Lecture Notes in Artificial Intelligence   6752
Edited by R. Goebel, J. Siekmann, and W. Wahlster

Subseries of Lecture Notes in Computer Science

Mohamed Kamel Fakhri Karray
Wail Gueaieb Alaa Khamis (Eds.)

Autonomous and Intelligent Systems

Second International Conference, AIS 2011
Burnaby, BC, Canada, June 22-24, 2011
Proceedings

Series Editors

Randy Goebel, University of Alberta, Edmonton, Canada
Jörg Siekmann, University of Saarland, Saarbrücken, Germany
Wolfgang Wahlster, DFKI and University of Saarland, Saarbrücken, Germany

Volume Editors

Mohamed Kamel
Fakhri Karray
University of Waterloo, Department of Electrical and Computer Engineering
Waterloo, ON, N2L 3G1, Canada
E-mail: {mkamel, karray}@uwaterloo.ca

Wail Gueaieb
University of Ottawa, School of Information Technology and Engineering (SITE)
800 King Edward Avenue, Ottawa, ON, K1N 6N5, Canada
E-mail: wgueaieb@site.uOttawa.ca

Alaa Khamis
German University in Egypt, Faculty of Engineering and Material Science
Department of Mechatronics Engineering
Main Entrance El Tagamoa El Khames, New Cairo City, Egypt
E-mail: alaa.khamis@guc.edu.eg

ISSN 0302-9743 e-ISSN 1611-3349
ISBN 978-3-642-21537-7 e-ISBN 978-3-642-21538-4
DOI 10.1007/978-3-642-21538-4
Springer Heidelberg Dordrecht London New York

Library of Congress Control Number: 2011929221

CR Subject Classification (1998): I.2, H.4, I.4, I.5, H.5

LNCS Sublibrary: SL 7 – Artificial Intelligence

© Springer-Verlag Berlin Heidelberg 2011
This work is subject to copyright. All rights are reserved, whether the whole or part of the material is
concerned, specifically the rights of translation, reprinting, re-use of illustrations, recitation, broadcasting,
reproduction on microfilms or in any other way, and storage in data banks. Duplication of this publication
or parts thereof is permitted only under the provisions of the German Copyright Law of September 9, 1965,
in its current version, and permission for use must always be obtained from Springer. Violations are liable
to prosecution under the German Copyright Law.
The use of general descriptive names, registered names, trademarks, etc. in this publication does not imply,
even in the absence of a specific statement, that such names are exempt from the relevant protective laws
and regulations and therefore free for general use.

Typesetting: Camera-ready by author, data conversion by Scientific Publishing Services, Chennai, India

Printed on acid-free paper

Springer is part of Springer Science+Business Media (www.springer.com)

Preface

AIS 2011, the International Conference on Autonomous and Intelligent Systems, held in Burnaby, British Columbia, Canada, June 22–24, 2011 was the second edition in the AIS series of annual conferences alternating between Europe and North America. The idea of organizing these conferences was to foster collaboration and exchange between researchers and scientists in the broad fields of autonomous design and intelligent systems, addressing recent advances in theory, methodology and applications. AIS 2011 was organized at the same time with ICIAR 2011, the International Conference on Image Analysis and Recognition. Both conferences are organized by AIMI—Association for Image and Machine Intelligence, a not-for-profit organization registered in Ontario, Canada, and co-sponsored by the Center for Pattern Analysis and Machine Intelligence of the University of Waterloo, the Kitchener Waterloo Chapters of the IEEE Computational Intelligence Society and the IEEE Systems, Man and Cybernetics Society.

For AIS 2011, we received a total of 62 full papers from 23 countries. The review process was carried out by members of the Program Committee and other reviewers; all are experts in various areas of the conference. Each paper was reviewed by at least two reviewers and checked by the Conference Chairs. A total of 40 papers were finally accepted and appear in one volume of the proceedings. The high quality of the papers is attributed first to the authors, and second to the quality of the reviews provided by the experts. We would like to sincerely thank the authors for responding to our call, and to thank the reviewers for their careful evaluation and feedback provided to the authors. It is this collective effort that resulted in the strong conference program and high-quality proceedings.

This year AIS provided some travel grants to a limited number of authors attending from outside North America.

We were very pleased to be able to include in the conference program keynote talks by well-known experts: Toshio Fukuda, Nagoya University, Japan; William A. Gruver, Simon Fraser University, Canada; Ze-Nian Li, Simon Fraser University, Canada; Andrew Sixsmith, Simon Fraser University, Canada; and Patrick Wang, Northeastern University Boston, United States of America. We would like to express our sincere gratitude to the keynote speakers for accepting our invitation to share their vision and recent advances in their specialized areas.

Special thanks are also due to Jie Liang, Chair of the local Organizing Committee, and members of the committee for their advice and help. We are thankful for the support and facilities provided by Simon Fraser University. We are also grateful to Springer's editorial staff, for supporting this publication in the *Lecture Notes in Artificial Intelligence* (LNAI) series.

We would like to thank Khaled Hammouda, the webmaster of the conference, for maintaining the Web pages, interacting with the authors and preparing the proceedings.

Finally, we were very pleased to welcome all the participants to AIS 2011. For those who were not able to attend, we hope this publication provides a good view of the research presented at the conference, and we look forward to meeting you at the next AIS conference.

Mohamed Kamel
Fakhri Karray
Wail Gueaieb
Alaa Khamis

AIS 2011 – International Conference on Autonomous and Intelligent Systems

General Co-chairs

Mohamed Kamel
Canada
mkamel@uwaterloo.ca

Fakhri Karray
Canada
karray@uwaterloo.ca

Program Co-chairs

Wail Gueaieb
Canada
wgueaieb@site.uottawa.ca

Alaa Khamis
Egypt
akhamis@pami.uwaterloo.ca

Track Chairs

Autonomous and Intelligent Systems
William Malek, Canada
Mohammad Biglarbegian, Canada

Signals and Intelligent Communication Systems
Slim Boumaiza, Canada

Haptics and Human–Machine Interaction
Abdulmotaleb El Saddik, Canada
Dana Kulic, Canada

Intelligent Data Analysis
Reda Alhajj, Canada
Richard Khoury, Canada

Local Organizing Committee

Jie Liang (Chair)
Simon Fraser University
Canada

Carlo Menon
Simon Fraser University
Canada

Faisal Beg
Simon Fraser University
Canada

Jian Pei
Simon Fraser University
Canada

Ivan Bajic
Simon Fraser University
Canada

Publicity Chair

Shahab Ardalan
Gennum Corp., Canada

Conference Secretary

Cathie Lowell
Canada

Webmaster

Khaled Hammouda
Canada

Co-Sponsorship

PAMI – Pattern Analysis and Machine Intelligence Group
University of Waterloo
Canada

Department of Electrical and Computer Engineering
Faculty of Engineering
University of Porto
Portugal

Department of Electrical and Computer Engineering
University of Waterloo
Canada

AIMI – Association for Image and Machine Intelligence

Simon Fraser University, Canada

Computational Intelligence Society Chapter, IEEE Kitchener Waterloo Section

Systems, Man, & Cybernetics (SMC) Chapter, IEEE Kitchener Waterloo Section

Advisory Committee

Pierre Borne	Ecole Centrale de Lille, France
Toshio Fukuda	Nagoya University, Japan
Elmer Dadios	De La Salle University, Philippines
Clarence de Silva	University of British Columbia, Canada
Mo Jamshidi	University of Texas, USA
Jong-Hwan Kim	Korea Advanced Institute for Science and Technology, South Korea
T.H. Lee	National University of Singapore, Singapore

Oussama Khatib   Stanford University, USA
Kauru Hirota   Tokyo Institute of Technology, Japan
Witold Perdrycz   University of Alberta, Canada

Technical Program Committee

Mohsen Afsharchi	Iran
Elmahdy Ahmed	Egypt
Salah AlSharhan	Kuwait
Farzad Aminravan	Canada
El-Badawy Ayman	Egypt
George Baciu	Hong Kong
Blake Beckman	Canada
Momotaz Begum	USA
Abder rezak Benaskeur	Canada
Pierre Borne	France
Rui Camacho	Portugal
Daniela Constantinescu	Canada
Vitor Costa	Portugal
Elmer Dadios	Philippines
Mehmet Önder Efe	Turkey
Mohamed Elabd	Kuwait
Arturo de la Escalera	Spain
Morteza Farrokhsiar	Canada
Guedea Federico	Mexico
Baris Fidan	Canada
Florin Filip	Romania
Hadi Firouzi	Canada
Toshio Fukuda	Japan
James Green	Canada
Jason Gu	Canada
Maki Habib	Egypt
Tarek Hamdani	Tunisia
Kauru Hirota	Japan
El-Sherif Hisham	Egypt
Morgan Elsayed Imam	Egypt
Mo Jamshidi	USA
Soo Jeoen	Canada
Jane You Jia	Hong Kong
Amor Jnifene	Canada
Jun Jo	Australia
Fakhri Karray	Canada
Oussama Khatib	USA
Richard Khoury	Canada
Keivan Kianmehr	Canada
Jong-Hwan Kim	South Korea

Dana Kulic	Canada
T.H. Lee	Singapore
Xuelong Li	UK
Honghai Liu	UK
Mohamed Masmoudi	Tunisia
Aníbal Castilho Coimbra de Matos	Portugal
Rene V. Mayorga	Canada
Vincent Myers	Canada
Homayoun Najjaran	Canada
Urbano Nunes	Portugal
Se-Young Oh	South Korea
Eugénio Oliveira	Portugal
Tansel Ozyer	Turkey
Witold Perdrycz	Canada
Emil Petriu	Canada
Kumaraswamy Ponnambalam	Canada
Luis Paulo Reis	Portugal
Mohamed Ben Rejeb	Tunisia
René Reynaga	Bolivia
Ana Paula Rocha	Portugal
Agos Rosa	Portugal
Ferat Sahin	USA
Rogelio Seto	Mexico
Clarence de Silva	Canada
Hooman Tahayori	Canada
Mehmet Tan	Turkey
Francisco José Rodríguez Urbano	Spain
Slawo Wesolkowski	Canada
Peter Won	Canada
Dongrui Wu	
James Yang	USA
Simon X. Yang	Canada
Jia Zeng	USA
Liangpei Zhang	China

Additional Reviewers

Kacem Abida	Canada
Masoud Alimardani	Canada
Jongeun Cha	Canada
Mohammad Azam Javed	Canada
Howard Li	Canada
Seyed Reza Miralavi	Iran
Mustapha Yagoub	Canada
Miao Yun Qian	Canada

Table of Contents

Autonomous and Intelligent Systems

Global Optimal Path Planning for Mobile Robots Based on Hybrid Approach with High Diversity and Memorization 1
Yun-Qian Miao, Alaa Khamis, Fakhreddine O. Karray, and Mohamed S. Kamel

Small Tree Probabilistic Roadmap Planner for Hyper-Redundant Manipulators 11
Eric Lanteigne and Amor Jnifene

A Fuzzy Logic Approach for Indoor Mobile Robot Navigation Using UKF and Customized RFID Communication 21
M. Suruz Miah and Wail Gueaieb

Toward Opportunistic Collaboration in Target Pursuit Problems 31
Soheil Keshmiri and Shahram Payandeh

A Multi-agent Framework with MOOS-IvP for Autonomous Underwater Vehicles with Sidescan Sonar Sensors 41
Liam Paull, Sajad Saeedi G., Mae Seto, and Howard Li

Three-Dimensional Path-Planning for a Communications and Navigation Aid Working Cooperatively with Autonomous Underwater Vehicles 51
Mae L. Seto, Jonathan A. Hudson, and Yajun Pan

Autonomous Haulage Systems – Justification and Opportunity 63
Juliana Parreira and John Meech

Building a No Limit Texas Hold'em Poker Agent Based on Game Logs Using Supervised Learning 73
Luís Filipe Teófilo and Luís Paulo Reis

Intelligent and Advanced Control Systems

Autonomous Manipulation Combining Task Space Control with Recursive Field Estimation 83
Soo Jeon and Hyeong-Joon Ahn

Accessibility of Fuzzy Control Systems 93
Mohammad Biglarbegian, Alireza Sadeghian, and William W. Melek

Thermal Dynamic Modeling and Control of Injection Moulding
Process 102
    Jaho Seo, Amir Khajepour, and Jan P. Huissoon

Analysis of Future Measurement Incorporation into Unscented
Predictive Motion Planning 112
    Morteza Farrokhsiar and Homayoun Najjaran

Nonlinear Maneuvering Control of Rigid Formations of Fixed Wing
UAVs 124
    Ismail Bayezit and Baris Fidan

Intelligent Control System Design for a Class of Nonlinear Mechanical
Systems 134
    S. Islam, P.X. Liu, and A. El Saddik

Trends in the Control Schemes for Bilateral Teleoperation with Time
Delay.. 146
    Jiayi Zhu, Xiaochuan He, and Wail Gueaieb

Bilateral Teleoperation System with Time Varying Communication
Delay: Stability and Convergence 156
    S. Islam, P.X. Liu, and A. El Saddik

Online Incremental Learning of Inverse Dynamics Incorporating Prior
Knowledge 167
    Joseph Sun de la Cruz, Dana Kulić, and William Owen

Experimental Comparison of Model-Based and Model-Free Output
Feedback Control System for Robot Manipulators 177
    S. Islam, P.X. Liu, and A. El Saddik

P-Map: An Intuitive Plot to Visualize, Understand, and Compare
Variable-Gain PI Controllers 189
    Dongrui Wu

Sufficient Conditions for Global Synchronization of Continuous
Piecewise Affine Systems 199
    Hanéne Mkaouar and Olfa Boubaker

Intelligent Sensing and Data Analysis

Question Type Classification Using a Part-of-Speech Hierarchy 212
    Richard Khoury

Exploring Wikipedia's Category Graph for Query Classification 222
    Milad Alemzadeh, Richard Khoury, and Fakhri Karray

Combination of Error Detection Techniques in Automatic Speech
Transcription 231
   Kacem Abida, Wafa Abida, and Fakhri Karray

Developing a Secure Distributed OSGi Cloud Computing Infrastructure
for Sharing Health Records 241
   Sabah Mohammed, Daniel Servos, and Jinan Fiaidhi

Extreme Learning Machine with Adaptive Growth of Hidden Nodes
and Incremental Updating of Output Weights 253
   Rui Zhang, Yuan Lan, Guang-Bin Huang, and Yeng Chai Soh

Face Recognition Based on Kernelized Extreme Learning Machine 263
   Weiwei Zong, Hongming Zhou, Guang-Bin Huang, and Zhiping Lin

Detection and Tracking of Multiple Similar Objects Based on
Color-Pattern 273
   Hadi Firouzi and Homayoun Najjaran

Argo Vehicle Simulation of Motion Driven 3D LIDAR Detection and
Environment Awareness 284
   *Mohammad Azam Javed, Jonathan Spike, Steven Waslander,
   William W. Melek, and William Owen*

Signal Processing and Pattern Recognition for Eddy Current Sensors,
Used for Effective Land-Mine Detection 294
   Hendrik Krüger and Hartmut Ewald

Velocity Measurement for Moving Surfaces by Using Spatial Filtering
Technique Based on Array Detectors 303
   Martin Schaeper and Nils Damaschke

Human-Machine Interaction

Towards Unified Performance Metrics for Multi-robot Human
Interaction Systems 311
   Jamil Abou Saleh and Fakhreddine Karray

Augmented HE-Book: A Multimedia Based Extension to Support
Immersive Reading Experience 321
   *Abu Saleh Md Mahfujur Rahman, Kazi Masudul Alam, and
   Abdulmotaleb El Saddik*

Human-Machine Learning for Intelligent Aircraft Systems 331
   Stuart H. Rubin and Gordon Lee

Haptic Data Compression Based on Curve Reconstruction 343
   *Fenghua Guo, Yan He, Nizar Sakr, Jiying Zhao, and
   Abdulmotaleb El Saddik*

Intelligent Circuit Analysis and Signal Processing

Diastolic Timed Vibrator: Applying Direct Vibration in Diastole to Patients with Acute Coronary Ischemia during the Pre-hospitalization Phase 355
    Farzad Khosrow-khavar, Marcin Marzencki, Kouhyar Tavakolian, Behrad Kajbafzadeh, Bozena Kaminska, and Carlo Menon

Low Noise CMOS Chopper Amplifier for MEMS Gas Sensor 366
    Jamel Nebhen, Stéphane Meillère, Mohamed Masmoudi, Jean-Luc Seguin, and Khalifa Aguir

Modeling Reliability in Wireless Sensor Networks 374
    Ahmed Ibrahim Hassan, Maha Elsabrouty, and Salwa Elramly

Fading Compensation for Improved Indoor RF Tracking Based on Redundant RSS Readings 384
    Andreas Fink and Helmut Beikirch

Redundant Reader Elimination Approaches for RFID Networks 396
    Nazish Irfan, Mustapha C.E. Yagoub, and Khelifa Hettak

NoC-MPSoC Performance Estimation with Synchronous Data Flow (SDF) Graphs 406
    Kamel Smiri and Abderrazak Jemai

Author Index 417

Global Optimal Path Planning for Mobile Robots Based on Hybrid Approach with High Diversity and Memorization

Yun-Qian Miao[1], Alaa Khamis[2],
Fakhreddine O. Karray[1], and Mohamed S. Kamel[1]

[1] Department of Electrical and Computer Engineering, University of Waterloo,
Waterloo, Ontario, Canada, N2L 3G1
{mikem,karray,mkamel}@pami.uwaterloo.ca
[2] Faculty of Petroleum and Mining Engineering, Suez Canal University, Egypt
akhamis@pami.uwaterloo.ca

Abstract. This paper presents a hybrid approach to the path planning problem of autonomous robots that combines potential field (PF) method and genetic algorithm (GA). The proposed PF+GA approach takes the strength of both potential field and genetic algorithm to find global optimal collision-free paths. In this integrated frame, the PF is designed as gradient-based searching strategy to exploit local optimal, and the GA is used to explore over the whole problem space. Different implementation strategies are examined through simulations in 2D scenarios. The conducted experiments show that global optimal path can be achieved effectively using the proposed approach with a strategy of high diversity and memorization.

Keywords: hybrid approach, path planning, potential field, genetic algorithm.

1 Introduction

Robot path planning is an interesting problem that has been studied extensively over the last two decades. This problem addresses how to find a collision-free path for a mobile robot from a start position to a given goal position, amidst a collection of obstacles. Efficient algorithms for solving problems of this type have important applications in areas such as: mobile robotics, intelligent transportation, computer animation, drug design, and automated surveillance [15].

Path planning can be seen as an optimization problem which aims at improving some system objectives with several constraints. System objectives may be the travelling distance, travelling time, and consuming energy. However, the travelling distance is the most common objective. Beside this, the necessity of considering collisions avoidance and path smoothness makes the problem becoming more challenging.

Different conventional approaches have been developed to solve the path planning problem [9], such as A* [14,12], the Rapidly Exploring Random Trees

(RRT) [11,16,13], and the Distance Transform approach [5]. These conventional approaches are grid-based and usually computationally expensive especially in case of high resolution. When lowering problem space's resolution, their precision is degraded. Other algorithms such as potential field approach and probabilistic roadmap methods [10,17] have been proposed to deliver more faster and practical solutions to this problem, and generalize it as possible for any robot with many degrees of freedom.

The potential field (PF) approach [6] is based on gradient of potential field that is built through a fitness function of considering both travelling distance and obstacles avoidance. The performance of PF method is largely associated with the initial setup, and may easily be trapped at local minima.

Genetic Algorithm (GA) has been recognized as one of robust searching techniques for complex optimization problem. There are several reports of applying GA to solve the path planning problem [4,18,3,1]. The most challenging problems of these GA approaches include needing domain specific operators (such as shortcutting and smoothening for local disturbance), which add extra computational load and can only achieve near optimal performance.

This paper examines the strengths and constraints of PF and GA algorithm, and proposes a hybrid approach that combines both algorithms to solve the path planning problem. In this novel integrated framework, the PF method is used to exploit local optimal solutions and the GA is designed to explore the problem space. Different implementation strategies are examined and analyzed through simulations in a 2D environment. It is demonstrated that the combining approach can effectively achieve global optimal in various situations when properly implemented.

The rest of this paper is organized as follows. Section 2 introduces theoretical background on Potential Field and Genetic Algorithm for path planning. Section 3 describes the details of the proposed hybrid approach. Section 4 presents our experimental results and analysis. Finally, conclusions and future directions are summarized in section 5.

2 Background

2.1 Potential Field (PF)

Potential field methods are gaining popularity in obstacle avoidance applications for mobile robots and manipulators. The idea of virtually forces acting on a robot has been suggested by Andrews and Hogan [2] and Khatib [6]. In these approaches obstacles produce repulsive forces onto the robot, while the target applies an attractive force to the robot. The sum of all forces determines the subsequent direction and speed of travel. One of the reasons for the popularity of this method is its simplicity and elegance. Simple PF method can be implemented quickly and initially provide acceptable results without requiring many refinements. Krogh and Thorpe [8] had applied the PF method to off-line path planning.

The problem of collision-free path planning may be treated as an optimization problem with two objectives. One objective is avoiding collision with obstacles. Another objective is to minimize the travelling distance. To integrate these two factors, we can formalize a path's fitness function that is comprised of two parts, energy function of distance and obstacles penalty function, respectively, as:

$$F(Path) = \omega_l E_l + \omega_o E_o \tag{1}$$

where E_l represents energy function of path distance, E_o represents collision penalty function, ω_l and ω_o are weights of the two parts.

The most significant problem of PF is the problem of trapped in local minima. There are recovery methods for remedying the trap situation to find feasible solution, which is most likely to result in non-global optimal solution [7].

2.2 Genetic Algorithm (GA)

GA has been recognized as a robust searching algorithm to optimize complex problems with the concept of evolution strategy. To use GA for solving the path planning problem, there are the following three steps to consider:

- *First:* Representing the problem and generating the initial population.
- *Second:* Selecting genetic operators for reproduction.
- *Third:* Using domain specific operators for considering constraints such as obstacles avoidance and path smoothness.

Problem Representation. *Initialization:* Create an initial population with a predefined population size. The population contains number of individuals (i.e., chromosomes). Each individual represents a solution for the problem under study. In our case, each solution is in fact a path between the start and end point in the search space. The initial population with size N can be presented as:

$$InitialPopulation = \langle S_1, S_2, \ldots, S_N \rangle \tag{2}$$

Each structure S_i is simply a float string of length M, in general. This structure represents a vector of node numbers in the space which contains "x" and "y" coordinates of a point in the search space. Normally, GAs individuals can contain any point value between the starting and ending point. Thus, the individual generated by GA is in the form of:

$$S_i = \langle P_{i1}, P_{i2}, \ldots, P_{iM} \rangle \tag{3}$$

where, M is the number of visited node in the search space. In order to include the starting and ending point, we need to make some modification to the individual structure so that we can add the starting point as the first point, P_{i1}, and ending point as the last point, P_{iM}.

Fitness function: Fitness function represents an important part of any evolutionary process using GA. Appropriate selection of the fitness function will lead the search towards the optimal solution.

The optimal path, in our case, is the shortest collision-free path between the starting and ending point. Thus, the fitness value for a complete solution S_i will be computed as:

$$Fitness(S_i) = \sum_{m=1}^{M-1} d_m \qquad (4)$$

where d_m is Euclidean distance between points P_{im} and $P_{i(m+1)}$.

Genetic operators. *Selection:* The selection procedure is applied to choose the individuals that participate in the reproduction process to give birth to the next generation. In order to make the whole population evolution towards good solutions, selection is executed probabilistically based on roulette wheel where the chance for individual to reproduce is proportional to its fitness function. Then the probability being chosen is:

$$Prob(S_i) = 1 - \frac{Fitness(S_i)}{\sum_{j=1}^{N} Fitness(S_j)} \qquad (5)$$

Crossover: This operator combines two selected paths (parents) to generate two offsprings as follows: a random mating position node is selected on each parent. Then, this node splits the path into two sections. The first offspring is generated by combining the first fraction of the first parent with the second fraction of the second parent. And, the second offspring is generated by combining the first fraction of the second parent with the second part of the first parent.

Mutation: This operator changes the node's coordinates of the selected path, except the start point and end point keep unaffected. The mutation happens at given level of possibility with a predefined range.

Immigration: In case of further increasing population diversity, immigration operator is introduced, which creates some random individuals and injects them into the next generation during the evolution progress.

Memory: This operator stores the best solution from the population of present generation, and clones it directly to the next generation. This operation is used to keep the evolution consistently converging towards the direction of better fitness, and avoid interference that comes from crossover and mutation.

Domain specific operators. *Repair:* This operation executes when infeasible path detected, which refers that there exists a segment of the planned path going across an obstacle. [3]'s repair operator is to generate a set of random points around the intersecting obstacle and connects these points to pull the segment around the obstacle.

Shortcut: This operator removes the intermediate points between two points if the direct connection line is collision-free.

Smooth: This operator is applied to a path with two connected segments with sharp turn around, which is not executable for actual mobile robots. The connection node is removed and replaced with two new nodes on each segment such that makes the path with smooth segments connections.

3 The Hybrid Approach

The PF method has advantages of finding a collision-free smooth solution effectively, but with a main drawback of easy trapping in local minima. The GA algorithm with proper strategy can explore complex problem space to search for global optimal, but in most cases, the solution is only near best. Furthermore, when applying pure GA algorithm to the problem of path planning, there need some extra domain specific operators as described in the previous section. These operations will degrade the usability of GA because of their extra computational burden.

In this section, a combining PF+GA approach is described, which takes the advantages of both methods and makes up their shortages.

3.1 Structure of PF+GA Path Planner

The pseudo code of PF+GA combining approach is described in Algorithm 1. The algorithm is terminated when a predefined number of generations is reached, or the results converge to a setting level.

Algorithm 1. PF+GA algorithm for path planning

1: Initialize the environment (setup parameters, obstacles representation, start point, goal point);
2: Initialize the first generation by randomly generating paths as individuals;
3: Applying PF method to each individual to make a local optimal solution;
4: **repeat**
5:   Evaluate the fitness of each solution;
6:   Applying GA operators: Selection, Mutation, Crossover, Immigration, Memory;
7:   Next generation produced;
8:   Applying PF method to individuals of new generation;
9: **until** termination condition reached
10: Output the result

3.2 Implementation Strategies

Diversity methods. While selection is an operator to lead the population converging to better fitness, diversity operations intend to explore the problem space. Crossover, mutation, and immigration operators and their different combinations are candidates for increasing diversity. Obviously, low level of diversity will make GA easily being trapped at a region, and GA is showing converging quickly to some sub-optimal prematurely.

Among these diversity operations, mutation is used to search locally. Because the proposed approach combining PF method for digging locally using gradient based method, the functionality of mutation is then duplicated and can be taken out from GA operation's pool, which reduces algorithm's computation cost.

The later experimental results also prove the idea of unnecessary of mutation operation in this hybrid approach.

Memory strategy. On the one hand, we need diversity for broad exploring. On the other hand, high level of diversity will lead the population oscillating and good solutions are easy to be disturbed by diversity interference.

In [18] and [3], authors introduced the concept of memory operator to GA, which memorizes the best solution so far and clones it directly to the next generation. The memory strategy is adopted in our combining PF+GA approach to keep the evolution progress stable.

Strategies composition. GA algorithm performance is affected by different level of each operator and different kind of combination of these operators. When using in PF+GA approach, the implementation strategy becomes more complex and related performance will be studied by experiments in the next section.

4 Experimental Results

4.1 Experimental Setup

The experiments were conducted in 2D environments with different size of obstacles. For each experiment condition, the results are collected by 5 runs to avoid randomness. The composition strategies of implementation to be examined are listed in Table 1.

Table 1. Implementation strategies and their parameters

Strategy	Description	Parameters
A0	Low diversity, no memory	Cross Rate = 0.1; Mutation Rate = 0.1; Immigrate Rate = 0; Memory = 0;
A1	Low diversity, has memory	Cross Rate = 0.1; Mutation Rate = 0.1; Immigrate Rate = 0; Memory = 1;
A2	High diversity, no memory	Cross Rate = 0.2; Mutation Rate = 0.2; Immigrate Rate = 0.2; Memory = 0;
A3	High diversity, has memory, no mutation	Cross Rate = 0.2; Mutation Rate = 0; Immigrate Rate = 0.2; Memory = 1;
A4	High diversity, has memory, has mutation	Cross Rate = 0.2; Mutation Rate = 0.2; Immigrate Rate = 0.2; Memory = 1;

4.2 Results

When using low diversity strategy (A0, see Table 1), Fig. 1 demonstrates the situation of prematurely converging to a sub-optimal solution.

On the contrary, Fig. 2 demonstrates the situation of achieving global optimal solution in the case of using high diversity with memory strategy (A4). As explained in Table 1, A0 and A4 represent the two extremes of the implementation strategies.

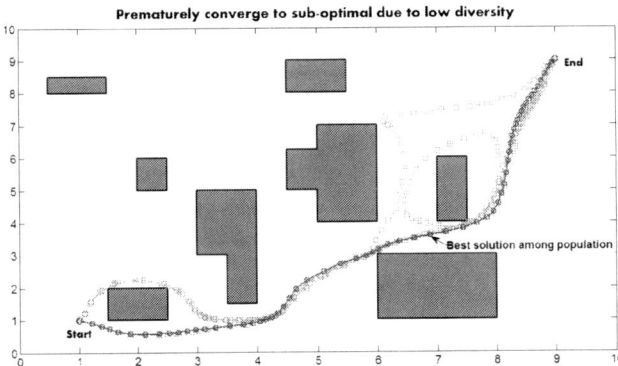

Fig. 1. An example shows the situation of prematurely converging to sub-optimal due to low diversity (A0)

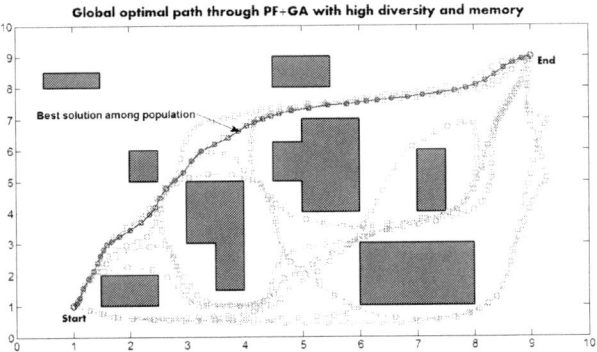

Fig. 2. An example demonstrates achieving global optimal solution through PF+GA approach with high diversity and memory strategy (A4)

Further, to investigate the effect of "memory" strategy, Fig. 3 gives us an inside look about the fitness evolving process as evolution proceeding by comparing strategy A0 and A1, which the only difference is "no memory" and "has memory".

Performance comparison of different strategies. The performance evaluation for each implementation strategy is determined by the best solution (shortest path) among the final population. Table 2 reports the average path length and standard deviation according to 5 runs of each settlement.

Results in Table 2 clearly indicates that strategies without memory operator perform worst (A0 and A2), while the strategy with high diversity and memorization performs best (A4). Comparing A3 and A4 ("no mutation" and "has

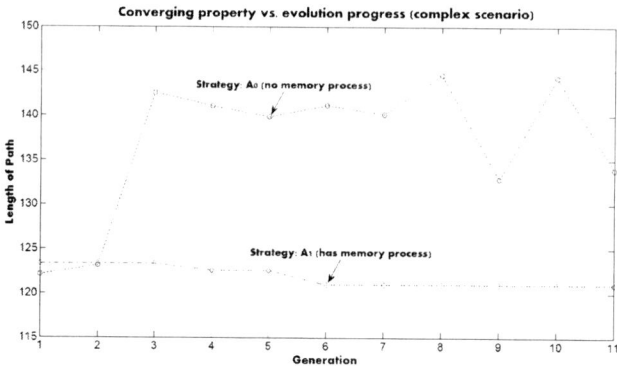

Fig. 3. Convergence property along the evolution progress with and without memorization (A0 vs. A1)

Table 2. Performance comparison between different implementation strategies

Strategy	A0	A1	A2	A3	A4
Average	131.42	122.12	124.93	122.14	**120.51**
Std. Dev.	4.43	0.70	2.78	**0.69**	1.86

mutation"), there is no distinguishable performance gap. When no mutation injected, the results are more consistent (standard deviation is smaller).

4.3 Discussion

PF method is a gradient based approach that demonstrates ability to find local optimal path and avoid obstacles in a harmony way. The solution path of PF is heuristically smooth and suitable for mobile robot navigation. However, PF solution's performance is decided by its initialization, and is lack of global optimal solution ability.

GA as a population-based meta-heuristic optimization method owns the ability of searching global optimal solution through evolution strategy. But, when applying pure GA to the path planning problem, the individuals of population are in most probably infeasible, especially in complex scenarios with high number of obstacles. To overcome this problem, some specific domain operators can give a hand with extra computation load. And the solutions are only near optimal.

The proposed hybrid approach is showing effective and global optimal to accomplish path planning with benefits that come from both PF and GA. Even so, different implementation strategy affects the algorithm's performance a lot. The interesting findings are:

- In general, high diversity with memory strategy performs better than any other combinations.

- Diversity plays a key role for exploring problem space to avoid prematurely converging to sub-optimal solution. The diversity of population can achieve by the operations of crossover, mutation and immigration. But, the mutation has less contribution in this case, because the PF executes local optimal searching and replaced the functionality of mutation.
- Memory is a very important operator in this problem, which colons the best solution so far to the next generation directly. It keeps the fitness of each generation decreasing monotonically. On the contrary, the solution of "no memory" oscillates along evolution progress because of possible interference by crossover.
- Diversity plays another key role for exploring problem space to avoid converging to sub-optimal prematurely.

5 Conclusion

This paper proposed a hybrid approach to mobile robot path planning problem combining PF and GA algorithms. This approach takes advantages of PF and GA methods and overcomes their inherent limitations. By studying different implementation strategies, it was revealed that the PF+GA combining framework with high diversity and memorization strategy is effective to find global optimal solutions.

Future work will focus on extension to dynamic environments with moving obstacles.

Acknowledgments. This work has been supported through the Collaborative Research and Development (CRD) project "DIScrimination of Critical Objects and EVents in PErvasive Multimodal SuRveillance Systems (DISCOVER)" funded by Natural Sciences and Engineering Research Council of Canada (NSERC) and Thales Canada Inc.

References

1. AL-Taharwa, I., Sheta, A., Al-Weshah, M.: A mobile robot path planning using genetic algorithm in static environment. Journal of Computer Science 4(4), 341–344 (2008)
2. Andrews, J.R., Hogan, N.: Impedance control as a framework for implementing obstacle avoidance in a manipulator. In: Hardt, D.E., Book, W. (eds.) Control of Manufacturing Processes and Robotic Systems, pp. 243–251. ASME, Boston (1983)
3. Elshamli, A., Abdullah, H.A., Areibi, S.: Genetic algorithm for dynamic path planning. In: Proc. of Canadian Conf. Elect. and Comput. Eng., Niagara Falls, vol. 2, pp. 677–680 (2004)
4. Gerke, M.: Genetic path planning for mobile robots. In: Proceedings of the American Control Conference, vol. 4, pp. 2424–2429 (1999)
5. Jarvis, R.A.: Collision-free trajectory planning using distance transforms. Mechanical Engineering Transactions, Journal of the Institution of Engineers ME-10(3), 187–191 (1985)

6. Khatib, O.: Real-time obstacle avoidance for manipulators and mobile robots. The International journal of Robotics Research 5(1), 90–98 (1986)
7. Koren, Y., Borenstein, J.: Potential field methods and their inherent limitations for mobile robot navigation. In: Proceedings of the IEEE International Conference on Robotics and Automation, vol. 2, pp. 1398–1404 (1991)
8. Krogh, B.H., Thorpe, C.E.: Integrated path planning and dynamic steering control for autonomous vehicles. In: Proceedings of the 1986 IEEE International Conference on Robotics and Automation, San Francisco, California, pp. 1664–1669 (1986)
9. Latombe, J.C.: Robot motion planning. Kulwer Academic Publisher, Boston (1991)
10. LaValle, S.: Planning Algorithms. Cambridge University Press, Cambridge (2006)
11. LaValle, S.M.: Rapidly-exploring random trees: A new tool for path planning. Technical Report 98-11, Dept. of Computer Science, Iowa State University (1998)
12. Lozano-Perez, T., Wesley, M.A.: An algorithm for planning collision-free paths among polyhedral obstacles. Commun. ACM 22(10), 560–570 (1979)
13. Masehian, E., Nejad, A.H.: A hierarchical decoupled approach for multi robot motion planning on trees. In: IEEE International Conference on Robotics and Automation (ICRA), pp. 3604–3609 (2010)
14. Nilsson, N.J.: Problem Solving Methods in Artificial Intelligence. McGraw-Hill, New York (1971)
15. Strandberg, M.: Robot Path Planning: An Object-Oriented Approach. PhD thesis, Automatic Control Department of Signals, Sensors and Systems Royal Institute of Technology (KTH) Stockholm, Sweden (2004)
16. Tedrake, R.: LQR-Trees: Feedback motion planning on sparse randomized trees. In: Proceedings of Robotics: Science and Systems, RSS (2009)
17. Teeyapan, K., Wang, J., Kunz, T., Stilman, M.: Robot limbo: Optimized planning and control for dynamically stable robots under vertical obstacles. In: IEEE International Conference on Robotics and Automation (ICRA), pp. 4519–4524 (2010)
18. Trojanowski, K., Michalewicz, Z., Xiao, J.: Adding memory to the evolutionary planner/navigator. In: Proc. of the 4th IEEE ICEC, pp. 483–487 (1997)

Small Tree Probabilistic Roadmap Planner for Hyper-Redundant Manipulators

Eric Lanteigne and Amor Jnifene

Royal Military College of Canada

Abstract. This article describes a single-query probabilistic roadmap planner, called Small Tree, for hyper-redundant manipulators. The planner incrementally builds two solution paths from small alternating roadmaps rooted at the two input configurations, until a connection linking both paths is found. In this manner, the solution path grows from both input queries. The process has the advantage of continuously replacing the two initial input queries, whose original connectivity in configuration space may have been difficult, with new target configurations. The method also has the ability to initiate the manipulator motion before the complete solution is found by using the partial solution path connected to the initial configuration.

Keywords: path planning, hyper-redundant manipulators, probabilistic roadmaps.

1 Introduction

In robotics, probabilistic roadmaps (PRM) are motion planning algorithms that operate in the configuration space $\mathcal{C}$ of the robot. Probabilistic roadmaps are graphs of connected nodes, where each node corresponds to a manipulator configuration located in the free-space $\mathcal{F}$ of the robot, where $\mathcal{F}$ is the obstacle-free and self-collision-free subspace of $\mathcal{C}$. There are two general types of PRM methods used in the motion planning of robotic devices: the single-query and the pre-processing methods. The single-query method solves a motion planning query for a given start and goal configuration, whereas the pre-processing method solves various motion planning problems using a pre-processed roadmap covering an adequate portion of the free space [5].

PRM planners using a single-query strategy were developed by a number of researchers [10,12,13,14]. The majority of the research focused on improving the efficiency of the local planner, since it has been shown that more than 95% of the processing time is spent on the node connection phase [2,6]. The lazy collision checking algorithm described in [3] assumes all the edges of the roadmap are collision-free. When the shortest path is found, the edges forming the path are checked for collisions. If a collision occurs at one of the edges, the nodes of the branch attached to the edge are removed from the graph and the graph is updated with new nodes, and the process is repeated until a collision-free path is found. The lazy collision checking algorithm was also adopted in the Single-query,

Bi-directional, Lazy in collision checking (SBL) probabilistic roadmap planner [12]. The bi-directional planner builds two trees rooted at the initial and final configurations of the manipulator. The expansion of the roadmap is achieved by randomly selecting a parent node from one of the trees, with a probability inversely proportional to the density of surrounding nodes. A new collision-free node is created in the neighbourhood of the parent node. If the distance between the closest node in the opposite tree and the newly added node is below a certain threshold, the edges connecting the branches of both trees are checked by the local planner. The expansion of both trees continues until a collision-free path is found.

Combinations of single-query PRMs and other methods have been developed for a number of specific problems. In [11], the bi-directional PRM algorithm was combined with a traveling-salesman tour algorithm to solve multi-goal path planning problems of redundant spot-welding and car painting manipulators. The single-query and preprocessing methods are combined in [1] by using sample-based tree methods as a subroutine for the PRM. The nodes of the PRM are trees are constructed using single query path planning methods such as expansive space trees, Rapidly-exploring Random Trees (RRTs), or others.

Similar to PRMs, the RRT algorithm builds a graph of nodes by incrementally selecting a random node in the configuration space and locating the nearest node in the tree. A new node is selected by moving from the nearest node in the direction of the random node by an incremental distance ϵ (assuming that the motion in any direction is possible) [9]. The new node and the edge connecting the new node to the nearest node is then added to the tree. Usually, an RRT alone is insufficient to solve a planning problem [9]. The RRT-Connect path planner developed by Kuffner and LaValle [7] works by incrementally building two RRTs rooted at the initial and final configurations using a simple greedy heuristic to explore the free space. Instead of extending the RRT by ϵ, the heuristic iterates in the direction of the random node until it reaches the random node or an obstacle surface [7].

In this article, we present a single-query bi-directional PRM planner that combines the properties of the SBL and the RRT-Connect algorithm. The planner combines small random walks and lazy collision checking to incrementally build a solution path from two input configurations. The following two sections describe the framework behind the proposed algorithm and examine the numerical results of simulations on a 17 degree of freedom (dof) planar manipulator.

2 Small Tree PRM

The Small Tree (ST) algorithm is a single-query lazy in collision checking PRM planner for manipulators with high-dimensional configuration spaces. The proposed method is a combination of a bi-directional and a uni-directional search algorithm. It is bi-directional in the sense that it builds roadmaps at each of the two input queries, and unidirectional in the sense that at any moment in time, there is only one roadmap being constructed from a single root configuration.

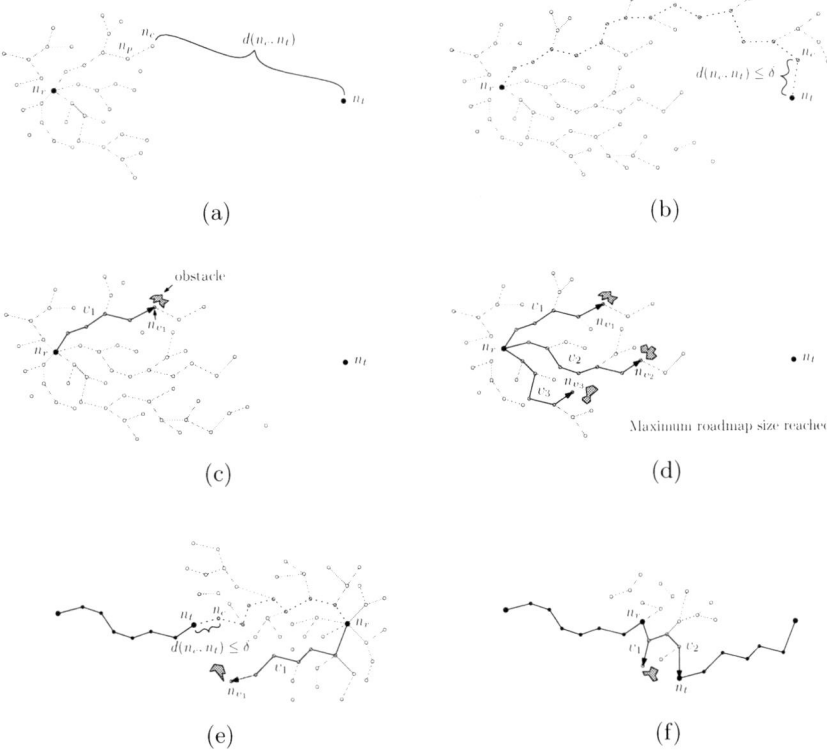

Fig. 1. ST roadmap: (a) illustrates the generation of the first roadmap; (b) illustrates a child node in proximity to the target node along with the potential solution path; (c) illustrates the search direction of the local planner and the elimination of the branch in collision; (d) illustrates the saved partial solution paths when the roadmap reaches its maximum size; (e) illustrates the generation of the second roadmap; and, (f) illustrates the generation of the third and final roadmap with the collision-free path

The roadmaps have a tree-like structure in configuration space. Each new node is linked to a single node in the existing roadmap, and no loops of nodes can occur. The process of adding nodes to the roadmap is summarized by the following three steps. First a parent node n_p is selected from the roadmap. Then a collision-free child node n_c is generated by mutating select dimensions of n_p, and connected to n_p via a hypothetically collision-free edge. Lastly, the distance between n_c and the target node n_t is computed using the following metric:

$$d(n_c, n_t) = \sum_{i=1}^{M} |(s_i)_c - (s_i)_t|, \qquad (1)$$

where s_i is the state of the ith module, and M is the total number of modules in the manipulator.

The algorithm first begins by setting initial manipulator configuration q_i to the root node n_r of the first tree, and the final manipulator configuration q_f to the target node n_t. The planner builds a uni-directional roadmap rooted at n_r using the method described above. For each new child node inserted in the graph, an attempt is made to find a path connecting it to the target node. This is achieved by comparing the distance d between n_c and n_t to a threshold δ that controls the frequency of calls to the local planner. This step is illustrated in Figure 1(a).

If the length of the edge connecting n_c and n_t is below the δ, the local planner checks the edges of the connecting path (highlighted by the dotted line in the example shown in Figure 1(b)). The edges of this path are checked sequentially from n_r to n_t. If an obstacle is found, as in Figure 1(c), the entire branch following the colliding edge is removed from the roadmap and the collision-free portion of the path is saved. In this example, the first partial solution path is v_1.

When the number of generated child nodes reaches the maximum graph size N of the uni-directional portion of the ST planner, the closest collision-free portion of the failed attempts at connecting the root node to the target node is inserted in the final solution path (based on (1)). In the case of the example in Figure 1(d), the node n_2 of the second path is closest to the target node (assuming the linear is representative of the distance in the configuration). A new roadmap is then constructed from the opposite query towards the tip of this new path, as shown in Figure 1(e). The method alternates between the two branches, saving the closest collision-free portions of each attempt, until a connection fuses both ends and completes the solution.

3 Simulations

3.1 Implementation

The parent node selection process is similar to the indexation method used by the SBL planner [12]. The method is based on the minimum distribution of nodes in a select subspace of the configuration space. The nodes of the roadmap are indexed in an h-dimensional array A, which partitions an h-portion of the configuration space into a grid of equally sized cells corresponding to the number of actuator states. A parent node is then randomly selected from the within one of these cells with a probability inversely proportional to the number of nodes contained in that cell.

Simulations are performed on a serial manipulator composed of $M = 17$ discretely-actuated modules. The states of the discrete actuators are divided into five equally-spaced orientations thereby facilitating the application of the distance metric(1). The five states $s_i = 1, 2, \ldots, 5$, illustrated in Figure 2, approximate the capabilities the discrete manipulator concept described in [8].

The child node n_c is created by randomly modifying S states within any of the M dimensions of the parent node n_p, where one state change is defined as moving one module from state i to state $i + 1$ or $i - 1$. The manipulator configuration contained in n_c is checked for collisions by computing the forward

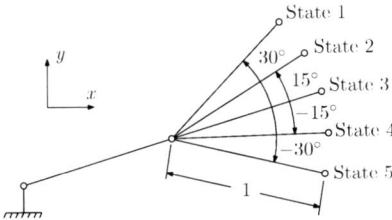

Fig. 2. Discrete actuator states

kinematics of the manipulator and comparing the module positions with the obstacle locations.

The algorithm executes η attempts to form a collision-free child node. If the initial η attempts fail, the search area is reduced by reducing S, and the process is reiterated. If a child node cannot be produced after η^S attempts, it is assumed that the parent node is trapped by the surrounding obstacles and a path cannot be formed from this node. The planner then simply selects a new parent node and resumes the search.

For each new node inserted in the roadmap, the corresponding cell in A is updated. For the present simulations, the array samples two dimensions of the configuration space at a given time ($h = 2$) and has a grid size is equal to the number of discrete states of the module actuator. These dimensions are periodically changed to ensure a diffusion along all dimensions of the configuration space, and at each cycle a new array is constructed to reflect the new dimensions [12]. The re-construction rate of A was fixed for all simulations as different rates did not noticeably affect the overall performance of the algorithm. The two dimensions depicted in A are selected at random and changed after the generation of every 10 child nodes to ensure a maximum diffusion rate on the even the smallest roadmaps.

3.2 Simulation Results

The Small Tree planner was tested using MATLAB on a 2.5 GHz AMD Athlon 64 X2 processor. Over 10^3 tests were performed for each simulation and individual tests were terminated if the total number of nodes generated by the planner exceeded 5×10^4. The ST planner was first evaluated by subjecting the manipulator to the simple path planning scenario shown in Figure 3. The circular obstacle has a diameter of 4 units and is centered at $x = 15$ in scenario 2 (distances and lengths with respect to the unit length of one module).

The ST planner is compared to the SBL planner described in [12] as SBL displayed a number of similarities to the proposed planner, and was capable of solving high dof planning problems in highly complex environments. Architecturally, both methods are single-query bi-directional planners. Both use lazy collision checking, and use the same adaptive node generation technique. The experimental results in [12] demonstrated that the method was capable of

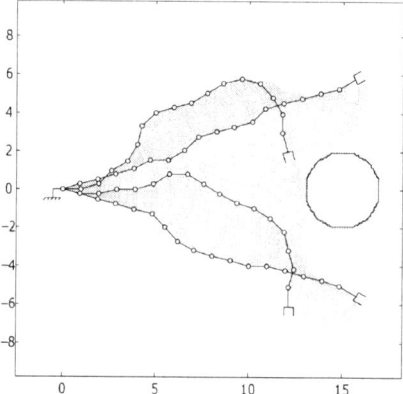

Fig. 3. Sample solution path to a simple scenario

Table 1. Simulation results for the: (a) narrow passage; and, (b) multiple obstacles

(a)

	ST	SBL
Success rate (%)	100	69
Time (seconds)	5.8	33.6
Nodes ($\times 10^3$)	5.3	19.5

(b)

	ST	SBL
Success rate (%)	100	29
Time (seconds)	11.3	29.1
Nodes ($\times 10^3$)	8.8	17.0

successfully solving the narrow passage path planning problem of a 6-dof manipulator, and the simultaneous path planning problems of 6 manipulators with 6-dof in an automotive welding station.

The parameters controlling the node generation of the two methods were standardized as much as possible. Both methods use similar threshold values δ and search areas S. The size of the uni-directional graph of the ST planner was fixed at $N = 25$ nodes. The success rate and average execution times of ST and SBL for the problem in Figure 3 are shown in Figures 4 and 5 for a range of δ and S values.

There are a number of factors that increase the complexity of the planning algorithm. Most notably the difficulty of the path planning problem, but also the number of dofs of the system and the number of states of each dof. The two bi-directional planners were tested on the two scenarios shown in Figure 6. The average results are given in Tables 1 using the midrange values of $S = 10$ and $\delta = 10$ for both planners, and a graph size of $N = 25$ for the ST planner

The data of Tables 1, although limited in scope, show that the ST planner is capable of handling difficult obstacle scenarios with 100% reliability. Both the total number of nodes generated by the ST planner and the average execution

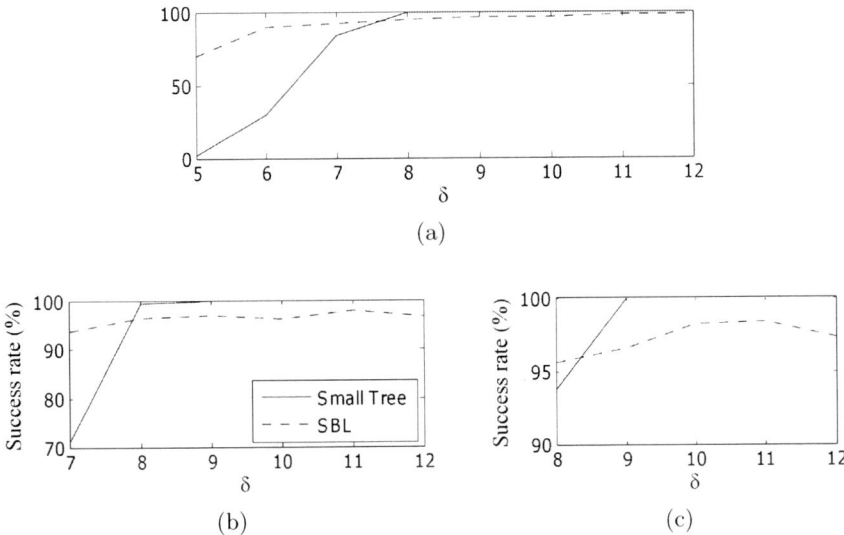

Fig. 4. Success rate comparison between the ST and SBL planners for: (a) $S = 4$; (b) $S = 8$; (c) $S = 12$

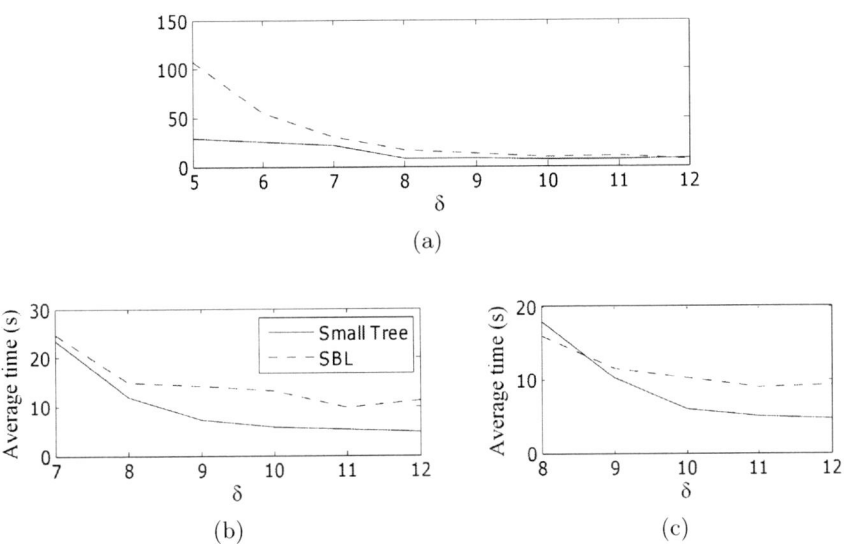

Fig. 5. Average execution times of the ST and SBL planners for: (a) $S = 4$; (b) $S = 8$; (c) $S = 12$

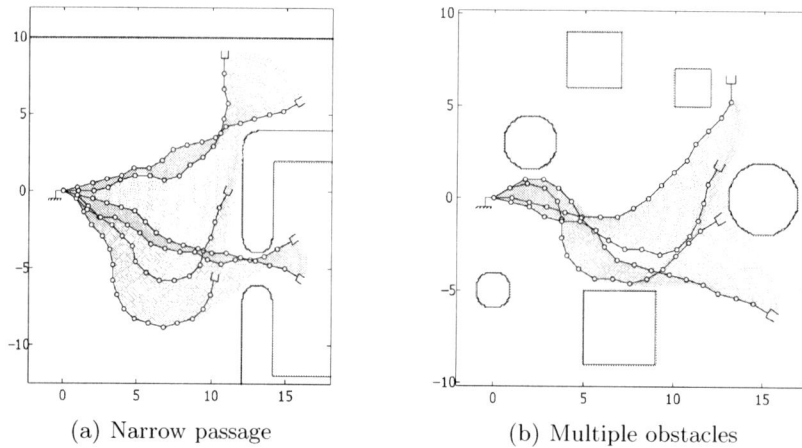

(a) Narrow passage (b) Multiple obstacles

Fig. 6. Sample solution path to two difficult scenarios

times were only slightly higher than the circle obstacle scenario. Various values of δ and S did not considerably improve the performance of the SBL planner.

4 Discussion

At threshold values above $\delta = 8$, the ST planner achieves a 100% success rate for all values of the number of state mutations. While both planners generated approximately the same number of nodes before reaching a solution, the average execution time of ST was between 25 to 50% lower than the SBL method.

The differences in the average times are linked to the differences in the tree connection phase of each method. The SBL planner attempts to connect the child node to a number of close nodes in the opposite roadmap. At high thresholds, a large number of candidate connections are available. The child nodes of the ST method are compared to a single node: the last node of the opposite partial solution path, and the local planner is called only if the resulting edge is lower than the threshold δ. It was also found that the performance of the ST planner diminished rapidly when $\delta < 8$. At these thresholds, a smaller number of partial solution paths were produced. This limits the number of choices available to the planner at the selection stage (shown in Figure 1(d)) and stalls the growth of the solution path. For the case where $S = 4$ and $\delta = 4$, it was found that a significant amount of roadmaps did not produce any partial solution paths. However, further testing of ST on the circle obstacle revealed even higher success rates and faster execution times when lower tree sizes were used. At $N = 15$, the average execution time could be reduced by 25% (for $S = 4$ and $\delta = 8$) compared to Figure 5(a).

The architecture of the ST planner does have a tendency to generate solution paths containing a large number of nodes. Above $\delta = 8$ in Figure 5, the ST planner generated solution paths containing an average of 310 nodes, or 7.6%

of the total number of generated nodes by the planner, and the SBL planner generated solution paths containing an average of 92 nodes, or 2.7% of the total number of generated nodes by the planner. However, reducing the number of configurations in the solution path using the local planner did not significantly affect the total execution time. This was achieved by connecting two nodes at random from the solution path and checking the resulting edge using the local planner. If no collisions occur, all the nodes between the two configurations would be removed from the solution. The process stops when no further nodes can be eliminated.

There exists a number of different optimization methods for improving a given path in high dimensional $\mathcal{C}$ spaces. An example is the variational approach described in [4]. The method improves a given path in terms of both the path length and the safety against collisions using a set of differential equations. Although this method is capable of generating smoother paths, the purpose of the aforementioned method is to provide a quick and concise solution path.

5 Conclusion and Future Work

In this article, we describe and evaluate a new PRM planner architecture capable of solving the path planning problem of serial hyper-redundant manipulators. The planner also has ability to utilize its growing paths before the complete solution is found. The partial solution path connected to the initial manipulator configuration could be optimized separately, while the planning is still in progress, and sent to the robot controller. Although the resulting overall solution would be less than optimal in terms of the distance traveled, the execution time could be theoretically reduced by up to half in static environments, and the algorithm could potentially be applied to dynamic environments. Other directions for future research include the on-line modification of planner's parameters based on the node generation success rate when subjected to new obstacle environments and other manipulators, and comparing the ST planner to other high-dimensional path planners such as RRTs.

Acknowledgment. The authors would like to acknowledge the scholarship from the Natural Sciences and Engineering Research Council of Canada.

References

1. Akinc, M., Bekris, K.E., Chen, B.Y., Ladd, A.M., Plaku, E., Kavraki, L.E.: Probabilistic roadmaps of trees for parallel computation of multiple query roadmaps. In: Dario, P., Chatila, R. (eds.) Robotics Research. Springer Tracts in Advanced Robotics, vol. 15, pp. 80–89. Springer, Heidelberg (2005)
2. Amato, N.M., Bayazit, O.B., Dale, L.K., Jones, C., Vallejo, D.: OBPRM: An obstacle-based PRM for 3D workspaces. In: Proceedings of the third workshop on the algorithmic foundations of robotics on Robotics: the algorithmic perspective, pp. 155–168 (1998)

3. Bohlin, R., Kavraki, L.E.: Path planning using lazy PRM. In: Proceedings of the IEEE International Conference on Robotics and Automation, April 2000, vol. 1, pp. 521–528 (2000)
4. Dasgupta, B., Gupta, A., Singla, E.: A variational approach to path planning for hyper-redundant manipulators. Robotics and Autonomous Systems 57(2), 194–201 (2009)
5. Geraerts, R., Overmars, M.: A comparative study of probabilistic roadmap planners. Springer Tracts in Advanced Robotics: Algorithmic Foundations of Robotics V7, 43–58 (2003)
6. Kavraki, L.E., Švestka, P., Latombe, J.C., Overmars, M.H.: Probabilistic roadmaps for path planning in high-dimensional configuration spaces. IEEE Transactions on Robotics and Automation 12(4), 566–580 (1996)
7. Kuffner, J.J., LaValle, S.M.: RRT-Connect: An efficient approach to single-query path planning. In: Proceeding of the IEEE International Conference on Robotics and Automation, April 2000, vol. 2, pp. 995–1001 (2000)
8. Lanteigne, E., Jnifene, A.: Experimental study on SMA driven pressurized hyper-redundant manipulator. Journal of Intelligent Materials Systems and Structures 16(9), 1067–1076 (2008)
9. LaValle, S.M.: Rapidly-exploring random trees: A new tool for path planning. Tech. Rep. TR 98-11, Iowa State University (October 1998)
10. Nielsen, C.L., Kavraki, L.E.: A two level fuzzy PRM for manipulation planning. In: Proceeding of the IEEE International Conference on Intelligent Robots and Systems, November 2000, vol. 3, pp. 1716–1721 (2000)
11. Saha, M., Roughgarden, T., Latombe, J.C., Sánchez-Ante, G.: Planning tours of robotic arms among partitioned goals. International Journal of Robotics Research 25(3), 207–223 (2006)
12. Sánchez, G., Latombe, J.C.: A single-query bi-directional probabilistic roadmap planner with lazy collision checking. In: International Symposium on Robotics Research, pp. 403–417 (2001)
13. Švestka, P.: Robot motion planning using probabilistic roadmaps. Ph.D. thesis, Utrecht University, Netherlands (1997)
14. Vallejo, D., Remmler, I., Amato, N.M.: An adaptive framework for single shot motion planning: A self tuning system for rigid and articulated robots. In: Proceedings of the IEEE International Conference on Robotics and Automation, vol. 4, pp. 21–26 (2001)

A Fuzzy Logic Approach for Indoor Mobile Robot Navigation Using UKF and Customized RFID Communication

M. Suruz Miah and Wail Gueaieb

School of Information Technology and Engineering
University of Ottawa, Ottawa, Ontario, Canada
suruz.miah@uOttawa.ca, wgueaieb@site.uOttawa.ca

Abstract. Most common techniques pertaining to the problem of mobile robot navigation are based on the use of multiple sensory data fusion and of excessive number of reference Radio Frequency (RF) stations. However, spatial layout, high navigation accuracy, and cost problems limit the applicability of those techniques in many modern robotic applications. The current manuscript outlines a novel computationally inexpensive indoor mobile robot navigation system with an intelligent processing of received signal strength (RSS) measurements of a customized Radio Frequency IDentification (RFID) system. The high navigation accuracy is achieved by incorporating a conventional stochastic Uncented Kalman Filter (UKF), which integrates RSS measurements from a customized RFID system and sensory data from the robot's wheel encoder. The workspace is defined by placing a number of RFID tags at 3-D locations. The orthogonal 2-D projection points of those tags on the ground define the target points where the robot is supposed to reach. The customized RFID system is simulated using a comprehensive electromagnetic commercial software, FEKO. The validity and suitability of the proposed navigation system are demonstrated by conducting extensive computer simulations in order to provide high-performance level in terms of accuracy and scalability.

Keywords: Customized RFID reader, fuzzy logic controller, mobile robot navigation, perception, received signal strength, UKF.

1 Introduction

RFID technology has been extensively deployed recently in the field of robotics thanks to its simplicity, compact size, cheap price, and flexibility [1]. The navigation problem of a mobile robot has been traditionally solved by several approaches suggested in the literature. One of the most popular techniques is the RSS-based navigation/localization using RFID technology where the line-of-sight (LoS) distance (or range) between an RFID tag and a reader is computed by Friss transmission formula [2]. In general, RSS is a feature of choice for estimating range in wireless networks because it can be obtained relatively effortlessly and inexpensively (i.e., no extra hardware is required) [3] which makes it good candidate to solve a localization problem. However, it has lately been discovered that achieving accurate mapping of RSS information to range is almost impossible in real-world reverberant environment [4], depleting the advantages listed above.

Hence, our current work contributes to the development of a mobile robot navigation system where accurate mapping of RSS measurements to LoS distance is not absolutely necessary. The robot is assumed to be equipped with a customized RFID reader and the environment is defined by placing a set of RFID tags at 3-D points whose orthogonal projection points on the ground define the target points to be reached by the robot. The RFID system architecture is discussed later in the paper. The majesty of this work is that the direction of the target point is determined using RSS values which are provided by the customized RFID reader. The robot then updates its orientation towards the target point followed by a set of control actions to reach that point which are determined by the processing of RSS measurements using a fuzzy logic controller (FLC). The robot's position accuracy is refined through a conventional stochastic filter, UKF.

The rest of the manuscript is outlined as follows. Some of the most commonly used RFID-based robot navigation systems are presented in section 2. Section 3 illustrates the proposed navigation framework. The robot's noisy positions and orientations are filtered using the UKF which is described in section 4. The performance of the proposed navigation system is then evaluated numerical experiments reported in section 5. Finally, section 6 concludes with some key advantages of the current research in real-life.

2 Related Study

Due to the recent advancements in RFID technology, and its applications in the field of robotics, localization systems have been used to deliver instantaneous position information in an indoor and/or outdoor environment. A large body of research has considered localizing a mobile robot using the emerging RFID technology. Since an RFID system can recognize objects at high-speed and send data within various distances, RFID technology has been gaining popularity within the robot community [5]. Hahnel et al. studied to improve robot localization with a pair of RFID antennas [6]. They presented a probabilistic measurement model for RFID readers that allow them to accurately localize the RFID tags in the environment.

In addition, robot's position estimation techniques can be classified as range-based and bearing-based [4]. The main idea behind range-based techniques is to trilaterate the robot's position using some known reference points and the estimated distances at those points in the environment. Distances can be estimated from either RSS measurements or time-based methods. A small subset of such works have explored the use of Time of Flight (ToF) [7] or Time Difference of Arrival (TDoA) measurements [8]. On the other hand, bearing-based schemes use the direction of arrival (DoA) of a target. However, these schemes require multiple range sensors in order to be better suited for mobile robot applications [4].

Despite the significant contributions of RFID systems and RSS measurements stated in the literature, the navigation problem of a mobile robot still faces technical challenges. For instance, the performance of RSS-based localization and/or navigation systems highly degrades in a reverberant environment and such systems are not applicable where approximating RSS measurements to LoS distance with a high precision is required. As such, the current work is devoted to the development of an RSS-based navigation system which might be a promising alternative to the existing robot navigation systems to date. It is noteworthy to mention that approximating RSS measurements to LoS distance, which most of the RSS-based navigation systems depend on, is not absolutely necessary in the proposed navigation system. To the best of the authors knowledge, this is the first attempt of finding direction using RSS measurements of a customized RFID system for the purpose of robot navigation in an indoor environment.

3 Proposed Navigation System

A high level setup of the proposed navigation system with a customized RFID reader mounted on a mobile robot is depicted in Fig. 1. The four RFID tags, Tag1, Tag2, Tag3, and Tag4, are attached to the 3-D points of the ceiling of an indoor office environment, for instance. In this configuration, the robot's target points, p1, p2, p3, and p4, are the orthogonal projection points of those tags on the ground. The mobile robot is pre-programed with these four tag IDs (a set of RFID tag IDs in general case) and is supposed to reach p1, then p2, after that p3, and finally p4. For it to do so, the robot continuously reads tag IDs using the customized RFID reader, and extracts and processes RSS values of the signal coming from the current target tag at a particular time instant. The communication between the tags and the reader is performed through the multiple receiving antennas mounted on the reader.

The current work is mainly composed of three components: a mobile robot, a customized RFID system, and a guiding principle. These components rely on the fact that an accurate data processing and control actions are performed by the robot while navigating in its workspace. In the following, we describe these components in detail.

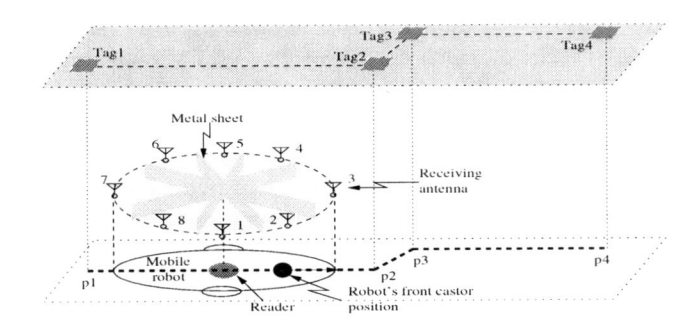

Fig. 1. High level system architecture of the proposed navigation system

3.1 Robot Model

The current work deploys a differential drive mobile robot. To derive its kinematic model, let us consider (x, y) and γ being the position and the heading angle of a robot with respect to a ground-fixed inertial reference frame X-Y. The rotational velocities of the robot's left and right wheels are characterized by the sole (scaler) axis angular velocities ω_L and ω_R, respectively; under the assumption that the front castor wheel is passive and simply provides necessary support for its balance. The robot's position is the midpoint of the wheelbase of length l_b connecting the two lateral wheels along their axis [9].

The discrete-time equivalent of the robot's kinematic model at time k can be derived as

$$q(k) = \begin{bmatrix} x(k) \\ y(k) \\ \gamma(k) \end{bmatrix} = q(k-1) + T\frac{r}{2} \begin{bmatrix} \cos\gamma(k-1) & \cos\gamma(k-1) \\ \sin\gamma(k-1) & \sin\gamma(k-1) \\ \frac{2}{l_b} & -\frac{2}{l_b} \end{bmatrix} \begin{bmatrix} \omega_R(k-1) \\ \omega_L(k-1) \end{bmatrix}, \quad (1)$$

where r is the radius of each wheel, T is the sampling time in seconds, $q(k) \in \mathbb{R}^3$ represents the robot's current pose (position and orientation), and $[\omega_R(k)\ \omega_L(k)]^T$ is the control input u_k for its actuator. It is worth to mention, however, that the low sampling frequency and high velocities of the robot can be a significant source of odometric error. We employ the FLC and the UKF to resolve odometric problem during the robot's navigation.

3.2 Customized RFID System

Most of the commercially available RFID systems on the market provide only static information to its data processing unit which limits its applicability in many real-world proximity-based applications. As such, we pioneer the current work with a customized RFID reader that consists of eight receiving antennas [10]. Note that the tag architecture of the customized RFID system remains the same as that of a typical commercial RFID system. As can be seen in Fig. 1, we customize an RFID reader by designing an antenna array system where eight antennas are placed at fixed angular positions, δ_i with $i = 1, \ldots, 8$, of a circle of radius r_a centered on a horizontal rectangular metal plate. Note that this metal plate (known as ground plane) of dimension $l \times l$ needs to be mounted on the RFID reader which acts as an RF front-end. As noted in the literature, RF signals can easily be distorted due to reverberation, and therefore the error in RSS measurements drastically increases in reverberant environment. As a result, each receiving antenna on the customized RFID reader is spatially separated from the other antennas by vertical rectangular metal shields of dimension $l \times h_p$. The vertical metal shield also reflects the fact that antennas' RSS values of the signal coming from a tag can easily be distinguished from each other.

The receiving antennas are interfaced with the reader's back-end through a 8-to-1 multiplexer and an RSS power detector in addition to its existing analog and digital front-end interface circuits for receiving static binary IDs from RFID tags. The RFID reader can extract RSS values from eight receiving antennas through the multiplexer using time-division multiple-access (TDMA) technique. The eight RSS values along with the current tag ID are then simply passed to the robot's central processing unit for guiding itself in the pre-specified workspace.

3.3 Guiding Principle

The purpose of this section is to compute the direction of the current target RFID tag (DOCT) and then navigate to reach the 2-D point of that tag. The robot mainly remains at two stages: *rotation* and *navigation*, while reaching a set of target points in the workspace. Initially (*source* state), the robot starts its actuators to rotate itself towards the current target point (*rotation* stage) based on its reader's RSS measurements of the target tag. Once the current orientation of the robot is updated, it passes the control to the *navigation* stage and remains in this state by generating appropriate control actions until the target point is reached. If this target point is not the final point in the set of pre-defined target points, then the robot goes back to the *rotation* stage to update its current heading to the next target point. This process continues until it reaches the final destination point (*destination* state). In the following, we illustrate how the two main stages can be implemented.

Tuning Robot's Direction. Suppose that $\bar{m}_i(k)$ is the (noisy) RSS value received by antenna i at time k, $i = 1, \ldots, 8$, where $m_i(k)$ is the ideal value of that signal (without reverberation nor noise). In other words, $\bar{m}_i(k) = m_i(k)G(\delta_i) + v_i(k)$, where $v_i(k)$ is the noise associated with the RSS value received at antenna i and $G(\delta_i)$ is the gain of the antenna in the direction of δ_i. A preliminary study has been conducted to ensure that reverberated noisy signal can be mostly filtered out by a simple moving average filter, which is what we used in this work.

The DOCT is approximated by identifying the antenna with the maximum RSS value. That is,

$$\text{DOCT} = \arg\max_{\delta_i, 1 \leq i \leq 8}(\bar{m}_i(k)). \tag{2}$$

Note that the DOCT can be approximated more precisely by comparing the differences between the RSS values of various antenna combinations, i.e., $\bar{m}_i(k) - \bar{m}_j(k)$, with $i, j = 1$ to 8 and $i \neq j$. The robot then updates its current orientation based on the DOCT.

Reaching Target Points. Once the direction of the current target on the robot's desired path is detected, it rotates itself towards the target tag as described earlier. Its mission is then to reach the target point which is the tag's orthogonal projection point on the ground. For it to do so, a fuzzy logic controller is employed for generating adequate control actions which are based on stochastic RSS measurements provided by the reader.

Fuzzy logic control is of a paramount importance in many control problems due to its appropriateness in various complex or ill-defined systems. It uses a set of well-structured if-then rules to represent system's input-output relationships as opposed to complex mathematical models defining the system. Fig. 2 represents the controller used by the mobile robot. The perception block of the controller aggregates RSS values from eight receiving antennas. However, RSS values from the antenna 2 and 4 are passed to the preprocessing block of the controller for it to reflect that RSS values of these two antennas are sufficient to control speeds of the robot's right and left wheels, respectively. The differential input of the FLC, $\Delta\bar{m}_{24}$, is defined as $\Delta\bar{m}_{24} = \bar{m}_2 - \bar{m}_4$.

The normalized values of the input $\Delta \bar{m}_{24} \in [-1200, 1200]$ nW, are fed to the FLC for producing normalized left and right wheel speeds, ω_{Ln} and ω_{Rn}, respectively. The post-processing block is used to scale the normalized speeds to engineering units. In this work, it contains the output speed gain that is tuned to best fit to the wheel actuators. Hence, the actual speeds of both wheels generated by the post-processing block are ω_R and ω_L which are directly applied to the robot's actuators to obtain its new pose using (1).

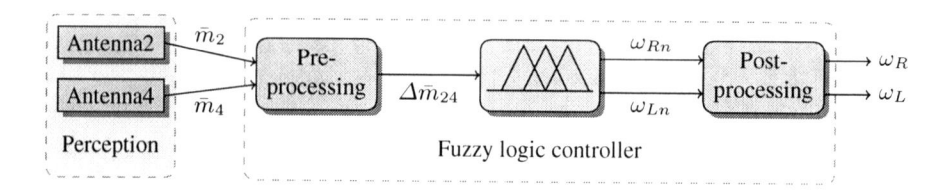

Fig. 2. Robot's speed control mechanism

4 Pose Estimation Using UKF

Although Extended Kalman Filter (EKF) is a widely used state estimation algorithm, it has several drawbacks that can easily be resolved using the Uncented Kalman Filter, UKF. The main advantages of the UKF over the EKF can be sought in [11]. The UKF used in the current work fuses the data from the wheel encoders and RSS measurements in order to better estimate the robot position. It deterministically extracts so-called sigma points from the Gaussian statistics and passes them through the nonlinear model defined in (1).

The UKF localization approach for the robot to approximate its position recursively can be sought in [12]. It integrates the RSS measurements of the RFID reader with the previous pose $q(k-1)$ and the error covariance matrix $P(k-1)$ to estimate its pose $q(k)$ and covariance $P(k)$ at time k.

5 Simulation Results

In this section, we evaluate the performance of the proposed navigation system using a circular shaped differential drive virtual mobile robot with a wheel base of 30 cm and a radius 5 cm of each wheel. The maximum output speed gain of the robot's actuator is 1.5 rad/s. The standard deviation of the left and right wheel rotational speeds are chosen as 0.1 rad/s; and that of the measurement noise associated with the robot's position (x, y) is 0.2 m and the robot's orientation γ is 0.1 rad, respectively. The configuration of the proposed RFID reader antenna system using the commercial simulation software FEKO[1] is shown in Fig. 3. Eight receiving dipole antennas are placed on a circle $45°$ apart from each other. A vertical metal shield is placed between adjacent antennas to

---
[1] www.feko.info

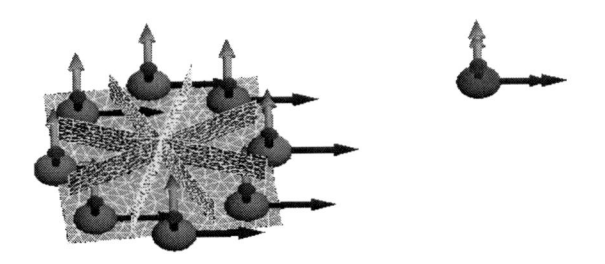

Fig. 3. Antenna system design for an RFID reader using FEKO

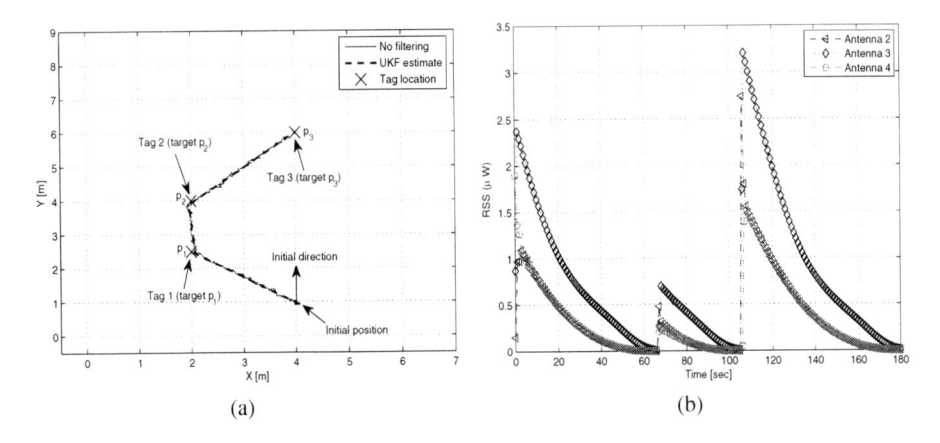

Fig. 4. (a) Robot's actual vs. desired trajectory for a path with three target points, and (b) RSS measurements received during navigation

reduce the effect of RF signal reverberation. The operating frequency of the RFID system used in the current work is 2.4 GHz. The performance metric adopted is the robot's position error e_k which is defined as $e_k = \sqrt{x_e^2 + y_e^2}$, where (x_e, y_e) is the error in the robot's local coordinate system.

The first test bed was adopted by placing three RFID tags with IDs, 1, 2, and 3, at the 3-D positions $(2, 2.5, 3)$ m, $(2, 4, 3)$ m, and $(4, 6, 3)$ m, respectively. Hence, the target points where the robot is supposed to reach are the 2-D points $p_1 = (2, 2.5)$ m, $p_2 = (2, 4)$ m, and $p_3 = (4, 6)$ m on the ground, respectively. Fig. 4(a) represents the actual and desired trajectories of the mobile robot during navigation. Initially, the robot is placed at $(4, 1)$ m with an initial orientation of 90°. It then updates its position and orientation based on the RSS measurements of tag 1 to reach point p_1. The RSS values during navigation are shown in Fig. 4(b). As can been seen, antenna 3 provides the maximum RSS value once the robot's heading is aligned with the target IDs, 1, 2, and 3, respectively. This is because antenna 3 is placed at the front castor position of the robot. As the robot moves towards target points, p_2 and p_3, its corresponding position errors and actual orientations are recorded as shown in Figs. 5(a) and 5(b), respectively. The superiority of applying UKF is clearly observed from the figures in reducing both position and orientation errors. Notice that the position error is enveloped between ≈ 3

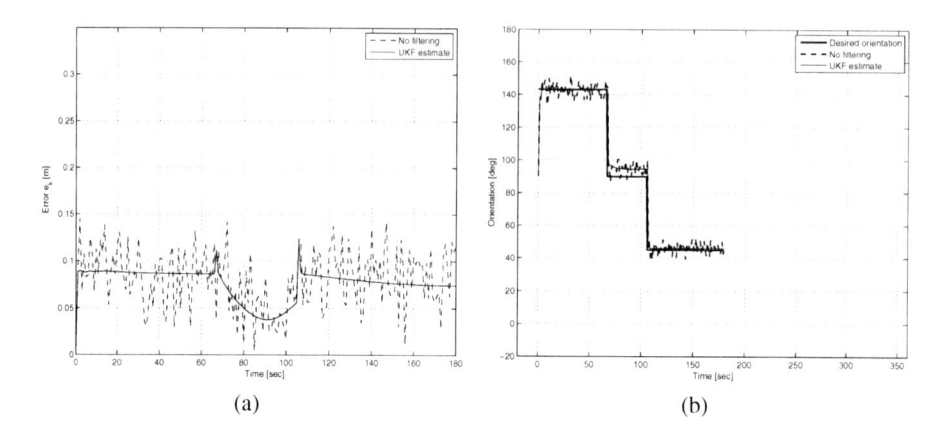

Fig. 5. Navigation system's performance for a path with three target points, (a) position error, and (b) desired vs. actual orientations

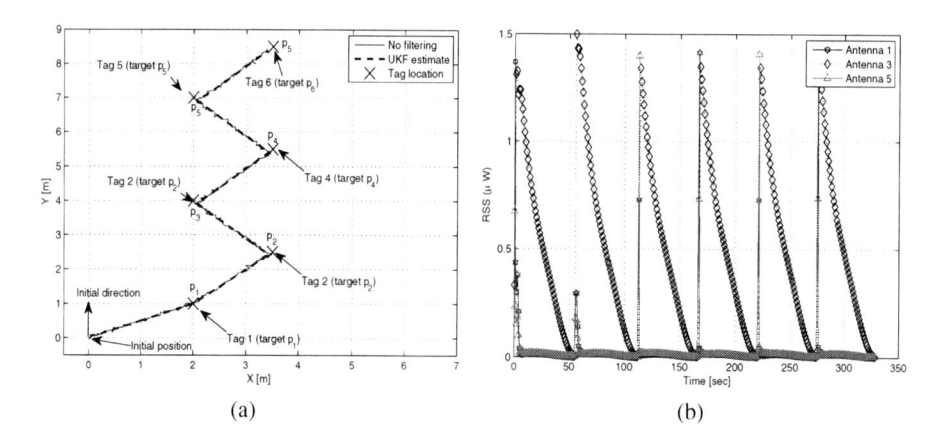

Fig. 6. (a) Robot's actual vs. desired trajectory for a path with six target points, and (b) RSS measurements received during navigation

and 15 cm. As can be seen from Fig. 5(b), the robot's current orientation is highly fluctuating when no filtering technique is applied. The smooth updated orientations, however, can be achieved by using UKF, as expected. The percentile error in Fig. 5(a) reveals that during 98% of the total simulation time, the robot's position error is less than 8 cm when UKF is used whereas it is 15 cm or less when it is not.

To better illustrate the performance of the proposed navigation system, six tags were placed in the environment representing the target points p_1, p_2, p_3, p_4, p_5, and p_6, as shown in Fig. 6(a). The corresponding position errors are presented in Fig. 7(a). The performance of the current navigation system is not affected much by choosing the relative positions of the tags in the sense that the robot could still reach the target points with a satisfactory position error. However, as expected, the orientation errors are higher at the sharp corners of the path which is natural. This is clear from Fig. 7(b). The RSS

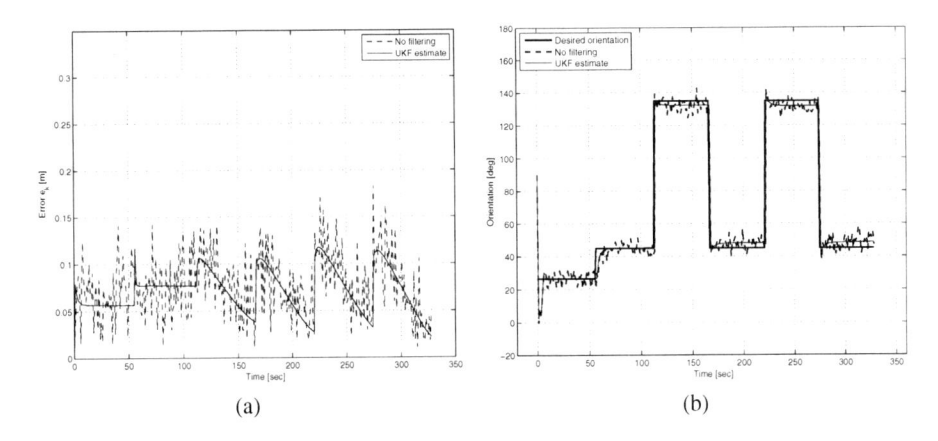

Fig. 7. Navigation system's performance for a path with six target points, (a) position error, and (b) desired vs. actual orientations

values provided by the receiving antennas are presented in Fig. 6(b) which ensures that the front antenna (antenna 3) always provides the maximum RSS values when the robot's front castor is aligned with the target tag as desired. As for the percentile error in Fig. 7(a), 100% of the simulation time the position error is less than or close to 12 cm. However, without the UKF position estimation technique, the position error reaches up to 18 cm. This clearly demonstrates the benefit of using UKF for the proposed navigation technique.

6 Conclusion

In this paper, we proposed a novel RSS-based mobile robot navigation system using a customized RFID reader which opens up a possibility of applying it in an indoor environment. The RSS measurements are employed as key observations since they can be easily obtained without specialized hardware or extensive training. The robot uses RSS measurement to update its position and orientation as it approach a current target point in the workspace. The positions and orientations are then rectified through UKF. The customized RFID reader is simulated using the electromagnetic software, FEKO. The performance of the proposed system is well demonstrated through computer simulations that reflect the validity and suitability of the proposed navigation system to provide high-performance level in terms of navigation accuracy and scalability.

The proposed navigation system has three major advantages that make it good candidate in many real world indoor navigation systems. Firstly, it is independent of the environment where the robot is deployed. Secondly, the odometry information is not absolutely necessary since it accumulates position and orientation errors over time. Thirdly, an accurate mapping of RSS information to distance is not required as it is the case for most RFID based localization system.

References

1. Finkenzeller, K.: RFID Handbook, 2nd edn. Wiley, Swadlincote (2004)
2. Bekkali, A., Matsumoto, M.: RFID indoor tracking system based on inter-tags distance measurements. In: Wireless Technology. LNEE, vol. 44, pp. 41–62. Springer, Heidelberg (2009)
3. Youssef, M.: The Horus WLAN location determination system. Ph.D. dissertation, University of Maryland, Maryland (2004)
4. Kim, M., Chong, N.Y.: Direction sensing RFID reader for mobile robot navigation. IEEE Transactions on Automation Science and Engineering 6(1), 44–54 (2009)
5. Kulyukin, V., Gharpure, C., Nicholson, J., Pavithran, S.: RFID in robot-assisted indoor navigation for the visually impaired. In: 2004 IEEE/RSJ IROS, Sendai, Japan, pp. 1979–1984 (2004)
6. Hahnel, D., Burgard, W., Fox, D., Fishkin, K., Philipose, M.: Mapping and localization with RFID technology. In: Proceedings - IEEE ICRA, New Orleans, LA, United States, April 2004, vol. 1, pp. 1015–1020 (2004)
7. Lanzisera, S., Lin, D.T., Pister, K.S.J.: RF time of flight ranging for wireless sensor network localization. In: Proceedings of the Fourth Workshop on Intelligent Solutions in Embedded Systems, WISES 2006, Vienna, Austria, June 2006, pp. 165–176 (2006)
8. Ni, L.M., Liu, Y., Lau, Y.C., Patil, A.P.: LANDMARC: indoor location sensing using active RFID. In: Proceedings of the First IEEE International Conference on Pervasive Computing and Communications, Fort Worth, TX, USA, pp. 407–415 (2003)
9. Miah, S., Gueaieb, W.: A stochastic approach of mobile robot navigation using customized rfid systems. In: International Conference on Signals, Circuits and Systems, Jerba, Tunisia (November 2009)
10. Miah, S., Gueaieb, W.: On the implementation of an efficient mobile robot navigation system: An RFID approach. In: Proceedings of the International Conference on Intelligent Autonomous Systems (IAS 2010), Ottawa, Canada (August 2010)
11. Julier, S.J., Uhlmann, J.K.: New extension of the Kalman filter to nonlinear systems. In: Signal Processing, Sensor Fusion, and Target Recognition VI, Orlando, FL, USA, vol. 3068, pp. 182–93 (1997)
12. Thrun, S., Burgard, W., Fox, D.: Probabilistic Robotics. The MIT Press, London (2005)

Toward Opportunistic Collaboration in Target Pursuit Problems

Soheil Keshmiri and Shahram Payandeh

Experimental Robotics Laboratory, School of Engineering Science,
Simon Fraser University, 8888 University Drive, Burnaby, B.C. Canada V5A 1S6
ska61@sfu.ca, shahram@cs.sfu.ca

Abstract. This paper proposes an opportunistic framework to modeling and control of multi-robots moving-target pursuit in dynamic environment with partial observability. The partial observability is achieved via the introduction of third party agent (referred to as mediator) that transforms the target's as well as group members' positioning information into the robotic agents' common knowledge, thereby eliminating the necessity of direct inter-robots communication. The robotic agents involved are modeled as fully autonomous entities, capable of determining their corresponding action profiles, using a strategy inference engine. The robot's inference engine consists of two sub-rating components viz. fixed or predefined sub-rating, and the variable or the opportunistic sub-rating. The action profiles at individual level are further analyzed by the mediator that finalizes the agent's action assignment at every execution cycle. It has been proven that addition of the third party mediator guarantees the optimality of group level performance.

Keywords: Multi-agent systems, Cooperative intelligent systems, Autonomous and intelligent robotics.

1 Introduction

One of the most fascinating and yet highly challenging problem in such areas as robotics and computer games is undoubtedly the *target pursuit* problem. Police car-chase [1], animal-hunting [2], etc. are among scenarios in which the target-pursuit capability plays an important role. What, in its general sense, is refereed to as multi-robots target pursuit problem, however, is in fact combination of several sub-problems namely, navigation control of the robotic agents involved [3], tracking the target [4,5], and coordinating the robotic group formation [6,7]. The problem may even sometimes consists in searching for the target [8] as well. The pursuit problem was originally proposed in [9] where a centralized coordination approach for pursuing a prey with four predators in a non-hazy grid world was presented. The agents were allowed to move only in horizontal and vertical directions whereas the prey followed a random movement. The game-theoretic approach in [10] attempts to address the pursuit problem in a non-hazy grid world, where the predators are coordinated implicitly by incorporating the global goal of a group of predators into their local interests using a payoff function. A coordination algorithm for an obstacle-free environment with static target is presented in [11] which is based on the multi-agent version of [12]. The generic algorithm in an obstacle-free grid world is used in [13] to evolve (and co-evolve) neural network controllers, rather than program controllers. In their work, the authors used the term co-evolution to refer to maintaining and evolving individuals for taking different roles in a pursuit task. In [14], the decision mechanism of the agents is divided into two hierarchical levels viz. the agents' learning level for target selection and the action selection level that is directed towards agent's respective target. Algorithm in [15] implements a fuzzy cost function controller in a game-theoretic framework for multi-robots pursuit of a moving target in an obstacle-free environment. Authors in [16] present a multi-agent real-time pursuit algorithm, MAPS, which employs two coordination strategies called blocking escape directions (BES) and using alternative proposals (UAL). The paper also contains a very useful literature review on the topic of multi-agents target pursuit.

The proposed approach in present work views the problem of action selection among group of autonomous robots as an instance of distributed probabilistic inference in which the joint probability distribution over the set of available actions is constructed as per individual robots' observations. By *autonomy* here, we refer to a control strategy in which no external mechanism may involve in robot's locomotion and decision-making processes at the individual level other than the robot itself. The environment in which the robotic team is situated is highly cluttered and is, in its very nature, dynamic, since the robotic agents are required to prevent any contingent collision with not only the obstacles but one another as well. In present approach, no explicit inter-robots communication is necessary, instead, the robots attain necessary information such as other group members' as well as target's relative positioning information via communicating with a third party agency, the *mediator*. The mediator analyzes the ranked rating of the individual robots on the available action set and resolve any contingent conflict among agents.

The remaining of the paper is organized as follows: Problem description is given in section 2. Overall architecture of the proposed system is explained in Section 3. Sections 4 and 5 elaborate the opportunistic inference engine and the mediation units. Simulation result and performance analysis are provided in section 6. Conclusion and some insight on future direction are presented in section 7.

2 Problem Description

In a dynamic target pursuit mission, there are several crucial factors that may significantly constrain the agent's probability of success. Robot-target distance, energy level of the robot, presence of the obstacle(s) along the robot's path, terrain's condition, etc. are among such factors. These factors might be pondered as elements of the set χ that represents *robot's state* with regards to its pursuing target as well as the environment in which it is situated. Problem of the multi-robots, real-time moving-target pursuit can then be expressed as:

Definition 1 (Multi-Robots Pursuit). Given a group of robots' positioning information $r_i \in \mathbb{R}^d$, $(d \geq 2)$, i=1...n, a set χ_i that represents i^{th} robot's state with regards to its pursuing target, and the target's positioning information τ, how to find set of actions $a_i \in A$, where A represents the set of all available actions to the robotic group while pursuing the target, such that the set of all assigned actions $A' = \{a_i\}, A' \subseteq A$ to the robotic members be a conflict-free set i.e. $\forall a_i, a_j \in A', a_i = a_j \leftrightarrow i = j$, $i, j \leq n$ and that A' maximizes the group level probability of success. Other word:

$$\pi(A') \in argmax_{a_i \in A'} \sum_{i=1}^{n} \pi_i(a_i) \qquad (1)$$

Here, $\pi_i(a_i)$ refers to the ith robot probability of success, given the action a_i and the robot's state χ_i, i.e.

$$\pi_i(a_i) \equiv \pi_i(a_i|\chi_i), \ a_i \in A \qquad (2)$$

3 System Architecture

Fig. 1 shows the overall architecture of the proposed system. The architecture consists of two main components, namely the *robotic agent* and the *mediation unit*. The mediation unit comprises two data structures: *public Information Board (PuIB)* where the agents' common knowledge data such as their own as well as target's positioning information are stored and shared with all the robotic members. The other data structure in this module is the *private Information Board (PrIB)* where the individual agents'

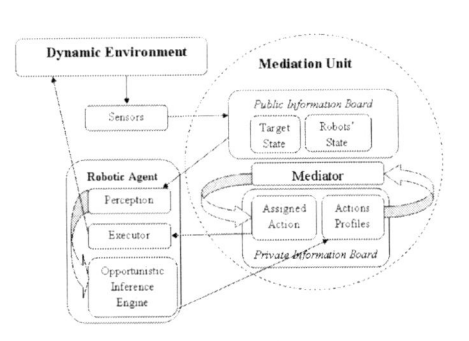

Fig. 1. Overall system architecture

action profiles and the agent's specific robot-action assignment by the mediator are stored. The entries of the private portion is only accessible to the designated robot and the final decisions are not shared by the members at different execution cycles. The other module of the mediation unit is the *mediator* whose functionality will be elaborated in section 5. The actual physical robot (second component of the architecture) obtains its own, its group members' and the target's positioning information through the sensory devices as well as the *PuIB* and passes them on to its *opportunistic inference engine*. The *opportunistic inference engine* is the module in which individual agent's reasoning and rating regarding the actions available to the agent takes place. The outcome of this module is the robot's *ordered action profile* that contains the rating i.e. robot's own estimates on its probability of success, performing either of the available actions. The profiles are ordered incrementally and as per their corresponding probability of success from the agent's perspective and are the inputs to the actions profile sub-module of the *PrIB*.

4 Robot's Opportunistic Inference Engine

As an autonomous entity, every agent would try to increase its probability of success, given the factors such as computed target's approaching points by the robot (i.e. directions from which robot may approach the target while pursuing it, given the current robot and target locations), its present energy level, presence or absence of the obstacles, etc. Such probability is, in essence, the robot's gain, that has been achieved through the execution of the robot's proposed action i.e.

$$\pi_i(a_i) = \sum_{a_i \in A_i} \sum_{x \in \chi_i} \pi_i(a_i|x) P(a_i) \qquad (3)$$

Where π_i, a_i, and χ_i are the gain, the action, and the robot's state with regards to the pursuing target as well as the robot's environment, respectively. A_i is the complete set of actions that are available to i^{th} agent. The ith agent's probability of success over the available action set to the agent i.e. A_i are computed such that:

$$0 \leq \pi_i(a_i) \leq 1$$
$$\sum_i \pi_i(a) = 1, \ \forall a \in A_i \qquad (4)$$

4.1 Action Selection through Two-Level State-Rating System

As identified in [17], there are two types of knowledge available to the agents viz. externally ascribed knowledge i.e. the default type of knowledge, hardwired into the agent (e.g. robot-task pre-assignment, etc.), and the explicit knowledge or knowledge that is acquired through the agent-environment interaction. Despite the fact that default knowledge may not thoroughly capture the dynamic and the uncertainty of the robotic agent's environment, it may still be considered as a valuable resource while calculating agents' gains and hence the agents' confidence over their proposed action profiles. Other word, the agent may incorporate their deterministic views of the surrounding environment into the action selection mechanism [19], thereby taking advantage of their default perspective of their tasks and environment dynamic once no further decision might be concluded. Letting the default and the explicit states of the robotic agent be represented by x^d and x^o, respectively, (3) can be restated as:

$$\pi_i(a_i) = \sum_{a_i \in A_i} \sum_{x \in \chi_i} \pi_i(a_i|x^d) P(a_i) + \pi_i(a_i|x^o) P(a_i) \qquad (5)$$

The default state, x^d, of the robotic agents are fixed-valued ratings of the agents' state and is initialized at the mission commencement and remains the same throughout the course of operation. Whereas the opportunistic state, x^o, is updated upon robots' conclusions on their action profiles at each iteration.

Definition 2. An action $\hat{a}_r$ selected by robot **r** is an optimal action if:

$$\pi_r(\hat{a}_r) \geq \pi_r(a), \ \forall a \in A_r \qquad (6)$$

i.e. $\hat{a}_r \in argmax \ \pi_r(a)$, $a \in A_r$, where A_r represents the set of all available actions to **r**.

Theorem 1. Action profile computed by every individual robot's opportunistic inference engine, is a profile that is ordered by the robot's probability of success for every available action and that contains the individual's optimal action, given the robot's state χ.

Proof. Let A_i be the set of all available actions to the ith robot, $i = 1 \ldots n$, with n being the total number of the robots. Then for every action $a_i \in A_i$ we have:

$$\pi_i(a_i) = \sum_{a_i \in A} \sum_{x \in \chi_i} \pi_i(a_i|x^d)P(a_i) + \pi_i(a_i|x^o)P(a_i) \qquad (7)$$

Let Π_i be the set of all such values for the ith robot in which $\pi_i(a_i) \geq 0$, i.e.

$$\Pi_i = \{\pi_i(a_i)|a_i \in A_i \ \& \ \pi_i(a_i) \geq 0\} \qquad (8)$$

if Π_i is a singleton set, then the theorem is trivial. If $\|\Pi_i\| > 1$ where $\|\Pi_i\|$ denotes the number of elements in Π_i, however, there exists at least one element $\pi_i^{'}(a_i) \in \Pi_i$ that:

1. $\pi_i^{'}(a_i) > \pi_i(a_i)$, $\forall \pi_i(a_i) \in \Pi_i$, then $\pi_i^{'}(a_i)$ strongly dominates all the members and is the one that maximizes the ith agent's probability of success.
2. $\pi_i^{'}(a_i) \geq \pi_i(a_i)$, $\forall \pi_i(a_i) \in \Pi_i$, then $\pi_i^{'}(a_i)$ weakly dominates all the members and is the one that maximizes the probability of success of the agent i.
3. $\exists \pi_i(a_i) \in \Pi_i$, $\pi_i(a_i) = \pi_i^{'}(a_i)$, then the agent i is indifferent between the two and either choice would maximize the agent's probability of success.

Theorem 2. For a robot **r** whose optimal action is denoted by $\hat{a}_r$, any set of actions A^*, $\|A^*\| \geq 1$, $\hat{a}_r \in A^*$, and $\pi_r(a^*) > 0$, $a^* \in A^*$, it is true that: $\pi_r(a^*) \leq \pi_r(\hat{a}_r)$, $\forall a^* \in A^*$, given the state of the robot χ_r.

Proof. If A^* is a singleton set, then the result is trivial. Let $\|A^*\| > 1$, $\hat{a}_r \in A^*$. If $\exists a^* \in A^*$ whose gain is higher than that of the $\hat{a}_r$, using (3) we have:

$$\pi_r(\hat{a}_r) = argmax_{x \in \chi_r} \sum_{a \in A^*} \pi_r(a|x)P(a)$$

$$= argmax_{x \in \chi_r} \sum_{a \neq a^*} \pi_r(a|x)P(a) + \pi_r(a^*|x)P(a^*)$$

$$< argmax_{x \in \chi_r} \sum_{a \neq a^*} \pi_r(a^*|x)P(a) + \pi_r(a^*|x)P(a^*)$$

$$= \pi_r(a^*) \qquad (9)$$

A contradiction to the original assumption of $\hat{a}_r$ being optimal.

Theorem 1 states that every robot is capable of determining its corresponding action profile that would comprise the robot's individual optimal action. Whereas theorem 2 expresses that for a robot, determining one ordered action profile that contains its optimal action would suffice as any other possible action profile's optimal action would essentially be at most as good as the one that has already been calculated.

5 Mediator

Group level optimality is essentially analogous to maximization of the joint probability of success of the entire robotic team, i.e. maximizing the sum of the gains of all the members under the restriction of no two teammates doing the same task [18]. To ensure aforementioned requirement, every robotic agent communicates its set of ordered actions profile with the mediator. The mediator then analyzes all the robots' sets of actions to ensure no conflict among the selected actions exists. Every action $a_i \in A_i$ for the ith robot, is examined against the rest of the group members' action sets and will be assigned to the ith robot if it is conflict-free, i.e. if no other group member has chosen it. Otherwise, the action will be assigned to the robot with the highest probability of success and will be removed from the available action entries. The robot whose action has been eliminated, will be assigned with the action in its updated action set that has the highest probability of success. The procedure will be repeated till all the robots are assigned with an action.

Proposition 1 Mediator's action assignment preserves group level action selection optimality.

Proof. 1. **Non − conflicting action selection**: No conflict among agents means:
$$\nexists a_i, a_j \in A, \ a_i = a_j, \ i \neq j, \ i,j = 1 \ldots n \tag{10}$$
With n and A being total number of agents and the full action set available to all the agents, respectively. This implies that $\forall i, j = 1 \ldots n, \ a_i \neq a_j \ \& \ \hat{a}_i = argmax_{a \in A} \ \pi_i(a), i = 1 \ldots n$, is satisfied which would in turn satisfy (1).
2. **Conflicting action selection**: In such a case, agents with no conflicting choice would perform their preferred, maximizing actions, respectively. For the conflicting agents, however, it is true that:
$$\pi_i(\hat{a}) \geq \pi_j(\hat{a}), \ i \neq j \tag{11}$$
with $\hat{a}$ being a joint maximizing action that causes the conflict between the ith and j^{th} agents. As per (11), conflicting action would be assigned to the jth agent due to higher gain obtained, hence preserving the agent's gain maximization action. For the ith agent, it is true that:
$$\exists a' \in A - \{\hat{a}\} \ \& \ a' \in argmax_{x_i \in \chi_i}$$
$$\sum \pi_i(a_i|\chi_i)P(a_i), \ a_i \in A_i - \{\hat{a}\} \tag{12}$$

That is, every robot action profile and due to result obtained in Theorem 1 would comprise an action that would maximize the robot's probability of success after the elimination of the conflicting action. Equation (12) implies that, every robot selected action satisfies (1) at group level and the definition (2), given the available set of action to the robot.

6 Case Study

Following assumptions have been made while performing the experiments:

1. Target's positioning information is always known to the mediator.
2. Every robotic agent is capable of communicating with mediation unit. Individual agents update their corresponding positioning information in *PuIB* (Fig. 1) and provide the mediator with their finalized selected action profiles.
3. Positioning information of the target as well as those of the robots is accessible with all the robotic members of the group, i.e. they form the common knowledge of the group.
4. Robotic teammates do not share their decisions over their action profiles.

Target is equipped with a kernel-based navigation controller [20] and is assumed to move with constant velocity. While performing the pursuit mission, every robot calculates series of approaching points (i.e. points based on which robot may approach the target). To calculate the approaching points, robots use the methodology adapted in [7] where the target is considered to be the point that minimizes the sum of the distances of all the robots to it. As per number of approaching points calculated (three approaching points, for instance, in current implementation) every robot calculates its action profile, using its *opportunistic inference engine*. The action profile of the ith robot is in fact an ordered tuple $(a_i{}^1, \ldots, a_i{}^m)$, $a_i \in A$, $i = 1 \ldots n$, $n \leq m$ where m represents the action selected by the ith agent, would the agent select the mth computed approaching point to pursue the target. Set A is the complete set of actions available to all agents, n represents the total number of robotic agents and m is total number of calculated approaching points ($n \leq m$), respectively. As shown in Fig. 2(a), robot's action profile indicates the corresponding agent's estimates of its probability of success (i.e. robot's gain) as per different calculated target's approaching points (*TAP*). Once calculated, every robot communicates its corresponding ordered action profile with the mediation unit. At group level and referring to Fig. 2(b), mediator analyzes robots' action profiles information and resolve any possible conflict that may arise among agents while carrying out their selected action. Fig. 2(c) shows robots' communicated action profiles (left column, plots 1, 3, 5, and 7) and the mediator's finalized approaching point assignments (right column, plots 2, 4, 6, and 8) at different stage. As shown in the plots 1 through 4, robots two and three are competing over the target's approaching point (TAP) 3, as they both achieve their highest gain i.e. probability of success by pursuing target, choosing $TAP3$ over the other available choices. Robot 1, however, achieves its highest gain by choosing target's approaching point (TAP) 1. Robot

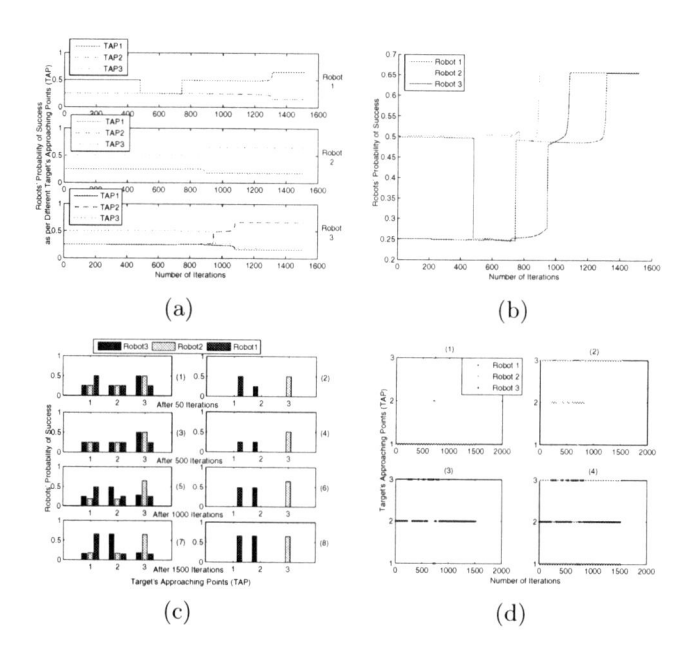

Fig. 2. (a) Individual robot's inference engine estimates of success as per different target's approaching points(TAP). Different curve corresponds to evolution of different robot-TAP probability of success i.e. robot's gain on selecting that particular approaching point over the other possible choices. The robots' gain over all possible approaching points satisfy $0 \leq \pi_i(a_i) \leq 1$ and $\sum_i \pi_i(a) = 1$, $a \in A_i$. (b) Mediator's finalized robot-TAP assignment at every iteration and as per group level optimality. Colors are referring to assigned actions' probability of success at each iteration for robot1 (red), robot2 (green), and robot3 (blue), respectively. (c) Probability of success for different target's approaching points calculated by the individuals (left column) and the finalized approaching point assignment to the individuals by the mediator (right column) after 50 (1 & 2), 500 (3 & 4), 1000 (5 & 6), and 1500 (7 & 8) iterations. (d) Robot-TAP assignments by the mediator to (1) robot1, (2) robot2, (3) robot3 at every execution cycle and the overall group performance as per robots-target approaching point assignment (4).

one's choice has no conflict with other teammates and hence is approved by the mediator (plots 2 and 4). For the robots with conflicting choices, the mediator assigns the concerned conflicting point (i.e. TAP number 3) to robot 2 and instructs robot 3 to pursue the target using TAP number 2. Fig. 2(d) represents the robot-target approaching point assignment throughout the pursuit mission and at every iteration cycle.

7 Conclusions and Future Works

An attempt towards collaborative target pursuit mission with a group of mobile robots has been presented. It has been shown that individual agent's perspective regarding the task and the environment in which they are situated in, might be considered as the basis upon which the agent probability of success in the mission can be calculated. Such probabilistic view of the agent towards its task might be interpreted as the agent's probability of success and hence the level of commitment to different possible actions that are available to agent. It has been proven that, such approach may provide every individual robot with the opportunity of calculating its own action profile and as per agent's perspective regarding the task. Such action profile is essentially the profile that contains robot's optimal action at individual level. It has further been shown that, group level optimality of individual action assignment can be achieved with the aim of a mediator whose primary goal is to resolve the conflict-of-interest among robotic agents. Such approach would not only preserve the autonomy of the individuals while interacting with their environment but would also provide the group with a decision-making opportunity that would involve every members of the team and their perspective while delivering the designated task.

The pursuing target's motion in present work, however, lacks any randomness and/or uncertainty that might be exhibited in real-life pursuit mission. The simulated environment also does not include any moving obstacles that would, in turn, increase the complexity of decision-making process. Exploration and addressing aforementioned limitations of the present approach in a very complex environment are subject of future research.

References

1. Cliff, D., Miller, G.F.: Co-Evolution of Pursuit and Evation II: Simulation Methods and Results, Technical Report CSRP377, COGS, University of Sussex (1995)
2. Cliff, D., Miller, G.F.: Co-Evolution of Pursuit and Evation I: Biological and Game-Theoretic Foundations, Technical Report CSRP311, COGS, University of Sussex (1994)
3. Keshmiri, S., Payandeh, S.: An Overview of Mobile Robotic Agents Motion Planning in Dynamic Environments. In: Proceedings of the 14th IASTED International Conference, Robotics and Applications (RA20), MA, Boston, pp. 152–159 (2009)
4. Chung, C.F., Furukawa, T.: Coordinated Pursuer Control using Particle Filters for Autonomous Search-and-Capture. Robotics and Autonomous Systems J. 57, 700–711 (2009)
5. Meng, Y.: Multi-Robot Searching using Game-Theory Based Approach. Int. J. of Advanced Robotic Systems 5(4), 341–350 (2008)
6. Balch, T., Arkin, R.C.: Behavior-Based Formation Control for Multi-Robot Teams. IEEE Trans. on Robotics and Automation 14(6), 926–939 (1998)
7. Keshmiri, S., Payandeh, S.: Isogonic Formation with Connectivity Preservation for a Team of Holonomic Robots in a Cluttered Environment. In: IEEE/WIC/ACM International Conference on Web Intelligence and Intelligent Agent Technology (WI-IAT), Toronto, ON, vol. 3, pp. 346-349 (2010)
8. Bourgault, F., Furukawa, T., Durrant-Whyte, H.F.: Coordinated Decentralized Search for a Lost Target in a Baysian World. In: Proceedings of the 2003 IEEE/RSJ International Conference on Intelligent Robotics and Systems, Las Vegas, NE, pp. 48–53 (2003)

9. Benda, M., Jagannathanand, V., Dodhiawalla, R.: On optimal cooperation of knowledge sources, Technical Report No.BCS-G2010-28, Boeing Advanced Technology Center (1986)
10. Levy, R., Rosenschein, J.: A game theoretic approach to the pursuit problem. In: 11th international workshop on distributed artificial intelligence (1992)
11. Kitamura, Y., Teranishiand, K., Tatsumi, S.: Organizational strategies for multi-agent real-time search. In: Proceedings of international conference on multi-agent systems (ICMAS 1996), pp. 150–156 (1996)
12. Korf, R.: A layered approach to learning coordination knowledge in multiagent environments. Artificial Intelligence J. 42(2-3), 189–211 (1990)
13. Yong, C., Miikkulainen, R.: Cooperative coevolution of multi-agent systems, Technical report: AI01-287, University of Texas at Austin (2001)
14. Erus, G., Polat, F.: A layered approach to learning coordination knowledge in multiagent environments. Applied Intelligence 27(3), 249–267 (2007)
15. Harmati, I., Skrzypczyk, K.: Robot Team Coordination for Target Tracking using Fuzzy Logic Controller in Game Theoretic Framework. Robotics and Autonomous Systems J. 57, 75–86 (2009)
16. Undeger, C., Polat, F.: Multi-agent real-time pursuit. Autonomous Agent and Multi-Agent Systems J. 21, 69–107 (2010)
17. Halpern, J.Y., Moses, Y., Vardi, M.Y.: Algorithmic Knowledge. In: Proceedings of the Fifth Conference on Theoritical Aspects of Reasoning about Knowledge (TARK), pp. 255–266 (1994)
18. Dias, M.B., Zlot, R., Kalra, N., Stentz, A.: Market-based multirobot coordination: A survey and analysis. Proceedings of IEEE 94(7), 1257–1270 (2006)
19. Ordean, A., Payandeh, S.: Design and Analysis of an Enhanced Opportunistic System for Grasping through Evolution. In: The Third ECPD International Conference on Advanced Robotics, Intelligent Automation and Active Systems, pp. 239–244 (1997)
20. Keshmiri, S., Payandeh, S.: Robot Navigation Controller: a Non-Parametric Regression Approach. In: IFAC Workshop on Intelligent Control System(WICS 2010), Sinaia, Romania, September 29-October 2 (2010)

A Multi-agent Framework with MOOS-IvP for Autonomous Underwater Vehicles with Sidescan Sonar Sensors

Liam Paull[1], Sajad Saeedi G.[1], Mae Seto[2], and Howard Li[1]

[1] COllaborative Based Robotics and Automation (COBRA) Group
University of New Brunswick
[2] Mine and Harbour Defense Group
Defense R&D Canada - Atlantic

Abstract. In this research, a framework for building Autonomous Underwater Vehicle (AUV) control algorithms that is based on the MOOS-IvP middleware is presented. The Sidescan Sonar Sensor (SSS) is commonly used to generate sonar images in which mine-like objects can be identified. A common mission specification would be to cover an entire area of seabed up to a specified confidence with the SSS. Here, a base station community is implemented that maintains a map of coverage confidence of the SSS, and provides the user with 2D and 3D simulations and the ability to implement advanced control schemes to achieve this mission. The development happens in two stages: 1) A minimalist configuration where only the necessary applications are used to develop and test outer loop control, and 2) A configuration that includes simulated hardware. The benefits are ease of use, faster development, and reduced hardware testing and cost.

Keywords: Multi-Agent Systems, Autonomous Underwater Vehicles, Sidescan Sonar Sensor, MOOS-IvP.

1 Introduction

Robotics systems and missions are becoming more complex. As a result, it is important for agents and sub-systems to work in parallel and share information to achieve difficult tasks more efficiently. One such task is underwater Mine Counter-Measures (MCM). The US Navy has referred to underwater mine removal as the most problematic mission facing unmanned undersea vehicles and the Navy at large [6]. Using multiple vehicles increases the efficiency and quality of results through data redundancy and increased robustness, and also allows for missions that would be infeasible with only one vehicle through heterogeneous vehicle configurations [12].

The Mine and Harbor Defense group at Defense R&D Canada - Atlantic (DRDC-A) conducts research and development on concepts which increase the autonomous decision-making capabilities of unmanned systems. One of these techniques is to use collaborative systems for countering undersea threats. The use and coordination of multiple vehicles tends to increase the complexity and the number of operators required. The use of a Multi-Agent System (MAS) can mitigate this to an extent. In particular, DRDC-A and the Canadian Navy are experimenting with concepts involving the use of multiple Autonomous Underwater Vehicles (AUVs) for underwater exploring, mapping and mine-hunting surveys. Such AUVs have on-board Sidescan Sonar Sensors (SSS) that generate sonar imagery of the seabed.

Architectures developed for ground or air based systems often do not apply directly to the undersea environment because of the extremely limited communication capabilities and lack of access to GPS.

In [12], a multi-agent architecture is developed for AUVs that uses a blackboard-based broadcast communication system called DELPHIS. Here, the system has three main components: the mission controller, the AUV database, and the mission model. The potentially limiting aspect of this framework is that distinct plans are generated for each AUV in the 'collective' and, as such, it will be difficult for them to achieve global goals efficiently.

In [11], an auction-based system is developed for multi-robot naval MCM called DEMiR-CF. As is common in many multi-robot undersea MCM schemes, one agent is responsible for performing a fast area coverage, and one agent is responsible for investigating further each mine-like object (MLO) detected. The 'cover' and 'identify' tasks are optimally allocated to available resources.

The Mission-Oriented Operating Suite (MOOS) is a middleware that was developed specifically for underwater robots by Paul Newman. The Interval Programming (IvP) Helm was written by Michael Benjamin and works as a MOOS application that implements multi-objective behavior-based control [4].

Eickstedt, et al. have used MOOS-IvP to support sensor driven missions where the robotics platform must adapt its mission plan based on real-time measurements [1]. Their platform is also capable of joint control with other sensing agents. Jiang et al. have developed an auction-based multiple AUV system with MOOS-IvP through the development of bidder and auctioneer applications. However, this auction system is heavily dependent on reliable communication between the auctioneer and the bidder [2].

In general, most previous underwater MASs suffer from overly simplistic and inefficient mission specifications, heavy reliance on reliable communication of unrealistic amounts of data, and system configurations that are not intuitive for users who are not expertly trained in the system.

In this research, a multiple AUV framework is built on top of the existing MOOS-IvP modules that is extensible, reliable, scalable, and easy to use. A base station community is implemented that will allow researchers to: 1) Visualize the AUV in the environment in 2D and 3D, 2) Monitor the coverage confidence over the environment, and 3) Implement long term path planning strategies more easily to achieve missions more efficiently. Only the pose of the vehicles and the waypoints generated at the base station need to be transmitted. The inner loop control, sensing, actuation, and navigation is performed on-board the AUVs.

This framework allows for rapid development of more complex missions. In addition, the need for expert knowledge of hardware programming is eliminated.

The remainder of the paper will be structured as follows: Section 2 will provide some background on the SSS and MOOS-IvP, Section 3 will describe the MOOS-IvP multi-agent framework, and Section 4 will discuss some conclusions and future work.

2 Background of Research

This section will give some background on the sidescan sonar sensor and the MOOS-IvP middleware.

2.1 Sidescan Sonar Sensor

As discussed in [10], most path planning research has assumed that the robot is equipped with a forward-looking sensor with a two-dimensional footprint. However, in this case, the area covered by the sensor is assumed to be effectively one dimensional perpendicular to the direction of motion. As a result, the sensor only covers an area of the map while the AUV is in motion. In addition, data obtained from the SSS is only considered meaningful while the AUV is in rectilinear motion. Fig. 1 shows an example of the area covered by the sensor for a given path.

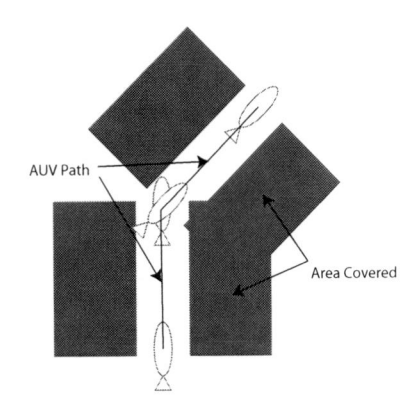

Fig. 1. An example of the AUV path and corresponding area covered by its sidescan sonar sensor

One of the most important considerations of the SSS is that the area directly underneath the AUV is not covered. It should be noted that the entire area covered by the sensor will not be covered with the same sensor performance. The sensor performance is a function of many factors. The sensor performance is defined by the $\mathcal{P}(y)$ curve which defines the confidence that a target is correctly classified as a function of lateral distance from the AUV. Three sample $\mathcal{P}(y)$ curves are shown in Fig. 2, one each for sand, cobble and clay seabed types. The curves are computed using the method described in [5].

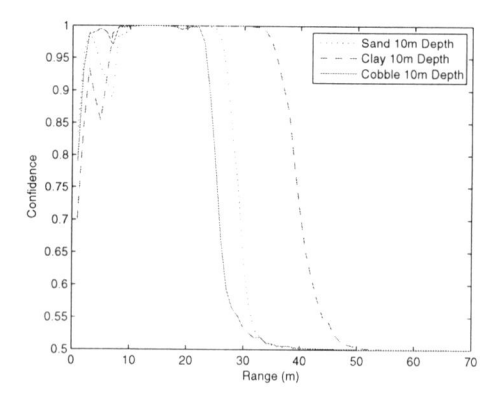

Fig. 2. $\mathcal{P}(y)$ curves for three different seabed conditions

2.2 MOOS-IvP

MOOS is a middleware used by the US navy to implement multi-agent systems. The *IvPHelm* application was later built so that higher level behaviors can be defined.

MOOS-IvP utilizes three main philosophical frameworks: the backseat driver, publish-subscribe, and behavior-based control.

Backseat Driver: In the backseat driver paradigm, the vehicle control and autonomy are decoupled. Essentially, MOOS does not worry about how the vehicle's navigation and control system function, as long as it receives accurate information from the sensors, and the vehicle can act upon the autonomy decisions that are sent back.

Publish-Subscribe: All communication goes through the *MOOSDB*. MOOS applications publish events, which are variable-value pairs and applications also subscribe to different variables. Most of the MOOS applications will normally be sensors, actuators or the visualizers etc., but one is special: the *pHelmIvP* application. The *pHelmIvP* is responsible for reconciling all the desired behaviors.

Behavior-Based Control: Different desired behaviors are defined in the mission behavior file. Behaviors can be selected from amongst predefined behaviors, or can be defined from scratch. On each iteration, the *pHelmIvP* generates an objective function from each behavior, and generates a 'decision' consisting of variable-value pairs that define the action that will optimize the multiple objectives. In addition, attributes of the behaviors can be assigned dynamically using mode variables and hierarchical mode declarations.

More details are provided in the MOOS-IvP documentation[4].

3 MAS Framework with MOOS-IvP

In the proposed MAS framework, there are separate communities for each AUV and one for the base station. Several applications run on each community, some of which are included in a MOOS-IvP distribution and some of which were developed. It should be emphasized that even user-developed MOOS applications have a heavy reliance on pre-existing libraries that are provided with the distribution.

3.1 System Overview

In the proposed development framework, there are two configurations: 1) A minimalist configuration that accelerates the development of advanced control and planning strategies, 2) A simulation configuration substitutes variables for actual sensor and actuator data at the lowest level [9].

Configuration 1 - Minimalist: A diagram showing the minimalist configuration with no sensor or actuator processes is shown in Fig. 3.

In this setup, there are three communities, two of which represent actual AUVs and eventually will run on-board the vehicles, and one that is a base station and intended to run on a ship or shore based computer.

In each AUV community there is a *MOOSDB*, a *pLogger*, a *pNodeReporter*, a *pHelmIvP*, a *pMarinePID*, *uMVS*, and a *pMOOSBridge*.

- *MOOSDB*: No community can exist without a *MOOSDB*. It is the center of the architecture and controls publishing and subscribing of variables.
- *pLogger*: Writes MOOS variables to a log file [4].
- *pNodeReporter*: Maps pose into a report that is consumed by the *pMarineViewer* 2D map [4].
- *pHelmIvP*: Responsible for taking a set of behaviors and resolving them into desired values for the inner loop controller [4].
- *pMarinePID*: An inner loop PID controller that tracks the references generated by pHelmIvP by generating the appropriate actuator values (thrust, rudder, and elevator) [4].
- *uMVS*: A multiple vehicle kinematics and dynamics simulator that is also capable of simulating acoustic communications [9].
- *pMOOSBridge*: Responsible for the communication of variables with other communities. The *pMOOSBridge* can also rename local variables and send them to specific communities. The major advantage of this configuration is that the other applications in the AUV do not need to worry about what community they are in, they can all be identical. The pMOOSBridge appends the 'AUV_N_' prefix onto the variables so that they can be handled appropriately by the base station [7].

The *BaseStation* community does the long term planning through an interface with MATLAB® called *iMatlab* [8], maintains the confidence map through the ConfidenceMap application, has 2D map visualization through *pMarineViewer*

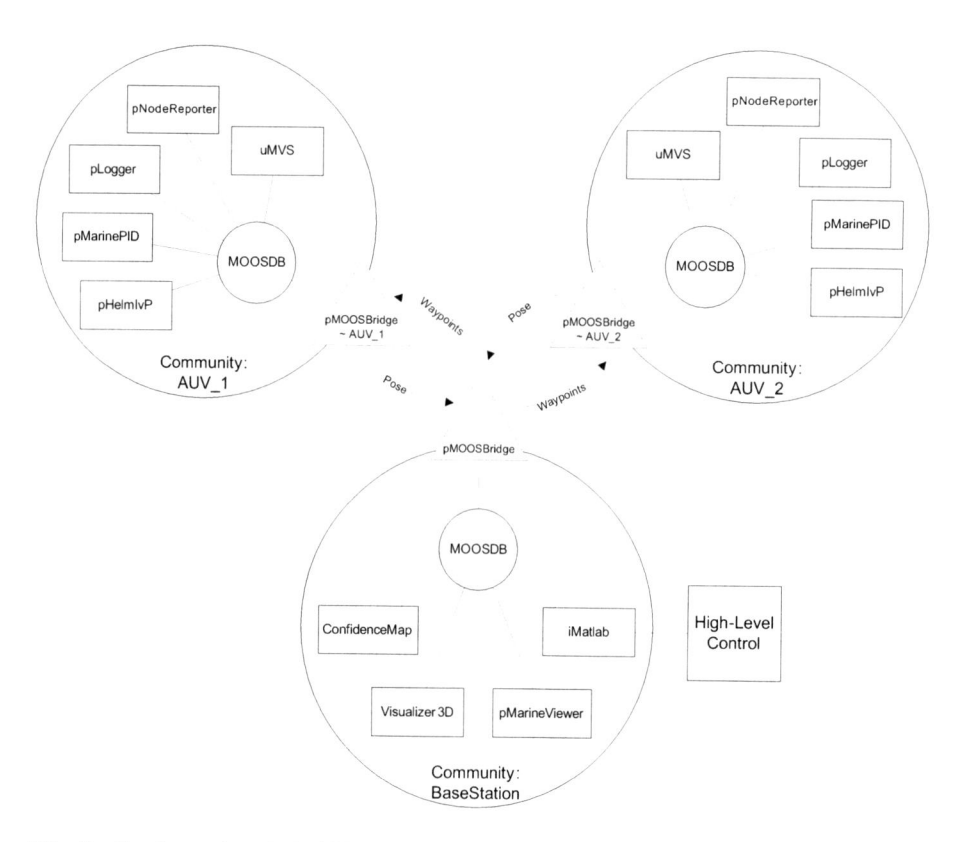

Fig. 3. Configuration 1: 2 AUV communities and one base station community with minimal applications

([4] sec.10), has 3D visualization, as well as requiring its own *MOOSDB* and *pMOOSBridge* applications. To summarize, the long range planner, *ConfidenceMap* (Sec. 3.2), and *Visualizer3D* (Sec. 3.3) are applications that were developed. The other applications were appropriately configured and their purposes will be discussed here:

– *MOOSDB*: Same as above, handles all published and subscribed data.
– *pMOOSBridge*: Does the inverse of the *pMOOSBridge* applications on the AUVs. Specifically, it maps community specific variables generated on the base station, such as waypoints, and sends them to the appropriate community with generic names.
– *pMarineViewer*: A 2D mapping application that is also capable of displaying multiple vehicles at one time [4]. A screenshot of two AUVs.
– *iMatlab*: An application that allows a matlab script to enter a MOOS community and therefore publish and subscribe to variables. It is based on using 'mex' to compile some MOOS code that MATLAB® can call directly [8].

Configuration 2 - Hardware Simulation: In this next configuration, the hardware sensing and actuation modules are added to the AUV communities as shown in Fig. 4. By setting the *simulator* variable to *true* in the mission file, the sensing instruments subscribe to simulation variables (SIM_*) generated by the simulator, and publish sensor data. Similarly, the actuator applications subscribe to simulated desired actuator variables and re-publish the same values as actual actuator values. For more details, refer to [9]. The *pNAV* application is responsible for taking the asynchronous sensor readings and producing accurate estimates of vehicle pose using Extended Kalman Filtering.

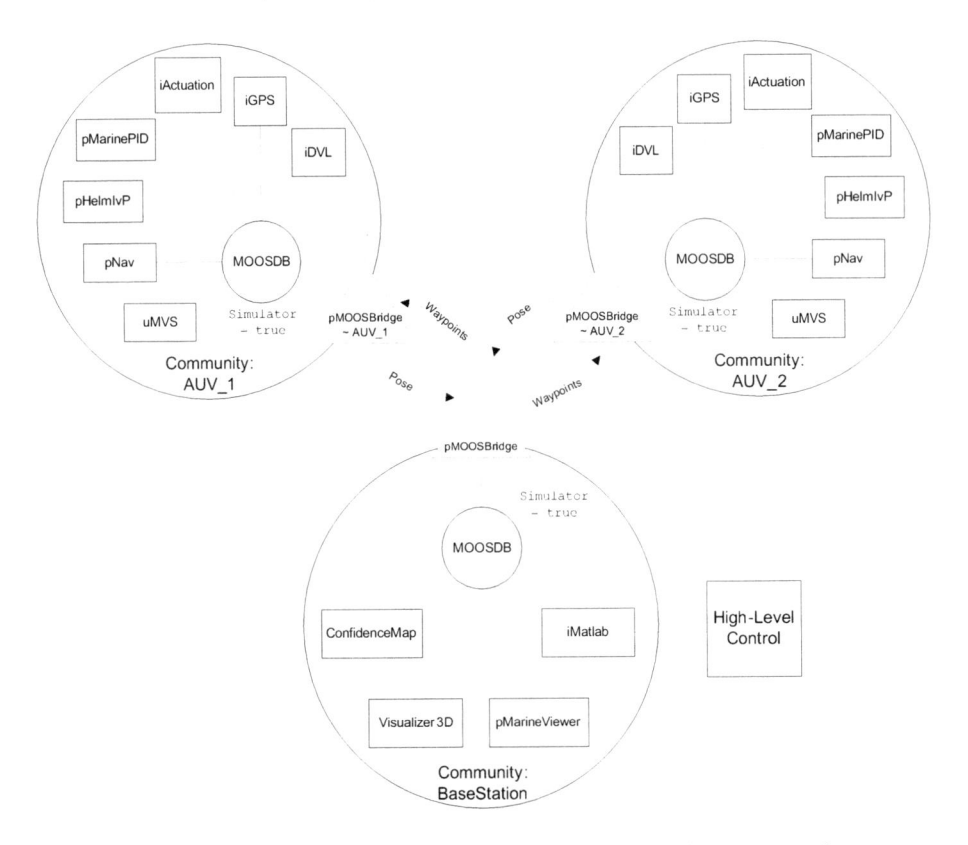

Fig. 4. Configuration 2: Hardware simulation includes all sensor and actuator applications

3.2 ConfidenceMap

The mission for which this framework was designed is seabed coverage with a SSS. As mentioned, SSS performance is crucial to being able to identify MLOs. The SSS performance is dependent on seabed type, depth and other factors. The mission is considered as completed when the average coverage confidence over the environment reaches a specified threshold.

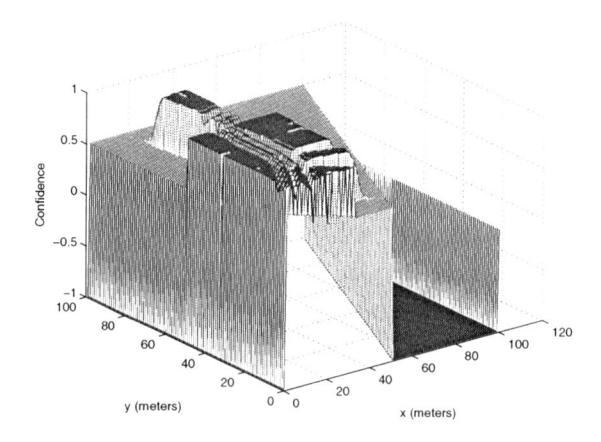

Fig. 5. A map of confidence over a test environment

The *CoverageMap* application on the base station registers for the pose variables of all AUVs and any environmental factors that they report. A map of the confidence over the entire area of interest is maintained and used by the long range planner. The *CoverageMap* application is responsible for signaling the completion of the mission. A confidence plot that was generated by MATLAB® is shown in Fig. 5.

In the figure, the coordinates of the environment are $((0,0), (0,100), (100,100), (100,40))$. Anything outside of the environment is assigned a confidence of -1. The confidences over the entire environment are initialized to 0.5 and then updated as the AUVs move about. In this case, there is one AUV that began at $(0,0)$ and then traveled parallel to the y axis and made a right turn.

The application maintains current and previous poses and updates the map accordingly based on intermittent and asynchronous AUV pose data. If the change in yaw is above a threshold, the update is not made because the SSS data is not valid in the case that the AUV is turning. Otherwise, a line is drawn between the previous and current (x, y) locations and is stepped along as perpendicular locations are updated.

3.3 3D Visualizer

A major challenge of AUV development is that it is difficult and costly to observe the vehicles in the environment. To mitigate this problem, an OpenGL based 3D visualizer originally designed to work with the MIRO middleware has been adapted and ported to MOOS-IvP [3]. Seabed data must be provided from previous bathymetric surveys. The 3D visualizer subscribes to the poses of all AUVs and displays them as shown in Fig. 6. One AUV is always at the center of the screen, while others will appear if they are in view.

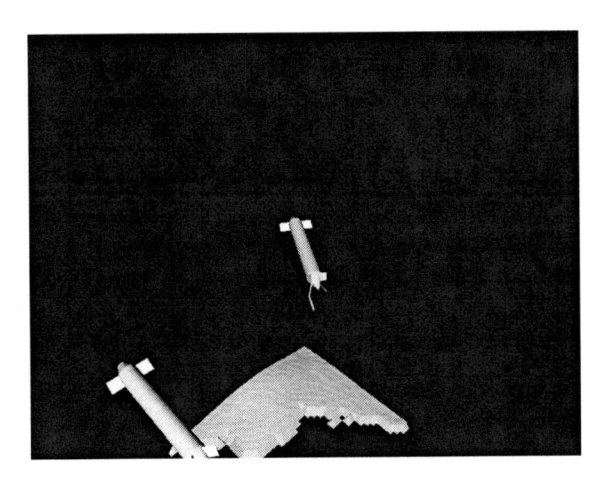

Fig. 6. A screenshot of the developed 3D Visualizer displaying 2 AUVs and bahymetric data

4 Conclusion

A framework for development of advanced control algorithms for Autonomous Underwater Vehicles with Sidescan Sonar Sensors has been presented. In the proposed approach, there are three levels of development: one with minimal overhead, one with hardware simulation, and then the final stage with actual hardware and acoustic communication. A base station community is used to simulate the vehicles in 2D and 3D, as well as maintain coverage confidences over the search space and generate paths to achieve the global mission goals. The implementation of this system will allow for reduced development time, impact on hardware, and cost.

Acknowledgment

This research is supported by Natural Sciences and Engineering Research Council of Canada (NSERC) and Canada Foundation for Innovation.

References

1. Eickstedt, D., Benjamin, M., Curcio, J.: Behavior based adaptive control for autonomous oceanographic sampling. In: IEEE International Conference on Robotics and Automation, pp. 4245–4250 (2007)
2. Jiang, D., Pang, Y., Qin, Z.: Coordination of multiple auvs based on moos-ivp. In: 8th IEEE International Conference on Control and Automation (ICCA), pp. 370–375 (2010)

3. Li, H., Popa, A., Thibault, C., Trentini, M., Seto, M.: A software framework for multi-agent control of multiple autonomous underwater vehicles for underwater mine counter-measures. In: 2010 International Conference on Autonomous and Intelligent Systems (AIS), pp. 1–6 (2010)
4. Benjamin, M., Newman, P., Schmidt, H., Leonard, J.: An overview of moos-ivp and a brief users guide to the ivp helm autonomy software. In:, http://dspace.mit.edu/bitstream/handle/1721.1/45569/MIT-CSAIL-TR-2009-028.pdf (June 2009)
5. Myers, V., Pinto, M.: Bounding the performance of sidescan sonar automatic target recognition algorithms using information theory. IET Radar Sonar Navig. 1(4), 266–273 (2007)
6. Navy, U.: The navy unmanned undersea vehicle (uuv) master plan (tech rep. a847115). Tech. rep., U.S. Navy (2004)
7. Newman, P.: Bridging communities with pmoosbridge, http://www.robots.ox.ac.uk/~pnewman/MOOSDocumentation/Essentials/Bridging/latex/MOOSBridge.pdf (June 2009)
8. Newman, P.: Moos meets matlab - imatlab, http://www.robots.ox.ac.uk/~pnewman/MOOSDocumeatation/tools/iMatlab/latex/iMatlab.pdf (March 17, 2009)
9. Newman, P.: Using the marine multivehicle simulator: umvs, (March 2009), http://www.robots.ox.ac.uk/~pnewman/MOOSDocumentation/tools/Simulation/Marine/latex/MarineMultiVehicleSimulator.pdf
10. Paull, L., Saeedi, S., Li, H., Myers, V.: An information gain based adaptive path planning method for an autonomous underwater vehicle using sidescan sonar. In: IEEE Conference on Automation Science and Engineering (CASE), pp. 835–840 (2010)
11. Sariel, S., Balch, T., Erdogan, N.: Naval mine countermeasure missions. IEEE Robotics Automation Magazine 15(1), 45–52 (2008)
12. Sotzing, C.C., Lane, D.M.: Improving the coordination efficiency of limited-communication multi-autonomus underwater vehicle operations using a multiagent architecture. Journal of Field Robotics 4, 412–429 (2010)

Three-Dimensional Path-Planning for a Communications and Navigation Aid Working Cooperatively with Autonomous Underwater Vehicles

Mae L. Seto[1], Jonathan A. Hudson[2], and Yajun Pan[3]

[1] Defence R&D Canada, Dartmouth, Nova Scotia, Canada
mae.seto@drdc-rddc.gc.ca
[2] Dalhousie University, Halifax, Nova Scotia, Canada
jn620053@dal.ca
[3] Dalhousie University, Halifax, Nova Scotia, Canada
yajun.pan@dal.ca

Abstract. Simulations for optimized path-planning of an autonomous communication and navigation aid (CNA) working cooperatively with submerged autonomous underwater vehicles (AUVs) in mine hunting surveys are presented. Previous work largely ignores optimizing the CNA path. Since AUVs in mine hunting missions maintain constant altitude this more general, and optimized, path-planner allows for variable vertical separations between the CNA and the survey AUVs. The path-planner also allows the CNA to be an AUV, not just a surface craft. The ability of the CNA to calculate on-line the cost of a heading decision L time steps into the future reduces the survey AUVs' position errors. Increase computation effort from such 'look-ahead' is offset by imposed maximum and minimum distances between the CNA and each survey AUV. The distance penalty also decreases computation effort and bounds the AUV position errors. An optimized CNA path-planner so described has not been previously reported.

Keywords: path-planning, communications and navigation aid, AUV.

1 Introduction

Under water robots like autonomous underwater vehicles (AUVs) are common considerations for underwater tasks like surveys, inspections, pollution monitoring, and a variety of naval operations. An example of a naval operation is mine hunting surveys. In these surveys, the AUV performs a lawn mower or ladder pattern so its on-board side scan sonar can efficiently collect seabed images over an area (Fig. 1). The lane spacing is adjusted to be smaller than the sonar swath width so neighboring swaths overlap. In these surveys, the AUV maintains prescribed altitude as it looks for targets on the seabed and, as a result, its depth varies constantly. The acquired sonar images are analyzed to detect targets such as mine-like objects. In these surveys it is useful to re-acquire a target's image from a different perspective to confirm its location and classification. For example. a circular cylinder in a side scan sonar image appears different depending on the direction (aspect) that the sonar images it.

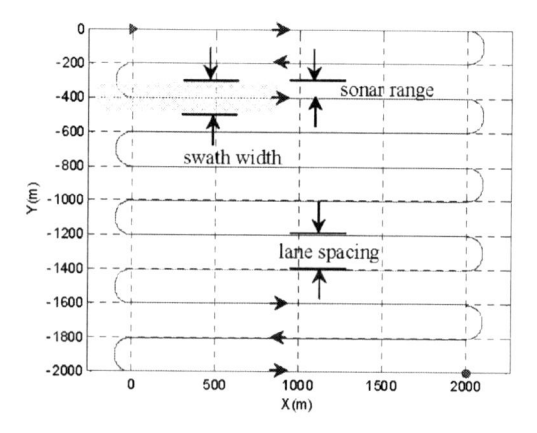

Fig. 1. Side scan sonar mission geometry (start ▶, recovery ●) with 200m lane spacing

Prior to using AUVs in mine hunting surveys, surface ships towed the sonar through the water to obtain the seabed images. The use of AUVs with on-board side scan sonar mitigates the risk of high valued assets and people in potential minefields. However, there are trade-offs in replacing surface ships with AUVs for mine hunting surveys: (1) AUVs do not typically transit faster than 4 knots for extended times – this is several times slower than a surface ship; (2) when a surface ship tows the sonar the images are accessible (through the tow cable) to the operator for target recognition analysis, and (3) when a surface ship tows the sonar there is a bound on the sonar images' position error since the sonar depth and tow cable length are known. These trade-offs are aggravated by limitations in working underwater.

One way to potentially recover the ship's higher mission speed using AUVs is to deploy multiple AUVs that cooperate / collaborate. If two AUVs were deployed on orthogonal headings one could acquire a target image and the other could re-acquire it from the orthogonal direction in the same time. Further, if the AUVs collaborate to confirm their findings or alter their mission to pick-up a target perceived missed by the other this could increase the confidence that all targets are acquired. Such robot cooperation / collaboration requires some level of inter-robot communications.

Acoustic means provide the best range for underwater communications. However, water has inherently low bandwidth, high attenuation, and multi-path returns for acoustic signals. This compromises reliable and long range acoustic propagation. Acoustic communications in an ocean environment is only on the order of kilometers. This means it is not possible to transmit large amounts of high bandwidth data, like sonar images, reliably or over long distances in a timely manner. To reduce the required bandwidth, the AUVs could analyze the images on-board for target recognition and transmit a summary to another AUV or operator. Consequently, on-board target recognition is emerging as a required and new capability. The value of the sonar images and their subsequent target recognition analysis is related to how well these images can be accurately positioned or geo-referenced. Ambiguity in a target's position means it might not be re-acquired and have its identity confirmed. Thus the requirement for accurate positioning is high in mine hunting surveys.

The under water environment has no absolute references for positioning and navigation as there is no underwater global positioning system (GPS). AUVs dead-reckon for positioning and navigation with on-board inertial navigation units (INUs) and Doppler velocity logs (DVL). With dead-reckoning, position errors grow with time without bound. One way to enable more accurate positioning is to use communication and navigation buoys, but these require time to deploy and manned boats in a minefield. The AUV can surface periodically for a GPS fix to put a bound on its position error. However, periodic surfacing impacts the mission since it takes longer to perform the mission. These limitations with AUVs can be addressed with a communication and navigation aid (CNA) that works with the submerged AUVs.

1.1 Surveys with AUVs and CNAs

The CNA concept uses a dedicated vehicle(s) to support submerged AUVs in their positioning, navigation, and communications. The CNA can be a manned surface ship or an autonomous system like an AUV or an unmanned surface craft (ASC).

An area of current AUV research is in the use of multiple AUVs working together and with heterogeneous platforms such as unmanned ASCs and unmanned air vehicles [1]. For cooperation or collaboration among multiple manned and autonomous platforms to be reliable and effective, the communication and navigation issues inherent in AUV operations must be addressed.

The CNA aids communication by linking the submerged AUVs to each other and to the surface control station which can be a surface ship, aircraft or on shore (Fig. 2). The CNA uses acoustic modems to exchange information with the submerged survey AUVs and conventional radio for in-air communication with the control station.

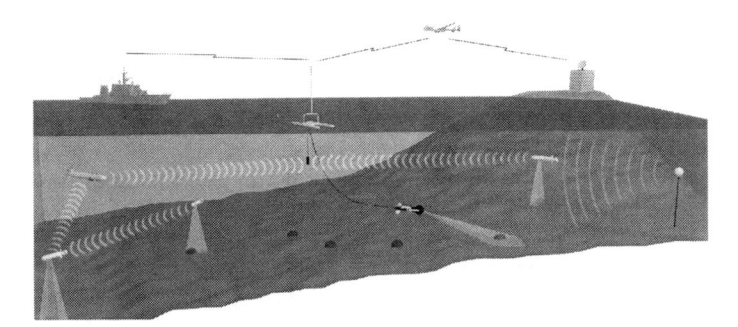

Fig. 2. AUVs assisted by a communication and navigation aid in the form of an autonomous surface craft

A CNA aids AUV navigation by providing its own global position (from GPS) to the survey AUVs. The survey AUVs use the CNA's global position, and their position relative to it, to refine estimates of their own global positions. This is more efficient than localization methods where the survey AUV either periodically surfaces for its own GPS position or employing a field of communication and navigation buoys.

This paper reports on the use of autonomous systems as a CNA for submerged AUVs. Specifically, the original contribution looks at optimized three-dimensional path-planning for the CNA when the AUV and CNA are changing depths. This path planner is further optimized through a distance penalty that bounds the AUV position errors and without increasing the computation burden. If individual AUVs can navigate well between the regular CNA updates, an AUV CNA could also re-acquire a target after it has been initially acquired by another AUV. It would do this in between surfacing for GPS fixes. This way the CNA is not a dedicated CNA and expands on the general CNA concept defined earlier.

Section 2 reviews recent developments in using autonomous systems as CNAs for cooperative / collaborative AUVs. Section 3 describes the contributions to develop a more general optimized path-planning capability for an AUV or ASC CNA. Section 4 presents results from the study and Section 5 reports on work towards in-water validation of the algorithm. Section 6 concludes with a few remarks.

2 Review of Recent CNA Developments

The use of a single vehicle (submerged or surface) as a communication [2]–[4] or navigation [5]–[9] aid to an underwater vehicle has been considered. One, or two, vehicles can aid both communication *and* navigation [10]–[16]. Cooperative AUV navigation [17] for survey and surveillance missions differs from the aforementioned general CNA concept in two ways. *First,* it does not use a dedicated CNA vehicle for multi-AUV surveys. All but one AUV is equipped with an expensive INU. The role of the AUV with the modest INU is to surface periodically for a GPS fix that it shares with the others. The other AUVs update their own position based on their relative position from it. The AUV with the modest INU contributes to the survey / surveillance mission between its GPS fixes. *Secondly,* all AUVs in the group exchange their position and range estimates to improve their navigation filters. Each AUV's filter combines measurements from sensors (weighted by accuracy) to yield the best position estimate.

In another application [10] for large area searches, two CNA vehicles maintain a specific distance from the survey AUV and in the second scenario the two CNA vehicles maintain a right-angle formation with the survey AUV (CNA at apex of right angle). This right angle formation while not an optimized path does provide a positioning solution but employs more assets to do so.

Among researchers who use a single CNA vehicle, few optimize its path. A CNA zig-zag path was used for mine hunting and oceanographic surveys as well as disaster response [13]. This zig-zag path is not an optimized path. The submerged AUVs meanwhile performed lawn mower maneuvers (Fig. 1). A zig-zag and an encirclement pattern was also employed by a single ASC CNA [18] where both patterns adapted to the changing survey AUV path.

Another approach [19] determines a discrete set of M positions that can be reached by the CNA prior to the next update. The CNA selects the position that minimizes the survey AUV's position error and in the process, optimizes its own path. The AUV position error, determined from the trace of its position covariance matrix, is sent to the CNA. Permutations of one or more AUVs working with one or more CNAs are studied and the CNA speeds are varied. In the case of multiple CNAs each makes its own decision to minimize the AUVs' position errors. Then, each CNA takes the other CNAs' selected best position into account to optimize its own path.

Chitre [11],[15] considers a two-dimensional situation to optimize the CNA path through minimizing the survey AUV(s) position error estimates. Two-dimensional path-planning was possible by assuming constant depths for both CNA vehicle and survey AUV(s). Simulation were for one and two survey AUVs aided by one CNA vehicle. The CNA vehicle path could be determined through *underway* or *pre-deployment* modes. In pre-deployment mode, the CNA vehicle knows the survey AUVs' planned paths and the survey AUVs know the CNA's planned path a priori. In underway mode, the CNA vehicle still knows a priori the survey AUVs' planned paths and is also able to accept updates of the survey AUVs' position error estimates. This information allows for adaptive path-planning by the CNA vehicle. This adaptation requires increase communication and computation for all vehicles.

The CNA vehicle plans its path with knowledge of the survey AUVs' paths and uses an optimization that minimizes the survey AUVs' position errors. Limits are imposed on the CNA vehicle's distance from the survey AUVs. The result is a reduction in the radial direction error but not the tangential one. In Fig. 3, the blue ellipse near the survey AUV represents its unaided position error and the green ellipse its error when aided by CNA range measurements.

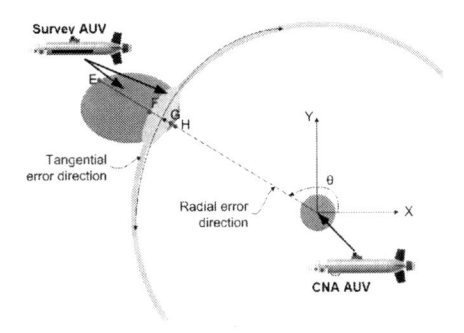

Fig. 3. Bounded survey AUV position error, adapted from [15]

In Fig. 3, the blue circle near the CNA AUV is its position error. Notice the survey AUV's position error reduces significantly in the radial direction from E-H to F-G but the tangential error is unchanged. The green ellipse is used to determine the CNA's next waypoint. Ideally, the CNA vehicle selects its next waypoint on the major (longest) axis of the green ellipse since this minimizes the survey AUV's overall position error.

As per [11], the CNA vehicle transmits an update to the N survey AUVs every τ seconds and determines its own heading, ψ, for the next τ seconds. At each time step, t, the CNA vehicle increments its heading by δ_t^{CNA} to minimize the survey AUVs' position errors. This heading change must be less than the CNA's maximum turn rate, i.e. $\delta_t^{CNA} \leq |\psi_{max}^{CNA} \times \tau|$. For implementation ease, the range of possible turn angles is a discrete set of A values where A is an odd integer greater than 1. At each time step the CNA vehicle's heading and position are described by Eq. (1) and (2) from [11], where x_t^{CNA} is the CNA vehicle's position in the x and y directions and s^{CNA} is the CNA vehicle's speed. Also, the angle between the CNA vehicle and any survey AUV, j, (out of a group of N survey AUVs) is described by $\theta_{t+1}^j = \angle \left(x_{t+1}^j - x_{t+1}^{CNA} \right)$ (Fig. 3).

$$\psi_{t+1}^{CNA} = \psi_t^{CNA} + \delta_t^{CNA} \quad (1)$$

$$x_{t+1}^{CNA} = x_t^{CNA} + \tau \times s^{CNA} \begin{pmatrix} \cos \psi_{t+1}^{CNA} \\ \sin \psi_{t+1}^{CNA} \end{pmatrix} \quad (2)$$

As shown in Fig. 3, the update from the CNA vehicle leads to survey AUV position errors in two orthogonal directions – radial and tangential. The radial error (along θ_{t+1}^j) is modeled as zero-mean Gaussian so the radial error is $\varepsilon_t^j = \sigma$. The tangential error, based on survey AUVs dead-reckoning from velocity estimates is described by Eq. (3) [11]. α represents the survey AUVs' velocity estimate accuracy.

$$\varepsilon_{t+1}^{-j} = \frac{\left(\varepsilon_t^j \varepsilon_t^{-j}\right)^2}{\left(\varepsilon_t^j \cos \gamma_t^j\right)^2 + \left(\varepsilon_t^{-j} \sin \gamma_t^j\right)^2} + \alpha\tau : \quad \gamma_t^j = \theta_{t+1}^j - \theta_t^j \quad (3)$$

At each time step t, the CNA vehicle chooses one of A possible headings. Each heading has error ε_{t+1}^j and ε_{t+1}^{-j} for each survey AUV. The CNA chooses the heading that minimizes the cost function, Eq. (4), which sums position errors over all AUVs.

$$C(S_t, a_t) = \sum_j \left[\left(\varepsilon_{t+1}^j\right)^2 + \left(\varepsilon_{t+1}^{-j}\right)^2 \right] \text{ [11]} \quad (4)$$

To improve the CNA path-planning algorithm's performance, a look-ahead function, V_t, (Eq. 5, [11]) considers the cost of current heading decisions L time steps into the future. S_t is the decision space and a_t is the decision at time t applied to each survey AUV. Low cost correlates directly to low AUV position error. As in the cost function without look-ahead, the CNA vehicle chooses the heading option that minimizes the survey AUVs' position errors. Simulations with different CNA vehicle heading values, vehicle speeds, look-ahead levels, and either one or two survey AUVs show the positioning error reduces with increase number of heading values, (A), CNA speed, and look-ahead time steps. Not unexpectedly, this increases computation effort. This may be a concern if the CNA plans its path underway rather than prior to the mission. Note, time at look-ahead level L = (current time + τ) + $L \times \tau$.

$$V_t(S_t) = \min \sum_{t=t'}^{t+L-1} C(S_t, a_t) \quad (5)$$

3 New Contributions to Optimized CNA Path-Planning

3.1 Three-Dimensional Path-Planning – Variable Depths

Previous assumptions [11] of constant depth are valid for some applications, but mine hunting side scan sonar missions require the survey AUV to maintain constant altitude above a varying sea bed. Dramatic changes in underwater topography is not unusual in littoral waters where mine hunting is considered. This paper reports on work that includes variable survey AUV and CNA vehicle depth in optimizing the CNA's path.

A third dimension is added by decomposing θ into θ_{xy} and $\theta_{x'z}$. $\theta_{x'z}$ is the angle between the line joining the vehicles and the horizontal plane. θ_{xy} is the bearing relative to an arbitrary origin. With the AUV CNA, ϕ, represents absolute CNA vehicle pitch angle (for an ASC CNA vehicle, $\phi = 0$). At each time step, the CNA AUV determines its new pitch (Eq. (6)) and heading angle (Eq. (1)). The pitch increment, β_t^{CNA} (positive for AUV nose up), for which there are B possible discrete values, is constrained to not exceed the maximum vehicle pitch rate, $\dot{\phi}_{max}$, i.e. $\beta_t^{CNA} \leq |\dot{\phi}_{max}^{CNA} \times \tau|$. Thus the CNA vehicle position at time t is shown in Eq. (7).

$$\phi_{t+1}^{CNA} = \phi_t^{CNA} + \beta_t^{CNA} \tag{6}$$

$$\begin{bmatrix} x_{t+1}^{CNA} \\ y_{t+1}^{CNA} \\ z_{t+1}^{CNA} \end{bmatrix} = \begin{bmatrix} x_t^{CNA} + \tau \cdot s^{CNA} \cos\left(\psi_{t+1}^{CNA}\right) \cos\left(\phi_{t+1}^{CNA}\right) \\ y_t^{CNA} + \tau \cdot s^{CNA} \sin\left(\psi_{t+1}^{CNA}\right) \cos\left(\phi_{t+1}^{CNA}\right) \\ z_t^{CNA} - \tau \cdot s^{CNA} \sin\left(\phi_{t+1}^{CNA}\right) \end{bmatrix} \tag{7}$$

The CNA vehicle still plans its path to minimize the survey AUVs' error. The error is calculated in the same manner as before since both vehicles maintain (and regularly exchange) accurate estimates of their depths (easily measured on-board). However, it is still beneficial to consider depth changes in both the survey AUVs and the CNA vehicle to maintain required maximum distances. In the case of an AUV CNA, changes in CNA depth place the vehicle in x-y locations that may otherwise be impossible with constant speed. Increased number of waypoint values, $A \times B$ over A, allows the AUV CNA to reduce the survey AUVs' error further than an ASC CNA.

Constraints of maximum and minimum three-dimensional distance from the survey AUVs are imposed on the CNA vehicle position, and with an AUV CNA, maximum and minimum depths as well as maximum absolute pitch angle.

3.2 Three-Dimensional Path-Planning – Distance Penalty

Distance limitations are imposed to penalize position values, x_t^{CNA}, the CNA may choose in its path-planning optimization if the position values violate maximum and minimum distances between the CNA and any survey AUV. The maximum distance between the CNA and a given survey AUV is maintained to ensure reasonable quality acoustic communication between the two. The minimum distance between the CNA and a given survey AUV is maintained to avoid collisions. The effect of distance limitations on optimizing the CNA path-planning has not been previously reported.

Every τ seconds (CNA update interval) the path-planning algorithm (Section 3.1) evaluates Eq. (4) for each heading value based on its ability to reduce the position error in all survey AUVs. At the same time, the distance between the CNA and any given AUV is also calculated. If this distance is less than the minimum distance or greater than the maximum distance, a revised cost is calculated (Eq. (8) and Eq. (9)) which penalizes that heading value so that it is less likely to be chosen as optimal.

The consequences of vehicles coming too close is more serious than having them too far apart. Therefore, the distance penalty penalizes 'too close' heading values more severely than 'too far' ones as shown in Eq. (8) and Eq. (9) for a given δ_t^{CNA}.

if distance < min_dist: $C'(S_t, a_t) = 2 \times \left(C(S_t, a_t) + 2 \times (\text{min\_dist} - \min(x_t^j - x_t^B)) \right)$ (8)

if distance > max_dist: $C'(S_t, a_t) = 2 \times \left(C(S_t, a_t) + \max(x_t^j - x_t^B) - \text{max\_dist} \right)$ (9)

In the case of multiple survey AUVs, it is possible that a heading option may be too close to one vehicle and too far from another. When this happens, the distance penalty function applies the 'too close' penalty rather than the 'too far' penalty.

In the simulations here, the maximum allowed distance between the CNA and any survey AUV was 1000 m while the minimum allowed distance was 100 m. These distances are arbitrary and chosen for consistency with previous results [11].

4 Results

4.1 3D Path-Planning – Variable Depth

The simulations conducted thus far indicate that constraints on distance limitations between the CNA and any one AUV, CNA depth, and CNA pitch (to change depth quicker) can be met while minimizing the survey AUVs' position errors (not shown). The addition of the third dimension yields more meaningful path-planning for AUVs tasked to altitude-keep over varying terrain.

The implementation of look-ahead significantly increases the computational effort. This limits the practical use of look-ahead, L, to only a few time steps, τ, into the future. Up to 5 look-ahead time steps were explored since it was computationally feasible. Even with $L = 4 \tau$ look-ahead steps, the reduction in position error was 20% for a CNA with 2 survey AUVs. For the case of $L = 4 \tau$ with 5 AUVS, the reduction in position error was 22%.

While adding the third dimension and incorporating look-ahead takes more computational effort, the path-planning on-board the CNA is still able to calculate a new waypoint every τ seconds. This is based on the MATLAB® simulations performed to date. That way, the CNA does not have to pause between waypoints to calculate its next destination. In a dynamic environment, a large number of look-ahead levels may not reduce survey AUV position error because the survey AUVs may not follow the path predicted by the look-ahead function far into the future.

Tables 1 shows a comparison of the errors in the survey AUV's position as calculated by the CNA. As expected, this error decreases consistently with increase CNA speed. Table 2 shows the error between the CNA and 3 individuals AUVs working cooperatively in 3 side-by-side areas as per Fig. 1. As per the case with one AUV in Table 1, the AUV position error calculated by the CNA decreases with speed. The error is larger since the CNA is now optimizing over 3 AUVs instead of 1. The simulation parameters are shown in Table 3.

Table I. AUV or ASC CNA with 1 survey AUV

CNA type	s^{CNA} (m/s)	distance from CNA to AUV (m)			error (m) - Eq. (3)	
		min	mean	max	max	rms
AUV	1.5	100.3	236.6	456.2	4.75	3.58
	2.0	101.2	262.0	566.6	5.04	3.51
ASC	1.5	100.4	240.9	461.6	4.81	3.60
	2.0	100.2	235.6	556.3	4.89	3.39
	3.0	100.0	108.0	160.7	2.94	2.45

Table 2. AUV or ASC CNA with 3 cooperative survey AUVs

CNA type	s^{CNA} (m/s)	max error (m) from CNA to AUV #			rms error (m) from CNA to AUV #		
		1	2	3	1	2	3
AUV	1.5	6.80	6.14	6.81	5.87	4.77	5.94
	2.0	6.47	5.58	5.58	5.34	4.27	5.40
ASC	1.5	6.80	6.10	6.10	5.87	4.77	5.94
	2.0	6.47	5.60	5.60	5.35	4.26	5.40
	3.0	5.91	4.86	4.80	4.61	3.80	4.67

In these simulated missions, each survey AUV is assigned a 500 m × 750 m area and transits at 1.5 meters / second (~ 3 knots). The survey AUVs begin the mission at 60 m depth and increase to 75 m depth half-way through the 2 hour mission. The CNA vehicle is constrained to remain within 1 kilometer (max distance) of every survey AUV and may come no closer to them than 100 meters (min distance).

In the case of an AUV CNA, its pitch may not exceed ± 60° and its depth may not be greater than 55 m (separation of 5 m from survey AUV). The minimum depth for an AUV CNA (and assumed modem depth on an ASC CNA) is set to 0.25 m.

Table 3. Simulation parameters for Tables 1 and 2 cases

parameter		value
time step	τ	10 sec
survey AUV j error in radial direction at $t = 0$	ε_0^j	1 m
survey AUV j error in tangential direction at $t = 0$	ε_0^{-j}	1 m
certainty of survey AUVs' velocity estimate	α	0.1 m² / sec
CNA maximum turng rate	ψ_{max}^{CNA}	4.0 deg / sec
CNA maximum pitch rate	ϕ_{max}^{CNA}	1.5 deg / sec
total discrete CNA heading increments	A	5
total discrete CNA pitch angles incremnts	B	5

As shown in Tables 1 and 2, increasing the CNA vehicle speed generally reduces the distance between the CNA and individual AUVs and each survey AUV's position error. However, this improvement is not uniform. In the case of a single survey AUV, increasing the speed from 1.5 m/s to 2 m/s reduced performance for both the ASC and AUV CNAs. This reversed trend is not observed in the case of 3 AUV surveys. Results for an AUV CNA travelling 3 meters / second are not shown since this speed is not typical for most AUVs. The ASC/AUV comparisons shown seem to indicate there is no significant advantage to using an AUV as a CNA. In fact, for these missions, an ASC CNA capable of 3 m/s (or perhaps more) would be the ideal solution, assuming no stealth requirement. On the other hand (in these simulated missions), the use of an AUV as the CNA vehicle is not noticeably worse than an ASC at similar speeds, so if stealth were a consideration, it appears that the positioning accuracy would not suffer dramatically.

4.2 3D Path-Planning – Distance Penalty

The use of the distance penalty does not prevent all violations of maximum and minimum distances between the CNA vehicle and survey AUVs (not shown). In some instances, the path-planner's optimizer ends up in a less-than-optimal portion of the decision space and selects a heading value that is the 'best out of a bad lot'.

It was shown previously [11] and in Section 4.1 that increase CNA speed reduces the survey AUVs' position error whether the distance penalty was applied or not. However, no correlation between the magnitude of the benefit (in terms of survey AUV position error) of the distance penalty and the speed of the CNA vehicle could be inferred from this.

As previously discussed [11], look-ahead reduces the overall survey AUV position error by adding the cost of future decisions to the cost of current decisions each time a decision is made (every τ) is made. Theoretically, and if there were no external forces perturbing the survey AUVs, with enough look-ahead a distance penalty would not be necessary. This hypothesis could not be shown in total costs with five survey AUVs but is apparent in total and average costs with two survey AUVs if a look-ahead level of four (or higher) were used (Fig. 5). If a high enough look-ahead level were used.

Five (or more) survey AUVs could be supported without a distance penalty. However, with two survey AUVs and no distance penalty, the distance and position error versus time trends (not shown), suggest the distances and errors are not bounded when using a look-ahead level of four and are bounded with a look-ahead of five (Fig. 5).

For the case of two AUVs, with a look-ahead level of four, the distance penalty bounds the error and the distance between the CNA and survey AUVs. Without the distance penalty the error and distance are not bounded even though the costs are very similar between with and without distance penalty (Fig. 5), though as mentioned earlier, there is a practical limit to the value of large look-ahead in a dynamic environment.

In the case of one survey AUV (look-ahead level = 0) the initial distance between the CNA vehicle and AUV determines whether the distance penalty makes a difference in reducing or bounding the AUV position error. In the example of a CNA and AUV initially separated by approximately 140 m, the distance penalty makes no difference in reducing or bounding AUV position error. In a second example, the CNA and survey AUV are initially over 1 kilometer apart (including depth). In this example, the distance penalty is needed to bound the survey AUV's position error.

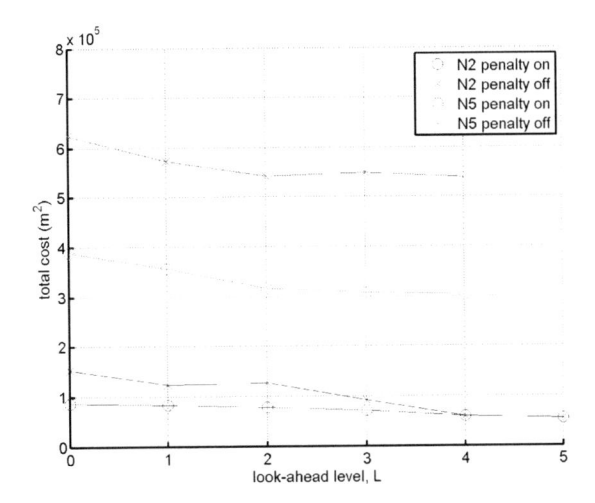

Fig. 5. Total cost (≈ survey AUV position error) of look-ahead level for two and five survey AUVs with and without distance penalty

Therefore, the distance penalty is of value when multiple autonomous vehicles must be deployed far apart. For any number of look-ahead levels or number of AUVs the addition of the distance penalty bounds the survey AUV position error.

5 Current and Future Work

Work is underway to validate the optimized CNA path-planner on an AUV CNA. It is being implemented on an IVER2 AUV that DRDC uses for autonomous behavior development in collaborative / cooperative AUVs. The survey AUVs are also IVER2's. All the IVER2's are equipped with WHOI micro-modems for communications. The middle-ware used in the AUV control architecture is MOOS-IVP [20]. This optimized path-planning is working in the pre-deployment mode. Effort will be applied to implement it in the underway mode. Then, the system will be tested on sea trials. Future work looks at integrating target re-acquisition and other secondary tasks into the optimized path-planner for an AUV CNA. The relative merits of using an AUV versus using an ASC for the CNA will also be considered.

6 Conclusions

The original work presented is an optimized CNA path-planner that accounts for variable depth differences between the CNA and survey AUVs. This allows the possibility of the CNA to be an AUV and not just an ASC as in earlier optimizers. Variable depth differences occur during mine-hunting surveys where the AUVs altitude-keep. The path-planner optimizes for minimum survey AUV position errors. This optimization was further improved by applying a distance penalty on the cost function which also had the added benefit of reducing the optimization computational effort. This CNA path-planner so optimized allows these autonomous systems to be applied to a wider range of missions.

References

1. Aguiar, A., et al.: Cooperative Autonomous Marine Vehicle Motion Control in the Scope of the EU GREX Project: Theory and Practice. In: Proc. Oceans 2009 - Europe, pp. 1–10 (2009)
2. Ferreira, H., et al.: SWORDFISH: An Autonomous Surface Vehicle for Network Centric Operations. In: Proc. Oceans 2007 - Europe, pp. 1–6 (2007)
3. Marco, D.B., Healey, A.J.: Current Developments in Underwater Vehicle Control and Navigation: The NPS ARIES AUV. In: Proc. Oceans 2000 MTS/IEEE Conference and Exhibition, pp. 1011–1016 (2000)
4. Marr, W.J., Healey, A.J.: Acoustic Data Transfer Limits using an Underwater Vehicle in Very Shallow Water. In: Proc. of OMAE 2004, 23rd International Conference on Offshore Mechanics and Arctic Engineering, Vancouver, pp. 711–718 (2004)
5. Alleyne, J.C.: Position Estimation from Range Only Measurements, Naval Postgraduate School, Monterey, CA., Master's Thesis (2000)
6. Hartsfield, J.C.: Single Transponder Range Only Navigation Geometry (STRONG) Applied to REMUS Autonomous Under Water [sic] Vehicles, Massachusetts Institute of Technology and Woods Hole Oceanographic Institution, Woods Hole, Mass., Thesis (2005)
7. Kinsey, J.C., Eustice, R.M., Whitcomb, L.L.: www.whoi.edu (2006), http://www.whoi.edu/cms/files/jkinsey-2006a_20090.pdf
8. Martins, A., Almeida, J.M., Silva, E.: Coordinated Maneuver for Gradient Search using Multiple AUVs. In: Proc. of Oceans 2003, pp. 347–352 (2003)
9. Pascoal, A., et al.: Robotic Ocean Vehicles for Marine Science Applications: European ASIMOV Project. In: Proc. Oceans 2000 Conference and Exhibition, Providence, R.I., pp. 409–415 (2000)
10. Bahr, A., Leonard, J.J., Fallon, M.F.: Cooperative Localization for Autonomous Underwater Vehicles. International Journal of Robotics Research 28(6), 714–728 (2009)
11. Chitre, M.: Path Planning for Cooperation Underwater Range-Only Navigation using a Single Beacon. In: International Conference on Autonomous and Intelligent Systems, 2010, Povoa de Varzim, Portugal, pp. 1–6 (2010)
12. Eustice, R.M., Whitcomb, L.L., Singh, H., Grund, M.: Experimental Results in Synchronous-Clock One-Way-Travel-Time Acoustic Navigation for Autonomous Underwater Vehicles. In: 2007 IEEE International Conference on Robotics and Automation, Roma, Italy, pp. 4257–4264 (2007)
13. Fallon, M.F., Papadopoulos, G., Leonard, J.J.: The 7th International Conference on Field and Service Robotics (July 2009), http://www.rec.ri.cmu.edu/fsr09/papers/FSR2009_0012_d9cc6671b367dcc393de3a4a60ded0fd.pdf
14. Perrier, M., Brignone, L., Drogou, M.: Communication Constraints and Requirements for Operating Multiple Unmanned Marine Vehicles. In: Proc. of the Seventeenth (2007) International Offshore and Polar Engineering Conference, pp. 1053–1058 (2007)
15. Rui, G., Chitre, M.: Cooperative Positioning using Range-only Measurements between Two AUVs. In: Proc. Oceans 2010, Sydney, Australia, pp. 1–6 (2010)
16. Yoon, S., Qiao, C.: Cooperative Search and Survey using Autonomous Underwater Vehicles (AUVs). To appear in IEEE Transactions on Parallel and Distributed Systems, 1–14 (2010)
17. Fallon, M.F., Papadopoulos, G., Leonard, J.J.: A Measurement Distribution Framework for Cooperative Navigation using Multiple AUVs. In: 2010 IEEE International Conference on Robotics and Automation, Anchorage, pp. 4256–4263 (2010)
18. Fallon, M.F., Papadopoulos, G., Leonard, J.J., Patrikalakis, N.M.: Cooperative AUV Navigation using a Single Maneuvering Surface Craft. The International Journal of Robotics Research (2010)
19. Bahr, A.: Cooperative Localization for Autonomous Underwater Vehicles. WHOI/MIT, Ph.D. Thesis (2009)
20. Benjamin, M.R., Newman, P.M., Schmidt, H., Leonard, J.J.: A Tour of MOOs-IvP Autonomy Software Modules. In: MIT-CSAIL-TR-2009-006, 40 pages (2009)

Autonomous Haulage Systems – Justification and Opportunity

Juliana Parreira and John Meech

Norman B. Keevil Institute of Mining Engineering,
The University of British Columbia, Vancouver, BC, Canada, V6T1Z4

Abstract. Driverless haulage trucks have recently been developed for open pit mines. To predict the benefits of an Autonomous Haulage System (AHS), a deterministic/stochastic model has been created to compare AHS to a manual system by estimating benchmarked Key Performance Indicators (KPIs) such as productivity, safety, breakdown frequencies, maintenance and labor costs, fuel consumption, tire wear, and cycle times. As this model is very complex and still in the process of development, the goal of this paper is to describe the model developed to describe vehicle motion and fuel consumption.

Keywords: autonomous haulage truck, simulation, vehicle motion, fuel consumption.

1 Introduction

The model described in this paper consists of a deterministic approach that describes truck movement in an open pit mine. It involves stochastic simulation to account for different events such as dumping, loading, queuing, breakdowns, downtime, shift changes, acceleration, braking, turning, and vehicular interactions. The model is developed in ExtendSim to project conventional shovel/truck simulation into a variety of truck sub-systems such as system controllers, fuel use, tire wear, and human performance in order to capture mechanical complexities and physical interactions of these sub-systems with a mine environment on a 24/7 time basis. Time during the shift and time during a work period also have major effects on performance. Running these two models in identical scenarios allows for the comparison of benchmarked KPIs that can demonstrate the adaptability and utilization of an AHS.

2 Layout of the Model

The model consists of eight CAT 793D trucks (each having a GVW of 846,000 lb., a net power of 1743 kW, and standard tires 40.00R57), one waste and one ore shovel, one crusher and one dump area. The model can be run either fully-autonomously or fully-manually. A probabilistic distribution is used to calculate loading/unloading times, truck and shovel maintenance, and shift changes. The Probabilistic or Monte Carlo simulation approach, in contrast to a deterministic one, is based on a variable possessing a degree of variation that cannot be attributed to known cause and effect relationships [1]. Since there is a lack of data regarding autonomous haulage trucks, data such Mean Time between Failures (MTBF) and Time to Repair (TTR) are assumed to have an empirical distribution. Once the model is validated and verified, the goal is to run a variety of fleet sizes in order to discover the impact of autonomy on the most effective fleet size under different circumstances.

3 Vehicle Motion

The model is a deterministic approach combined with a small increment of time used to calculate truck movement. The deterministic approach is used when the system variables are discrete, i.e., the system changes state as events occur, and only when an event occurs does the passing of real time have a direct effect on model activity or performance [2].

The data required for vehicle motion include speed-rimpull characteristics of the CAT 793D, together with haul road specifications such as section length, lane width, road bed quality and maximum speed and acceleration.

The mine model consists of 161 links, totaling 23 routes. For each link, acceleration is calculated by considering the physical limitations of each truck (motor power/road condition/loaded/empty), route speed, and safety distances established between other equipment. Acceleration is also a function of driver experience, personality, stress, and fatigue level. By examining different driver behaviors, variances can be captured, studied, and used to validate the model. As a result, the "optimum driver" can be determined and set as input for the AHS fleet model which will essentially be able to run with the best driver on all trucks all the time.

3.1 Small Increment Approach

The main objective of this technique is to calculate the speed of haulage trucks by small increments of time where during each interval; acceleration and force are assumed constant. It is important to know the weight of the truck, its rolling resistance, grade resistance, traction coefficient, and drive axle to calculate rimpull forces; thus, to output acceleration and speed.

The variable that accounts for the weight of trucks is changed dynamically according to cycle time (loaded/unloaded). Additionally, the weight distribution is set as a constant according to the Caterpillar truck handbook: a loaded truck has 67% of its weight on the drive axle while an empty truck has 54% of its weight on the drive axle. The haul road profiles such as section length, maximum speed, maximum acceleration, and grade resistance depend on the layout of the mine to be simulated.

3.2 Rolling Resistance

Rolling Resistance (RR) is defined as:

"When driving, energy is absorbed by the deformation of tires and ground. An example of this is rutting. The restraining effect this has on the vehicle is called rolling resistance"[3].

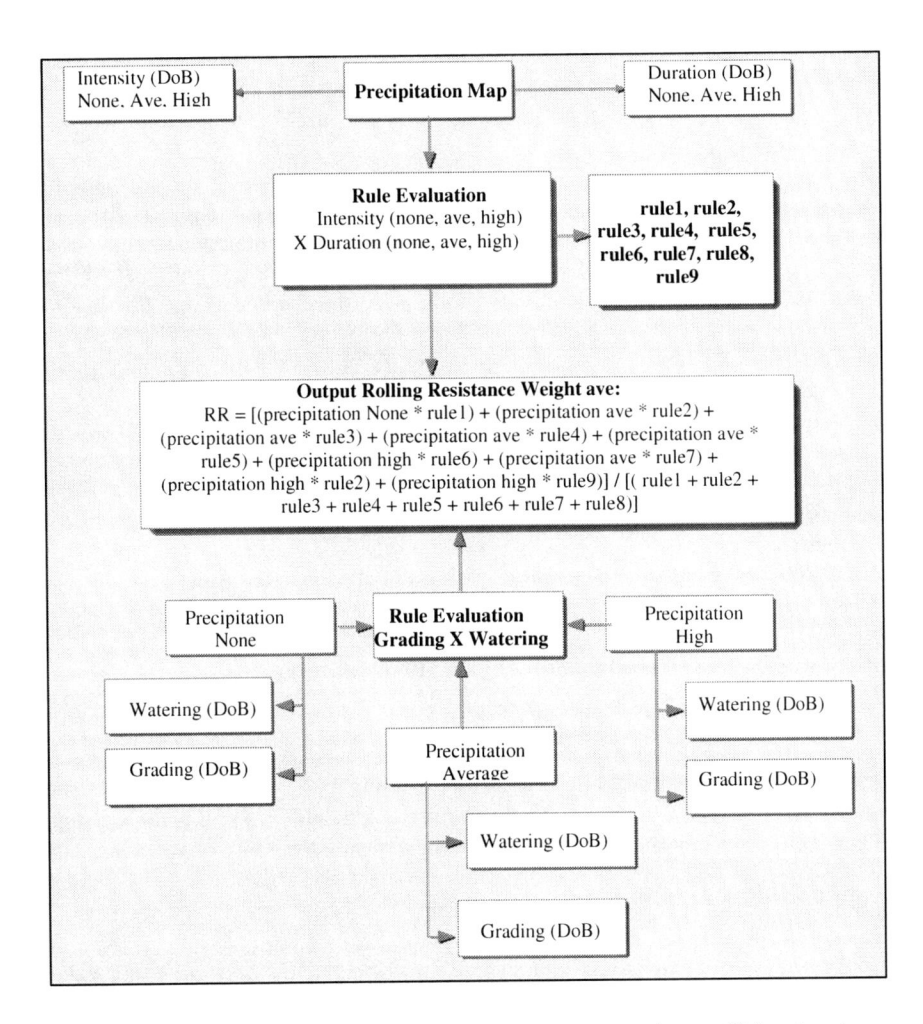

Fig. 1. Rolling Resistance – Fuzzy Logic Chart. Note that the traction coefficient has the same structure as the rolling resistance fuzzy logic

Truck speed depends on the rolling resistance at the ground-wheel interface [3]; if the road section currently has a rolling resistance higher than the mine average, the efficiency of a truck is decreased, and failure frequencies in equipment and tire wear are increased. The categories of "cut" and "impact" failures of tires are directly related to road conditions [4]. RR is difficult to estimate, and every mine sets its own rolling resistance value, which considers the combination of road conditions and equipment [5]. For example, for a watered, well-maintained, hard, smooth, stabilized surfaced roadway with no penetration under load, rolling resistance is about 2%.

In order to keep RR in the proximity of this number, water truck and grader schedules must be adjustable according to changes in road quality and the amount of water on the road. In addition, water trucks act to decrease dust while graders also minimize the effects of bumps, holes, and spillage [4].

Fuzzy logic, a form of multi-valued logic based on principles of approximate reasoning [6], has been used to calculate the rolling resistance and traction coefficient, which is the ratio of maximum pull developed by the machine to the total weight on the axle drivers [7] – see Figure 1.

The logic consists of a map of precipitation factors: intensity (mm) and duration (hours) of rain (or snow). These elements are passed through fuzzy sets that relate these discrete inputs to the degrees of belief (DoBs) in the linguistic terms "none", "average", and "high". For each scenario, i.e., precipitation (none, average and high), a second fuzzy map is applied together with the time since watering and grading were done (none, low, medium and high). Once these DoBs are known, IF-THEN rules are applied to all combinations - precipitation low (grading x watering), precipitation average (grading x watering), precipitation high (grading x watering).

To determine rolling resistance and traction coefficient of the road, all rules are fired and combined (defuzzified) using a MIN-MAX weighted average based on the input DoBs. This approach allows weather, which is not often accurately measured or known), to be "guestimated" with reasonable and acceptable accuracy. Every time a truck enters a segment, these rules are fired and the RR and traction coefficient are applied. Note that watering and grading variables depend on the time since the water truck and motor grader have passed. A count is used for this purpose in the model - when this variable achieves a specific value, it is reset to zero. Maintenance of these ancillary vehicles is also used to introduce delays in this reset.

3.3 Rimpull Forces

The model calculates rimpull forces to define machine performance capability. These forces are defined as available, usable, or required. Available rimpull is the amount of echanical force that the engine transfers to the transmission and final drives that is distributed to the place where the drive tires contact the ground [8]. The available rimpull can be determined from the rimpull-velocity curve provided in the manufacturer's handbook.

The curve below was approximated using 14 straight lines for each interval. By using these equations and knowing the weight of the truck (machine + payload) and effective grade (rolling resistance + grade resistance), the model calculates the maximum velocity and gear range for each step change in the next time interval.

Not all the available rimpull is used; the actual usable rimpull is defined as the amount of pull that the truck offers at the point where the drive tire contacts the ground. This force is a function of weight on the drive wheels and the friction between the tire and ground [8]. Knowing the traction coefficient and weight of the truck, available rimpull can be calculated.

Newton's law states that a body starts moving when forces are equal to motion resistance; thus, the rimpull required to move the truck is equal to total resistance plus rolling resistance.

$$\text{Usable rimpull} = \text{traction coefficient} + \text{weight of truck.} \quad (1)$$

$$\text{Required rimpull} = \text{rolling resistance} + \text{grade resistance.} \quad (2)$$

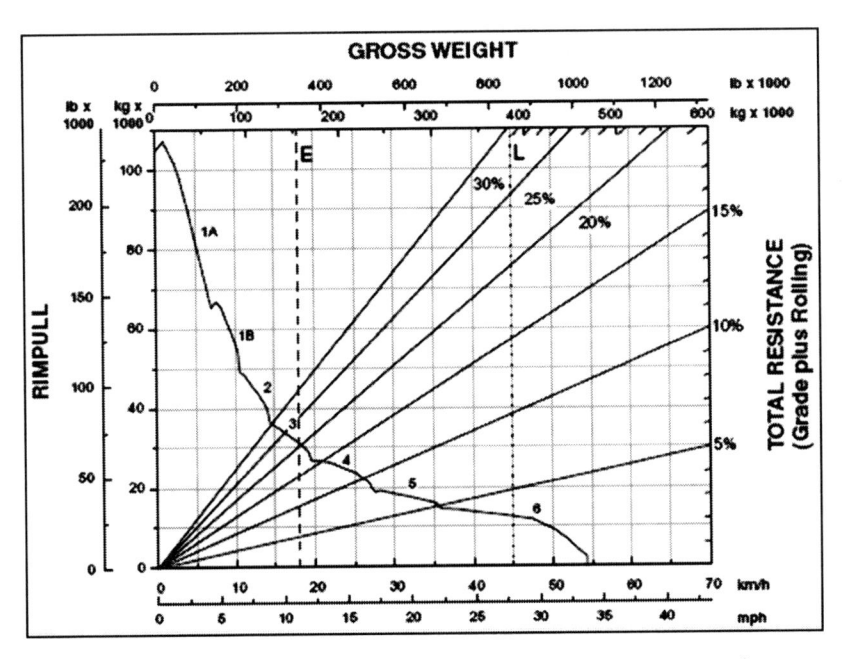

Fig. 2. Curve Speed x Rimpull – 793D CAT (from Caterpillar Handbook 38th edition)

In order to calculate acceleration, the model takes the lower value of available and usable rimpull applied to the following equation [1]:

$$A = [(R^* - Rreq) * 9.81] / W . \quad (3)$$

where: A= vehicle acceleration (m/s²)
R = the lower value between available and usable rimpull (tonnes)
Rreq = the required rimpull (tonnes)
W = the weight of payload and machine (tonnes)
9.81 = gravity acceleration (m/ s²)
When a truck is descending on a downhill section, available rimpull is 0.

Knowing the initial velocity and the small time increment value (Δt), velocity and motion equations can be used to find the actual velocity and distance traveled in Δt:

$$Vf = Vi + a * \Delta t . \quad (4)$$

$$D = Vi * \Delta t + (1/2)*a * (\Delta t)^2 . \quad (5)$$

For every interval Δt, the new rimpull force, acceleration, speed and traveled distance are calculated. At the end of each haul road segment, data is initialized for the next road segment and the simulation continues until the truck arrives at its destination.

For each time increment, the model verifies that the truck is driving at the speed limit and whether the acceleration/deceleration is higher than the maximum allowed. For example, if truck A is approaching other vehicles or an intersection, it decelerates.

Table 1. Driver Performance

Speed	Deviation of speed (%)					
	< ±1	±1	±2	±3	±4	> ±5
15 km/h	Excellent	Excellent	Good	Good	Regular	Bad

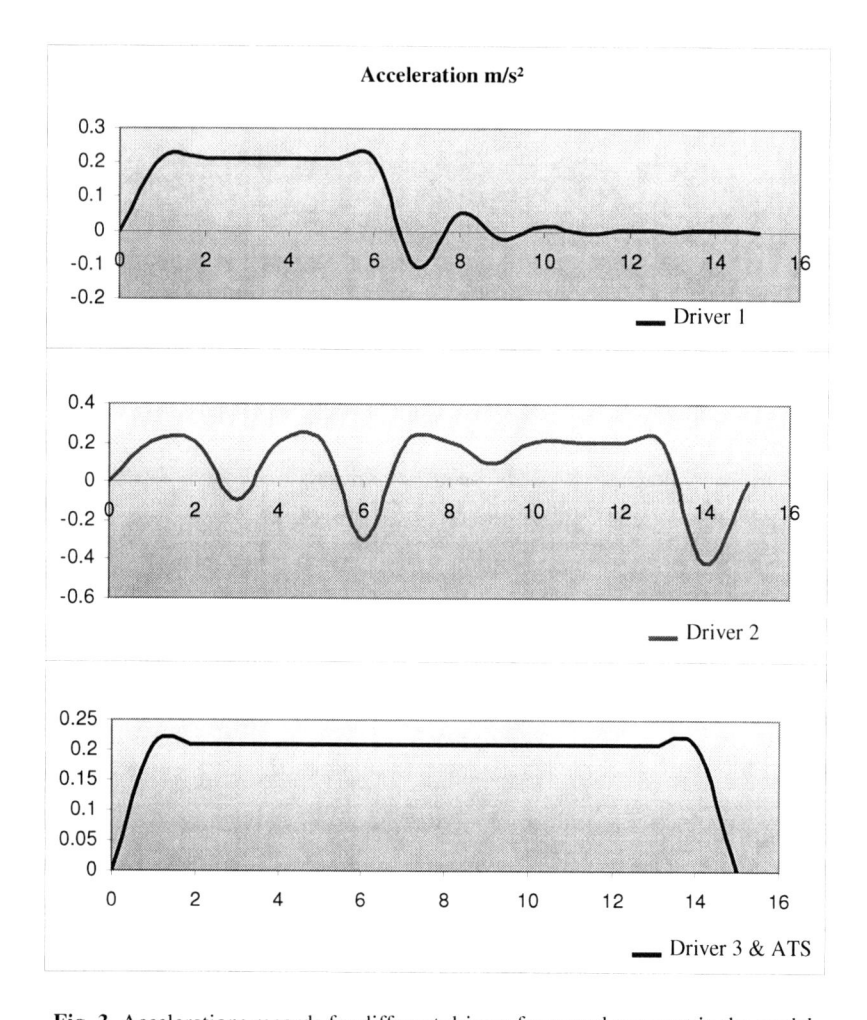

Fig. 3. Accelerations records for different drivers for a road segment in the model

If the front truck is stopped; then truck A breaks until it reaches a full stop. When trucks drive on a downhill slope, the safe speed limit is determined from manufacture retarding performance charts. After the effective grade and weight of the truck are known, a safe speed limit is determined. For example, if a truck that is driving downhill has a safe mechanical speed of 14 km/h and the speed limit on this road is 15km/h; then, the safe mechanical speed will be the maximum speed allowed for this truck, and the speed limit of the road is disregarded.

The act of accelerating/decelerating depends on a driver's abilities and personality; as a result, it is important to accurately estimate driver efficiency in order to compare the manual and autonomous systems. Many attributes such as experience, fatigue, training and gender result in varieties of efficiencies. For example, an aggressive person may drive beyond speed, accelerate, decelerate or brake sharply; on the other hand, an overly-careful employee may drive below the road limit, may accelerate, decelerate and break before the optimum time. As a result, driver performance can run the gamut from aggressive behavior, which be a driver who was trained a long time ago, who is inexperienced, tired, hungry, or under stress to a careful driver who is slow and inexperienced and perhaps, has not received enough training. A properly and recently trained experienced drive is considered to be the best driver.

Human behavior limits the efficiency of the machine; as a result, mechanical and tire wear, fuel consumption, cycle time, and productivity all change according to the driver. In contrast, autonomous haulage trucks can be set to operate at the optimum point of efficiency such that mechanical failures, fuel consumption, and tire wear are decreased. Furthermore, changes in acceleration are smoother with autonomous haulage trucks when compared to manually-driven trucks. Their operation is less variable.

4 Fuel Consumption

In order to calculate fuel consumption, the gear efficiency must be known; however, this data is very hard to obtain from manufacturer's handbooks as competitors are very interested in these numbers. According to the Caterpillar handbook, maximum power for a 793D is 1801kW, rated RPM is 1750, tire radius is 1.778 m. At this engine speed, efficiency is 100%. It is important to find engine speed to estimate the true efficiency.

From the 14 linear relationships used to approximate the Caterpillar Speed-Rimpull Curve, the efficiency can be estimated. Each equation has a range from minimum to maximum rimpull allowed for each interval. From the available rimpull and speed, the following equation is used to determine power and efficiency [9]:

$$\text{power (kW)} = \text{speed (m/s)} * \text{rimpull (kg)} . \tag{6}$$

$$\text{efficiency} = \text{power(kW)} / \text{maximum power (kW)} . \tag{7}$$

In the example below, the higher efficiency for gear#1 is 83.98%, and is given when available rimpull is 57,510 kg and velocity is 9.26 km/h.

Table 2. Estimating Efficiency

Equation for Gear 1	Km/h	Rimpull (X1000)	Power (kW)	Efficiency	Reduction Ratio	RPM	BSFC
R(v) = 255,36 - 21,43*V	8.05	67.23	1473	81.81%	121.40	1458	201
	8.45	64.80	1491	82.80%		1531	202
	8.86	62.37	1504	83.49%		1604	203
	9.26	59.94	1511	83.88%		1677	206
	9.66	**57.51**	**1513**	**83.98%**		1750	212
	10.06	55.08	1509	83.78%		1823	219
	10.47	52.65	1500	83.29%		1896	227
	10.87	50.22	1486	82.50%		1969	228

After estimating the gear efficiency, the reduction or gear ratio can be calculated according to the following equation [9]:

$$\text{Reduction ratio} = \text{rated RPM} * 2\pi * r / (60 * \text{speed}). \tag{8}$$

Where: 2π is used to convert angular into horizontal velocity, r is the radius of the tire, and the constant 60 is used to convert RPM into RPS.

In the example above, reduction ratio = $(1750*2\pi*1.778)/(60*9.66) = 121.40$.

After estimating reduction ratio and efficiency, the engine speed in RPM for every gear can be determined by using the formula [9]:

$$\text{RPM} = \text{reduction} * \text{speed} / (2\pi * r). \tag{9}$$

For the above optimum example, RPM = $121.40*8.05/(2\pi*1.778) = 1458$.

For every point that falls within any of the ranges of the linear equations; efficiency, reduction, engine speed and Break Specific Fuel Consumption(BSFC) are known. BSFC is used to measure fuel efficiency within the piston engine. Since Caterpillar provides only BSFC at 100% efficiency, BSCF also is estimated.

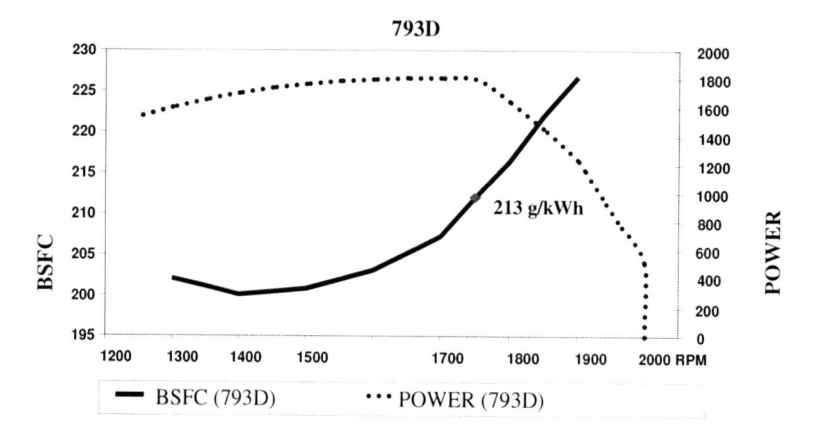

Fig. 4. Caterpillar BSFC chart

The Caterpillar curve above was divided into 6 linear equations to associate BSFC with engine speed (RPM). From these equations, the BSCF is calculated.

The chart above does not represent equipment in the real world as there are many additional losses such as heat and noise; as a result, a truck does not achieve 1801 kW, as seen in the chart. The model shows that in reality, efficiency peaks at an engine speed of 1750 rpm. The BSFC at this point is ~213 g/kWh which compares well to Caterpillar's manual range of 210-215.

With the approach, it is possible to determine the highest gear efficiency, the engine speed and BSFC. These variables will be very important when human behavior is eliminated from the system. Suggestions regarding road design can be made to account for this increased efficiency and decreased variance. Autonomous haulage trucks could drive at speeds with the highest gear efficiency all the time. For example, when an AHT uses gear #1, it would drive proximally at 9.66km/h if the variables remain constant. With increased equipment efficiency, engine performance also improves, and lower levels of fuel consumption and mechanical wear are noted.

By knowing the BSCF for each small increment of time, the model calculates the fuel consumption rate at each step change by using the following formula [10]:

$$C_f = (C_{sf} * P) / \rho_f . \qquad (10)$$

Where: C_f is fuel consumption rate (L/h), C_{sf} is BSFC at full power (g/kWh), P is the rated brake power (bkW), ρ_f is the fuel density (g/L).

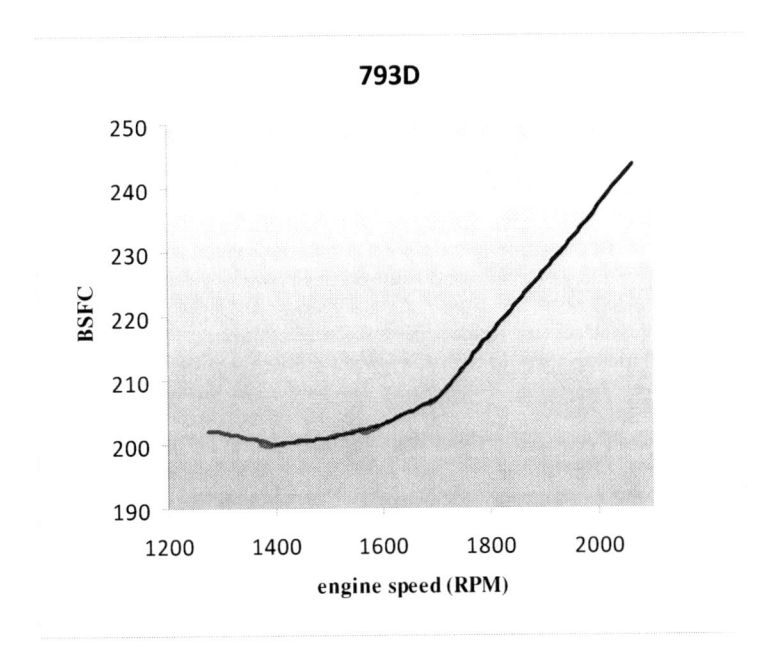

Fig. 5. Estimated BSFC for all Gear (1-6)

5 Conclusions

Manual processes involve variations since how a person drives differs according to his or her skills, moods and abilities; therefore, with a human driver, a truck is less likely to work at maximum efficiency. In contrast, autonomous haulage trucks can be programmed to operate at optimum braking and acceleration rates throughout the haul. Additionally, AHT can maintain speed at the highest gear efficiency. As such, significant savings in tire wear and fuel consumption are expected. Although the model described in this paper is still under development, when completed, a reasonably accurate comparison of AHT with human-drivers systems can be done using benchmarked Key Performance Indicators such as tire wear, fuel consumption, productivity, and maintenance. The anticipated results are a reduction of ~10% in fuel consumption, ~14% in maintenance costs, ~12% in tire wear, 5-50% in labor costs, ~8% in mechanical availability, ~5% in productivity, and ~12% in truck life. The cost-benefit analysis is complex, but if variables are estimated accurately and based on First-Principles modeling techniques, a combination of deterministic and probabilistic simulation should provide a proper assessment of the improvements.

Acknowledgment. The authors wish to express their appreciation to BHP-Billiton Nickel West Division for the funding of this research and for access to important data and information regarding autonomous haulage systems.

References

1. Fytas, K. G.: An Interactive Computer Simulation Model of Open Pit Haulage systems (1983)
2. Lewis, M.W., Werner, J., Sambirsky, B.: Capturing Unrealized Capacity. CIM Bulletin 97(1076), 57–62 (2004)
3. Smith, S.D., Wood, G.S., Gould, M.: A New Earthmoving Estimating Methodology. Construction Management and Economics 18, 219–228 (2004)
4. Monroe, M.D.: Optimizing Truck and Haul Road Economics. Journal of Mines, Metals, and Fuels, 255–261 (1999)
5. Karaftath, L.L.: Rolling Resistance of Off-Road Vehicles. Journal of Construction Engineering and Management 114(3), 458–470 (1988)
6. Von, C.A.: Fuzzy Logic and NeuroFuzzy applications explained. Prentice Hall PTR, Upper Saddle River (1995)
7. University of South Australia.: Civil Engineering Practice 1. Course intro to civil and mining engineering (2009)
8. King Fahd University of Petroleum & Minerals, Construction Engineering & Management Department: Construction Equipment & Methods. Course: CEM 530 (Spring 2004/2005)
9. Perdomo, J.L.: Detailed Haul Unit Performance Model. Virginia Polytechnic Institute, Civil Engineering (2001)
10. Kennedy, B.A.: Surface Mining, 2nd edn., p. 93 (1990)

Building a No Limit Texas Hold'em Poker Agent Based on Game Logs Using Supervised Learning

Luís Filipe Teófilo and Luís Paulo Reis

Departamento de Engenharia Informática, Faculdade de Engenharia da Universidade do Porto,
Portugal
Laboratório de Inteligência Artificial e Ciência de Computadores, Universidade do Porto,
Portugal
{luis.teofilo,lpreis}@fe.up.pt

Abstract. The development of competitive artificial Poker players is a challenge toArtificial Intelligence (AI) because the agent must deal with unreliable information and deception which make it essential to model the opponents to achieve good results. In this paper we propose the creation of an artificial Poker player through the analysis of past games between human players, with money involved. To accomplish this goal, we defined a classification problem that associates a given game state with the action that was performed by the player. To validate and test the defined player model, an agent that follows the learned tactic was created. The agent approximately follows the tactics from the human players, thus validating this model. However, this approach alone is insufficient to create a competitive agent, as generated strategies are static, meaning that they can't adapt to different situations. To solve this problem, we created an agent that uses a strategy that combines several tactics from different players.By using the combined strategy, the agentgreatly improved its performance against adversaries capable of modeling opponents.

Keywords: Poker; Machine Learning; Supervised Learning; Opponent Modeling; Artificial Intelligence.

1 Introduction

Poker is a game that is increasingly becoming a field of interest for the AI research community, on the last decade. Developing an agent that plays Poker presents a completely different challenge than developing agents for other games like chess or checkers. In these games, the players are always aware of the full state of the game. This means that these problems can be solved by using deepMinimax decision trees and game state classification heuristics. Notable results were achieved in this type of games such as the Deep Blue, which was the first to computer to beat a Chess champion in a series of games [1].

Unlike Chess, the Poker's game state is partially hidden because each player cannot see the opponents' cards.Therefore, it's much more difficult to create and analyze a decision tree. Poker is also stochastic game i.e. it admits the element of chance, making the outcome of the game unpredictable.

The announced characteristics of Poker make it essential to define mechanisms to model the opponents, before making a decision. By identifying the opponents playing style, it is possible to predict their possible actions and therefore make a decision that has better probability of success [2, 3].

The main goal of this work is to determine on how strategies used in the past by good human players can be used to support the creation of an artificial Poker player. This work is divided in the following steps:

- Extract a significant amount of games between human players;
- Extract information about the players;
- Define the classification problem: game state variables that can influence player's decision and the possible outcomes;
- Create a strategy by classifying game state instances, using supervised learning algorithms;
- Create and test an agent that uses the classifier and follows the strategy.

The rest of the paper is organized as follows. Section 2 briefly describes the game of poker. Section 3 describes related work about opponent modeling in Poker and some approaches followed earlier to create Poker agents. Section 4 describes the characteristics of the dataset that was used in this work. Section 5 presents the classification problem and the comparison of the supervised learning algorithms that were used to solve it. Section 6 presents some details about the implementation of the agent that uses the human strategy. Section 7 describes the results obtained by the agent. Finally, section 8 presents the paper main conclusions and some pointers for future work.

2 Texas Hold'em Poker

Poker is a gambling game where the players bet that their hand is stronger than the hands of their opponents. Every bet made by a player goes into the pot and at the end the winning player collects the pot.The winner is the last standing player or the player that has the best hand. Poker is a generic name for literally hundreds of games, with similar rules [4], called variants. This work is focused on the variant Texas Hold'em Poker, which is nowadaysthe most popular one.

2.1 Hand Ranking

A poker hand is a set of five cards that identifies the score of a player in a game of poker. The hand rank at any stage of the game is the score given by the 5 card combination composed by player cards and community cards that has the best possible score. This means that if more than 5 cards are available, some will be ignored.

The possible hand ranks are (from stronger to weaker ranks): Royal Flush (top sequence of same suit), Straight Flush (sequence of same suit), Four of a Kind (4 cards with same rank), Full House (Three of a Kind + Pair), Straight (sequence of cards), Three of a Kind (3 cards with same rank), Two Pair, One Pair (2 cards with same rank) and Highest Card (when does not qualify to any other ranks).

2.2 No Limit Texas Hold'em

No Limit Texas Hold'em is a Poker variation that uses community cards. At the beginning of every game, two cards are dealt for each player – pocket cards. A dealer player is assigned and marked with a dealer button. The dealer position rotates clockwise from game to game. After that, the player on the left of the dealer position posts the small blind (half of the minimum bet) and the player on the left of that one posts the big blind (minimum bet). The first player to talk is the one on the left of the big blind position.

After configuring the table, the game begins. The game is composed by four rounds (Pre-Flop, Flop, Turn, River) of betting. In each round the player can execute one of the following actions: Bet, Call, Raise, Check, Fold or All-In.

At any game round, the last standing player wins the game and therefore the pot. If the River round finishes, the player that wins is the one with the highest ranked hand.

This variation of Poker is also No Limit, which means that the players are allowed to bet any amount of chips, above the Bet or Raise value.

3 Related Work

Most noticeable achievements in Computer Poker are from the Poker Research Group [5] from University of Alberta mostly on the variant Limit Texas Hold'em. One of the most significant publications of the group is Darse Billings PhD thesis [6] where hepresents a complete analysis of the evolution of artificial poker agent architectures, demonstrating strengths and weaknesses of each architecture both in theory and in practice.

Other significant publication is [7] where a perfect strategywas defined for a very simple variant of Poker (Rhode Island). Another near-perfect strategy was achieved in [8] for Heads-Up No Limit variant, using the Nash Equilibrium Theory. More recently, it should be emphasized the article [9], which describes a new and more effective technique for building counter strategies based on Restricted Nash Response. Some achievements were also made using machine learning classifiers like in [10] where were studied evolutionary methods to classify in game actions.

Despite all the breakthroughs achieved by known research groups and individuals, no artificial poker playing agent is presently known capable of beating the best human players.

3.2 Opponent Modeling in Poker

A large number of opponent modeling techniques are based on real professional poker players' strategies, such as David Sklansky, who published one of the most renowned books on poker strategy [4]. In his book Sklansky says that Poker players can be classified in terms of tightness and aggressiveness. A player is tight if it plays 28% or less hands, and loose if it plays more than 28% of the times. With regard to aggressiveness, the player is aggressive if it has an aggression factor (1) over 1.0; otherwise he is a passive player.

$$Agression\ Factor = \frac{Num\ Raises}{Num\ Calls} \qquad (1)$$

With these two parameters, Sklansky defined groups of players (Sklansky' groups)that play similar hands in a similar way.

3.2 Poker Hand Rank Evaluation

Hand rank evaluation consists in checking if a given Poker hand is better than another. Usually, the Hand Rank Evaluator takes a poker hand and maps it to a unique integer rank such that any hand of equal rank is a tie, and any hand of higher rank wins. For a poker AI, it is absolutely critical to have the fastest hand evaluator possible [11]. Any poker agent may have to evaluate thousands of hands for each action it will take.

The fastest known evaluator is TwoPlusTwo Evaluator [12], which can evaluate about 15 millions of hands per second.

3.3 Odds Calculation

Evaluating the rank of the hand, which was discussed above, is about giving a score to a set of cards. A Poker AI in reality does not directly use the hand evaluators, because it does not know the opponent's cards, so there is no group of hands to compare. A Poker AI performs odds calculation, which consists on the prediction of its own hand success. For that purpose, it evaluates its own hand score and compares with possible opponent's hands scores. This prediction is used to help measuring the risk of an action.

There are various ways to determine the odds:

- Chen Formula [13]: this formula can determine the relative value of a 2 card hand;
- Hand Strength [6, 11]: determines how many hands are better than our, taking into account the number of opponents;
- Hand Potential [6, 11]: The hand potential is an algorithm that calculates PPOT and NPOT. The PPOT is the chance that a hand that is not currently the best improves to win at the showdown. The NPOT is the chance that a currently leading hand ends up losing. Therefore, they are used to estimate the flow of the game;
- Effective Hand Strength [6, 11]: Combines the hand strength and the hand potential formulas.

3.4 Agent Development Tools

There are some software tools that can aid the development of Poker agents. Most notable is the Meerkat API [11] which easily allows the creation of Poker agents. The Meerkat agents can be tested using game simulators that support this API, like Open Meerkat Test Bed [14].

4 Poker Data Extraction and Analysis

To create models of human poker players, the first step is to extract a great amount of Poker games to analyze the actions of the players. The chosen data source is composed by game logs from money games inonline casinos.

The Poker game logs represent the history of actions made by the players during the games on a given Poker table. These files don't represent the state of the game. Therefore to analyze the game it is necessary to replay the game with the same movements and card distribution.

Obtaining information from these documents is difficult since these files don't typically have an organized structure, making it difficult to parse the information. Moreover, there is no standard to represent game movements: each casino client has its own representation of logs. For this reason, to combine data from multiple sources, a new parser is needed for each game log format.

The package of game logs that was used in this work can be found here [15]. Some characteristics of the game logs are on table 2.

Table 1. Game logs characteristics

Characteristic	Value
Number of games	51.377.820
Number of players	158.035
Number of showdowns	2.323.538
Number of players with 500 or more showdowns	183

To characterize an action it is essential to know the cards that the player had. As it can be seen on table 1, the percentage of games in which card exposal occurs is very small, and only 183 players showed their cards more than 500 times. These were the players that were selected for this work.

After obtaining the game data, we selected the best players available in order to learn good tactics. For this task, a player list was generated containing some information about each player. The criteria used to choose a good player was its earnings as, for instance, a player with high negative earnings probably don't has a good tactic.

Table 2. Characteristics of the extracted players

Name	Game Count	Number of Shows	Earnings
John	15.763	638	1.003$
Kevin	20.660	838	30$
David	77.598	2.103	14.142$
Jeff	33.257	882	-4.945$

It should be noted that a player with negative earnings (Jeff) was included for testing purposes, to check if a tactic generated from a theoretical bad player loses against a tactic from a theoretical good player.

5 Learning Game Strategies

To learn the players' tactics supervised learning algorithms were used. In order to facilitate both the learning process and the implementation of the agent, we used WEKA[17]. The characteristics used to classify the players' tactics were based on game state variables that might influence the players' decisions during the game: position in table, money, last opponent's action and classification of last opponent. The following code represents the ARFF file structure associated with each player.

```
@relation poker
@attribute positionScore numeric
@attribute effectiveHandStrength numeric
@attribute percentageMoneyToBet numeric
@attribute percentageMoneyOnTable numeric
@attribute possibleWins numeric
@attribute didLastPlayerRaise numeric
@attribute didLastPlayerCall numeric
@attribute isLastPlayerAggressive numeric
@attribute isLastPlayerTight numeric
@attribute action {call, raise5%, raise10%, ...}
```

Only numerical attributes were used in this model. The boolean attributes were converted to 0 or 1 whenever the value was true or false. The last attribute is nominal and it represents the action that was taken by the player, being for that the class attribute. The defined model returns an action, based on all the other attributes. Another important fact to note is that each player has four ARFF files associated with it, each one representing a game round (Pre-Flop, Flop, Turn, River). This is because tactics used during the game tend to be different in each game round, due to factors such as the varying number of community cards available.

Different classifiers were tested to build this model. The classifiers that obtained a smaller average error, using tenfold cross validation, were search trees, more particularly Random Forest Trees (Fig 1). The errors were relative high as was expected as players tend to change tactic during the game, making it difficult to find patterns on the data.

The classifiers error rate was also analyzed per round (Fig. 2). It was found that the error is much higher in Flop and Turn rounds than in Pre Flop and River rounds. This was expected for Pre-Flop round, since there are no communities cards, the tactics tend to be simpler. As for River round, this can be explained by the lower number of players that reach this round.

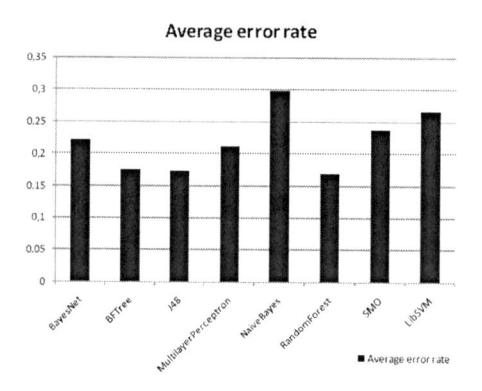

Fig. 1. Average classifier error rate

Error rate evolution

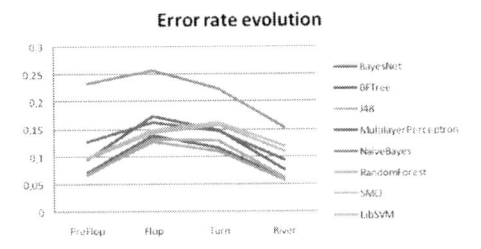

Fig. 2. Average classifier error rate evolution

6 Agent Implementation

The agent was implemented using theMeerkat API [11]. The way the agent chooses the action is based on the current game state and can be divided in the following steps.

- First, the agent determines which actions can be taken;
- The agent changes the tactic based on the game state;
- It uses the current tactic classifier to predict the action;
- If the prediction is null i.e. the classifier can't find any action that suits the situation, the agent checks if possible, otherwise it folds.

Regarding the action prediction, the classifiers weren't taught how to fold, because only hands with showdown were used. To solve this, we defined a criterion to fold some hands. The defined criterion was that if the Effective Hand Strength [11] is below 50%, then the agent has a probability of folding the hand equal to its tightness level.The agent will choose the class that has better acceptance level on the current game state. If that class acceptance level is below a defined minimum, the agent folds.

7 Tests and Results

Four types of tests were used to validate this approach: behavior tests, tests between generated tactics, tests against other agents previously developed and strategy tests. The tests were made on Open Meerkat Test Bed, with 1.000 games and table seat permutation.

The results of the behavior tests are on table 3. The agent approximately conserved the real human Sklansky's classification thus conserving its behavior.

Table 3. Behaviour tests

Name	Agent AF	Real player AF	Agent tightness	Real player tightness	Agent classification	Real player classification
John	4,77	5,12	0,32	0,31	Loose Aggressive	
Kevin	1,55	1,49	0,25	0,23	Tight Aggressive	
David	16,04	13,22	0,19	0,19	Tight Aggressive	
Jeff	9,42	8,40	0,44	0,33	Loose Aggressive	

Fig. 3 presents the bankroll evolution in a series of games between the agents. The humans that won more money in real life generated an agent with better results.

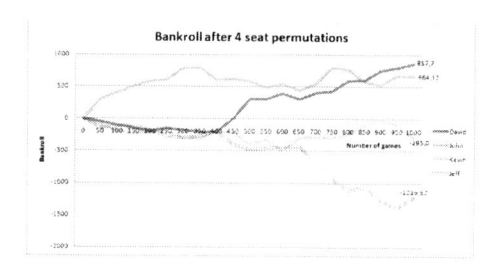

Fig. 3. Game between the players on table

Next, we tested John agent against HSB Bot (an agent that chooses his actions based on its pocket hand strength) (Fig. 4). John clearly won the match by a large margin. John was also tested against an always call agent, achieving similar results.

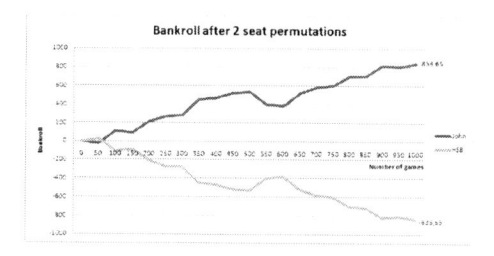

Fig. 4. John VS HSB

Finally, we tested John against MCTS Bot [18]. The agent was totally defeated (Fig. 5), because the MCTS is capable of opponent modeling.

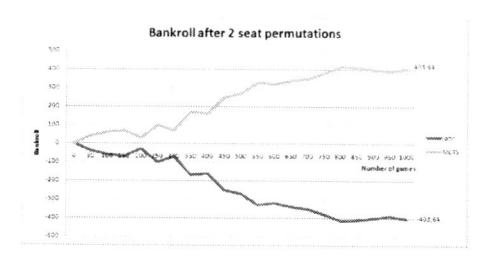

Fig. 5. John VS MCTS Bot

The results improved a lot by using a strategy that combines multiple tactics of generated agents (Fig. 6). The strategy changes the tactic when losing money, confusing the opponent modeling mechanisms. We can observe some cycles on the chart due to this fact. When the agent starts losing money, it changes its tactic, and after a little while it starts winning again, until the moment that the MCTS discovers the new tactic.

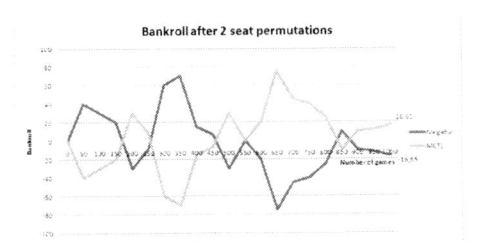

Fig. 6. Combined strategy VS MCTS

8 Conclusions and Future Work

There is still a long way to go to create an agent that plays poker at the level of the best human players. This research presented an approach based on supervised learning methodologies, where the agent relies on pre learned tactics based on past human experience to decide upon its actions. The created agent is not competitive against good poker players that are capable of opponent modeling. However the results greatly improved after combining various tactics, which means that an agent should not use a static tactic in a game like poker.

This approach can be promising for good human players, since they can create an agent based on their logs that will play like them, autonomously.

The generated agents could be improved in the future, by defining a more complex player model or specifying better heuristics to change tactic along the game. In the future, the system should also be tested against human players.

This approach might also be used in other incomplete information games by defining the game state variables of that particular game.

Acknowledgments. I would like to thank to Fundaçãopara a Ciência e a Tecnologia for supporting this work by providing my PhD Scholarship SFRH/BD/71598/2010.

References

1. Hsu, F.-H.: Behind Deep Blue: Building the Computer that Defeated the World Chess Champion. Princeton University Press, Princeton (2002)
2. Billings, D., Papp, D., Schaeffer, J., Szafron, D.: Opponent modeling in poker. In: AAAI 1998/IAAI 1998, pp. 493–499. American Association for Artificial Intelligence, Madison (1998)
3. Davidson, A.: Opponent modeling in poker. M.Sc., University of Alberta, Edmonton, Alberta, Canada (2002)
4. Sklansky, D.: The Theory of Poker: A Professional Poker Player Teaches You How to Think Like One. Two Plus Two (2002)
5. Computer Poker Research Group Homepage,
   http://webdocs.cs.ualberta.ca/~games/poker/
6. Billigs, D.: Algorithms and Assessment in Computer Poker. Ph.D., University of Alberta, Edmonton, Alberta (2006)

7. Gilpin, A., Sandholm, T.: Better automated abstraction techniques for imperfect information games, with application to Texas Hold'em poker. In: 6th international joint conference on Autonomous agents and multiagent systems, Honolulu, Hawaii, pp. 1–8. ACM, New York (2007)
8. Miltersen, P.B., Sørensen, T.B.: A near-optimal strategy for a heads-up no-limit Texas Hold'em poker tournament. In: 6th international joint conference on Autonomous agents and multiagent systems, Honolulu, Hawaii, pp. 1–8. ACM, New York (2007)
9. Johanson, M., Bowling, M.: Data Biased Robust Counter Strategies. In: Twelfth International Conference on Artificial Intelligence and Statistics, Clearwater Beach, Florida, USA, April 16-18, pp. 264–271 (2009)
10. Beattie, B., Nicolai, G., Gerhard, D., Hilderman, R.J.: Pattern Classification in No-Limit Poker: A Head-Start Evolutionary Approach. In: 20th conference of the Canadian Society for Computational Studies of Intelligence on Advances in Artificial Intelligence, Montreal, Quebec, Canada, pp. 204–215. Springer, Heidelberg (2007)
11. Felix, D., Reis, L.P.: An Experimental Approach to Online Opponent Modeling in Texas Hold'em Poker. In: 19th Brazilian Symposium on Artificial Intelligence: Advances in Artificial Intelligence, Savador, Brazil, pp. 83–92. Springer, Heidelberg (2008)
12. Poker Bot Artificial Intelligence Resources, http://spaz.ca/poker/
13. Chen, B., Ankenman, J.: The Mathematics of Poker. Conjelco (2006)
14. Félix, D., Reis, L.P.: Opponent Modelling in Texas Hold'em Poker as the Key for Success. In: ECAI 2008: 18th European Conference on Artificial Intelligence, pp. 893–894. IOS Press, Amsterdam (2008)
15. MeerkatOpentestbed, http://code.google.com/p/opentestbed/
16. Out Floppe Poker Q & A: Obfuscated datamined hand histories, http://www.outflopped.com/questions/286/obfuscated-datamined-hand-histories
17. Hall, M., Frank, E., Holmes, G., Pfahringer, B., Reutemann, P., Witten, I.H.: The WEKA data mining software: an update. SIGKDD ExplorNewsl. 11(1), 10–18 (2009)
18. Broeck, G., Driessen, K., Ramon, J.: Monte-Carlo Tree Search in Poker Using Expected Reward Distributions. In: 1st Asian Conference on Machine Learning: Advances in Machine Learning, Nanjing, China, pp. 367–381 (2009)

Autonomous Manipulation Combining Task Space Control with Recursive Field Estimation

Soo Jeon[1] and Hyeong-Joon Ahn[2]

[1] University of Waterloo, Waterloo ON N2L 3G1, Canada,
soojeon@uwaterloo.ca,
[2] SoongSil University, Seoul 156-143 Korea

Abstract. This paper proposes an autonomous manipulation strategy that enables an articulated manipulator to detect and locate a particular phenomenon occurring in the surrounding environment. More specifically, the sensor-driven task space control of an end-effector is combined with the field estimation and the target tracking in an unknown spatial field of interest. The radial basis function (RBF) network is adopted to model spatial distribution of an environmental phenomenon as a scalar field. Their weight parameters are estimated by a recursive least square (RLS) using the collective measurements from the on-board sensors. Then, the asymptotic source tracking has been achieved by the control law based on the gradient of the estimated field. Experimental results verified the effectiveness of the proposed scheme.

Keywords: autonomous manipulator, task space control, field estimation.

1 Introduction

Using robotic manipulators for autonomous inspection and repair has been a major concern in a number of industrial sectors [1], especially, in the operation of hazardous or remotely located facilities such as the nuclear reactor [2] and the offshore oil & gas platform [3]. They are often configured on moving platforms for mobility and usually equipped with a variety of sensors to perform wide range of monitoring and inspection operations [3]. Kinematic redundancy is desired in some cases for extra degrees of freedom to enable dexterous maneuvers in complex environments [4]. Without a target point nor the desired trajectory, the movement of the end-effector needs to be resolved by an instantaneous motion command determined while it navigates through an unknown scalar field of interest. For instance, in the emergency repair of a gas leak in a process plant, the robot needs to command its motion tasks while it traces out the origin of the leak.

The real time estimation of an unknown field has recently been an important issue in mobile sensor networks, especially, in their application to environmental monitoring and mapping [5, 6, 7]. The common approach is to progressively refine the knowledge of the unknown field based on new observations that are available at every instant. There exist several different approaches depending on how we assume the mathematical model of the unknown field and how to design the motion command. Martinez [6] employed the non-parametric inference method where the spatial field is represented by interpolation without a priori knowledge on the structure of the unknown field. In contrast, Choi et al. [5] assumed that the field can be represented by the weighted sum of a number of radial basis functions (RBF). Leonard et. al. [7] formulated the gradient estimation scheme considering an affine function as an approximation of the field. To trace a target source (e.g., the origin of toxic chemical leak) through the unknown spatial field, the most common approach has been using the gradient climbing [8, 5, 9, 10].

By combining the on-line field estimation with the task space control using artificial potential field, this paper presents a new sensor-driven control strategy for control of the robot end-effector in an unknown field. This paper is organized as follows. Section 2 presents the problem statement and basic formulations. Then the main results follow in Section 3 which describes the formulation of the recursive field estimation. Experimental results are given in Section 4 to verify the proposed scheme. The concluding remarks are presented in Section 5.

2 Problem Statement and Basic Formulation

Consider a robot manipulator deployed to a certain region where the unknown field of interest is distributed over the operational range of the manipulator. Without a priori knowledge on the location of the target source, the manipulator needs to run a *learning process* to make its way through the field to find the target source. As the sensors collect more data about the field, the manipulator will progressively build a better knowledge on the configuration of the field. Eventually, it needs to converge to the target source, i.e. the global maximum of the field. Such a concept is illustrated in Fig. 1 where the distribution of the unknown field is represented by the iso-surfaces and their gradient vectors. The manipulator will be instrumented with a number of field sensors at different

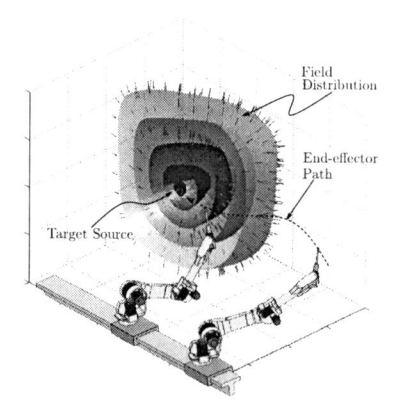

Fig. 1. A robot manipulator and an unknown field of interest

parts of the body to provide enough data to estimate the parameters characterizing the field and also to identify various objects to be avoided during the navigation. The main objective of this paper is to formulate a task space control strategy to enable the end-effector of the manipulator continuously directed toward the target source of the unknown field. Especially, the emphasis is on the experimental validation of this concept. In this paper, the term *field* refers to a scalar field representing a spatial distribution of an environmental phenomenon. It is assumed that the scalar value at each point is the only measurement available through local sensing. This excludes typical cognitive sensors such as vision, ultrasonic or haptic devices. Temperature distribution, diffusion of chemicals or magnetic/electric field can be such examples.

2.1 Manipulator Dynamics in Task Space

Consider a mechanical manipulator with n links, the dynamics of which is given by the well-known Euler-Lagrange equation as

$$M(q)\ddot{q} + L(q,\dot{q})\dot{q} + G(q) + F_f(\dot{q}) = \tau + F_d \qquad (1)$$

where $q(t) \in \mathbb{R}^n$ is the joint angle vector, $\tau(t) \in \mathbb{R}^n$ is the joint torque, $M(q) \in \mathbb{R}^{n \times n}$ is the inertia matrix (symmetric and positive definite), $L(q,\dot{q})\dot{q} \in \mathbb{R}^n$ is the Coriolis and centrifugal force, $G(q) \in \mathbb{R}^n$ is the gravitational force, $F_f(\dot{q}) \in \mathbb{R}^n$ is the nonlinear joint friction force, and $F_d \in \mathbb{R}^n$ is the vector of unknown external disturbance forces. Let $p(t) \in \mathbb{R}^\ell$ represent the configuration of the end-effector in the task space. Note that $\ell = n$ if we assume the fully-actuated and non-redundant manipulator. $p(t)$ is related to the joint space vector $q(t)$ by

$$p(t) = \psi(q(t)), \quad \dot{p} = v = J\dot{q}, \quad \ddot{p} = a = J\ddot{q} + \dot{J}\dot{q} \qquad (2)$$

where $\psi : \mathbb{R}^n \to \mathbb{R}^\ell$ is the forward kinematics that maps the joint space variable q to the task space variable p and $J(q) = \frac{\partial \psi}{\partial q}$ is the manipulator Jacobian. Consider a modified inverse dynamics control law of the form [13]

$$\tau = \hat{M}a_q + \hat{L}\dot{q} + \hat{G} + \hat{F}_f - \hat{F}_d \qquad (3)$$

where $\hat{M}, \hat{L}, \hat{G}, \hat{F}_f$ and $\hat{F}_d$ denote the estimated (computed) values for the corresponding time-varying parameters and a_q is an equivalent desired acceleration referenced to the joint space coordinate system. From Eq. (2), we can map the desired joint acceleration a_q to the desired task space acceleration a_x as

$$a_x = \hat{J}a_q + \dot{\hat{J}}\dot{q} \qquad (4)$$

where $\hat{J}$ and $\dot{\hat{J}}$ are the estimated values of Jacobian and its derivative respectively. The hat notation $(\hat{\bullet})$ will be used for the estimated value of the corresponding variable. Substituting Eqs. (3) and (4) to Eq. (1), we get the decoupled dynamic system in task space as

$$\dot{v} = a_x + \mathcal{F}_p \qquad (5)$$

where $\mathcal{F}_p$ is a small perturbation term given by

$$\mathcal{F}_p = JM^{-1}\{\Delta(MJ^{-1})a_x + \Delta(L - MJ^{-1}\dot{J})\dot{q} + \Delta(G + F_f - F_d)\}. \quad (6)$$

Δ denotes the parameter error for each term, i.e., $\Delta(\bullet) := (\hat{\bullet}) - (\bullet)$.

In this paper, the desired task space acceleration a_x will take the form of

$$a_x(t) = \Gamma_f - K_v \hat{v}(t) \quad (7)$$

where Γ_f is a control law to be derived in Section 3.2. The last term is the damping effect applied to the task space velocity $\hat{v}(t) = \hat{J}\dot{q}(t)$ with the diagonal matrix $K_v \in \mathbb{R}^{\ell \times \ell}$ representing damping coefficients.

Note that our control task does not presume any knowledge on the desired trajectory. A critical issue arising from the implementation of the suggested control scheme is how to handle the kinematic singularity that may be encountered during the navigation. If the manipulator is redundant, the gradient of additional potential functions can be augmented by the projection onto the null space of the Jacobian to keep the joint angles away from the singular configurations without compromising the task space control action [11]. While this method can prevent the redundant manipulator from entering singularities by taking advantage of its self motion, a problem still occurs if the task space motion command from the field estimation is persistently directing the end-effector toward the boundary of the workspace. To overcome such issues, we may use an ad-hoc approach. One method is to project the desired task space command onto the vector orthogonal to the singular direction. By doing this, we may drive the end-effector as close as possible to the target source while tolerating the singular configuration at or near the boundary of the work space. More details on this method can be found in [12].

3 Spatial Field Estimation and Tracking

In practice, the joint feedback control law given in Eq. (3) is implemented in a digital controller so is recommended to run at the sample rate fast enough to realize the desired motion control performance. However, the control action that will be given by the estimated field data (which will be implemented to a_x in Eq. (5) by recursive least square (RLS) method) does not need to run at such a fast rate. This is because that the RLS algorithm updates the parameter based on the field sensing which may not contain new information until the end-effector moves to a new location where the data is distinguishable from the previous measurement. Hence, in this subsection, we will formulate the RLS in discrete time domain to account for the different time scale of the RLS compared to other motion control signals.

3.1 Recursive Estimation of Spatial Field of Interest

The control law is based on the gradient information learned up to the present time. Specifically, we will adopt the approach using the radial basis function

(RBF) network as a mathematical model of the regional field [5]. Consider a differentiable spatial basis function vector $h(\xi) : \mathbf{R}^\ell \to \mathbf{R}^m$, where $\xi \in \mathbf{R}^\ell$ is the coordinate vector in ℓ dimensional space and m denotes the total number of basis functions used to approximate the field with reasonable accuracy within the surveillance region. Then, we can represent the true field of interest denoted by $\mathcal{T}(\xi) : \mathbf{R}^\ell \to \mathbf{R}$ as a weighted sum of basis functions as

$$\mathcal{T}(\xi) := \theta^T h(\xi) = \sum_{i=1}^{m} \theta_i h_i(\xi) \qquad (8)$$

where $\theta := \begin{bmatrix} \theta_1 & \theta_2 & \cdots & \theta_m \end{bmatrix}$ is a weight vector that parameterizes the field. $h_i(\xi) = \bar{h}_i(\|\xi - \mu_i\|)$ is the ith radial basis function centered at a location $\mu_i \in \mathbf{R}^\ell$ in the surveillance region. Note that we assume the linear form of the basis function network, which is intended to utilize the standard RLS technique as our learning tool. Any set of (differentiable) functions that their response changes monotonically with distance from the center point can be used as basis functions. In describing environmental or biological behaviors, a typical choice is the Gaussian RBF which can be written as

$$h_i(\xi) = \exp\left(-\frac{\|\xi - \mu_i\|^2}{\sigma^2}\right) \qquad (9)$$

where σ denotes the common width of basis functions.

A set of field data is assumed to be measured by on-board sensors mounted to the manipulator links. Denoting the number of sensors by m_s, the measurement equation for the $j^{th} \in \{1, 2, ..., m_s\}$ field sensor can be written as

$$\lambda_j(k) = \mathcal{T}(\xi^j(k)) + w_j(k) \qquad (10)$$

where k is the discrete time index, i.e., $t = kT_f$ with T_f as the sample time for the RLS routine. $\xi^j(k)$ is the coordinate of the location of the jth sensor at time k and $w_j(k)$ is the corresponding measurement noise which is assumed to be white and zero mean. The sensor signals and their noises are denoted as

$$\lambda(k) := \begin{bmatrix} \lambda_1(k) & \cdots & \lambda_{m_s}(k) \end{bmatrix}^T, \quad w(k) := \begin{bmatrix} w_1(k) & \cdots & w_{m_s}(k) \end{bmatrix}^T.$$

The main goal of the field estimation is to get the best estimate, $\hat{\theta}(n)$ at the present time $t = nT_f$ based on all the prior (noisy) observations $\lambda(1), ..., \lambda(n)$. Specifically, $\hat{\theta}(n)$ is obtained as a solution to minimize the least square error:

$$\hat{\theta}(n) = \min_{\theta \in \Theta} \sum_{k=1}^{n} \|\lambda(k) - H(k)^T \theta\|^2 \qquad (11)$$

where $\Theta \subset \mathbf{R}^m$ is a compact set for the parameter vector θ and $H(k) \in \mathbf{R}^{m \times m_s}$ is the basis function matrix evaluated at each sensor location, i.e,

$$H(k) := \begin{bmatrix} h(\xi^1(k)) & \cdots & h(\xi^{m_s}(k)) \end{bmatrix}. \qquad (12)$$

The solution to Eq. (11) can be obtained in a recursive way as [14]

$$\hat{\theta}(k) = \hat{\theta}(k-1) + K(k)\left(\lambda(k) - H^T(k)\hat{\theta}(k-1)\right) \tag{13a}$$

$$K(k) = P(k)H(k) = P(k-1)H(k)\left(I_{m_s} + H^T(k)P(k-1)H(k)\right)^{-1} \tag{13b}$$

$$P(k) = \left(I_m - K(k)H^T(k)\right)P(k-1) \tag{13c}$$

where $I_{(\bullet)}$ is the identity matrix of size $\bullet$ and

$$P^{-1}(k) := \mathcal{H}(k)\mathcal{H}^T(k) = \sum_{i=1}^{k} H(i)H^T(i). \tag{14}$$

Note that $P(k)$ is singular if $k \times m_s < m$. So $P_0 := P(0)$ must be selected as positive definite. Then $P(k) = \left(P_0^{-1} + \mathcal{H}(k)\mathcal{H}^T(k)\right)^{-1}$, and $P(k) \succ 0$ for all k. Also, note that $P(k)$ can be made arbitrarily close to $\left(\mathcal{H}^T(k)\mathcal{H}(k)\right)^{-1}$ by choosing P_0 sufficiently large. In practice, P_0 may be chosen as large as allowed by the initial learning effort and the noise sensitivity. Typically, P_0 is a diagonal matrix.

3.2 Field Tracking with Gradient Climbing

Given the estimated weight vector $\hat{\theta}(k)$ in Eq. (13a), the field tracking control law Γ_f in Eq. (7) can be implemented by the gradient:

$$\Gamma(k) = K_f \widehat{\nabla \mathcal{T}}(\hat{p}(k)) = K_f \left.\frac{\partial h(\xi)}{\partial \xi}\right|_{\xi=\hat{p}(k)}^T \hat{\theta}(k) \tag{15}$$

where $K_f \in \mathbf{R}^{\ell \times \ell}$ is a diagonal matrix for the gain of field tracking. K_f needs to be selected considering the intensity of the field measurements and the magnitude of the control force of the actuator. Note that our field model $\mathcal{T}(\xi)$ can be considered as a class of artificial potential functions parameterized by the unknown weight vector $\hat{\theta}(k)$ and known basis function $h(\xi)$. Thus, any additional control effort such as the obstacle avoidance can easily be augmented into the field model with required potential functions simply added to $\mathcal{T}(\xi)$.

Equation (15) may be regarded as the desired velocity in task space. $\mathcal{T}(\xi)$ is unknown a priori and the gradient vector in Eq. (15) may encounter a large control effort if there is a sudden change in the scalar field. To avoid such situation, the maximum allowable task space velocity can be imposed as follows.

$$a_x(t) = K_v \bar{v} \text{sat}\left(\bar{v}^{-1} K_v^{-1} K_f \left.\frac{\partial h(\xi)}{\partial \xi}\right|_{\xi=\hat{p}(k)}^T \hat{\theta}(k)\right) - K_v \hat{v}(t) \tag{16}$$

where $\bar{v} \in \mathbf{R}^{\ell \times \ell}$ is a diagonal matrix with its elements being the magnitude of the maximum allowable task space velocities. Note that Eq. (16) used notations for both the continuous time and the discrete time domain.

4 Experimental Results

The proposed control method is applied to a two link manipulator deployed to a planar region to search for a source location of the field. The first joint of the robot is assumed to be positioned at the origin of the spatial coordinate frame. The experimental setup is shown in Fig. 2 with two reflective color sensors mounted as shown in Fig. 2(a) (OPB780Z, *OPTEK Tech.*). Figure 2(b) illustrates that the (unknown) field has been arranged on top of the arm such that the color sensors can read the intensity of the gray map at each point.

The gray map representing the potential field is modeled as shown in Fig. 3. The original field model (color map) in Fig. 3(a) is converted into an equivalent gray scale map as shown in Fig. 3(b) which is then printed on a $4' \times 6'$ paper and pasted on a glass plate (See Fig. 2(b)). Parameters for the experimental tests are summarized in Table 1. The decoupling feedback control law is run at the base sample rate of $T_s = 1\,ms$ while the field estimation is run at $T_f = 50\,ms$. The real-time control is implemented using the NI PCIe RIO board (PCIe-7842R) which runs under the environment of a Realtime Desktop PC with LabView.

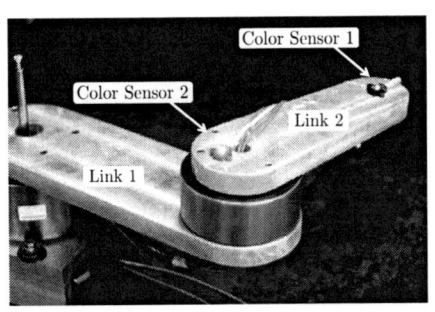

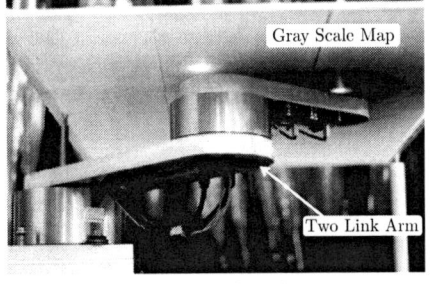

(a) Two link robot with color sensors (b) Configuration of scalar sensing

Fig. 2. Experimental setup

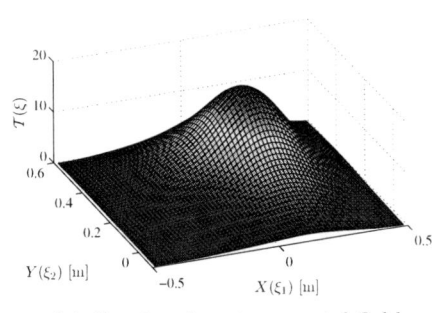

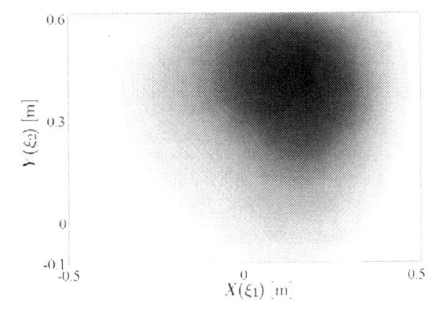

(a) Simulated environmental field (b) Equivalent gray map for color sensors

Fig. 3. Model of the unknown environmental field created for experiments

Table 1. Parameters for the experimental tests

Plant para.	Values ($1^{st}, 2^{nd}$)	Units	Field model para.	Value	Units
Link length	0.32, 0.21	$[m]$	# of bases, m	3×3	
Motor tq. const.	2.304, 1.112	$[Nm/V]$	RBF Width, σ^2	0.45	
Inertia, J	0.210, 0.021	$[kgm^2]$	Field region x	[-0.5, 0.5]	$[m]$
			Field region y	[-0.1, 0.6]	$[m]$

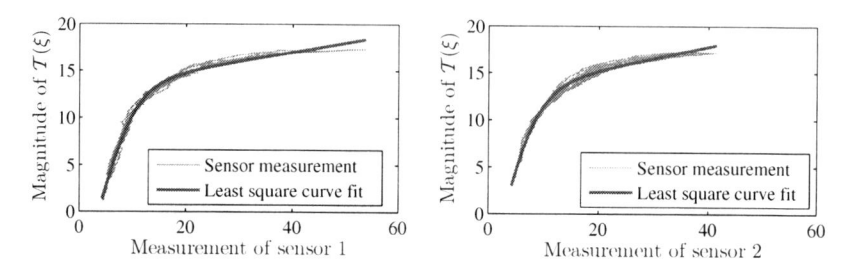

Fig. 4. Nonlinearity of color sensor and calibration

The reflective color sensors used in the test possess nonlinear characteristics with respect to the numerical value of the gray scale map. Thus, measurements from the color sensors are first collected off-line and processed to obtain the scaling function through the nonlinear least-square curve fit as shown in Fig. 4.

(a) End-effector trajectory.

(b) Convergence of source tracking

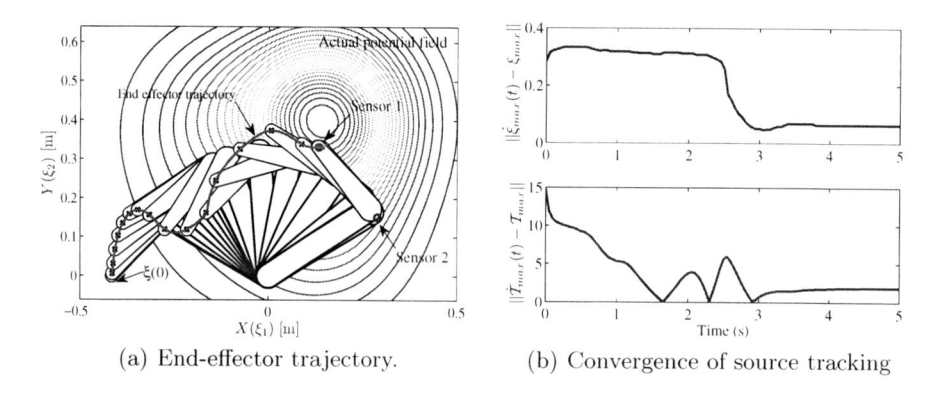

Fig. 5. End-effector trajectory and source tracking performance

The experimental results are summarized in Figs. 5 and 6. Figure 5(a) shows snap shots (0.25 second interval) of the manipulator started at $\xi(0) = (-0.41, -0.25)$. The true potential field is also shown in the background as a contour map. The end-effector (the end point of the second link) eventually converged to a point near the target. The related errors are plotted in Fig. 5(b). More specifically, the upper plot shows the vector norm of coordinate error of

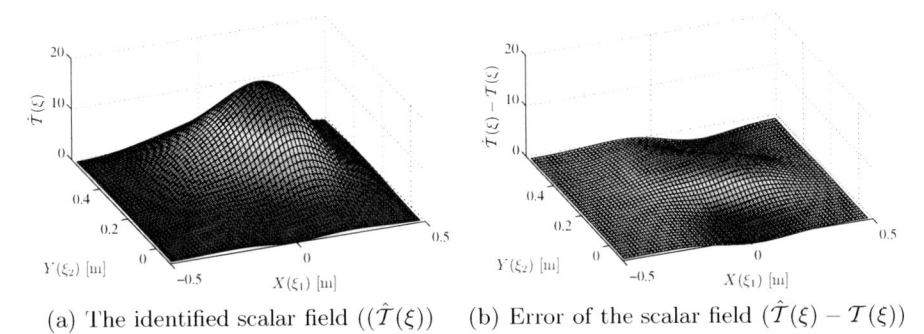

(a) The identified scalar field $(\hat{\mathcal{T}}(\xi))$ (b) Error of the scalar field $(\hat{\mathcal{T}}(\xi) - \mathcal{T}(\xi))$

Fig. 6. Performance of scalar field estimation

the maximum point. The localization error of the end-effector is around 0.063 m. The lower plot how the error of the maximum value of the potential field changes with respect to time.

Figure 6(a) shows the identified field as a result of the manipulator motion of Fig. 5(a). To see the error more clearly, the difference between the true field and the identified field, i.e. $\hat{\mathcal{T}}(\xi) - \mathcal{T}(\xi)$ is plotted in Fig. 6(b). The maximum value of the actual field is $\mathcal{T}_{max} = 17.3$ at $\xi_{max} = (0.135, 0.399)$ while that of the identified field is $\hat{\mathcal{T}}_{max} = 15.43$ at $\hat{\xi}_{max} = (0.128, 0.333)$. From Fig. 6, the error of the scalar field magnitude is within $-2.05 \leq \hat{\mathcal{T}}(\xi) - \mathcal{T}(\xi) \leq 4.6$ across the field region.

5 Conclusion

In this paper, we verified a new task space control design of a robot end-effector to achieve the source tracking in an unknown field of interest. Using the RBF network, the unknown field is estimated by the RLS scheme and the estimated field is used for the end-effector to navigate and track the gradient direction.

The results in this paper will be useful in extending the application of articulated manipulation to various autonomous operation tasks.

References

1. Parker, L.E., Draper, J.V.: Robotics applications in maintenance and repair. In: Handbook of Industrial Robotics, 2nd edn. (1998)
2. Vertut, J., Coiffet, P.: Teleoperation and Robotics: Applications and Technology. Prentice-Hall, Englewood Cliffs (1985)
3. Erik, K., Pål, L., Andreas, T.A.: A Robotic concept for remote inspection and maintenance on oil platforms. In: ASME 28th International Conference on Ocean, Offshore and Arctic Engineering, Honolulu, Hawaii, USA (2009)
4. Glass, K., Colbaugh, R., Seraji, H.: Real-time control for a serpentine manipulator. In: IEEE/RSJ International Conference on Intelligent Robots and Systems, Grenoble, France, pp. 1775–1780 (1997)

5. Choi, J., Oh, S., Horowitz, R.: Distributed learning and cooperative control for multi-agent systems. Automatica 45(12), 2802–2814 (2009)
6. Martinez, S.: Distributed interpolation schemes for field estimation by mobile sensor networks. IEEE Transactions on Control Systems Technology 18(2), 491–500 (2009)
7. Leonard, N., Paley, D.A., Lekien, F., Sepulchre, R., Fratantoni, D.M., Davis, R.E.: Collective motion, sensor networks, and ocean sampling. Proceedings of the IEEE 95(1), 48–74 (2007)
8. Grünbaum, D.: Schooling as a strategy for taxis in a noisy environment. Evolutionary Ecology 12, 503–522 (1998)
9. Christopoulos, V.N., Roumeliotis, S.I.: Adaptive sensing for instantaneous gas release parameter estimation. In: IEEE International Conference on Robotics and Aotumation, Barcelona, Spain, pp. 4450–4456 (2005)
10. Khatib, O.: Real-time obstacle avoidance for manipulators and mobile robots. The International Journal of Robotics Research 5(1), 90–98 (1986)
11. Sciavicco, L., Sicilino, B.: Modeling and Control of Robot Manipulators, 2nd edn. McGraw-Hill Companies, Inc., New york (2000)
12. Jeon, S.: Singularity Tolerant Robotic Manipulation for Autonomous Field Tracking in an Unknown Environment. In: ASME IDETC 2010, Montreal, QC, Canada (2010)
13. Spong, M.W., Hutchinson, S., Vidyasagar, M.: Robot Modeling and Control. John Wiley & Sons, Inc., Chichester (2006)
14. Åström, K.J., Wittenmark, B.: Adaptive Control, 2nd edn. Addison-Wesley, Reading (1995)
15. Patel, R.V., Shadpey, F.: Control of Redundant Robot Manipulators: Theory and Experiments. Springer-Verlag GmbH, Heidelberg (2005)

Accessibility of Fuzzy Control Systems

Mohammad Biglarbegian[1,2], Alireza Sadeghian[1], and William W. Melek[2]

[1] Computer Science Department, Ryerson University, Toronto
[2] Mechanical & Mechatronics Engineering Department,
University of Waterloo, Waterloo
mbiglarb@scs.ryerson.ca, asadeghi@ryerson.ca, wmelek@uwaterloo.ca

Abstract. This paper presents a systematic methodology for delineating the accessibility of fuzzy logic control systems (FLCSs). Accessibility is an essential prior step for determining the controllability of plants in which the state trajectories are verified if they are accessible to a full dimension set. Thus, accessibility is an important metric in control design. We propose a methodical approach of examining the accessibility of FLCSs through linearization. Two examples are provided to demonstrate the utility of the developed methodology. The approach presented herein is an attempt to derive the conditions for assessing the accessibility of FLCSs systematically and facilitate the design of accessible FLCSs which is of great interest to researchers in the control community.

1 Introduction

Accessibility is a very important concept in the process of designing control systems [3]. Accessibility determines if the trajectories of a plant can be driven from an initial point to a full dimension set (not necessarily the entire space) [4]. Thus, it provides a tool to determine beforehand if the system under investigation, can achieve the control design requirements. Accessibility is the main prior step in determining controllability, i.e., if a system is not accessible, it will not be controllable. Parallel to controllability, observability of systems is also a metric in the design of observers – observability is usually noted as a dual to controllability in the literature. Thus, in addition, deriving the controllability of a system will allow us to easily obtain the observability conditions paralleling a similar approach.

Fuzzy logic control systems (FLCSs) have been extensively used in many practical applications [1,2]. Although initially they were subjected to criticism for the lack of mathematical analysis, strong and rigorous studies carried out on these systems have further developed them into one of the widely-used nonlinear control tools [5]. Nevertheless, there still remain some important conecpts/preperties of FLCSs necessary to be investigated thoroughly to help understand the behavior of these complex nonlinear systems. One of the most important aspects is accessibility through which one can easily verify the controllability of the system under investigation.

To the best of our knowledge, there exists no work in the literature for analyzing the accessibility of FLCSs. This paper derives the mathematical accessibility conditions for FLCSs. We start with presenting a general methodology for determining the accessibility of FLCSs and derive a machinery to assess the derived conditions. Our approach is not limited to restricting assumptions and it also can be easily used for systems with a large number of rules. Furthermore, it can be extended for interval type-2 fuzzy logic systems (IT2 FLSs) indicating its strength. The organization of this paper is as follows: Section 2 reviews the structure of TS FLCSs and also presents the definition of accessibility. Section 3 presents the accessibility analysis, Section 4 provides examples, and Section 5 concludes the paper.

2 Background

This section reviews TS FLCSs' structure and presents the definition of accessibility.

2.1 TS FLCs

The general structure of a TS FLCS is as follows:

If r_1 is F_1^i and r_2 is F_2^i and $\cdots$ r_z is F_z^i, Then

$$\dot{\boldsymbol{x}}^i(t) = \boldsymbol{A}^i \boldsymbol{x}(t) + \boldsymbol{B}^i \boldsymbol{u}(t) \tag{1a}$$

$$\boldsymbol{y}(t) = \boldsymbol{C}^i \boldsymbol{x}(t) \tag{1b}$$

where $i = 1, ..., M$, F_j^i represents the fuzzy set (FS) of input state j in rule i, r_j is the jth input state, $\boldsymbol{y}$ and $\boldsymbol{x}(t)$ are the output and state vectors of the ith rule, respectively, $\boldsymbol{u}(t)$ is the input vector, M is the number of rules, $\boldsymbol{A}^i \in \mathbb{R}^{n \times n}$, $\boldsymbol{B}^i \in \mathbb{R}^{n \times m}$, $\boldsymbol{C}^i \in \mathbb{R}^{q \times n}$, $\boldsymbol{x}(t) \in \mathbb{R}^n$, $\boldsymbol{y}(t) \in \mathbb{R}^q$, and $\boldsymbol{u}(t) \in \mathbb{R}^m$. Getting the weighted average of all rules, $\dot{\boldsymbol{x}}(t)$ can be expressed as

$$\dot{\boldsymbol{x}}(t) = \frac{\sum_{i=1}^{M} f^i \left[\boldsymbol{A}^i \boldsymbol{x}(t) + \boldsymbol{B}^i \boldsymbol{u}(t) \right]}{\sum_{i=1}^{M} f^i} \tag{2}$$

Define

$$h^i(\boldsymbol{x}) \equiv \frac{f^i(\boldsymbol{x})}{\sum_{j=1}^{M} f^j(\boldsymbol{x})} \tag{3}$$

Using (3), (2) can be expressed as

$$\dot{\boldsymbol{x}}(t) = \sum_{i=1}^{M} h^i(\boldsymbol{x}) \left[\boldsymbol{A}^i \boldsymbol{x}(t) + \boldsymbol{B}^i \boldsymbol{u}(t) \right]$$

$$= \sum_{i=1}^{M} h^i(\boldsymbol{x}) \boldsymbol{A}^i \boldsymbol{x}(t) + \sum_{i=1}^{M} h^i(\boldsymbol{x}) \boldsymbol{B}^i \boldsymbol{u}(t) \tag{4}$$

For IT2 FLSs, a new inference engine with a closed-form, $m-n$ IT2 FLS, was introduced in [2] which its output is given by

$$Y_{\text{m-n}}(\boldsymbol{x}) = m\frac{\sum_{i=1}^{M} \underline{f}^i(\boldsymbol{x}) y_i(\boldsymbol{x})}{\sum_{i=1}^{M} \underline{f}^i(\boldsymbol{x})} + n\frac{\sum_{i=1}^{M} \overline{f}^i(\boldsymbol{x}) y_i(\boldsymbol{x})}{\sum_{i=1}^{M} \overline{f}^i(\boldsymbol{x})} \quad (5)$$

where y_i is the output of each rule (if $\underline{f}^i(\boldsymbol{x}) = \overline{f}^i(\boldsymbol{x})$, then $m+n=1$). Lower and upper firing strengths of the ith rule, $\underline{f}^i(\boldsymbol{x})$ and $\overline{f}^i(\boldsymbol{x})$, are given by

$$\underline{f}^i(\boldsymbol{x}) = \underline{\mu}_{\tilde{F}_1^i}(x_1) \star \cdots \star \underline{\mu}_{\tilde{F}_p^i}(x_p) \quad (6)$$

$$\overline{f}^i(\boldsymbol{x}) = \overline{\mu}_{\tilde{F}_1^i}(x_1) \star \cdots \star \overline{\mu}_{\tilde{F}_p^i}(x_p) \quad (7)$$

where $x_1, \cdots, x_p$ are state variables. Note that one can readily uses the $m-n$ IT2 FLSs for the following accessibility analyses by replacing it with the type-1 TS model structure. Hence, the methodology presented here can be extended for IT2 FLCSs as well.

2.2 Definition of Accessibility

Definition 1. *[4]. Consider $\phi(t, s, x, u)$ representing the trajectory at the time t, where x is the state, s, is the initial condition, and u is the allowable control input. A reachable set R is a set of states that can be reached from the origin, i.e., $\phi(t, 0, 0, u)$. A dynamic system is defined as accessible if the reachable set R has a nonempty interior.*

In other words, accessibility deals with the dimension of the reachable set.

3 Accessibility Analysis

3.1 Linearization

This section starts with analyzing the accessibility through linearization. If the linearized system is controllable, then the the nonlinear system is controllable. However, if the linearized system is not controllable no decisive conclusion can be made about the nonlinear system (it may or may not be controllable).

The equilibrium point[1] of the system given by (2) is $\boldsymbol{x}_0 = \boldsymbol{0}$, $\boldsymbol{u}_0 = \boldsymbol{0}$. The linearized system can be expressed as

$$\dot{\boldsymbol{x}}(t) = \boldsymbol{E}\boldsymbol{x} + \boldsymbol{F}\boldsymbol{u} \quad (8)$$

---

[1] Note that depending on the specific FLCS under investigation, other points *might* lead to $\dot{\boldsymbol{x}}(t) = 0$. However, since we are dealing with a problem in general which is not limited to a specific one, we consider an equilibrium point which is common for all FLCSs.

where

$$E = \frac{\partial \left[\sum_{i=1}^{M} h^i(\boldsymbol{x}) \boldsymbol{A}^i \boldsymbol{x}(t)\right]}{\partial \boldsymbol{x}} (\boldsymbol{x}_0, \boldsymbol{u}_0) = \sum_{i=1}^{M} \frac{\partial \left[h^i(\boldsymbol{x}) \boldsymbol{A}^i \boldsymbol{x}\right]}{\partial \boldsymbol{x}} \bigg|_{\boldsymbol{x}_0} \quad (9)$$

and

$$F = \frac{\partial \left[\sum_{i=1}^{M} h^i(\boldsymbol{x}) \boldsymbol{B}^i \boldsymbol{u}(t)\right]}{\partial \boldsymbol{u}} (\boldsymbol{x}_0, \boldsymbol{u}_0) = \sum_{i=1}^{M} h^i(\boldsymbol{x}) \boldsymbol{B}^i \bigg|_{\boldsymbol{x}_0} \quad (10)$$

We now obtain an expression for $\frac{\partial h^i(\boldsymbol{x})}{\partial x_j}$. To do so, we first derive $\frac{\partial f^i(\boldsymbol{x})}{\partial x_j}$.

$$\frac{\partial f^i(\boldsymbol{x})}{\partial x_j} = \frac{\partial}{\partial x_j} \left[\mu^i(x_1) \star \cdots \star \mu^i(x_n)\right] \quad (11)$$

Using algebraic product as the t-norm, (11) is calculated as

$$\frac{\partial f^i(\boldsymbol{x})}{\partial x_j} = \mu_1^i(x_1) \cdots \mu_{j-1}^i(x_{j-1}) \cdot \frac{\partial \mu_j^i(x_j)}{\partial x_j} \cdot \mu_{j+1}^i(x_{j+1})$$

$$\cdots \mu_n^i(x_n) = \left[\prod_{\substack{k=1 \\ k \neq j}}^{n} \mu_k^i(x_k)\right] \frac{\partial \mu_j^i(x_j)}{\partial x_j} \quad (12)$$

Next we calculate $\frac{\partial}{\partial x_j} \left[\sum_{i=1}^{M} f^i(\boldsymbol{x})\right]$

$$\frac{\partial}{\partial x_j} \left[\sum_{i=1}^{M} f^i(\boldsymbol{x})\right] = \frac{\partial}{\partial x_j} \left[\sum_{l=1}^{M} f^l(\boldsymbol{x})\right] = \sum_{l=1}^{M} \left[\frac{\partial f^l(\boldsymbol{x})}{\partial x_j}\right]$$

$$= \sum_{l=1}^{M} \left(\prod_{\substack{k=1 \\ k \neq j}}^{n} \mu_k^l(x_k)\right) \cdot \frac{\partial \mu_j^l(x_j)}{\partial x_j} \quad (13)$$

Using (12) and (13), $\frac{\partial h^i(\boldsymbol{x})}{\partial x_j}$ is given as follows:

$$\frac{\partial h^i}{\partial x_j} = \frac{\frac{\partial f^i}{\partial x_j} \cdot \left[\sum_{l=1}^{M} f^l\right] - f^i \cdot \left(\frac{\partial}{\partial x_j} \left[\sum_{l=1}^{M} f^l\right]\right)}{\left[\sum_{l=1}^{M} f^l\right]^2}$$

$$= \frac{\left(\prod_{\substack{k=1 \\ k \neq j}}^{p-1} \mu_k^i(x_k)\right) \frac{\partial \mu_j^i(x_j)}{\partial x_j} \left[\sum_{l=1}^{M} f^l\right]}{\left[\sum_{l=1}^{M} f^l\right]^2}$$

$$- \frac{f^i \cdot \sum_{l=1}^{M} \left[\left(\prod_{\substack{k=1 \\ k \neq j}}^{n-1} \mu_k^l(x_k)\right) \cdot \frac{\partial \mu_j^l(x_j)}{\partial x_j}\right]}{\left[\sum_{l=1}^{M} f^l\right]^2} \quad (14)$$

Note that $E_{i,j}, (i,j = 1, \cdots, n)$ is computed as

$$E_{i,j} = \sum_{k=1}^{n} \sum_{l=1}^{M} \frac{\partial h^l(\boldsymbol{x})}{\partial x_j} A^l_{ik} x_k \bigg]_{\boldsymbol{x}_0}^{\to 0} + \sum_{l=1}^{M} h^l(\boldsymbol{x}) A^l_{ij} \bigg]_{\boldsymbol{x}_0}$$

$$= \sum_{l=1}^{M} h^l(\boldsymbol{0}) A^l_{ij} \tag{15}$$

Similarly, $F_{i,j}, (i = 1, \cdots, n, j = 1, \cdots, m)$ is given by

$$F_{i,j} = \sum_{l=1}^{M} h^l(\boldsymbol{x}) B^l_{ij} \bigg]_{\boldsymbol{x}_0, \boldsymbol{u}_0} = \sum_{l=1}^{M} h^l(\boldsymbol{0}) B^l_{ij} \tag{16}$$

Therefore, matrix $\boldsymbol{E}$ in (9) is computed as

$$\boldsymbol{E} = \begin{bmatrix} \sum_{l=1}^{M} h^l(\boldsymbol{0}) A^l_{11} & \cdots & \sum_{l=1}^{M} h^l(\boldsymbol{0}) A^l_{1n} \\ \sum_{l=1}^{M} h^l(\boldsymbol{0}) A^l_{21} & \cdots & \sum_{l=1}^{M} h^l(\boldsymbol{0}) A^l_{2n} \\ \vdots & & \vdots \\ \sum_{l=1}^{M} h^l(\boldsymbol{0}) A^l_{n1} & \cdots & \sum_{l=1}^{M} h^l(\boldsymbol{0}) A^l_{nn} \end{bmatrix} \tag{17}$$

And matrix $\boldsymbol{F}$ in (10) is calculated as

$$\boldsymbol{F} = \begin{bmatrix} \sum_{l=1}^{M} h^l(\boldsymbol{0}) B^l_{11} & \cdots & \sum_{l=1}^{M} h^l(\boldsymbol{0}) B^l_{1m} \\ \sum_{l=1}^{M} h^l(\boldsymbol{0}) B^l_{21} & \cdots & \sum_{l=1}^{M} h^l(\boldsymbol{0}) B^l_{2m} \\ \vdots & & \vdots \\ \sum_{l=1}^{M} h^l(\boldsymbol{0}) B^l_{n1} & \cdots & \sum_{l=1}^{M} h^l(\boldsymbol{0}) B^l_{nm} \end{bmatrix} \tag{18}$$

Define

$$\boldsymbol{G} \equiv [\boldsymbol{F}, \boldsymbol{EF}, \boldsymbol{E}^2 \boldsymbol{F}, \cdots, \boldsymbol{E}^{n-1} \boldsymbol{F}] \tag{19}$$

If rank $\boldsymbol{G} = n$, then the linearized system is controllable which means the nonlinear system is controllable. If rank $\boldsymbol{G} \neq n$, then the linearized system is not controllable, but no conclusive decision can be made about the nonlinear system in general, i.e., it may or may not be controllable.

4 Examples

This section presents benchmark examples for the developed methodologies on the accessibility/controllability analyses of FLCSs.

Example 1. This example is adopted from [6]. Consider the nominal plant given as follows:

Plant rules: (antecedent MFs[2] are shown in Figure 1)

---
[2] We have used Gaussian MFs corresponding to the original MFs.

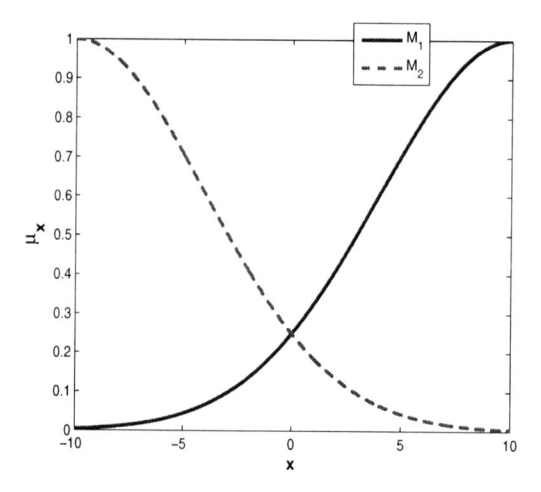

Fig. 1. Example 1–MFs of the nonlinear plant

Rule 1: If $x_1(k)$ is M_1, then $\dot{x} = A_1 x(k) + B_1 u(k)$
Rule 2: If $x_1(k)$ is M_2, then $\dot{x} = A_2 x(k) + B_2 u(k)$

where

$$A_1 = \begin{bmatrix} 0.2 & 0.5 \\ 5 & 0.6 \end{bmatrix}, \quad A_2 = \begin{bmatrix} -0.4 & -0.8 \\ 3.5 & 0.6 \end{bmatrix}$$

$$B_1 = \begin{bmatrix} 0 & 0.1 \\ 0.1 & 0 \end{bmatrix}, \quad B_2 = \begin{bmatrix} 0 & 0.1 \\ 0.2 & 0 \end{bmatrix}$$

Using (17) and (18), matrices E and F are computed as

$$E = \begin{bmatrix} 0.0833 & 0.1806 \\ 0.2083 & 0 \end{bmatrix}, \quad F = \begin{bmatrix} 0 & 0 \\ -0.0139 & 0 \end{bmatrix}$$

Consequently, G is given by

$$G = \begin{bmatrix} 0.0833 & 0.1806 & -0.0025 & 0 \\ 0.2083 & 0 & 0 & 0 \end{bmatrix}$$

in which rank G is 2, showing the controllability of the linearized system. Hence, the nonlinear system is controllable.

Example 2. In this case study, we consider the well-known inverted pendulum which is a benchmark control in the literature. An inverted pendulum is shown in Figure 2.

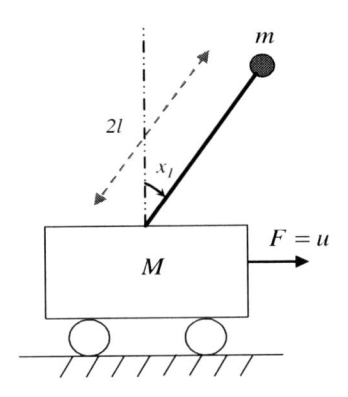

Fig. 2. Example 2–inverted pendulum

The inverted pendulum system has nonlinear dynamics and its equations of motion are given as

$$\dot{x}_1 = x_2 \tag{20}$$

$$\dot{x}_2 = \frac{g \sin x_1 - amlx_2^2 \sin(2x_1)/2 - au.\cos x_1}{4l/3 - aml \cos^2 x_1} \tag{21}$$

where $x_1(t)$ and $x_2(t)$ are the angular position and velocity of the pendulum, respectively, $u(t)$ is the control input, m is the pendulum mass, M is the cart mass, $2l$ is the length of the pendulum, and $a \equiv \frac{1}{m+M}$.

Define $\boldsymbol{x}(t) \equiv [x_1(t), x_2(t)]^T$ where $x_1(t)$ and $x_2(t)$ are the state variables, i.e., angular position and velocity of the pendulum. In [7] an inverted pendulum using a TSK FLS was modeled where the structure of the plant is given by

Plant rules: (see Figure 3 for antecedent MFs)[3]

Rule 1: If $x_1(t)$ is "about 0", then $\dot{\boldsymbol{x}} = \boldsymbol{A}_1\boldsymbol{x}(t) + \boldsymbol{b}_1 u(t)$

Rule 2: If $x_1(t)$ is "about $\frac{\pi}{2}$ or $-\frac{\pi}{2}$", then $\dot{\boldsymbol{x}} = \boldsymbol{A}_2\boldsymbol{x}(t) + \boldsymbol{b}_2 u(t)$

in which

$$\boldsymbol{A}_1 = \begin{bmatrix} 0 & 1 \\ 17.3118 & 0 \end{bmatrix}, \quad \boldsymbol{A}_2 = \begin{bmatrix} 0 & 1 \\ 9.3696 & 0 \end{bmatrix}$$

$$\boldsymbol{b}_1 = \begin{bmatrix} 0 \\ -0.1765 \end{bmatrix}, \quad \boldsymbol{b}_2 = \begin{bmatrix} 0 \\ -0.0052 \end{bmatrix}$$

Matrices $\boldsymbol{E}$ and $\boldsymbol{F}$ are computed as

$$\boldsymbol{E} = \begin{bmatrix} 0 & 0 \\ 1.0544 & 0 \end{bmatrix}, \quad \boldsymbol{F} = \begin{bmatrix} 0 \\ 0.0196 \end{bmatrix}$$

---

[3] x_1 is given in radians.

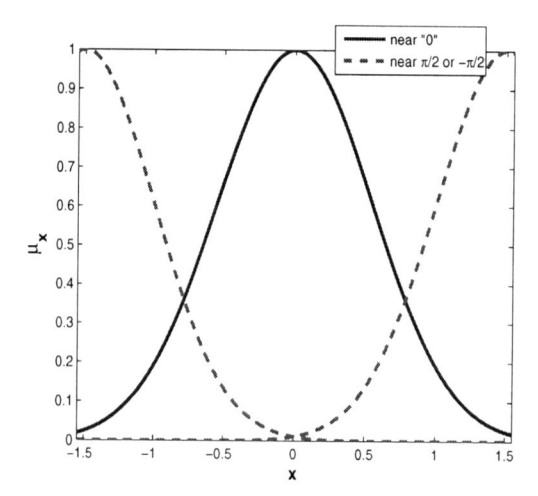

Fig. 3. Example 2–inverted pendulum MFs

Consequently, G is given by

$$G = \begin{bmatrix} 0 & 0 & 0 \\ 1.0544 & 0 & 0 \end{bmatrix}$$

where rank $G = 1$, which indicates that the linearized system is not controllable. Consequently, no decisive conclusion can be made on the controllability of the nonlinear system. This case study demonstrates the limitation of the linearization approach. To draw a conclusion on the accesibility/controllability of the nonlinear system, we must perform a more general approach, parallel to nonlinear control theory, which is the future work of the present paper.

5 Conclusion

In this paper we rigorously analyzed the accessibility of FLCSs, a very important parameter for control design. The analyses developed herein will subsequently help mathematically investigate the controllability of fuzzy systems that will ultimately enable their utility in various applications through verifying these important properties (verification). The presented approach can be easily used for FLCS's analysis as well as the design of accessible fuzzy control systems. The future work of this work will be dedicated on more examples and computational efforts associated with the proposed approach. Future work is geared toward developing stronger accessibility/controllability analyses for FLCSs, parallel to nonlinear control theory.

References

1. Feng, G.: A Survey on analysis and design of model-based fuzzy control systems. IEEE. Tran. Fuzzy. Syst. 14(5), 676–697 (2006)
2. Biglarbegian, M., Melek, W.W., Mendel, J.M.: On the stability of interval type-2 TSK fuzzy logic control systems. IEEE Tran. Sys., Man, and Cyb.: Part B 4(3), 798–818 (2010)
3. Sastry, S.: Nonlinear systems: analysis, stability, and control, 1st edn. Springer, Heidelberg (1999)
4. Sontag, E.D.: Kalman's controllability rank condition: from Lineaer to nonlinear. Math. Sys. Th.: The Influence of R. E. Kalman, 453–462 (1991)
5. Hsueh, Y.-C., Su, S.-F., Tao, C.W., Hsiao, C.-C.: Robust L2-gain compensative control for direct-adaptive fuzzy-control-system design. IEEE Tran. Fuzzy Sys. 18(4), 661–673 (2010)
6. Chang, W.-J., Sun, C.-C.: Constrained fuzzy controller design of discrete Takagi-Sugeno fuzzy models. Fuzzy Sets and Sys. 133(1), 37–55 (2003)
7. Tanaka, K., Wang, H.O.: Fuzzy control systems designs and analysis, 1st edn. John Wiley & Sons, Chichester (2001)

Thermal Dynamic Modeling and Control of Injection Moulding Process

Jaho Seo, Amir Khajepour, and Jan P. Huissoon

Department of Mechanical and Mechatronics Engineering
University of Waterloo
j7seo@engmail.uwaterloo.ca

Abstract. Thermal control of a mould is the key in the development of high efficiency injection moulds. For an effective thermal management system, this research provides a strategy to identify the thermal dynamic model for the design of a controller. Using neural networks and finite element analysis, system identification is carried out to deal with various cycle-times for moulding process and uncertain dynamics of the mould. Based on the system identification, a self-adaptive PID controller with radial basis function (RBF) is designed to tune controller parameters. The controller's performance is studied in terms of tracking accuracy under different moulding processes.

Keywords: Plastic injection moulding; thermal dynamic modeling; cycle-times; neural networks; finite element analysis; RBF based self-adaptive PID control.

1 Introduction

Injection moulding is a primary manufacturing process to produce parts by injecting molten plastic materials into a mould. Thermal control is a key issue in this process since uniform temperature in the moulds contributes to production quality by reducing problems such as shrink porosity, poor fill and prolonged cycle-times for part solidification [1, 2].

Many approaches have been proposed to deal with the thermal control in mould (or die) systems. A PI [3] and PID algorithms [1] were applied to manage the cavity temperature on a plastic injection moulding and high-pressure die-casting, respectively. To improve limitation of PID control in presence of uncertain or nonlinear dynamics, a Dahlin controller [4] and the Model Predictive Control (MPC) have been utilized in diverse range of mould systems for thermal control [5, 6, 7, 8]. Despite of improved performance compared to PID control, the addressed controllers are not robust in some circumstances with more complex nonlinearities Specifically, because these controllers are based on a linear "best-fit" approximation (e.g., ARX and ARMAX), the performance of the controllers is affected largely by modeling errors arisen from uncertain dynamics.

Although accurate modeling of the thermal dynamics of moulds is a prerequisite to successful thermal control, it is a difficult task in practice due to mould uncertainties. For example, a mould is a complex continuous system with cooling and heating channels causing the modeling and control to be quite complicated. Unmodeled thermal dynamics of moulds (such as convection and radiation) also provide further modeling challenges.

To deal with inherent challenges of modeling in a mould system, this paper considers a neural network (NN) approach. By applying NN techniques, the thermal dynamics with uncertainties in a plastic injection mould is modeled. In addition, our modeling covers various cycle-times for plastic moulding process that has not been studied (i.e., most documented approaches for thermal management of mould systems have only considered a fixed cycle-time). In this study, the system identification is conducted using the temperature distribution obtained through a finite element analysis (FEA). Based on this system identification, a controller is designed using the RBF based self-adaptive PID control.

Section 2 describes the mould system. Section 3 presents a methodology for modeling thermal dynamics using FE simulation and NN techniques. In Section 4, a controller is designed and its performance is discussed. Finally, Section 5 provides concluding remarks.

2 Injection Mould and Various Cycle-Times

Figure 1 shows a plastic injection mould used for the analysis in this study. Hot polymer injected into the mould cavity is cooled to the demoulding temperature by heat transfer to coolant through the cooling channel in close proximity to the cavity. Since the dominant heat for the injection moulding process is transferred by means of the conduction and convection by the coolant [3, 9], the flow rate and temperature of the coolant are chosen as the control parameters. Finite element simulations for thermal dynamic analysis of the mould are carried out based on this input-output model.

In the plastic injection moulding process, the cycle-time is allocated for the main phases of injection, packing, holding, cooling and ejection. Since the cooling phase among these phases takes a large portion of the cycle-time to cool the polymer down to solidification temperature [10], the cooling time plays an important role in the cycle. The previous studies dealing with the system identification and thermal control in the injection moulding process have used a predetermined cooling time (thus cycle-time). However, the system identification using a fixed cycle-time cannot cope with the wide range of thermal dynamics variation when the cycle-time is not fixed. The control strategy based on this limited identification cannot make full use of the benefits of the cost-effective manufacturing which can be achieved by reducing the cooling time (or cycle-times). In this study, various cycle-times of the moulding process are considered in the process of thermal dynamics modeling.

3 Thermal Dynamics Modeling

3.1 Finite Element Analysis

The objective of the FEA is to identify the thermal dynamics (i.e., temperature distribution) during the moulding process in the injection mould and to obtain the input-output data set required for the system identification.

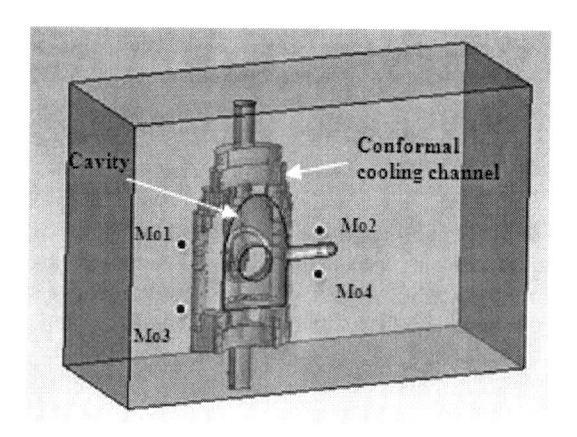

Fig. 1. Injection mould with cooling channel

The injection moulding process of the mould can be briefly described as follows: Polymer is injected into the cavity at a temperature of 205 ^oC before the injection moulding cycle starts. During the cycle, the heat removal from the melt and mould is achieved through heat conduction with the coolant and natural convection with air. After the cooling phase, the polymer part is ejected from the cavity of opened mould.

For the FEA, ANSYS CFX 11 software package was used with 3D CAD files from SolidWorks. For the polymer and mould, material properties of Santoprene 8211-45 [11] and Cast Iron GG-15 [12] were applied, respectively. The properties of water were used for the coolant. Flow of the coolant in the cooling channel was assumed to be turbulent. The initial temperature of the mould was assumed as the room temperature (25 ^oC). For natural convection with the stagnant air, the value of 6 W/m^2 ^oC was chosen as the convective heat transfer coefficient. A transient state analysis in the FEA was carried out to observe the temperature distribution over the cycle-times of injection moulding process at some nodes near the cavity. Four nodes (Mo1, Mo2, Mo3 and Mo4) located inside the mould in Fig. 1 were chosen to monitor the temperature distribution. For the FEA meshing, the convergence study for mesh refinement was conducted until a convergence in the temperature distribution was achieved.

3.2 System Identification Using NARX Model and FEA

To capture a dynamical model to describe the temperature distribution in the mould using the input-output data from the FEA, NARX (nonlinear autoregressive with exogenous inputs) model as a neural network approach is provided. NARX is a powerful class of nonlinear dynamical model which enables to deal with uncertain dynamics of the mould and in which embedded memory is included as temporal changes to reflect the dynamic response of the system. Due to this feature, the NARX has the following structure:

$$y(t) = y_N(t) + e(t)$$
$$= f(y(t-1), y(t-2), ..., y(t-n_y), u(t), u(t-1), u(t-2), ..., u(t-n_u)) + e(t) \quad (1)$$

where $y(t)$, $y_N(t)$, $u(t)$ and $e(t)$ stand for the system output, neural network output, system input and error between $y(t)$ and $y_N(t)$ at time t, respectively. The inputs are composed of past values of the system inputs and outputs. n_y and n_u are the number of past outputs and past inputs, respectively and are referred to the model orders.

When applying the NARX model to our modeling, the outputs in the NARX model are the temperatures at 4 nodes. However, the rapid change of the temperature during the cycle makes it difficult to use this temperature profile as a set-point and thus alternative variable is required as an output. Instead, cycle average temperature [4] defined in Eq. (2) serves as an output.

$$T_a = \frac{\int_0^P T dt}{P} = \frac{\sum_0^M T_n}{M}. \tag{2}$$

where t is time, T is the instantaneous temperature at a specific location, P is the cycle-time, M is the total number of samples during one cycle, and T_n is the temperature at the nth sample time. The use of T_a has an advantage of its insensitiveness to process noise.

The past states of temperatures at each node (i.e., past outputs) and the flow rate (input) of the coolant are used as inputs to the NARX model. A real plastic moulding process uses the coolant with constant temperature (15.5 °C) and thus, the coolant temperature was utilized as a boundary condition instead of an input in the NARX model. As mentioned before, our modeling is extended to deal with various cycle-times rather than a predetermined cycle-time that previous studies have considered. Therefore, the cycle-time for the moulding process is additionally included as input variable in the model. The NARX model to cover all above inputs (flow rate, cycle-time, past values of outputs at 4 nodes) and outputs (each T_a at 4 nodes) is a multi-input multi-output (MIMO) system. The following table shows inputs conditions for flow rate and cycle-time to generate the steady states of outputs used for training and validation of the NARX modeling. A part of training data sets is demonstrated in Fig. 2, which is generated for node Mo2 using all flow rates in Table 1 and one cycle-time (91 sec) among 4 types of cycle-times.

From the figure, it can be seen that for data gathering, each flow rate (input) with a given cycle-time keeps constant until a corresponding steady state of temperature at the node (output) is reached over many cycles [3]. Additional data sets for modeling at the same node (Mo 2) were generated by using the same flow rate ranges (0, 1, 3, 4, 5, 6, 8 gpm) with each different cycle-time (i.e., 81, 71, 61 sec). This method was also applied for data generation at the remaining three nodes (Mo1, Mo3, Mo4).

Table 1. Input conditions of FEA for NARX modeling

Data type	Input variables	Range
Training	Flow rate (gpm)	0, 1, 3, 4, 5, 6, 8
	Cycle-times (sec)	61, 71, 81, 91
Validation	Flow rate (gpm)	0, 2, 7.13
	Cycle-times (sec)	66, 86

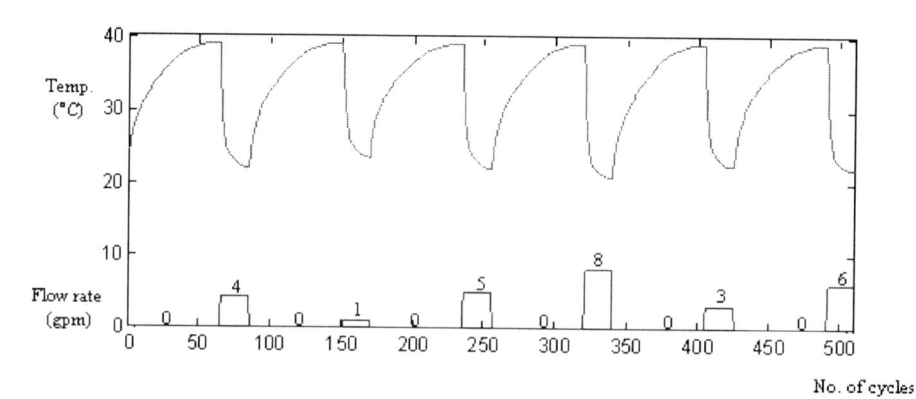

Fig. 2. Training data using all flow rate ranges (0, 1, 3, 4, 5, 6, 8 gpm) and one cycle-time (91 sec) at node Mo2

Next step was to determine the model orders in the NARX model of Eq. (1) by applying the Lipschitz criterion [13] and using the training data sets. The obtained model orders are presented in the schematic of NARX model of Fig.3. For better results and reduction of the calculation time, data was normalized by Eq. (3) before they were used for training and validation process.

$$y_{Nor,i}(t) = \frac{y_i^{max} - y_i(t)}{y_i^{max} - y_i^{min}}; \quad u_{Nor,m}(t) = \frac{u_m(t)}{u_m^{max}}. \quad (3)$$

where $y_{Nor,i}$, y_i^{max}, y_i^{min} are the normalized value, maximum, minimum of the output (temperature), y at i^{th} node, respectively. $u_{Nor,m}$ and u_m^{max} are the normalized value and maximum of the mth input, u_m (cycle-time or flow rate), respectively.

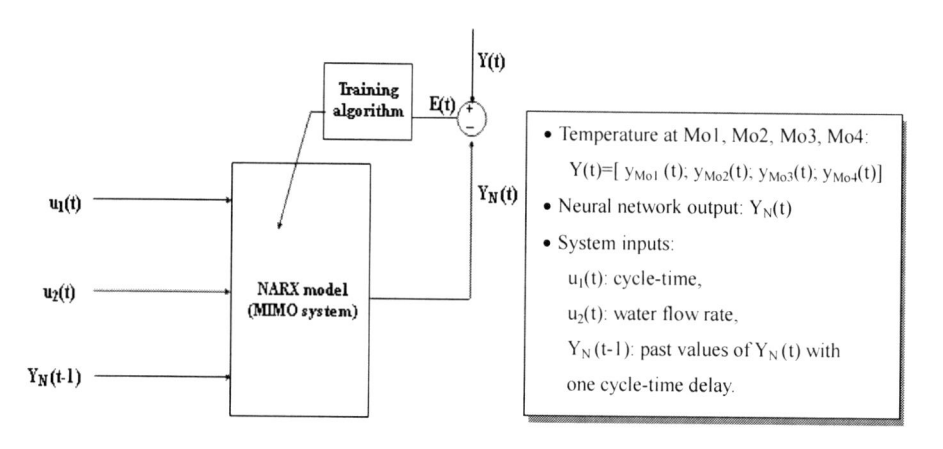

- Temperature at Mo1, Mo2, Mo3, Mo4:
  $Y(t) = [\ y_{Mo1}(t);\ y_{Mo2}(t);\ y_{Mo3}(t);\ y_{Mo4}(t)]$
- Neural network output: $Y_N(t)$
- System inputs:
  $u_1(t)$: cycle-time,
  $u_2(t)$: water flow rate,
  $Y_N(t-1)$: past values of $Y_N(t)$ with one cycle-time delay.

Fig. 3. Schematic of NARX model

The Levenberg-Marquardt (LM) algorithm was adopted as a training algorithm, and the number of hidden layer nodes was determined by trial and error, based on the smallest mean square error (MSE). After training with the training data with 200 epochs, the NARX model's performance (i.e., accuracy) was tested with the validation data sets. Using the validation input conditions (from Table 1), the NARX model is validated (see Fig.4). From Fig. 4, it can be noted that our NARX model has a good performance of temperature estimation at all nodes for new input conditions that are not utilized for training data.

4 Controller Design

4.1 Control of Cavity-Wall Temperature

The cavity-wall temperature after part ejection can indicate the quality of the final product. However, because the cavity wall is an area where locating sensors is challenging, the temperature measured at 4 nodes can be used instead of the cavity-wall temperature by utilizing the relation between them. From the training data sets used for NARX modeling, a strong linear relation (R=0.787) between an average of four T_a at 4 nodes and cavity-wall temperature right after ejection is found.

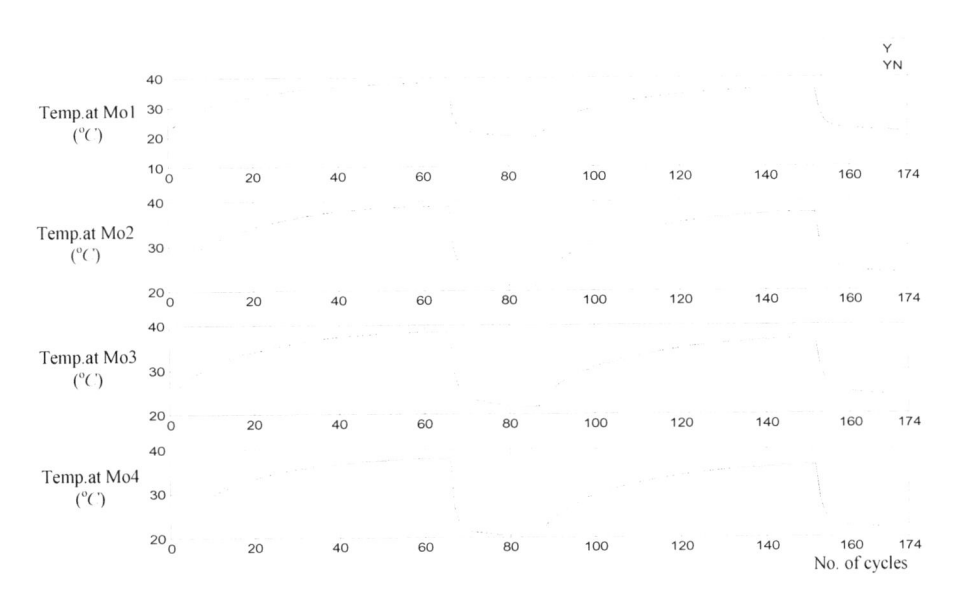

Fig. 4. Validation results (comparison between model (Y_N) and actual outputs (Y))

Since an average of four T_a is relatively insensitive to noise or unique response which can be generated at specific node, this value is more suitable to use for analyzing the relation with cavity-wall temperature rather than each T_a. When the part reaches the desired demoulding temperature for this material (Santoprene 8211-45: 85 ^{o}C), the cavity-wall temperature is around 25 ^{o}C, which corresponds to 22 ^{o}C for an average of four T_a from the linear relationship. Therefore, 22 $^{\circ}C$ will be used as an objective value to control the quality of moulding process (or product quality).

4.2 Self-tuning PID with RBF Neural Network

As an effective control strategy, neural networks using the radial basis function (RBF) is applied for the design of a self-tuning PID controller. The advantages of RBF neural network include simple structure, faster learning algorithms and effective mapping between a controlling system's input and output [14]. This PID controller overcomes a limitation of conventional PID controller that the determination of control parameters is not easy in nonlinear or uncertain systems, by the self-learning ability of tuning PID parameters online.

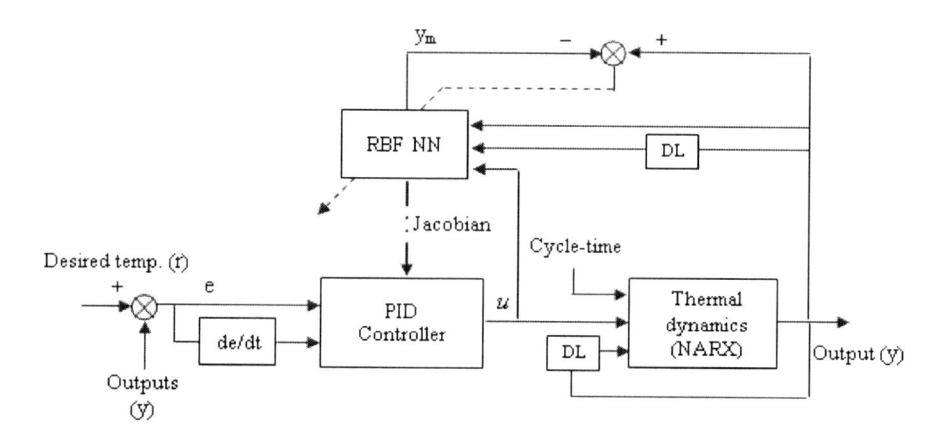

Fig. 5. Self-adaptive PID control with RBF neural network

Figure 5 describes the structure of self-adaptive PID controller using the RBF neural network. The main function of this control technique is to first identify the plant dynamics and secondly adjust the PID parameters adaptively based on RBF neural network identification.

The input vector, output vector in hidden layer and weight vector between hidden layer and output layers in RBF network are respectively,

$$X = [x_1, x_2, ..., x_n]^T ; H = [h_1, h_2, ..., h_m]^T ; W = [w_1, w_2, ..., w_m]^T . \quad (4)$$

where X, H and W are the input vector, output vector in hidden layer and weight vector between hidden and output layers. Each element (h_j) of vector H has a format of the Gaussian kernel function given by

$$h_j = \exp\left[\left(\|X - C_j\|^2\right)/2b_j^2\right], j = 1, ..., m; \quad C_j = [c_{1j}, c_{2j}, ..., c_{ij}, ..., c_{nj}]^T . \quad (5)$$

where h_j is the Gaussian kernel function, C_j and b_j are the center and width of h_j.

The output of RBF network, y_m is expressed as:

$$y_m = w_1 h_1 + w_2 h_2 + \cdots + w_m h_m. \tag{6}$$

The incremental PID control algorithm is given as follows:

$$\begin{aligned} u(k) &= u(k-1) + \Delta u(k) \\ &= u(k-1) + K_p \Delta e(k) + K_I e(k) + K_D \Delta^2 e(k). \end{aligned} \tag{7}$$

where $u(k)$ is the control variable at the time instant k. K_p, K_I and K_D (PID parameters) are proportional, integral and derivative gains, respectively. $e(k)=r(k)-y(k)$ is the control error; $\Delta e(k)=e(k)-e(k-1)$; $\Delta^2 e(k)=e(k)-2e(k-1)+e(k-2)$.

The W, C_j, b_j and PID parameters are updated by iterative algorithms using the gradient-descent technique addressed in [15] in which the Jacobian information (i.e., sensitivity of output to controlled input, $\partial y_m / \partial \Delta u(k)$) identified by RBF network is used for adjusting the PID parameters online as seen in Fig. 5.ï

4.3 Controller Performance

By adopting the aforementioned control technique, the performance of designed controller was evaluated using MATLAB/Simulink. The simulation results in Fig. 6 show the variation of average of four T_a (at 4 nodes) in NARX model with the

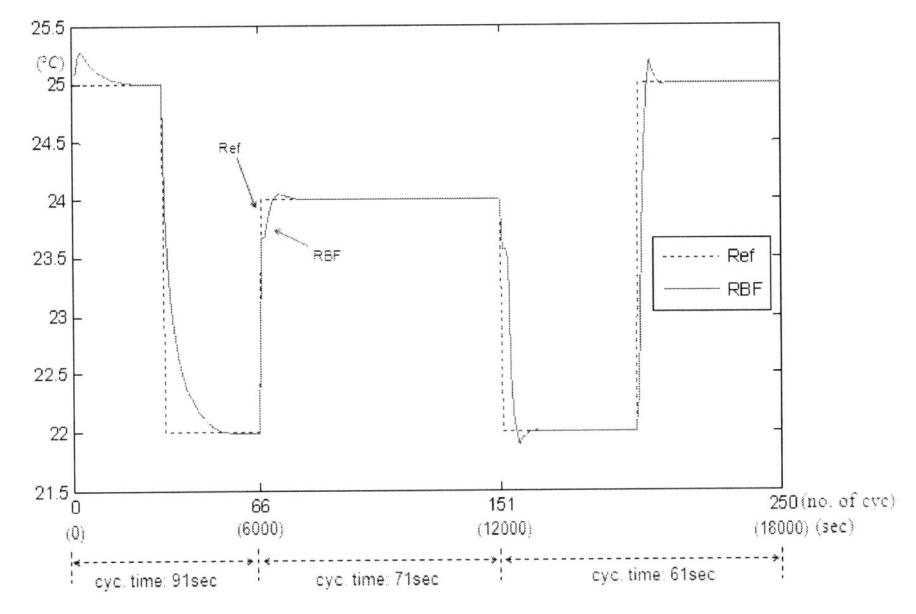

Fig. 6. RBF based self-adaptive PID Controller's performance for multi-setpoints with various cycle-times

controller. For the first range (0-6000 sec), the cycle-time of 91 sec (thus 66 no. of cycles) is considered with multi-setpoints of 25 °C and 22 °C (which corresponds to the demoulding temperature). Then, set-point is increased to 24 °C and kept constant during 6000 sec – 12000 sec with a different cycle-time of 71 sec (total 85 cycles).

Then multi-setpoints of 22 °C and 25 °C are used during 12000 sec – 18000 sec with the cycle-time of 61 sec (total 99 cycles).

The self-tuning PID controller with RBF shows a good tracking performance for all cycle-time ranges.

5 Conclusion

This study proposed an effective thermal control strategy for the plastic injection moulding process. The thermal dynamics was modeled using FEA and NN techniques allowing to tackle the problem of various cycle-times and uncertain dynamics of an injection mould system. Based on the model, a self adaptive PID controller with RBF neural network was introduced. By utilizing on-line learning algorithms to tune control parameters, the self-adaptive PID controller showed accurate control performance for the moulding process with diverse cycle-times.

Acknowledgements

The author would like to acknowledge the financial support of Natural Sciences and Engineering Research Council of Canada (NSERC) and Ontario Graduate Scholarship (OGS).

References

1. Bishenden, W., Bhola, R.: Die temperature control. In: 20th International Die Casting Congress and Exposition, North American Die Casting Association, pp. 161–164 (1999)
2. Kong, L., She, F., Gao, W., Nahavandi, S., Hodgson, P.: Integrated optimization system for high pressure die casting processes. Journal of Materials Processing Technology 201(1-3), 629–634 (2008)
3. Dubay, R., Pramujati, B., Hernandez, J.: Cavity temperature control in injection molding. In: 2005 IEEE International Conference on Mechatronics and Automation, vol. 2, pp. 911–916 (2005)
4. Gao, F., Patterson, W., Kamal, M.: Dynamics and Control of Surface and Mold Temperature in Injection Molding. International Polymer Processing 8, 147–157 (1993)
5. Dubay, R.: Self-optimizing MPC of melt temperature in injection moulding. ISA Transactions 41(1), 81–94 (2002)
6. Lu, C., Tsai, C.: Adaptive decoupling predictive temperature control for an extrusion barrel in a plastic injection molding process. IEEE Transactions on Industrial Electronics 48(5), 968–975 (2001)
7. Tsai, C., Lu, C.: Multivariable self-tuning temperature control for plastic injection molding process. IEEE Transactions on Industry Applications 34(2), 310–318 (1998)
8. Vetter, R., Maijer, D., Huzmezan, M., Meade, D.: Control of a die casting simulation using an industrial adaptive model-based predictive controller. In: Proceedings of the Symposium Sponsored by the Extraction and Processing Division of the Minerals, Metals and Materials Society, pp. 235–246 (2004)
9. Hu, H., Chen, F., Chen, X., Chu, Y., Cheng, P.: Effect of cooling water flow rates on local temperatures and heat transfer of casting dies. Journal of Materials Processing Technology 148, 439–451 (2004)

10. Potsch, G., Michaeli, W.: Injection molding: An introduction. Hanser/Gardner Publications (2007)
11. http://www.santoprene.com
12. http://www.matbase.com/material/ferrous-metals/cast-iron/gg-15/properties
13. He, X., Asada, H.: A new method for identifying orders of input–output models for nonlinear dynamic systems. In: Proceedings of American Control Conference, pp. 2520–2523 (1993)
14. Zhang, M., Wang, X., Liu, M.: Adaptive PID Control Based on RBF Neural Network Identification. In: 17th IEEE International Conference on Tools with Artificial Intelligence, pp. 681–683 (2005)
15. Jiang, J., Wen, S., Zhou, Z., He, H.: Fuzzy barrel temperature PID controller based on neural network. In: Proceedings of the 2008 Congress on Image and Signal Processing, vol. 1, pp. 90–94 (2008)

Analysis of Future Measurement Incorporation into Unscented Predictive Motion Planning

Morteza Farrokhsiar and Homayoun Najjaran

UBC, Okanagan School of Engineering,
3333 University way, Kelowna, Canada
{Morteza.Farrokhsiar,Homayoun.Najjaran}@ubc.ca

Abstract. An analysis of the future measurement incorporation into the unscented predictive motion planning algorithm for nonholonomic systems is presented. A two-wheeled robot is selected as the example nonholonomic system. A predictive motion planning scheme is used to find the suboptimal control inputs. In addition to the nonholonomic constraint, state estimation and collision avoidance chance constraints are incorporated to the predictive scheme. The closed form of the probabilistic constraints is solved by utilizing the unscented transform of the motion model. In order to evaluate the effect of future measurement incorporation into the planning algorithm, two different types of the unscented predictive planner, UPP-1 and UPP-2, are developed. In UPP-2 the future measurement is incorporated to the planning algorithm, whereas in UPP-1 the future measurement is ignored. Numerical simulation results demonstrate a high level of robustness and effectiveness of the both proposed algorithms in the presence of disturbances, measurement noise and chance constraints. Also, simulation results indicate that UPP-1 is a more conservative planner than UPP-2.

Keywords: Motion planning; Model predictive control; Chance constraints; Nonholonomic systems.

1 Introduction

THE main objective of this paper is to develop a motion planning scheme for a nonholonomic robot in unknown and dynamic environments and analyze the effect of the future measurement incorporation into the developed algorithm. In the robotics literature, motion planning refers to the construction of the control policies, i.e., system inputs, that drives the robot from an initial state to the goal state [1].

By growing interest in an increased level of autonomy of service robots, robot motion planning in complex and dynamic environments has become considerably important [2, 3]. In the classical motion planning it is assumed that planned actions can be implemented without any uncertainty [1, 4]. The planning algorithm for stochastic system has been modeled mainly as a dynamic programming (DP) problem. Dynamic programming deals with decision making under uncertainty. In the DP approach, the problem is solved by finding the optimal solution backward [5]. Initially, in the motion planning problem, only the process noise has been considered and it is been assumed that the environment can be sensed continuously. This case is known as perfect state information (PSI). The more general case of motion planning in which the assumption of PSI can be dropped is known as "imperfect state information" (ISI). In this case, the control signal is subjected to disturbance and system states cannot be perfectly measured.

Most often, the motion planning problem is subjected to different constraints on the system states and control signal. Avoiding obstacles and control action limitation are well-known examples of such constraints. Receding horizon control (RHC), or also model predictive control (MPC), is a suboptimal control scheme which is able to handle the control of stochastic systems subjected to explicit states and control signal constraints. In the RHC scheme, the problem is solved using information from current states, prediction of future measurements and future control signal over a finite horizon. Then a small portion of plan is executed. By obtaining the new measurement, the system states are updated and the planning problem is resolved. An extensive review on this topic can be found in [6, 7]. The RHC approach has been widely used in the uncertain robot location problems [8, 9]. In the stochastic systems, the RHC state constrains need to be reformulated as chance constrains, which ensure that the probability of collision with surrounding obstacles is below a given threshold [10]. Later, Blackmore used RHC-based probabilistic particle control to avoid the assumptions related to system linearity and Gaussian noise [11]. An important issue in RHC motion planning, which is the main attention of this paper, is incorporating the future information into the planning algorithm. Otherwise uncertainty grows with the open loop prediction of the system states. The unbounded growth of uncertainty will then require an exceedingly cautious planning solution to avoid obstacles [12]. In all of the stated stochastic RHC motion planning methods, the motion model has been assumed linear. However, in the real practice, mobile robots are usually nonlinear and nonholonomic. On the other hand, during the past decade, while there has been an increasing interest in deterministic MPC of nonholonomic systems [13-15] and related stability issues [16-18], the MPC design for nonholonomic systems with stochastic constraints has never been studied.

In this paper, the problem of motion planning of nonholonomic mobile robot using stochastic model predictive control has been discussed. The motion planning algorithm has to tackle both deterministic nonholonomic constraints of the nonlinear motion model and collision avoidance chance constraints. Blackmore has proposed a sequential Monte Carlo based method, particle control, to solve the chance constraint problem [11]. The main problem of sequential Monte Carlo methods is that the outcome of the control is function of the sample number. The number of samples needed for convergence grows proportional to the square of the state size [19]. Also, there is no guarantee to avoid the problem of degeneracy [20]. *If state distributions are Gaussian, or they can be approximated by Gaussian distributions it is possible to find a closed solution for the chance constraints* [10]. Different methods have been developed to approximate non-Gaussian function outputs with Gaussian distributions [19]. Some of these methods, such as well-known method of Taylor expansion rely on the analytical linearization, whereas some other methods, such as unscented transformation (UT), rely on the statistical linearization [21]. It has been reported that the statistical linearization methods outperforms analytical linearization methods [22]. In unscented predictive motion planning, the closed solution of chance constraints based on the unscented transform of the nonlinear motion models is developed. Finally, two types of MPC motion planning algorithm has been developed which are able to tackles both deterministic and stochastic constraints. In the first unscented predictive planner (UPP-1) future measurements are ignored. In the second unscented predictive planner (UPP-2), future measurements are incorporated into the unscented predictive planner.

The problem statement, including the system model, the motion planning formulation and chance constraints are discussed in the next Section. Different versions of the unscented predictive motion planning algorithms are discussed in the Section 3. The numerical simulation results and performance comparison of the two different algorithms are presented and discussed in Section 4, which is followed by concluding remarks and future work.

2 Problem Statement

2.1 The Nonholonomic System Model

In this paper a two-wheeled mobile robot, shown in Fig. 1, has been considered as an example of nonholonomic systems. Robot states are defined using robot position and rotation angle, $X = [x, y, \theta]^T$. The configuration space, C, has been considered as the state space; $X \in C = \mathbb{R}^2 \times S$. Considering linear and angular velocities as the robot input, $u = [v, \omega]$, the robot motion model can be defined as,

$$\begin{cases} \dot{x} = v\cos\theta \\ \dot{y} = v\sin\theta \\ \dot{\theta} = \omega \end{cases} \tag{1}$$

In Eq. (1), the deriving signals include both control signal and disturbances that act on the system in the same way. If the robot moves without slipping, the robot kinematics satisfies a nonholonomic constraint:

$$\dot{x}\sin\theta - \dot{y}\cos\theta = 0 \tag{2}$$

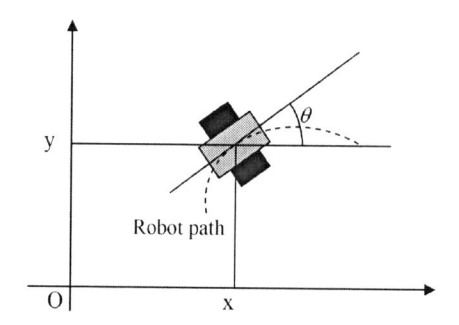

Fig. 1. Schematic of the two-wheeled mobile robot

The control system design for the stated nonholonomic system is a challenging task. Based on the well-known work of Brockett (1983), the nonholonomic robot model is considered non-controllable for the time invariant feedback control laws [23]. Also, in the real situations, saturation of the control inputs is inevitable. The focus of this paper is on the discrete-time control of the nonholonomic system. By discretizing the Eq. (1) based on the sampling time T, motion equations can be rewritten as:

$$\begin{cases} x(k+1) = x(k) + v(k)\cos\theta(k)T \\ y(k+1) = y(k) + v(k)\sin\theta(k)T \\ \theta(k+1) = \theta(k) + \omega(k)T \end{cases} \quad (3)$$

where $x(k), y(k)$ and $\theta(k)$ compose the state vector $X(k)$. Eq. (3) can be summarized as:

$$X(k+1) = f(X(k), u(k)) \quad (4)$$

It can be assumed that in the motion planning problem, the pose error is defined based on the difference of the final destination to the current pose:

$$\chi(k) = X_f - X(k) \quad (5)$$

Therefore, by assuming $\chi(k) = [x_e, y_e, \theta_e]^T$, the error dynamics can be defined as:

$$\begin{cases} x_e(k+1) = x_e(k) - v(k)\cos\theta(k)T \\ y_e(k+1) = y_e(k) - v(k)\sin\theta(k)T \\ \theta_e(k+1) = \phi(\theta_e(k) - \omega(k)T) \end{cases} \quad (6)$$

where $\phi: \mathbb{R} \to [-\pi, \pi)$ provides a unique value for all the same physical angles having different numerical values. Now, we can define the motion planning problem for the system defined in Eq. (6).

2.2 The Motion Planning Formulation

The problem considered in this paper is to find an optimal sequence of control inputs that ensures that the robot reaches the goal region in the finite time and avoids obstacles during the planning process. The optimality criterion is defined in terms of minimizing a cost function which is usually an indicator of the control effort. In the deterministic systems with perfect state information, the optimization can be defined as:

$$\min_u J(N, \chi, u_n^{n+N-1}) = \chi_{n+N}^T P_N \chi_{n+N} \\ + \sum_{i=0}^{N-1} (\chi_{n+i}^T Q_i \chi_{n+i} + u_{n+i}^T R_i u_{n+i}) \quad (7)$$

subject to

$$\chi_{n+i+1} = f(\chi_{n+i}, u_{n+i}) \quad (a)$$
$$\chi_{n+i} \in X_i \quad (b) \tag{8}$$
$$u_{n+i} \in U_i \quad (c)$$

The optimization result, $u^* = \arg\min(J)$, is the sequence of the optimal control inputs over the control horizon N. The robot motion model is introduced to the optimization equation as one of the constraints. Also, Eqs. (8.b and 8.c) indicate that the robot states and control inputs must be inside the feasible region of the optimization. In the motion planning in unknown dynamic environments, different source of uncertainties are introduced to the problem. These uncertainties may come from different sources [11]:

1. Due to imperfect state information, the robot states are known as a probability distribution over possible states specially, in unknown environments where landmark location is known as a probability distribution.
2. The system model is unknown which is a common limitation of mathematical modeling.
3. Unmodeled disturbances are inevitable.

By considering the existing disturbances, the system has to be modeled as:

$$\chi_{n+i+1} = f(\chi_{n+i}, u_{n+i}, v_{n+i}) \quad (a)$$
$$y_{n+i} = g(\chi_{n+i}, u_{n+i}, \upsilon_{n+i}) \quad (b) \tag{9}$$

where y_n the measurement output is related to system states and control inputs through the zero-order Markov chain g. Also, v_{n+i} and υ_{n+i} are the motion disturbances and measurement noise, respectively. It is assumed that the disturbances act on the robot in a similar way to the control input, u_{n+i}. In order to handle the stated uncertainties, a stochastic dynamic programming (SDP) is defined as:

$$\min_{u} J(N, \chi, u_n^{n+N-1}) = E(\chi_{n+N|I_n}^T P_N \chi_{n+N|I_n}$$
$$+ \sum_{i=0}^{N-1} (\chi_{n+N|I_n}^T Q_N \chi_{n+N|I_n} + u_{n+i|I_n}^T R u_{n+i|I_n})) \tag{10}$$

subject to

$$\chi_{n+i+1} = f(\chi_{n+i}, u_{n+i}, v_{n+i})$$
$$y_{n+i} = g(\chi_{n+i}, u_{n+i}, \upsilon_{n+i})$$
$$P(\chi_{n+i|I_n} \in X_i) \geq p_i \tag{11}$$
$$u_{n+i} \in U_i$$

where $E(.)$ is the conditional expectation value operator, and I_i is the set of information, $I_i = \{u_1 \cdots u_i, y_1 \cdots y_i\}$. Yan and Bitmead showed that solving the stated SDP is equivalent of solving an equivalent deterministic dynamic programming [10]:

$$\min_{u} J(N, \hat{\chi}, u_n^{n+N-1}) = \hat{\chi}_{n+N}^T P_N \hat{\chi}_{n+N}$$
$$+ \sum_{i=0}^{N-1} (\hat{\chi}_{n+i}^T Q_i \hat{\chi}_{n+i} + u_{n+i|I_n}^T R u_{n+i|I_n}) \quad (12)$$

subject to

$$P(\chi_{n+i|I_n} \in X_i) \geq p_i$$
$$u_{n+i} \in U_i \quad (13)$$

where $\hat{\chi}_i$ is defines as the mean of the expected value of the robot states conditioned to the available information, $\hat{\chi}_i = E(\chi_i | I_i)$. The probabilistic constraint $P(\chi_{n+i|I_n} \in X_i) \geq p_i$ is a general expression for the chance constraints. This constraint would be discussed elaborately in the next subsection.

2.3 Chance Constraints

In addition to uncertainty of the robot states, in unknown dynamic environments, the information of obstacles and other robots comes from uncertain estimation process. Therefore, the collision avoidance constraints in Eq. (8) has to be expressed in terms of chance constraints, $P(x_{i+1} \notin X_{free} | I_i) \leq \partial$, where X_{free} denotes the free space not occupied by obstacles and ∂ denotes confidence level, respectively. In this paper it is assumed a disk robot with radius ε must avoid point obstacles. The collision condition can be expressed as:

$$C : x^O \in B(x^R, \varepsilon) \quad (14)$$

where x^O and x^R are the obstacle and robot location. Also $B(x^R, \varepsilon)$ is the disk ball with the center of x^R and radius ε. The probability of collision can be defined as [12]:

$$P(C) = \int_{x^R} \int_{x^O} I_C(x^R, x^O) p(x^R, x^O) dx^R dx^O \quad (15)$$

where I_C, the indicator function is defined as:

$$I_C(x^O, x^R) = \begin{cases} 1 & x^O \in B(x^R, \varepsilon) \\ 0 & Otherwise \end{cases} \quad (16)$$

Using definition of Eq. (16) can be rewritten as:

$$P(C) = \int_{x^R} \left[\int_{x^O \in B(x^R, \varepsilon)} p(x^O | x^R) dx^O \right] p(x^R) dx^R \quad (17)$$

By assuming a small size for the robot, Eq. (17) can be approximated as:

$$P(C) \approx V_B \int_{x^R} p(x^O = x^R \mid x^R) p(x^R) dx^R \tag{18}$$

where V_B is the area of the robot. *If the robot state and obstacle distributions are Gaussian*, i.e., $x^R \simeq \mathcal{N}(\hat{x}^R, \Sigma_{x^R})$ and $x^O \simeq \mathcal{N}(\hat{x}^O, \Sigma_{x^O})$, the integral in Eq. (18) can calculated as [12]:

$$\int_{x^R} p(x^O = x^R \mid x^R) p(x^R) dx^R = \left(\sqrt{\det(2\pi\Sigma_C)}\right)^{-1} \exp\left[-0.5(\hat{x}^R - \hat{x}^O)^T \Sigma_C^{-1} (\hat{x}^R - \hat{x}^O)\right] \tag{19}$$

where $\Sigma_C = \Sigma_{x^R} + \Sigma_{x^O}$. Using the constraint of $P(C) \le \partial$ on the mean of the states, the chance constraint of Eq. (12) is equal to:

$$\left(\hat{x}^R - \hat{x}^O\right)^T \Sigma_C^{-1} \left(\hat{x}^R - \hat{x}^O\right) \ge -2\ln\left(\sqrt{\det(2\pi\Sigma_C)} \frac{\partial}{V_B}\right) \tag{20}$$

3 Unscented Model Predictive Motion Planning

The As it is discussed in the previous section, collision avoidance constraints are introduced to SDP as chance constraints. In this section, we will elaborate on the unscented transform that provides Gaussian approximation for state distributions, and thus, closed form solution for the chance constraints. At the end of this Section, the final formulation of the unscented predictive motion planning would be presented.

3.1 Unscented Transformed Motion Equations

The unscented transform relies neither on analytical approximation methods such as the Taylor expansion nor on the Monte Carlo random sampling. In this transform, the so called *sigma points* are extracted deterministically. By passing these points through the nonlinear function, the Gaussian approximation would be constructed.

Assume that χ_{n+i} is an n-dimensional Normal distribution with mean and covariance matrix of μ_χ and Σ_χ, respectively. The $2n+1$ sigma points $\chi^{[j]}$ are calculated by the following rule:

$$\chi^{[0]} = \mu_\chi$$
$$\chi^{[j]} = \mu_\chi + \left(\sqrt{(n+\lambda)\Sigma_\chi}\right)_j \quad \text{for } j = 1, \ldots, n \tag{21}$$
$$\chi^{[j]} = \mu_\chi + \left(\sqrt{(n+\lambda)\Sigma_\chi}\right)_{j-n} \quad \text{for } j = n+1, \ldots, 2n$$

where $\lambda = \alpha^2(n+\kappa) - n$, with α and κ that are scaling parameters [22]. In order to calculate the approximated mean and covariance, each sigma, $\chi^{[j]}$, point has two weights, one, $w_m^{[j]}$, related to the mean and the other one, $w_c^{[j]}$ related to the covariance.

$$w_m^{[0]} = \frac{\lambda}{n+\lambda}$$

$$w_c^{[0]} = \frac{\lambda}{n+\lambda} + (1 - \alpha^2 + \beta) \qquad (22)$$

$$w_m^{[i]} = w_m^{[i]} = \frac{\lambda}{2(n+\lambda)} \quad \text{for } j = 1, \ldots, 2n$$

The parameter β is used to incorporate the knowledge about the distribution which is being approximated by the Gaussian distribution. For an exact Gaussian distribution, $\beta = 2$ is the optimal choice [22]. Passing through the nonlinear function, the estimation mean, $\overline{\chi}_{n+i+1}$, and covariance, $\overline{\Sigma}_{n+i+1}$, of χ_{n+i+1} is estimated as:

$$\overline{\chi}_{n+i+1} = \sum_{j=0}^{2n} w_m^{[j]} \overline{\chi}^{[j]} \quad (a)$$

$$\overline{\Sigma}_{n+i+1} = \sum_{j=0}^{2n} w_c^{[j]} (\overline{\chi}^{[j]} - \overline{\chi}_{n+i+1})(\overline{\chi}^{[j]} - \overline{\chi}_{n+i+1})^T + R_{n+i} \quad (b) \qquad (23)$$

where R_{n+i} is the noise covariance of v_{n+i}, the system disturbance and also

$$\overline{\chi}^{[j]} = f(\chi^{[j]}, u_{n+i}) \qquad (24)$$

It is possible to summarize the Eq. (22.a) as a nonlinear mapping approximating the system equation and write it as:

$$\overline{\chi}_{n+i+1} = h(\overline{\chi}_{n+i}, u_{n+i}) \qquad (25)$$

The mapping h is Gaussian and hence, can be used to find the closed solution of the chance constraints. It should be noted that $\overline{\chi}_{n+i}$ is the estimation of χ_{n+i} in which future measurements are ignored.

3.2 Unscented Model Predictive Motion Planning Using Output Feedback

The last piece of the unscented model predictive motion planning algorithm is incorporation the future measurement. The idea of utilizing the future measurement information can be traced back to the *dual control theory* [24]. Incorporating the future measurement information into the model predictive algorithm prevents the control system to be overly conservative while preserves system cautiousness [12]. In the unscented model predictive motion planning algorithm, the future information is incorporated to the system through the unscented Kalman filter (UKF). Assume that g in Eq. (9. b) is a nonlinear function with additive noise covariance, Q_{n+i}. In the UKF algorithm, incorporating the measurement information commences with computing χ^*_{n+i+1}, the vector of sigma points of the normal distribution of $\mathcal{M}(\overline{\chi}_{n+i+1}, \overline{\Sigma}_{n+i+1})$. The algorithm proceeds as:

$$\overline{Y}_{n+i+1} = g(\chi^*_{n+i+1}), \quad \hat{y}_{n+i+1} = \sum_{j=0}^{2n} w_m^{[j]} \overline{Y}_{n+i+1}^{[j]}$$

$$S_{n+i+1} = \sum_{j=0}^{2n} w_c^{[j]} (\overline{Y}_{n+i+1}^{[j]} - \hat{y}_{n+i+1})(\overline{Y}_{n+i+1}^{[j]} - \hat{y}_{n+i+1})^T + Q_{n+j+1}$$

$$\Sigma^{x,y}_{n+i+1} = \sum_{j=0}^{2n} w_c^{[j]} (\chi^{*[j]}_{n+i+1} - \overline{\chi}_{n+i+1})(\overline{Y}_{n+i+1}^{[j]} - \hat{y}_{n+i+1})^T \quad (26)$$

$$K_{n+i+1} = \Sigma^{x,y}_{n+i+1} S^{-1}_{n+i+1}$$

$$\hat{\chi}_{n+i+1} = \overline{\chi}_{n+i+1} + K_{n+i+1}(y_{n+i+1} - \hat{y}_{n+i+1})$$

$$\Sigma_{n+i+1} = \overline{\Sigma}_{n+i+1} - K_{n+i+1} S_{n+i+1} K^T_{n+i+1}$$

In reality, the future measurement y_{n+i+1} is not available; therefore, *it is assumed that the future measurement is equal to the most likely measurement*. This assumption does not change the predicted state means, $\hat{\chi}_{n+i+1} = \overline{\chi}_{n+i+1}$ while reduces the estimation covariance, Σ_{n+i+1} and increases the feasibility region for the optimization Eq. (11). Finally, it is possible to express the unscented predictive planner with future measurement incorporation (UPP-2) in its final form:

$$\min_u J(N, \hat{\chi}, u_n^{n+N-1}) = \hat{\chi}^T_{n+N} P_N \hat{\chi}_{n+N}$$

$$+ \sum_{i=0}^{N-1} (\hat{\chi}^T_{n+N} Q_N \hat{\chi}_{n+N} + u^T_{n+i|l_n} R u_{n+i|l_n}) \quad (27)$$

subject to Eqs. (18, 25, 26) and $u_{n+i} \in U_i$.

The unscented predictive planner in which future measurements are ignored (UPP-1) is defined as an optimization similar to Eq. (27) but not subjected to Eq. (26). In the next section, numerical evaluations of the UPP-1 and UPP-2 for three different cases are discussed.

4 Simulation Results and Discussions

Efficiency of the proposed unscented predictive motion planning algorithms are evaluated through numerical simulations. In the simulation scenario, the robot is moved inside an environment in which obstacles are known only as a probabilistic distribution. The measurement equation is assumed to be linear:

$$g(\chi_{n+i}, \upsilon_{n+i}) = \chi_{n+i} + \upsilon_{n+i} \qquad (28)$$

The unscented predictive motion planning algorithm is applicable to both linearity and nonlinear measurement equation. The deriving signal disturbance and measurement noise are considered as a zero mean Gaussian with the covariance matrices of $R = I_2$ and $Q = 0.1 \times I_3$, respectively. Also I is the identity matrix. Control signals have the saturation magnitude equal to 5. The initial robot pose covariance is equal to P_o, where $P_0 = I_3$. The unscented transform parameters are selected as $\alpha = -1.1$ and $\kappa = 0$. It is assumed that the robot is posed at $X_0 = \begin{bmatrix} -25 & 0 & 0 \end{bmatrix}^T$. The obstacle distribution is selected as $x^O = \mathcal{N}([-8\ -1]^T, I_2)$. Other parameters are selected as, $\delta = 0.01$ and $V_B = 1$. The robot paths planned and executed using UPP-1 and UPP-2 are shown in Figs. 2 and 3, respectively. The evolution of the heading angle versus time for UPP-1 and UPP-2 are shown in Figs. 4 and 5.

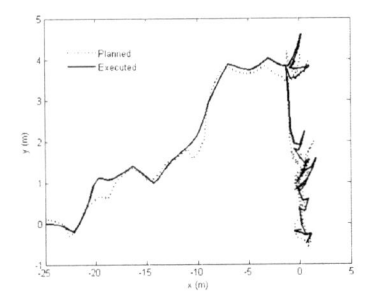

Fig. 2. Robot path with UPP-1; The planned path (dashed line) and exucted path (solid line)

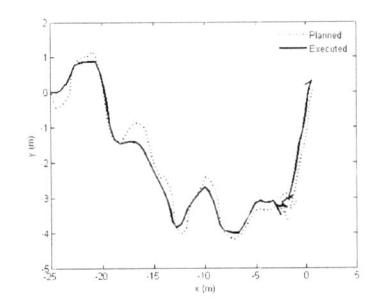

Fig. 3. Robot path with UPP-2; The planned path (dashed line) and exucted path (solid line)

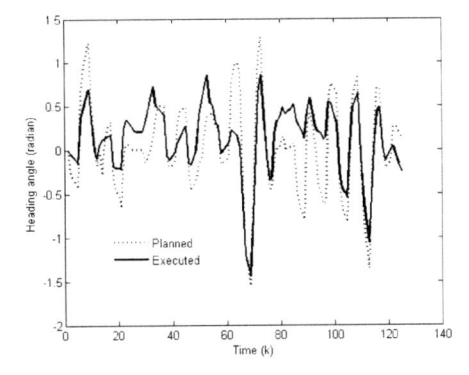

Fig. 4. Robot heading anlge, θ, with UPP-1, planned (dashed line) and exucted (solid line)

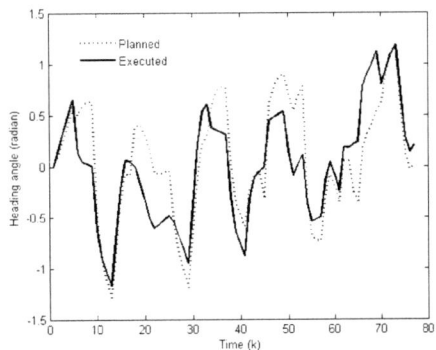

Fig. 5. Robot heading anlge, θ, with UPP-2, planned (dashed line) and exucted (solid line)

Simulation results indicate that the both unscented predictive planning motion planning algorithms are robust against the disturbances including uncertain initial location and severe driving signal disturbances and also measurement noise. The path oscillation close to the origin is related to the nonholonomic constraint. Also, simulation results indicates that while both motion planning algorithms are able to handle the uncertainty, UPP-2 drive the robot to the origin faster than UPP-1, whereas UPP-1 shows more cautious response in managing the uncertainty.

5 Conclusion

The problem of motion planning of a nonholonomic robot with chance constraints is discussed. A model predictive control has been used to solve the motion planning scheme. The nonholonomic constraint, state estimation and collision avoidance chance constraints are incorporated to the MPC scheme. The unscented transformed motion equations are used to find unscented the deterministic equivalent of the chance constraints. Two different types of the unscented predictive planner algorithm, UPP-1 and UPP-2 have been developed. In UPP-1, future measurements are ignored whereas in UPP-2 they are incorporated into the planning algorithm. Numerical simulations demonstrate that while both algorithms are robust, UPP-1 is more cautious than the UPP-2.

References

[1] LaValle, S.M.: Planning Algorithms. Cambridge Univ. Pr., Cambridge (2006)
[2] Roy, N., Gordon, G., Thrun, S.: Planning under uncertainty for reliable health care robotics. In: Field and Service Robotics, pp. 417–426 (2006)
[3] Tadokoro, S., Hayashi, M., Manabe, Y., Nakami, Y., Takamori, T.: On motion planning of mobile robots which coexist and cooperate with human. In: 1995 IEEE/RSJ International Conference on Human Robot Interaction and Cooperative Robots, Intelligent Robots and Systems 1995, vol. 2, pp. 518–523 (1995)
[4] Latombe, J.C.: Robot Motion Planning. Springer, Heidelberg (1990)
[5] Bertsekas, D.P.: Dynamic programming and optimal control, vol. I (1995)
[6] Mayne, D.Q., Rawlings, J.B., Rao, C.V., Scokaert, P.O.M.: Constrained model predictive control: Stability and optimality. Automatica 36(6), 789–814 (2000)
[7] Morari, M., Lee, H.: Model predictive control: past, present and future. Comput. Chem. Eng. 23, 667–682 (1999)
[8] Kuwata, Y.: Trajectory planning for unmanned vehicles using robust receding horizon control (2007)
[9] Li, P., Wendt, M., Wozny, G.: A probabilistically constrained model predictive controller. Automatica 38(7), 1171–1176 (2002)
[10] Yan, J., Bitmead, R.R.: Incorporating state estimation into model predictive control and its application to network traffic control. Automatica 41(4), 595–604 (2005)
[11] Blackmore, L.: A probabilistic particle control approach to optimal, robust predictive control (2006)
[12] Du Toit, N.E.: Robot Motion Planning in Dynamic, Cluttered, and Uncertain Environments: the Partially Closed-Loop Receding Horizon Control Approach (2010)
[13] Wan, J., Quintero, C.G., Luo, N., Vehi, J.: Predictive motion control of a mirosot mobile robot. In: Proceedings of the World Automation Congress, pp. 325–330 (2004)

[14] Shim, D.H., Chung, H., Sastry, S.S.: Conflict-free navigation in unknown urban environments. IEEE Robotics & Automation Magazine, 28 (2006)
[15] Zhu, Y., Ozgüner, U.: Constrained model predictive control for nonholonomic vehicle regulation problem. In: Proceedings of the 17th IFAC World Congress, pp. 9552–9557 (2008)
[16] Grimm, G., Messina, M.J., Tuna, S.E., Teel, A.R.: Nominally robust model predictive control with state constraints. IEEE Transactions on Automatic Control 52, 1856–1870 (2007)
[17] Michalska, H., Mayne, D.Q.: Robust receding horizon control of constrained nonlinear systems. IEEE Transactions on Automatic Control 38, 1623–1633 (1993)
[18] Gu, D., Hu, H.: Receding horizon tracking control of wheeled mobile robots. IEEE Trans. Control Syst. Technol. 14, 743–749 (2006)
[19] Thrun, S., Burgard, W., Fox, D.: Probabilistic Robotics. MIT Press, Cambridge (2006)
[20] Doucet, A., de Freitas, J.F.G., Gordon, N.J.: An intoduaction to sequential monte carlo methods. In: Doucet, A., de Freitas, J.F.G., Gordon, N.J. (eds.) Sequential Monte Carlo Methods in Practice, Springer, New York (2001)
[21] Julier, S.J., Uhlmann, J.K.: A new extension of the kalman filter to nonlinear systems. In: Int. Symp. Aerospace/Defense Sensing, Simul. and Controls, p. 26 (1997)
[22] Van Der Merwe, R., Wan, E.: The square-root unscented kalman filter for state and parameter-estimation. In: IEEE International Conference On Acoustics Speech And Signal Processing, pp. 3461–3464 (2001)
[23] Brockett, R.W.: Asymptotic stability and feedback stabilization (1983)
[24] Filatov, N.M., Unbehauen, H.: Survey of adaptive dual control methods. In: IEE Proceedings of Control Theory and Applications, vol. 147, pp. 118–128 (2000)

Nonlinear Maneuvering Control of Rigid Formations of Fixed Wing UAVs

Ismail Bayezit and Baris Fidan

University of Waterloo,
Department of Mechanical and Mechatronics Engineering,
200 University Avenue West,
Waterloo, Ontario, Canada
{ibayezit,fidan}@uwaterloo.ca
http://www.mme.uwaterloo.ca/

Abstract. This paper is on autonomous three-dimensional maneuvering of teams of fixed-wing *unmanned aerial vehicles* (UAVs), cooperating with each other during the mission flight. The goal is to design a distributed control scheme that preserves the shape of the UAV team formation by keeping the inter-agent distances constant during arbitrary maneuvers. The paper considers the dynamic characteristics and constraints of the UAVs resulting from being fixed-wing; and proposes a Lyapunov analysis based individual UAV control design to complement the distributed control scheme. After presentation of formal design of the distributed control scheme and individual UAV controllers, simulation results are provided demonstrating effectiveness of the proposed control design.

Keywords: Rigid formations, formation control, cooperative control, cohesive motion, autonomous vehicles, fixed-wing UAV.

1 Introduction

Distributed control and cooperation of dynamical systems has become a popular research area due to emerging real-life applications in various fields in recent years [5, 7–12]. A particular general problem in this research area is motion coordination of systems of autonomous mobile agents (robots or vehicles), which is considered in this paper as well.

The goal of the paper is to design a distributed control scheme together with low-level individual UAV controllers, to move a team of fixed-wing UAVs from an arbitrary initial setting to an arbitrary final setting without corrupting the formation geometry of the team. The initial and final settings are assumed to have the same formation geometry, which is defined by a set of inter-agent (UAV) distance constraints to be satisfied. The general *cohesive motion control problem* of moving a formation of autonomous agents in two and three dimensions have been considered in [2, 6, 14], where the distributed control scheme is similar to the one in this paper. The main contribution of this paper is implementation of such a control scheme in teams of *fixed-wing* UAVs via appropriate individual UAV control design.

2 Modeling and Problem Definition

Consider a team T of m UAV agents $A_1, ..., A_m$ moving in $\Re^3$ (where $m \geq 4$). Before formally defining the *cohesive motion control problem*, we present modeling of the multi-UAV system: Section 2.1 presents the modeling of the formation of the system. Dynamics of UAV agents are presented in Section 2.2 and the cohesive motion problem is defined in Section 2.3.

2.1 Modelling the Formation

In this paper, we consider asymmetric formation control structures [6, 13] where for each inter-agent distance keeping task only one agent (in the corresponding agent pair) is responsible. We represent the multi-UAV system, which is required to maintain a certain formation using such an asymmetric control architecture, by a directed graph $G_F = (V_F, E_F)$, called the directed underlying graph of the multi-UAV system or the formation. The directed underlying graph $G_F = (V_F, E_F)$ has a vertex set V_F and an edge set E_F where each vertex $i \in V_F$ corresponds to an agent A_i in T and each directed edge $\overrightarrow{(i,j)}$ from i to j represents the control and information link between neighboring agents A_i and A_j [6]; more specifically existence of $\overrightarrow{(i,j)}$ implies that A_i can sense its distance from A_j and is responsible to keep a desired distance d_{ij} from A_j. In this case, we also say that A_i follows A_j, or A_i is a follower of A_j. In the sequel, we call $T = \{A_1, A_2, ..., A_M\}$ together with the underlying graph $G_F = (V_F, E_F)$ and the distance set $D_F = \{d_{ij} \mid \overrightarrow{(i,j)} \in E_F\}$ a *formation* and represent by $F = (T, G_F, D_F)$, similarly to [14].

A *rigid* formation is a formation in which the distance d_{ij} between each agent pair (A_i, A_j) remains constant, which implicitly means that the formation shape is maintained, during any continuous motion of the formation. In a formation, if each agent in the formation is always able to satisfy the distance constraints it is responsible for, then this formation is called *constraint-consistent*. A formation that is both rigid and constraint-consistent is called *persistent*. Furthermore, a persistent formation maintaining its persistence with the minimum possible number of links, which is $|E_f| = 3|V_f| - 6$ in three dimensions [6, 13], is called *minimally persistent*. More formal definitions of *rigidity*, *constraint-consistence*, and *persistence* in three dimensions can be found in [13]. For further convenience, we make the following assumption about the formation and control architecture of the UAV team T:

A1: The formation $F = (T, G_F, D_F)$ is in a special minimally persistent form, called the *leader-follower* form, where $\overrightarrow{(2,1)}, \overrightarrow{(3,1)}, \overrightarrow{(3,2)} \in E_F$ and for all $i = \{4, ..., m\}$ $\overrightarrow{(i, i-3)}, \overrightarrow{(i, i-2)}, \overrightarrow{(i, i-1)} \in E_F$.

Assumption A1 is a widely used assumption in the literature [1, 6, 8, 12]. It poses that A_1 is a free agent with no distance constraints, A_2 has only one distance keeping constraint, A_3 has 2, and the other agents 3.

2.2 UAV Agent Dynamics

For the fixed-wing UAV dynamics model, we adopt the one for the Piccolo-controlled UAVs given in [4]:

$$\dot{x}_i = s_{lat,i}\cos\theta_i$$
$$\dot{y}_i = s_{lat,i}\sin\theta_i \qquad (1)$$
$$\dot{s}_{lat,i} = \frac{1}{\tau_s}\left(-s_{lat,i} + s_{lat,i_{cmd}}\right), \underline{s} \le \dot{s}_{lat,i} \le \overline{s}$$

$$\dot{\theta}_i = \omega_i$$
$$\dot{\omega}_i = \frac{1}{\tau_\omega}\left(-\omega_i + \omega_{i_{cmd}}\right) \qquad (2)$$

$$\dot{z}_i = \frac{1}{\tau_z}\left(-z_i + z_{i_{cmd}}\right), \underline{z} \le \dot{z}_i \le \overline{z} \qquad (3)$$

where $p_i(t) = [x_i(t), y_i(t), z_i(t)]^T \in R^3$, $s_{lat,i}$, θ_i and ω_i denote the position in three-dimensional earth fixed frame, lateral speed, lateral heading and lateral turn rate of the UAV agent A_i, respectively. The angle θ_i is defined within $-\pi < \theta_i \le \pi$. $s_{lat,i_{cmd}}$, $\omega_{i_{cmd}}$ and $z_{i_{cmd}}$ are the control signals. These three external inputs are generated with the help of inner control loop applied on mechanical actuators such as elevators, ailerons and rudders.

2.3 The Cohesive Motion Control Problem

Having the modeling of the multi-UAV system of interest presented, we present the formal control problem next. The following assumptions are made in the definition of the problem and implementation of the control task:

A2: Agents in the formation can perfectly measure their own positions as well as the (relative) positions of the agents they follow.

A3: The distance-sensing range for a neighbor agent pair (A_i, A_j) is sufficiently larger than the desired distance d_{ij} to be maintained.

A4: The agents A_1, A_2, A_3 know their final desired positions p_{1f}, p_{2f}, p_{3f}.

Next, we formally state the cohesive motion control problem of interest:

Problem 1: Consider a team T of m UAV agents $A_1, ..., A_m$ ($m \ge 4$) that are initial at positions, respectively, $p_{10}, ..., p_{m0} \in \Re^3$. Let $F = (T, G_F, D_F)$ be a minimally persistent formation in leader follower structure, where D_F is consistent with the initial positions p_{i0}. Each agent A_i moves with the dynamics (1)–(3). The control task is, under Assumptions $A_1 - A_4$, to generate the control signals $s_{lat,i_{cmd}}, \omega_{i_{cmd}}, z_{i_{cmd}}$ ($i \in \{1, ..., m\}$) to move T to a final desired setting defined by a set of desired final positions, $p_{1f}, ..., p_{mf}$, which is consistent with D_F, without deforming the shape of the formation, i.e. forcing the inter-agent distances track the values defined in D_F, during motion.

In the next section, we approach Problem 1 considering single-velocity integrator kinematic models, and design a distributed control scheme. Later, we discuss extension of this control scheme for fixed-wing UAV models in Section 4.

3 High-Level Distributed Formation Control Design

The high-level distributed control scheme is composed of an on-line trajectory generation and a distributed control law for generation of ideal desired velocities to follow these trajectories.

3.1 On-Line Trajectory Generation

It can be observed that the first three agents A_1, A_2, A_3 are the main agents coordinating the position and orientation of the formation F. The other agents just have the task of maintaining distances from the 3 agents they follow. Based on this observation and the definition of Problem 1, we pose the following desired positions for the m agents:

$$p_{1d}(t) = p_{1f} \tag{4}$$
$$p_{2d}(t) = p_1(t) + p_{21}, \; p_{21} = p_{2f} - p_{1f} \tag{5}$$
$$p_{3d}(t) = p_1(t) + p_{31}, \; p_{31} = p_{3f} - p_{1f} \tag{6}$$
$$p_{id}(t) = \bar{p}_{jkl}(t, p_i(t)), \forall i \in 4, ..., m \tag{7}$$

where $\bar{p}_{jkl}(t, p_i(t))$ denotes the intersection point of the spheres $S(p_j(t), d_{ij})$, $S(p_k(t), d_{ik})$ and $S(p_l(t), d_{il})$ that is closer to $p_i(t)$. In the notion $S(.,.)$, the first argument indicates the center of sphere and the second argument indicates the radius of sphere. The detailed mathematical formula for $\bar{p}_{jkl}(t, p_i(t))$ and its derivation can be found in [3]. $p_i(t) = [x_i(t), y_i(t), z_i(t)]^T$ and $p_{id}(t)$ denote the actual and the desired position of the centre of mass of the agent A_i at time t, respectively. Besides, denoting the relative position of a vector $\xi(t)$ in local coordinate frame of A_i (having the same axis directions with the global one) as $\xi^{(i)}(t)$, these desired positions can be expressed in local coordinates by

$$p_{id}^{(i)}(t) = p_{id}(t) - p_i(t) \tag{8}$$

3.2 Distributed Control Law Generating Desired Velocities

In this subsection, we model each agent A_i as a point agent, ignoring its geometric properties for the sake of simplicity as described in Section 2. We present our base distributed control design considering the following single-velocity-integrator kinematic model for the agents:

$$\dot{p}_i(t) = V_i(t) \tag{9}$$

where $p_i(t) = [x_i(t), y_i(t), z_i(t)]^T$ and $V_i(t) = [v_{x_i}(t), v_{y_i}(t), v_{z_i}(t)]^T \in \Re^3$ denote the position and velocity of the centre of mass of the agent A_i at time t. Considering Problem 1 and the control architecture of minimally persistent formations in leader-follower structure, the individual controller of each agent needs to be designed separately according to the number of distance constraints for that agent. As a result, we need four different types of individual desired velocity generators, generating the desired values $V_{id}(t)$ of $V_i(t)$ at each time t. For generation of desired velocities, we use essentially the same distributed control scheme as [2], generating the desired velocity vector.

$$V_{id}(t) = \bar{v}_i \sigma_i(t) p_{id}^{(i)}(t) / \|p_{id}^{(i)}(t)\| \tag{10}$$

$$\sigma_i = \begin{cases} 0, & \|p_{id}(t)\| < \varepsilon_k \\ \frac{\|p_{id}(t)\| - \varepsilon_k}{\varepsilon_k}, & \varepsilon_k \leq \|p_{id}(t)\| < 2\varepsilon_k \\ 1, & \|p_{id}(t)\| \geq 2\varepsilon_k \end{cases} \tag{11}$$

for each $i \in \{1, 2, ..., m\}$, where $\varepsilon_k > 0$ is a small design constant; the term $\sigma_i(t)$ is to avoid chattering due to small but acceptable errors in the reaching target position of A_i; $\bar{v}_i$ is a design constant for maximum speed. Note that V_{id} can be applied as is to the kinematic model (9). However, to apply it to the actual dynamic model (1)–(3), further control design consideration is needed for low level motion control. Our approach to this end is presented in the next section.

4 Low Level Motion Control Design for Fixed-Wing UAV Agents

Having established the high level control scheme generating the desired positions for the UAV agents on-line in Section 3, we present the low level individual UAV motion controller design to generate the control signals to track the desired positions generated by the high level formation control scheme in this section. Let us consider the following partitioning of the desired velocity $V_{id}(t)$ in (10)

$$V_{id} = \begin{bmatrix} V_{lat,id} \\ V_{z,id} \end{bmatrix} \in \Re^3 \tag{12}$$

$$V_{lat,id} = \begin{bmatrix} V_{x,id} \\ V_{y,id} \end{bmatrix} \in \Re^2 \tag{13}$$

Consider the Lyapunov function

$$P_{v_i}(t) = \frac{1}{2} e_{v_i}^T(t) e_{v_i}(t) \tag{14}$$

$$e_{v_i}(t) = \dot{p}_i(t) - V_{id}(t) \tag{15}$$

We approach design of the low level motion controllers to generate the command signals $s_{lat,i_{cmd}}, \omega_{i_{cmd}}, z_{i_{cmd}}$ based on forcing P_{v_i} in (14) to decay to zero. From (1)–(3), (15) we have

$$e_{v_i} = \begin{bmatrix} s_{lat,i} \cos\theta_i - V_{x,id} \\ s_{lat,i} \sin\theta_i - V_{y,id} \\ \frac{1}{\tau_z}(-z_i + z_{i_{cmd}}) - V_{z,id} \end{bmatrix} = \begin{bmatrix} s_{lat,i} e_{\theta_i} - V_{lat,id} \\ \frac{1}{\tau_z}(-z_i + z_{i_{cmd}}) - V_{z,id} \end{bmatrix} \tag{16}$$

where for any angle α, e_α is defined as $e_\alpha = [\cos\alpha, \sin\alpha]^T$. Selecting

$$z_{i_{cmd}} = z_i + \tau_z V_{z,id} \tag{17}$$

(14) reduces to

$$P_{v_i}(t) = \frac{1}{2}\|s_{lat,i} e_{\theta_i} - V_{lat,id}\|^2 \tag{18}$$

(1) and (18), together with the fact

$$\dot{e}_{\theta_i} = e_{\theta_i}^\perp \omega_i, \quad e_{\theta_i}^\perp \triangleq [-\sin\theta_i, \cos\theta_i]^T$$

leads to

$$\dot{P}_{vi} = (s_{lat,i} e_{\theta_i} - V_{lat,id})^T (\dot{s}_{lat,i} e_{\theta_i} + s_{lat,i} \dot{e}_{\theta_i} - \dot{V}_{lat,id})$$
$$= (s_{lat,i} e_{\theta_i} - V_{lat,id})^T (-\frac{1}{\tau_s}(s_{lat,i} e_{\theta_i} - V_{lat,id}) - \frac{V_{lat,id}}{\tau_s}$$
$$+ \frac{s_{lat,i_{cmd}} e_{\theta_i}}{\tau_s} + \omega_i s_{lat,i} e_{\theta_i}^\perp - \dot{V}_{lat,id})$$

which can be written in the form

$$\dot{P}_{vi} = -\frac{2}{\tau_s}P_{vi} + (s_{lat,i}e_{\theta_i} - V_{lat,id})^T \left(\begin{bmatrix} \frac{e_{\theta_i}}{\tau_s}, s_{lat,i}e_{\theta_i}^\perp \end{bmatrix} \begin{bmatrix} s_{lat,i_{cmd}} \\ \omega_i \end{bmatrix} \right. \tag{19}$$
$$\left. - \left(\frac{V_{lat,id}}{\tau_s} + \dot{V}_{lat,id} \right) \right)$$

If ω_i was applicable as a control signal, choosing

$$\begin{bmatrix} s_{lat,i_{cmd}} \\ \omega_i \end{bmatrix} = \begin{bmatrix} \tau_s e_{\theta_i} \\ s_{lat,i}^{-1}(e_{\theta_i}^\perp)^T \end{bmatrix} \left(\frac{V_{lat,id}}{\tau_s} + \dot{V}_{lat,id} \right) \tag{20}$$

(19) would be simplified to

$$\dot{P}_{vi} = -\frac{2}{\tau_s}P_{vi}$$

from which, using standard Lyapunov analysis arguments, P_{vi} can be shown to converge to zero. Since ω_i is not a control signal, we replace (20) with

$$\begin{bmatrix} s_{lat,i_{cmd}} \\ \omega_{id} \end{bmatrix} = \begin{bmatrix} \tau_s e_{\theta_i} \\ s_{lat,i}^{-1}(e_{\theta_i}^\perp)^T \end{bmatrix} \left(\frac{V_{lat,id}}{\tau_s} + \dot{V}_{lat,id} \right) \tag{21}$$

and generate the command signal corresponding to ω_i using the compensator

$$\omega_{i_{cmd}} = \frac{k_\omega \tau_\omega (s + 1/\tau_\omega)}{s + k_\omega}[\omega_{id}], \quad k_\omega > 0 \tag{22}$$

so that (2) and (22) lead to

$$\omega_i = \frac{k_\omega}{s + k_\omega}[\omega_{id}] \tag{23}$$

Using standard Lyapunov analysis and backstepping arguments it can be shown that (21), (22) force P_{v_i} and $\omega_i - \omega_{id}$ to converge to zero exponentially for sufficiently large k_ω. Summarizing our Lyapunov analysis based design, our low level motion controller for agent A_i is given by (17), (21), (22).

5 Simulations

In this section, we present the results of a set of simulations for a 5-agent formation satisfying Assumptions A1–A4 using the control laws proposed in Sections 3 and 4. In the simulations, we have used the following parameters: $\varepsilon_k = 0.01$; desired inter-agent distances are assumed to be $d_{ij} = 50\ m$ for $(i,j) \in \{(1,2),(1,3),(2,3),(1,4),(2,4),(3,4),(1,5),(2,5),(3,5)\}$. The initial positions of the agents are $p_{10} = [50, 0, 0]^T$, $p_{20} = [25, 43.3, 0]^T$, $p_{30} = [0, 0, 0]^T$, $p_{40} = [25, 14.435, 40.825]^T$ and $p_{50} = [25, 14.435, -40.825]^T$.

5.1 Simulations for the High-Level Coordination Control Law

To check the effectiveness of the high-level coordination control scheme presented in Section 3, we have first performed simulations with the single integrator kinematics (9). As mentioned in Section 2.1, we need to keep $3 \cdot 5 - 6 = 9$ inter-agent distances constant in order to keep the persistence of a formation with 5 agents in 3 dimensions.

For (9) with using control design in (10),(11), and a constant maximum speed $\bar{v}_i = 10 \; m/sec$, the simulation results are shown in Figures 1 and 2. As it can be seen in these figures, the five agents of the team move in formation to their final positions cohesively, satisfying the desired distance constraints, with less than 5% error after transients. It is clear that the agents do not collide with each other at any time and successfully reach their final desired positions.

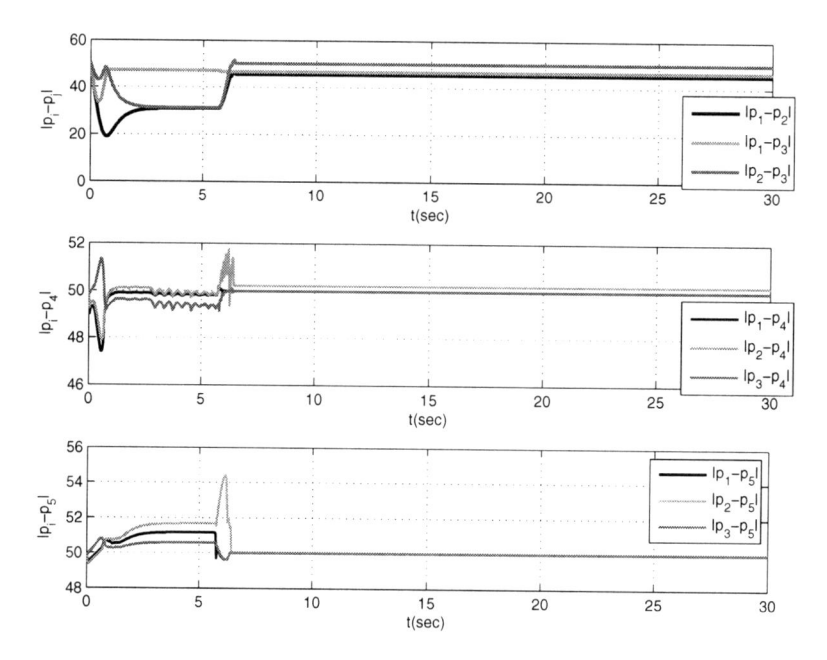

Fig. 1. Cohesive motion of a three-dimensional formation having leader-follower structure

5.2 Fixed-Wing UAV Formations

For the better understanding of the implementation on fixed-wing UAV formations, we present in this subsection the results of a set of simulations for the same 5-agent team under the same motion task with same initial and desired final positions, but considering motion of individual UAV agents with the dynamics (1)–(3) and the control laws (10)–(13),(17),(21),(22). The time constants τ_s, τ_w and τ_z in (1)–(3) are assumed to be 0.001 sec and the initial lateral speed is $s_{lat_0} = 10 \; m/sec$. Under these assumptions, as it can be seen in Fig.3 and 4, the five UAV agents can move in formation to their final positions cohesively. It is clear that the agents do not collide with each other and successfully

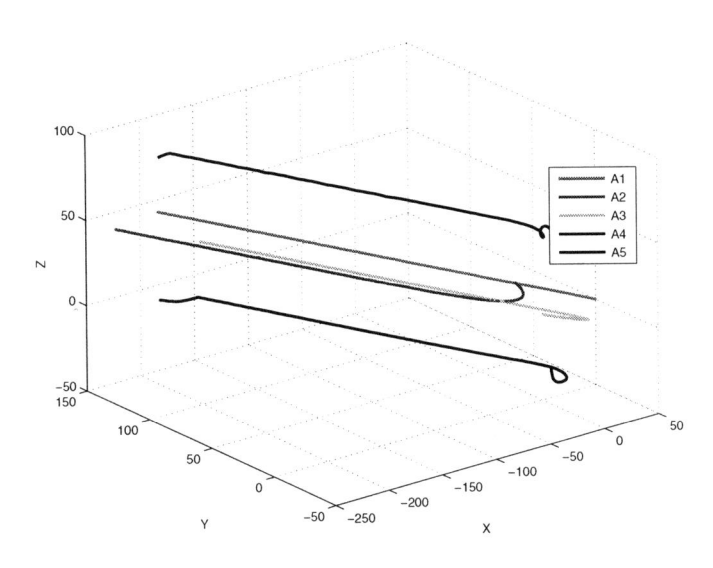

Fig. 2. Pathway of autonomous agents. Initial and final locations are indicated in colors red(dark) and green(light), respectively.

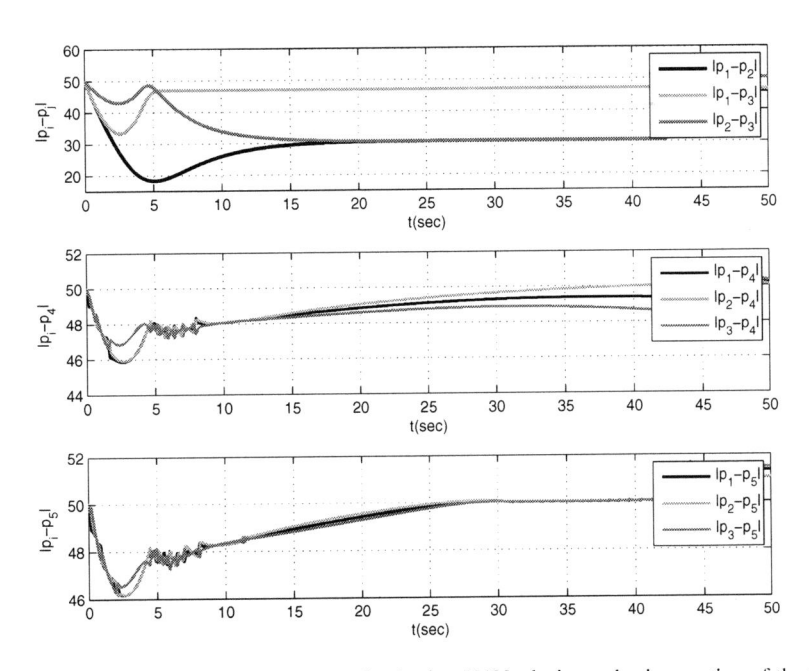

Fig. 3. The inter-agent distances between fixed-wing UAVs during cohesive motion of the three-dimensional formation in Section 4.

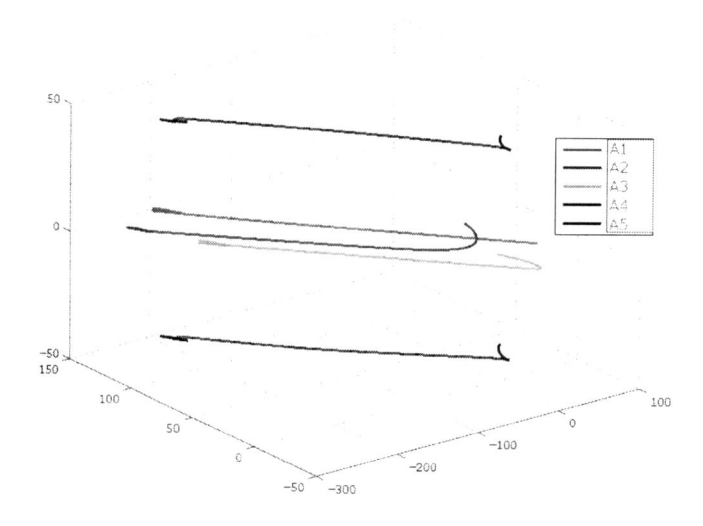

Fig. 4. Pathways of fixed-wing UAV in the simulation example of Section 4

reach their final positions. In this application as seen in Figure 4, the UAVs try to meet the altitude requirement and reach the desired x and y coordinates simultaneously.

6 Conclusion

In this paper, we have proposed a distributed formation control scheme for teams of fixed-wing UAVs preserving rigidity of the formation during maneuvers. The control scheme is composed of a high level distributed algorithm for generating desired trajectories of individual agents and low level motion controllers for individual UAVs considering their dynamics based on a Piccolo-type-UAV model. The high level distributed trajectory generation is based on geometric and graph theoretical analysis of the team formation, and the low level control design is based on Lyapunov analysis. The performance of the designed control scheme is verified via numerical simulations. Future research directions include consideration of dynamic uncertainties, external disturbances, and measurement noise effects.

References

1. Bai, H., Arcak, M., Wen, J.: Rigid body attitude coordination without inertial frame information. Automatica 44(12), 3170–3175 (2008)
2. Bayezit, I., Amini, M., Fidan, B., Shames, I.: Cohesive Motion Control of Autonomous Formations in Three Dimensions. In: Proc. Sixth International Conference on Intelligent Sensors, Sensor Networks and Information Processing, Brisbane, Australia (December 2010)
3. Bayezit, I., Fidan, B., Amini, M., Shames, I.: Distributed Cohesive Motion Control of Quadrotor Vehicle Formations. In: Proc. ASME 2010 International Mechanical Engineering Congress and Exposition, Vancouver, Canada (November 2010)

4. Bayraktar, S., Fainekos, G., Pappas, G.: Experimental cooperative control of fixed-wing unmanned aerial vehicles. In: 43rd IEEE Conference on Decision and Control, CDC 2004, vol. 4, pp. 4292–4298. IEEE, Los Alamitos (2005)
5. Drake, S., Brown, K., Fazackerley, J., Finn, A.: Autonomous control of multiple UAVs for the passive location of radars. In: Proc. 2nd International Conference on Intelligent Sensors, Sensor Networks and Information Processing, pp. 403–409 (2005)
6. Fidan, B., Anderson, B., Yu, C., Hendrickx, J.: Persistent autonomous formations and cohesive motion control. In: Ioannou, P.A., Pitsillides, A. (eds.) Modeling and Control of Complex Systems, pp. 247–275. CRC, Boca Raton (2007)
7. Gazi, V., Fidan, B.: Coordination and control of multi-agent dynamic systems: Models and approaches. In: Şahin, E., Spears, W., Winfield, A. (eds.) Swarm Robotics, pp. 71–102. Springer, Heidelberg (2007)
8. Olfati-Saber, R., Murray, R.: Distributed cooperative control of multiple vehicle formations using structural potential functions. In: Proc. of IFAC World Congress, pp. 346–352 (2002)
9. Paul, T., Krogstad, T., Gravdahl, J.: Modelling of UAV formation flight using 3D potential field. Simulation Modelling Practice and Theory 16(9), 1453–1462 (2008)
10. Sandeep, S., Fidan, B., Yu, C.: Decentralized cohesive motion control of multi-agent formations. In: Proc. 14th Mediterranean Conference on Control and Automation (June 2006)
11. Smith, R.S., Hadaegh, F.Y.: Distributed estimation, communication and control for deep space formations. IET Control Theory & Applications 1(2), 445–451 (2007)
12. Tanner, H.G., Pappas, G.J., Kumar, V.: Leader-to-formation stability. IEEE Transactions on Robotics and Automation 20(3), 443–455 (2004)
13. Yu, C., Hendrickx, J.M., Fidan, B., Anderson, B.D.O., Blondel, V.D.: Three and higher dimensional autonomous formations: Rigidity, persistence and structural persistence. Automatica 43(3), 387–402 (2007)
14. Zhai, S., Fidan, B., Oztürk, S., Gazi, V.: Single view depth estimation based formation control of robotic swarms: Obstacle avoidance, simulation, and practical issues. In: Proc. 16th Mediterranean Conference on Control and Automation, June 2008, pp. 1162–1167 (2008)

Intelligent Control System Design for a Class of Nonlinear Mechanical Systems

S. Islam[1,2,*], P.X. Liu[3], and A. El Saddik[1]

[1] School of Information Technology and Engineering (SITE),
University of Ottawa, 800 King Edward Road, Ottawa, Canada, K1N 6N5
[2] School of Engineering and Applied Sciences,
Harvard University, 29 Oxford Street, Cambridge, MA 02138 USA
[3] Department of Systems and Computer Engineering,
Carleton University, Ottawa, Ontario, Canada, K1S 5B6
sislam@sce.carleton.ca, sislam@seas.harvard.edu

Abstract. In this work, intelligent control technique using multiple parameter models is proposed for robust trajectory tracking control for a class of nonlinear systems. The idea is to reduce the controller gains so as to reduce the control efforts from the single model (SM) *certainty equivalence* (CE) principle based classical adaptive control approach. The method allows classical adaptive control to be switched into a candidate among the finite set of candidate controllers that best approximates the plant at each instant of time. The Lyapunov function inequality is used to identify a candidate that closely approximates the plant at each instant of time. The design can be employed to achieve good transient tracking performance with smaller values of controller gains in the presence of large scale parametric uncertainties. The proposed design is implemented and evaluated on 3-DOF Phantom $Premimum^{TM}$ 1.5 haptic robot device to demonstrate the effectiveness of the theoretical development.

Keywords: Adaptive Control; *Certainty Equivalence Principle* (CE); Switching Logic; Robotics; $Phantom^{TM}$ Haptic Device.

1 Introduction

The single model CE-based adaptive control technique has been attracted by control community for the last two decades [5, 6, 9, and 16]. Most adaptive control paradigms for nonlinear systems, however, only ensure asymptotic convergence property while transient tracking behavior was not well considered in stability analysis. If the unknown parameters and initial conditions belong to large compact set, then existing CE-based adaptive design provides poor transient tracking performance. This is due to the fact that the control system design with the CE principle assumed that uncertain parameters are assumed to be appeared linearly with respect to unknown nonlinear system dynamics. For example, in adaptive design for robotic systems, the control system stability

---

* Corresponding Author.

fully relies on the strict dynamical properties of the system (e.g., [1, 2, 4, 7, 8, 12, 15, 17, and many others]), where unknown parameters are required to be appeared linearly with respect to nonlinear robot dynamics such as, the inertial matrix $M(q) \in R^{n \times n}$ is symmetric, bounded, and positive definite and satisfies the following inequalities: $\|M(q)\| \leq M_M$ and $\|M^{-1}(q)\| \leq M_{MI}$ where M_M and M_{MI} are known bounded positive constant, the matrix $\dot{M}(q) - 2C(q,\dot{q})$ is skew symmetric and the norm of the gravity and centripetral-coriolis forces are upper bounded as $\|C(q,\dot{q})\| \leq C_M\|\dot{q}\|$ and $\|C(q,\dot{q}_d)\| \leq C_M\|\dot{q}_d\| \leq \alpha_c$, where C_M and α_c are known bounded positive constant. To improve transient tracking response in classical adaptive control framework, one may employ high controller gains to speed up the convergence rate of the state and parameter estimates. Particularly, state feedback based classical adaptive design demands high learning gains to obtain fast convergence of the parameter estimator. As a result, the existing CE-based adaptive design might not be practically applicable for many applications as, in practice, the control gains/efforts in most nonlinear control systems technology are limited. This is because high controller gains may excite unmodeled high-frequency dynamics as well as amplifies measurement noise associated with the joint position signals. To reduce controller gains from classical adaptive control (CAC) paradigm, we have introduced intelligent control strategy using multiple parameter models. The method extends CAC by allowing single model based classical controller to be switched instantaneously to a controller among finite set of candidate that improves overall tracking performance. The method first applies CAC over a pre-fixed small period of time. During this time, the logic checks whether or not the inequality for the Lyapunov function is satisfied. If the classical controller satisfies the inequality, then we apply it. If the CAC does not satisfy the inequality, then the logic switches to the controller among finite set of candidate controllers that satisfies the switching criterion. To demonstrate the theoretical development for real-life applications, experimental results on 3-DOF $Phantom^{TM}$ haptic device are presented.

The rest of the paper is organized as follows: In section 2, we derive CE-based CAC algorithm for a class of nonlinear mechanical systems. In section 3, intelligent control approach using multiple model parameters is then introduced to improve overall tracking response from the CAC design. Section 4 presents real-time implementation results on 3-DOF $Phantom^{TM}$ haptic device to illustrate the theoretical development of this paper. Finally, section 5 concludes the paper.

2 Classical Adaptive Control

In this section, we first derive the SM-based CAC approach for a class of multi-input multi-output nonlinear mechanical systems represented by the following compact differential equation

$$\dot{q} = \phi_f(q) + \phi_g(q)\tau \tag{1}$$

where $\tau = [\tau_1, \tau_2, \ldots, \tau_n]$, $q = \left[q_1, \dot{q}_1, \ldots, q_1^{(r_1-1)}, \ldots, q_n, \dot{q}_n, \ldots, q_n^{r_n-1}\right]^T$, $\phi_f(q) = [\phi_1(q) \ldots \phi_n(q)]^T$ and $\phi_g(q) = [\phi_{11}(q) \ldots \phi_{1n}(q); \ldots; \phi_{n1}(q) \ldots \phi_{nn}(q)]^T$. The functions $\phi_f(q)$ and $\phi_g(q)$ are smooth nonlinear that are unknown but belongs to known compact set. The system defined above can be represented by the equation of motion for n-link robot manipulators [1, 7, 11, 12, 17] as

$$M(q)\ddot{q} + C(q,\dot{q})\dot{q} + G(q) = \tau \qquad (2)$$

where $q \in \Re^n$ is the joint position vector, $\dot{q} \in \Re^n$ is the joint velocity vector, $\ddot{q} \in \Re^n$ is the joint acceleration vector, $\tau \in \Re^n$ is the input torque vector, $M(q) \in \Re^{n \times n}$ is the symmetric and uniformly positive definite inertia matrix, $C(q,\dot{q})\dot{q} \in \Re^n$ is the coriolis and centrifugal loading vector, and $G(q) \in \Re^n$ is the gravitational loading vector. Equation (2) can be written in compact form as given in (1)

$$\dot{q} = \phi_f(q,\dot{q}) + \phi_g(q)\tau \qquad (3)$$

where $\phi_f(q) = -M(q)^{-1}[C(q,\dot{q})\dot{q} + G(q)]$ and $\phi_g(q) = M(q)^{-1}$. In view of the dynamical properties of this system, we consider that $\phi_f(q,\dot{q})$ and $\phi_g(q)$ are smooth continuous nonlinear functions as $M(q) \in \Re^{n \times n}$, $C(q,\dot{q})\dot{q} \in \Re^n$ and $G(q) \in \Re^n$ are continuous over the given compact sets. The control objective is that the joint position $q(t)$ asymptotically tracks the desired joint position $q_d(t)$. We assume that the desired trajectory $q_d(t)$, its second and third derivatives are bounded such that $Q_d = [q_d \; \dot{q}_d \; \ddot{q}_d]^T \in \Omega_d \subset \Re^{3n}$ with compact set Ω_d. To begin with that, let us first represent the multi-input multi-output nonlinear dynamics in the following error state space form

$$\dot{e}_1 = e_2, \dot{e}_2 = \phi_1(e) + \phi_2(e_1)\tau - \ddot{q}_d \qquad (4)$$

where, $e = [e_1, e_2]^T$, $e_1 = (q - q_d)$, $e_2 = (\dot{q} - \dot{q}_d)$, $\dot{e}_2 = (\ddot{q} - \ddot{q}_d)$, $\phi_1(e) = -M(e_1 + q_d)^{-1}[C(e_1+q_d, e_2+\dot{q}_d)(e_2+\dot{q}_d)+G(e_1+q_d)]$ and $\phi_2(e_1) = M(e_1+q_d)^{-1}$. To meet the control objective, let us consider the following adaptive control law as state feedback (position-velocity) [18]

$$\tau(e, Q_d, \hat{\theta}) = Y(e, \dot{q}_r, \ddot{q}_d, \dddot{q}_d)\hat{\theta} - K_P e_1 - K_D e_2 \qquad (5)$$

$$\dot{\hat{\theta}} = -\Gamma Y^T(e, \dot{q}_r, \ddot{q}_d, \dddot{q}_d)S \qquad (6)$$

where $Y(e, \dot{q}_r, \ddot{q}_d, \dddot{q}_d)\hat{\theta} = \hat{M}(q)\ddot{q}_d + \hat{C}(q,\dot{q}_r)\dot{q}_d + \hat{G}(q)$, $\hat{\theta}$ is an estimate of uncertain parameters representing the inertial parameters of robot arms and its load, $K_P \in \Re^{n \times n}$, $K_D \in \Re^{n \times n}$, $S = (e_2 + \lambda e_1)$, $\dot{q}_r = (\dot{q}_2 - \lambda e_1)$, $\lambda = \frac{\lambda_0}{1+\|e_1\|}$, $\lambda_0 > 0$ and $\hat{M}$, $\hat{C}(.)$ and $\hat{G}(.)$ define the estimates of the $M(.)$, $C(.)$ and $G(.)$. Adaptation mechanism (6) may cause discontinuous property in the parameter estimates even after learning estimate converge to the actual parameter. To remove the discontinuous property from learning estimates, $\hat{\theta}$ can be adjusted with the smooth parameter projection scheme as $\dot{\hat{\theta}}_i = [Proj(\hat{\theta}, \Phi)]_i$ for

$\theta \in \Omega = \{\theta \mid a_i \le \theta_i \le b_i\}, 1 \le i \le p\}$ where, Φ_i is the i-th element of the column vector $-Y^T(e, \dot{q}_r, \dot{q}_d, \ddot{q}_d)S$ and $\delta > 0$ is chosen such that $\Omega \subset \Omega_\delta$ with $\Omega_\delta = \{\theta \mid a_i - \delta \le \theta_i \le b_i + \delta\}, 1 \le i \le p\}$. It is important to notice from the adaptive controller (5), (6) that the design is assumed uncertain parameter θ to be linearly parameterized with respect to nonlinear system dynamics $Y(e, \dot{q}_r, \dot{q}_d, \ddot{q}_d)$. To explore the convergence condition, we can consider the following Lyapunov-function candidate

$$V(e, \tilde{\theta}) = \frac{1}{2}S^T M S + \frac{1}{2}e^T K_P e + \frac{1}{2}\tilde{\theta}\Gamma^{-1}\tilde{\theta} \qquad (7)$$

with $\tilde{\theta} = (\hat{\theta} - \theta)$. We first take time derivative, $\dot{V}(e, \tilde{\theta})$, along with the closed loop trajectories, formulated by using (4), (5), (6) along with parameter projection mechanism. Then, after some manipulations, we can simplify $\dot{V}(e, \tilde{\theta})$ as

$$\dot{V}(e, \tilde{\theta}) \le -\lambda_{min.}(\Pi)\|e\|^2 \le 0 \qquad (8)$$

$\forall \hat{\theta}(0) \in \Omega, \forall \theta(0) \in \Omega, \forall e(0) \in \Omega_{co}, \forall \hat{\theta} \in \Omega_\delta, \Omega_c = \{e \mid e^T Q_{sm} e \le c\}, c > 0$ and $Q_{sm} = \begin{bmatrix} 0.5M & 0.5M\lambda \\ 0.5M\lambda & 0.5(\lambda^2 + K_P) \end{bmatrix}$, $\Pi = \Theta^T \Delta \Theta$ with Δ and Θ defined as $\Delta = \begin{bmatrix} K_1 I & 0 \\ 0 & K_2 I \end{bmatrix}$ and $\Theta = \begin{bmatrix} \frac{\lambda I}{2} & I \\ \frac{\lambda I}{2} & 0 \end{bmatrix}$ with $K_1 = [\lambda_{min.}(K_D) - 3\lambda_0 M_M - 2\lambda_0 C_M]$ and $K_2 = \left[\frac{4\lambda_{min.} K_P}{\lambda_0} - \lambda_{max.}(K_D) - 2\lambda_0 M_M - 2\lambda_0 C_M\right]$. The design derived above is assumed that system model dynamics is free from disturbances. Let us now remove this assumption by adding an unmodeled dynamics and unknown external disturbances term τ_d with the model (2) as

$$M(q)\ddot{q} + C(q, \dot{q})\dot{q} + G(q) = \tau + \tau_d \qquad (9)$$

We then introduce the single model based robust adaptive control law as

$$\tau(e, Q_d, \hat{\theta}) = \Upsilon(e, \dot{q}_r, \dot{q}_d, \ddot{q}_d)\hat{\theta} - K_P e_1 - K_D e_2 \qquad (10)$$

$$\dot{\hat{\theta}} = -\Gamma \Upsilon^T(e, \dot{q}_r, \dot{q}_d, \ddot{q}_d)S \qquad (11)$$

where $\Upsilon(e, \dot{q}_r, \dot{q}_d, \ddot{q}_d) = [\zeta(e, \dot{q}_r, \dot{q}_d, \ddot{q}_d)\; sgn(S)]$ $\hat{\theta}(t) = \begin{bmatrix} \hat{\theta}_l^T(t) & \hat{\theta}_{gd} \end{bmatrix}^T$ is an estimate of parameters $\theta(t) = \begin{bmatrix} \theta_l^T(t) & \theta_{gd} \end{bmatrix}^T$, $\theta_l(t)$ represent the parameters associated with inertial matrix $M(q_1)$ and centrifugal-coriolis matrix $C(q_1, q_2)$, $\|G(q) - d\| \le \theta_{gd}$ and $\zeta(e, \dot{q}_r, \dot{q}_d, \ddot{q}_d)\hat{\theta}_l$ can be defined as $\zeta(e, \dot{q}_r, \dot{q}_d, \ddot{q}_d)\hat{\theta}_l = \hat{M}(q)\ddot{q}_d + \hat{C}(q, \dot{q}_r)\dot{q}_d$. The tracking error bounds with the robust adaptive control law (10), (11) can be shown along the line of the stability analysis derived for control law (5), (6). To begin with that, let us derive the closed-loop model as $M(q)\dot{e}_2 = -C(q_1, q_2)e_2 - \lambda C(q_1, e_2)\dot{q}_d + \zeta(e, \dot{q}_r, \dot{q}_d, \ddot{q}_d)\tilde{\theta}_l + \tilde{\beta} - K_P e_1 - K_D e_2$ where $\tilde{\beta} = (\hat{\beta} - \beta)$ and $\zeta(e, \dot{q}_r, \dot{q}_d, \ddot{q}_d)\tilde{\theta}_l$ is defined as

$$\left[\hat{M}(q)\ddot{q}_d - M(q)\ddot{q}_d + \hat{C}(q_1, \dot{q}_r)\dot{q}_d - C(q_1, q_2 - \lambda e_1)\dot{q}_d\right]$$

Then, after some manipulations, we can show that $\dot{V}(e, \tilde{\theta}) \leq 0$. The switching function $sgn(.)$ in robust adaptive control design requires to estimate by using saturation function $sat(.)$ in order to reduce chattering phenomenon.

3 Intelligent Control System Using Multiple Model Parameters: A Hybrid Control Approach

The main practical problem with SM-based CAC technique is its poor transient tracking performance. If the initial conditions and parameter errors become large value, then the transient tracking errors as well as the joint oscillation will also become unacceptably large values. This is mainly because of the assumption that the nonlinear system dynamics represented by regressor model $Y(e, \dot{q}_r, \dot{q}_d, \ddot{q}_d)$ are assumed to be linear with respect to uncertain parameters θ. The common technique is to use high values of controller gains to achieve desired transient tracking performance. The problem, however, is that the controller gain requires to increase with the increase of the parametric uncertainty resulting very large control efforts. When the level of uncertainty is large, the gain Γ require to be very high to ensure good transient tracking performance. However, the use of large value of control gains is not a practical solution as, in practice, the designer cannot increase the control gains as well as saturation level as much required to achieve desired control objectives due to the presence of control chattering activity.

To deal with the problem associated with high control gains, we propose to use intelligent control strategy in multiple adaptive control framework. The method allows the designer to keep smaller values of Γ. Our interest here in this paper is to reduce the level of uncertainty via switching the CAC law into a candidate controller which best estimates the plant among a finite set of candidate. To identify best possible controller from a finite set, we consider on-line estimation of the Lyapunov-function candidate (5). We first equally divide the set $\theta \in \Omega$ into a finite number of smaller compact subsets such that $\theta_i \in \Omega_i$ with $\Omega = \bigcup_{i=1}^{N} \Omega_i$, $\theta \in \Omega_i$ and $i \in \mathcal{G} = \{1, 2, 3, 4, ..., N\}$. For a given compact set of the initial condition of interest $e(0) \in \Omega_{co}$, we then construct finite set of controller, bounded via saturating outside the region Ω_c, correspond to each of these smaller parameter subsets as

$$\tau^i(e, Q_d, \theta_i) = Y(e, \dot{q}_r, \dot{q}_d, \ddot{q}_d)\theta_i - K_P e_1 - K_D e_2$$

for all $(\theta, \theta_i) \in \Omega_i$, such that for every $\theta \in \Omega_i$ all the signals in the closed loop system under intelligent adaptive controller defined above started inside the set Ω_{co} are bounded and the output tracking error trajectories converge to zero as the time goes to infinity. The control gains K_P and K_D are common to all the N candidate controllers. The regressors model $Y(e, \dot{q}_r, \dot{q}_d, \ddot{q}_d)$ is also common to all controllers. Our aim is now to develop a control switching logic to identify $\tau(e, Q_d, \theta)$ from $\tau^i(e, Q_d, \theta_i)$. To identify which candidate will be used to generate the final control vector, a logic can be used in such a way ensuring that all the signals in the closed-loop systems are bounded, and the error trajectories converge to zero as time goes to zero. One may use pre-ordered control logic [14] to identify an appropriate candidate controller from a finite set of candidate. The difficulty associated with the pre-ordered algorithm is that if the number of controllers become large then the long switching search may produce unacceptable transient tracking errors and high-frequency control oscillation. This is because, in the presence of large number of controllers, the logic has to scan through a large number of controllers before converging to the one that satisfies the Lyapunov inequality. In this work, we allow the controller to be switched instantaneously among the finite set of candidates which improves overall tracking performance. To begin with this development, we consider Lyapunov-function candidate corresponding to each of these candidates $\tau^i(e, Q_d, \theta_i)$ as $\alpha_2^i \|e\|^2 \leq V_i(e, \tilde{\theta}_i) \leq \alpha_3^i \|e\|^2 \ \forall e \in \Omega_c^i = \{(e, \tilde{\theta}_i) \mid V_i(e, \tilde{\theta}_i) \leq c\}$ and $\forall (\theta, \theta_i) \in \Omega_i$, where $c > 0$, $\tilde{\theta}_i = (\theta_i - \theta)$ and α_2^i and α_3^i are known positive constants. Then, we proceed with the following Algorithm 1 to identify a candidate from the set of candidate which closely represents the plant at each instant of time such that the error trajectories asymptotically converge to zero.

Algorithm 1: Suppose that the finite set of controllers $i \in \mathcal{G} = \{1, 2, 3, ..., N\}$ as well as Lyapunov-function candidates, $V_i(e, \tilde{\theta}_i)$, are available at any time. Then we follow the following steps to obtain an appropriate candidate control law.

[A.] Define the initial time $t_o = 0$, the switching index $i \in \mathcal{G} = \{1, 2, 3, ..., N\}$ and a small positive time constant $t_d > 0$.

[B.] Put the classical adaptive controller (5), $\tau(e, Q_d, \hat{\theta})$, with standard adaptation mechanism (6) for a short period of time $t \in [t_o, t_o + t_d]$.

[C.] For $t \geq t_o + t_d$, we monitor the inequality for the Lyapunov-function candidates to see which control generates guaranteed decrease in the value of $V_i(t_s) - V_0(t) \leq 0$ with $V_0(t) = \frac{1}{2} e^T(t) Q_{sm} e(t) + \frac{1}{2} \tilde{\theta}^T(t) \Gamma^{-1} \tilde{\theta}(t)$ and $V_i(t_s) = \frac{1}{2} e^T(t_s) Q_{sm} e(t_s) + \frac{1}{2} \tilde{\theta}_i^T(t_s) \Gamma^{-1} \tilde{\theta}_i(t_s)$, where $\tilde{\theta}(t) = \left(\hat{\theta}(t) - \theta \right)$, $\tilde{\theta}_i(t_s) = (\theta_i(t_s) - \theta)$ and $t_s > t_o + t_d$ is the switching time. The logic stay with the CAC law (5), (6) in the loop until the moment of time $t_i \geq t_o + t_d$ when the switching condition is violated. If the classical controller does not satisfy the Lyapunov inequality at $t = t_i$, then it switches acting controller to the one that generates largest guaranteed decrease in the value of $V_i(t_s) - V_0(t) \leq 0$.

[D.] If the inequality $V_i(t_s) - V_0(t) \leq 0$ never violated, then there will not be any switching. This implies that the plant output tracks the desired trajectory $q(t) \to q_d(t)$ as the time goes to infinity.

[E.] If at some time, say t_i with $t_i \geq t_o + t_d$, the controller that acting in the loop does not satisfy $V_i(t_s) - V_0(t) \leq 0$, then another controller will be put in the system as there always exists a guaranteed minimum value of $V_i(t_s) - V_0(t) \leq 0$ at that instant of time.

Based on our above analysis, we now summarize the results in the following Theorem 1.

Theorem 1 : Consider the closed-loop system formulated by (1) with intelligent controller $\tau^i(e, Q_d, \theta_i)$ under the switching scheme introduced in Algorithm 1.

Then, there exists a time such that, according to Algorithm 1, the controller corresponding to the guaranteed decrease in the value of $V_i(t_s) - V_0(t) \leq 0$ is tuned to the plant which ensures that the tracking error signals in the closed-loop model are bounded and $e(t) \to 0$ when $t \to \infty$.

The logic designed in Algorithm 1 is based on disturbance free system dynamics. As a result, undesirable switching may arise from uncertainty causing disturbance, such as, unmodeled actuator dynamics, input and output disturbance. To remove the undesirable switching, one can modify controller $\tau^i(e, Q_d, \theta_i) = Y(e, \dot{q}_r, \ddot{q}_d, \dddot{q}_d)\theta_i - K_P e_1 - K_D e_2$ by $\tau^i(e, Q_d, \theta_i) = \Upsilon(e, \dot{q}_r, \ddot{q}_d, \dddot{q}_d)\theta_i - K_P e_1 - K_D e_2$ and then follow the same steps as introduced in Algorithm 1. Then convergence analysis under modified controller design can be derived straight forward manner along the line of the convergence analysis shown in Algorithm 1.

4 Experimental Results

In this section, we show the design and implementation process of the proposed design on real robotic systems. For this purpose, we use two 3-DOF $Phantom^{TM}$ Premimum 1.5A haptic manipulators as shown in Fig. 1 provided by SensAble Technologies, Inc.. The system is equipped with standard gimbal end-effector [3]. Each joint is attached with encoder for joint position measurement. The velocity signals are obtained by differentiating the joint position measurement as the system does not have velocity sensors. The equation of motion for $Phantom^{TM}$ robot can be defined as

$$\begin{bmatrix} m_{11} & m_{12} & m_{13} \\ m_{21} & m_{22} & m_{23} \\ m_{31} & m_{32} & m_{33} \end{bmatrix} \begin{bmatrix} \ddot{q}_1 \\ \ddot{q}_2 \\ \ddot{q}_3 \end{bmatrix} + \begin{bmatrix} c_{11} & c_{12} & c_{13} \\ c_{21} & c_{22} & c_{23} \\ c_{31} & c_{32} & c_{33} \end{bmatrix} \begin{bmatrix} \dot{q}_1 \\ \dot{q}_2 \\ \dot{q}_3 \end{bmatrix} + \begin{bmatrix} G_1 \\ G_2 \\ G_3 \end{bmatrix} = \begin{bmatrix} \tau_1 \\ \tau_2 \\ \tau_3 \end{bmatrix} \quad (12)$$

where m_{jk} are the elements of the inertia matrix, c_{jk} are the elements of the coriolis and centrifugal force matrix, G_j are the gravity vector and τ_j are joint torques with $j = 1, 2, 3$ and $k = 1, 2, 3$. We neglect the viscous and coulomb friction from the motion dynamics to make the nominal dynamics same as defined in (2). The components of motion dynamics given in (12) are modeled as $m_{11} = \theta_1 + \theta_2 \cos^2 q_2 + (\theta_3 + \theta_5) \sin^2 q_3 + 2\theta_6 \cos q_2 \sin q_3$, $m_{12} = 0$, $m_{13} = 0$, $m_{21} = 0$, $m_{22} = \theta_4 + \theta_5 - 2\sin(q_2 - q_3)$, $m_{31} = 0$, $m_{23} = M_{32} = \theta_5 - \theta_6 - 2\sin(q_2-q_3)$, $m_{33} = \theta_5$, $c_{11} = -(\theta_2 \sin q_2 \cos q_2 + \theta_6 \sin q_2 \sin q_3)\dot{q}_2 + ((\theta_3 + \theta_5)\sin q_3 \cos q_3 + \theta_6 \cos q_2 \cos q_3)\dot{q}_3$, $c_{12} = -(\theta_2 \sin q_2 \cos q_2 + \theta_6 \sin q_2 \sin q_3)\dot{q}_1$, $c_{13} = ((\theta_3 + \theta_5)\sin q_3 \cos q_3 + \theta_6 \cos q_2 \cos q_3)\dot{q}_1$, $c_{21} = (\theta_2 \sin q_2 \cos q_2 + \theta_6 \sin q_2 \sin q_3)\dot{q}_1$, $c_{22} = \theta_6 \cos(q_2 - q_3)(\dot{q}_3 - \dot{q}_2)$, $c_{23} = \theta_6 \cos(q_2 - q_3)(\dot{q}_2 - \dot{q}_3)$, $c_{31} = -(\theta_3 + \theta_5)\sin q_3 \cos q_3 \dot{q}_1 - \theta_6 \cos q_2 \cos q_3 \dot{q}_1$, $c_{32} = 0$, $c_{33} = 0$, $G_1 = 0$, $G_2 = \theta_7 \cos q_2$ and $G_3 = \theta_8 \sin q_3$. The model parameter sets are chosen as $\theta_1 = (I_{y1} + I_{y3} + I_{z2})$, $\theta_2 = (I_{y2} - I_{z2} + m_2 r_2^2 + m_3 l_1^2)$, $\theta_3 = (I_{z3} - I_{x3} - I_{y3})$, $\theta_4 = (I_{x2} + m_2 r_2^2 + m_3 l_1^2)$, $\theta_5 = (m_3 r_3^2 + I_{x3})$, $\theta_6 = m_3 r_3 l_1$, $\theta_7 = g(m_2 r_2 + m_3 l_1)$ and $\theta_8 = gm_3 r_3$ where m_j with $j = 1, 2, 3$ are the mass of the j-th link, I_{xj}, I_{yj}

Fig. 1. Experimental set-up for 3-DOF $Phantom^{Tm}$ haptic robot for real-time experimentation of CAC and intelligent control algorithm

and I_{zj} are the moment of inertia of each link and r_j are the center of masses for each link. We then design nonlinear regressor model as follows

$$Y(q,\dot{q},\ddot{q}) = \begin{bmatrix} Y_{11} & Y_{12} & Y_{13} & Y_{14} & Y_{15} & Y_{16} & Y_{17} & Y_{18} \\ Y_{21} & Y_{22} & Y_{23} & Y_{24} & Y_{25} & Y_{26} & Y_{27} & Y_{28} \\ Y_{31} & Y_{32} & Y_{33} & Y_{34} & Y_{35} & Y_{36} & Y_{37} & Y_{38} \end{bmatrix}$$

with $Y_{11} = \ddot{q}_1$, $Y_{21} = 0$, $Y_{31} = 0$, $Y_{12} = (\cos^2 q_2 \ddot{q}_1 - 2\cos q_2 \sin q_2 \dot{q}_1 \dot{q}_2)$, $Y_{22} = \cos q_2 \sin q_2 \ddot{q}_1^2$, $Y_{32} = 0$, $Y_{13} = (\sin^2 q_3 \ddot{q}_1 + 2\sin q_3 \cos q_3 \dot{q}_1 \dot{q}_3)$, $Y_{23} = 0$, $Y_{33} = -\sin q_3 \cos q_3 \dot{q}_1^2$, $Y_{14} = 0$, $Y_{24} = \ddot{q}_2$, $Y_{34} = 0$, $Y_{15} = (\sin^2 q_3 \ddot{q}_1 + 2\sin q_3 \cos q_3 \dot{q}_1 \dot{q}_3)$, $Y_{25} = (\ddot{q}_2 + \ddot{q}_3)$, $Y_{35} = (\ddot{q}_2 + \ddot{q}_3 - \sin q_3 \cos q_3 \dot{q}_1^2)$, $Y_{17} = 0$, $Y_{27} = \cos q_2$, $Y_{37} = 0$, $Y_{18} = 0$, $Y_{16} = (2\cos q_2 \sin q_3 \ddot{q}_1 - 2\sin q_2 \sin q_3 \dot{q}_1 \dot{q}_2 + 2\cos q_2 \cos q_3 \dot{q}_1 \dot{q}_3)$, $Y_{26} = -2\sin(q_2 - q_3)\ddot{q}_2 - \sin(q_2 - q_3)\ddot{q}_3 + \sin q_2 \sin q_3 \dot{q}_1^2 - \cos(q_2 - q_3)(\dot{q}_2 - \dot{q}_3)^2$, $Y_{36} = -2\sin(q_2 - q_3)\ddot{q}_2 - \cos q_2 \cos q_3 \dot{q}_1^2$, $Y_{28} = 0$ and $Y_{38} = \sin q_3$. The model parameters are identified via using least square method as; $\theta_1 = 50$, $\theta_2 = 60$, $\theta_3 = 48$, $\theta_4 = 80$, $\theta_5 = 9$, $\theta_6 = 8$, $\theta_7 = 235$ and $\theta_8 = 270$. The joint position is measured in radians and the control input torque is given in numerical values. The reference trajectory was defined as a square wave with a period of 8 seconds and an amplitude of ± 0.5 radians was pre filtered with a critically damped 2nd-order linear filter using a bandwidth of $\omega_n = 2.0$ rad./sec. Our main target was to employ a desired trajectory that usually uses in industrial robotic systems.

Let us first examine the tracking performance of CE-based CAC law (5), (6) on the given $Phantom^{TM}$ device. We first replace $\dot{q}$ in $c(q,\dot{q})$ by $\dot{q}_r = (\dot{q}_2 - \lambda e_1)$ to construct the nonlinear regressor model $Y(e, \dot{q}_r, \dot{q}_d, \ddot{q}_d)$. The experimental set up that used for the real-time application is depicted in Fig 1. The values of PD controller design parameters are selected in such away that provides acceptable transient tracking performance as $\lambda_0 = 2$, $K_P = diag(750, 1200, 1200)$ and $K_D = diag(500, 800, 800)$. The gains Γ are chosen to achieve faster parameter learning.

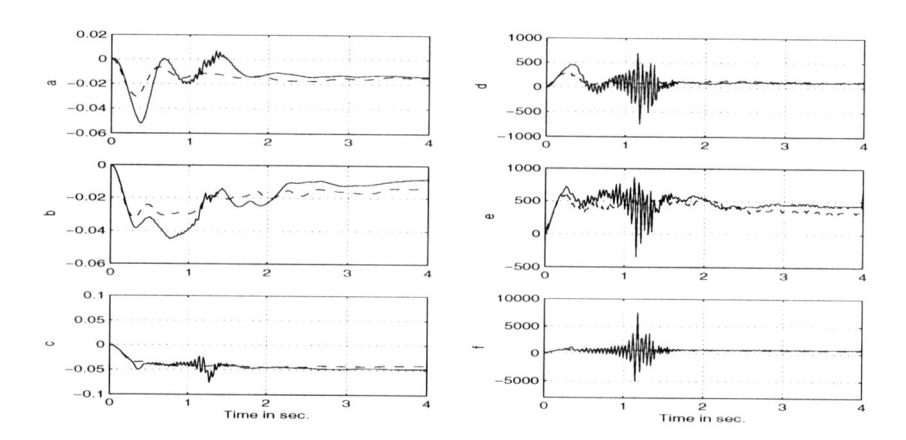

Fig. 2. Left Column: a, b & c) The tracking error (dash-line is for Theorem 1 and solid-line is for CAC algorithm (5), (6)) for joints 1, 2 and 3, Right Column: d, e & f) The control inputs (dash-line is for Theorem 1 and solid-line is for algorithm (5), (6)) for joints 1, 2 and 3

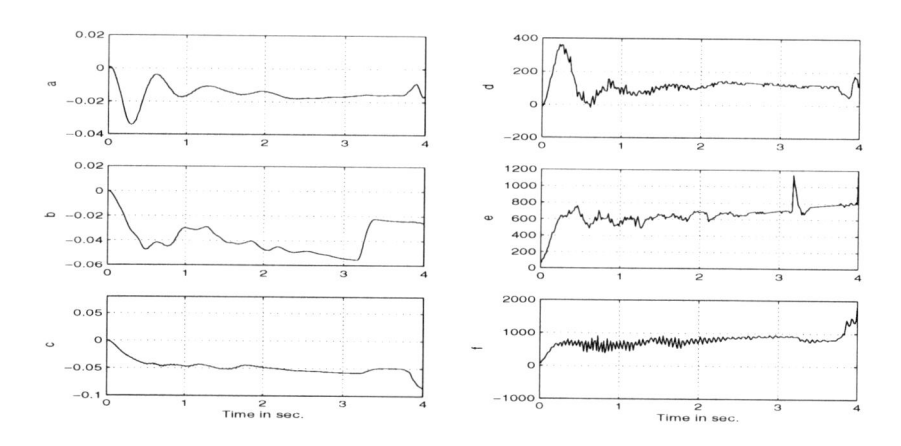

Fig. 3. Left Column: a, b & c) The tracking errors for joints 1, 2 and 3, Right Column: d, e & f) The control inputs for joints 1, 2 and 3

In our experimental test, the values of Γ are chosen as $\Gamma = diag(550, 550, 550, 500, 500, 500, 500, 500)$. With these parameter sets, we apply CAC algorithm on $Phantonm^{TM}$ device to track the given desired trajectory. The tested results are given as solid-line in Fig. 2. The control chattering phenomenon can be seen from this result.

We now implement multi-model based intelligent control on the same manipulator. The parameter $\theta = [\theta_1, \theta_2, \theta_3, ..., \theta_8]^T$ for $Phantom^{TM}$ device is

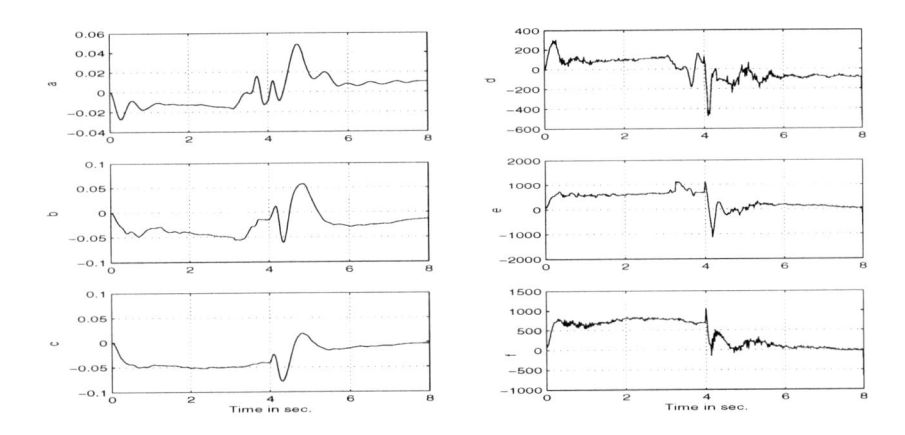

Fig. 4. Left Column: a, b & c) The tracking errors for joints 1, 2 and 3, Right Column: d, e & f) The control inputs for joints 1, 2 and 3 with an additional load attached to the end-effector of $Phantom^{TM}$ robot

arbitrarily equally distributed into a finite number as $\Omega = \bigcup_{i=1}^{11}\{\Omega_i\}$, that is, $\Omega = \bigcup_{i=1}^{11}\{\theta_i\} = \{0, 7, ..., 63, 70\} \times \{0, 8, ..., 72, 80\} \times \{0, 6, ..., 54, 60\} \times \{0, 12, ..., 108, 120\} \times \{0, 2, ..., 18, 20\} \times \{0, 2, ..., 18, 20\} \times \{0, 32, ..., 288, 320\} \times \{0, 32, ..., 288, 320\} \subset \Re^8$. The control design parameters λ_0, K_P and K_D are common to all $i = 11$ candidate controllers. The value of λ_0, K_P, K_D and Γ are chosen as $\lambda_0 = 2$, $K_P = diag(750, 1200, 1200)$, $K_D = diag(500, 800, 800)$ and $\Gamma = diag(550, 550, 550, 500, 500, 500, 500, 500)$. With these design sets, we now construct set of candidate controllers as a state feedback $\tau^i(e, Q_d, \theta_i) = Y(e, \dot{q}_r, \dot{q}_d, \ddot{q}_d)\theta_i - K_P e_1 - K_D e_2$ with $i = 11$. Then, we choose $t_d = 0.003$ and follow the steps of Algorithm 1 for Theorem 1. The tested results are shown in Fig. 2 as dash-line. In view of this Fig., we can observe that the control chattering phenomenon under CAC increases causing very large control efforts while chattering free control action with smaller control input under multi-model based adaptive design can be observed. In addition, the proposed intelligent approach provides better tracking performance than CAC design.

Let us now inspect the control performance of the multi-model based intelligent control system with respect to dynamical model parameter changes. To induce such model dynamic changes, we consider two cases in our next experimental evaluation. In first case, we increase the parametric error uncertainty by adding approximate 0.85 kg mass externally into end-effector of the $Phantom^{TM}$ manipulators. We then define the following set of controller design parameters as $\lambda_0 = 2$, $K_P = diag(750, 1200, 1200)$, $K_D = diag(500, 800, 1000)$ and $\Gamma = diag(700, 700, 700, 700, 700, 700, 700, 700)$. Aswe can notice from our design parameters that we increase the derivative and learning gains from our last evaluation to examine the effect of the control chattering phenomenon. The tested results with Theorem 1 under parameter changes are shown in Fig. 3. By comparing the dash-line of Fig. 2 with Fig. 3, we can observe that the tracking errors and control efforts are slightly higher than the tracking errors obtained in our last evaluation. This is due to the fact that we increase the modeling error uncertainty by adding an external loads to the system. It is also noticed from our various evaluations that, even with the presence of large modeling errors, the control efforts are relatively smaller and free from chattering activity. For comparison, we apply CAC design under the same set of design parameters defined above, but the control system goes unbounded. Due to space limit, we remove these experimental results. In second case, we change the operating dynamics while the $Phantom^{TM}$ manipulator is executing specified task. To create such situation in our experimental evaluation, we first operate the system with the above distributed parameter sets. Then, at approximately 4.075 sec., we add 0.85 kg mass into the second link to increase the modeling error uncertainty. In view of the parameter sets, there are two dynamics changes in the whole process dynamics. Then, we implement Theorem 1 on the given system with the same design constants that applied for our last implementation. The tested results are depicted in Fig.4. From these results, we can observe that the tracking errors as well as control efforts at the time of dynamical changes are little larger than the tracking errors and control efforts obtained under fixed dynamic case for Theorem 1. We also notice from these results that the control inputs with sudden dynamic changes are free from chattering activity. It is important to note that, with the same design parameter sets, we could not run the experiment with CAC algorithm (5), (6) due to the presence of excessive control chattering phenomenon.

5 Conclusion

We have presented multi-model based intelligent control system strategy for a class of nonlinear mechanical systems. Our interest was to improve overall tracking performance of the single model classical adaptive control technique with relatively smaller values of the control gains. The method reduced the controller gains from CE-based classical adaptive control scheme via reducing the level of parametric uncertainty through on-line estimation of the Lyapunov-function inequality. The design increased the convergence speed of standard adaptation mechanism by resetting the parameter estimate of CAC design into a family of candidate model that closely approximates the plant at each instant of time. The experimental evaluation on 3-DOF $Phantom^{TM}$ manipulators has been given to illustrate the theoretical development for the real-time applications.

Acknowledgment

This work is partially supported by the Natural Science and Engineering Research Council (NSERC) of Canada Postdoctoral Fellowship grant and Canada Research Chairs Program grant.

References

1. Hua, C., Liu, P.X.: Convergence analysis of teleoperation systems with unsymmetric time-varying delays. IEEE Transaction on Circuits and Systems-II 56(3), 240–244 (2009)
2. Erlic, M., Lu, W.: A reduced-order adaptive velocity observer for manipulator control. IEEE Transactions on Robotics and Automation 11, 293–303 (1995)
3. Polouchine, I.G.: Force-reflecting tele operation over wide-area networks, Ph. D. Thesis, Carleton University, Ottawa, Canada (2009)
4. Pomet, J.-B., Praly, L.: Adaptive nonlinear regulation estimation from the Lyapunov equation. IEEE Transactions Automatic Control 37, 729–740 (1992)
5. Astrom, K.J., Wittenmark, B.: Adaptive control, 2nd edn. Addision-Wesley, Reading (1995)
6. Narendra, K.S., Annaswamy, A.M.: Stable Adaptive Systems. Prentice Hall, New Jersy (1988)
7. Spong, M.W., Vidyasagar, M.: Robot Dynamics and Control. Wiley, New York (1989)
8. Spong, M.W.: On the robust control of robot manipulators. IEEE transactions on Automatic Control 37(11), 1782–1786
9. Ioannou, P., Sun, J.: Robust adaptive control. Prentice-Hall, Englewood Cliffs (1996)
10. Murray, R.M., Li, Z., Sastry, S.S.: A mathematical introduction to robotic manipulation. CRC Press, Boca Raton (1994)
11. Islam, S., Liu, P.X.: PD output feedback for industrial robot manipulators. To appear in IEEE/ASME Transactions on Mechatronics (2009)
12. Islam, S.: Adaptive output feedback for robot manipulators using linear observer. In: Proceedings of the international Conference on Intelligent System and Control, Orlando, Florida, November 16-18 (2008)
13. Islam, S., Liu, P.X.: Adaptive fuzzy output feedback control for robotic manipulators. In: Proceeding of the 2009 IEEE SMC, Texas, San Antonio, USA, October 11-14, pp. 2704–2709 (2009)
14. Islam, S., Liu, P.X.: Adaptive sliding mode control for robotic systems using multiple model parameters. In: IEEE/ASME AIM, Singapore, July 14-16, pp. 1775–1780.
15. Islam, S., Liu, P.X.: Robust control for robot manipulators by using only join position measurements. In: Proceeding of the 2009 IEEE SMC, San Antonio, TX, USA, October 11-14, pp. 4113–4118 (2009)
16. Sastry, S., Bodson, M.: Adaptive control: stability, convergence and robustness. Prentice-Hall, Englewood Cliffs (1989)
17. Lu, W.-S., Meng, M.Q.H.: Regressor formulation of robot dynamics: computation and applications. IEEE Transactions on Robotics Automation 9(3), 323–333 (1993)
18. Berghuis, H., Ortega, R., Nijmeijer, H.: A robust adaptive robot controller. IEEE Transactions on Robotics and Automation 9, 825–830 (1993)

Trends in the Control Schemes for Bilateral Teleoperation with Time Delay

Jiayi Zhu[1], Xiaochuan He[2], and Wail Gueaieb[1]

[1] Machine Intelligence, Robotics and Mechatronics (MIRaM) Laboratory
School of Information Technology and Engineering (SITE)
University of Ottawa, Ottawa ON K1N 6N5, Canada
[2] Studies on Teleoperation and Advanced Robotics (STAR) Laboratory
Department of Systems and Computer Engineering
Carleton University, Ottawa ON K1S 5B6, Canada

Abstract. In the last two decades, an increasing interest has been observed in the development of effective bilateral teleoperation systems. One of the biggest challenges encountered however is the presence of time delay, which can cause instability and poor transparency to the system. Researchers have been trying to tackle this problem from different angles and various control schemes have been proposed and tested. This paper aims to do an overview of those schemes and to compare them based on their assessment in order to give new researchers a summary of the current state of art. The schemes are divided into four main categories: passivity based, prediction based, sliding-mode control schemes and others. The comparison is based on transparency level, stability requirement, delay type, and delay range.

Keywords: bilateral teleoperation, time delay, delayed teleoperation, control scheme, passivity.

1 Introduction

Teleoperation is an action performed over distance where the human operator at the local site can control a robot at the remote site. When such an operation is qualified as bilateral, the operator can perceive the corresponding force feedback when the remote robot enters in contact with its environment. In the literature, the robot directly manipulated by the human operator is called "master" and the one located at the remote site is called "slave". A basic block diagram of such a system is shown in figure 1.

Such a system has a large potential with possible applications in nuclear, mining, space, medicine and many other fields. Unilateral teleoperation where the force feedback is non-existent is already in service in those fields; however as mentioned in [20], if force feedback was incorporated in those systems, it could ameliorate the completion time and quality of the overall teleoperated task. The incorporation of force feedback in the control system is unfortunately challenging, especially in the presence of a time delay which may cause inefficiency to the human operator and instability to the system.

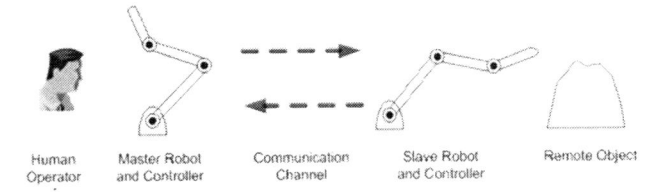

Fig. 1. An illustration of the bilateral teleoperation system

A few papers have done surveys of teleoperation control schemes in [2,4,8,11, 15,16,19]; however none was found to be specifically oriented towards both bilateral and time-delayed control schemes. Also, a sorting of the control schemes into categories is desirable since a growing number of controllers is being suggested. Furthermore, this work aims to provide a more conceptual explanation of the control schemes in order to ensure the provision of a comprehensive introduction suitable for new researchers in this field.

In this paper, a variety of proposed control schemes will be examined with the aim of introducing readers to the different control methods. Each concept will be presented and their pros and cons investigated.

This work will be divided into different parts, section 2 will give a background of time delay in teleoperation system, section 3 will talk about the different control schemes that have been proposed in the literature along with their features, section 4 gives a few remarks about the trends of the future works in this field, and section 5 will end with a conclusion.

2 Time Delay in Teleoperation

Time delay in teleoperation systems can be caused by the potentially large physical distance between the master and slave robots, the processing time of the controllers, the network congestion, and the communication bandwidth. It has been shown as early as 1966 in [5] that even a delay less than 50 ms can destabilize a bilateral teleoperation system. It must be mentioned however that time delay is not the only cause of instability of a teleoperation system. In fact, other sources include noise, external disturbances, force scaling, non passive environment, etc. But this paper will specifically focus on the problems caused by time-delay and the suggested solutions.

Beside instability, time delay may cause the human operator to be unable to instantaneously perceive her/his action and its corresponding reaction. This would decay the quality of a teleoperated task. One of the first methods that appeared to deal with time delayed teleoperation system uses the "move and wait" strategy where open-loop control is used as mentioned in [18]. The system is guaranteed to be stable in this case, however this method is very time consuming and is tiring for the operator.

In the literature, there are usually two categories of time delays considered: constant and time-varying. The constant time delay is usually in the order of less than one second, and it can find its application in master-slave robots that are physically linked, either by electrical wires or by optical fibers. On the other hand, the variable time delay is relevant to the applications using Internet as a communication channel. Due to the constantly changing condition of the network, time delay can vary between 0.1 s to 3 s [20].

3 Control Schemes for Time-Delayed Bilateral Teleoperation

In the literature, various control schemes are proposed for bilateral operation. All those presented in this paper appeared in the last two decades and have been experimentally verified. The comparison will be done for delay type and range, and also for the two most critical design criteria of a teleoperation control scheme: transparency and stability. Transparency refers to the accuracy in the position and force tracking. Ideally, a perfect transparency would make the human operator feel and believe that he or she is at the remote site. Stability on the other hand refers to the fact that the system output would always be bounded for a bounded input. In the close-loop control structures, this means that the poles should all be on the left hand side of the s-plane. Those two criteria often need to be traded off since a high bandwidth allow high transparency at the cost of degrading stability due to the time delay.

3.1 Passivity-Based Schemes

In order to ensure the overall system stability, its passivity can be used as the design objective. Passivity is a sufficient condition for stability. A passive system is one that can only consume energy without being able to produce any. In a two-port telerobotic system where the energy flow is characterized by $\dot{x}(t) = [\dot{x}_1(t), \dot{x}_2(t)]^T$ and effort by $F(t) = [F_1(t), F_2(t)]^T$ ($\dot{x}$ represents velocity, F represents force, 1 represents the master side, and 2 represents the slave side), this implies:

$$\int_0^\infty F^T(t)\dot{x}(t)dt \geq 0 \ .$$

The passivity-based control was first proposed by Anderson and Spong [1] and was considered a breakthrough by many researchers. In this method, the communication channel of the system was redefined in order to guarantee passivity independent of time delay. This was done by combining network theory, scattering transformation and passivity. Two years later, Niemeyer and Slotine further built into the idea by introducing wave variables [12]. The wave transformation converges the power variables (force and velocity) into wave variables as follow:

$$u = \frac{b\dot{x} + F}{\sqrt{2b}}, v = \frac{b\dot{x} - F}{\sqrt{2b}} \ ,$$

where u and v are the wave variables, while $\dot{x}$ and F are the power variables velocity and force, respectively. The parameter b is the wave impedance and can be tuned to modify the system's behaviour. Stability problem is solved under the assumption that each blocks of the overall system is passive, including human operator, environment, controllers, robots and communication channel.

This control scheme was experimented using the hard contact and back drivability tests. The results show that the constant time delay can be in the range of 40 ms to 2 s without destabilizing the system. But the tracking performance of the position and force commands (transparency) of the system was quite poor. When the time delay was increased to 2 s, the force and position tracking began to severely degrade.

Those problems may be caused by wave reflection, discrete sampling, numerical errors, and data loss. To solve them, Niemeyer and Slotine proposed two possible solutions in 2004 [13]: the first was to use wave integral, where no extra bandwidth was required, but the system can transmit position, velocity and force information (instead of just velocity and force); the second solution was the addition of a corrective term in response to the drift. It introduced no extra power, and slowly compensated for the position errors. The wave integral method has a potential to destabilize system under time-varying delays, therefore it was also proposed to include a reconstruction filter in the design. Another problem addressed by Niemeyer and Slotine was wave reflection, which was the issue of wave variables circulating several time in the system before dying out, compromising system's transparency. They suggested to use impedance matching where each subsystem is designed as a damper before the impedance is chosen and adjusted. This control scheme was tested using a constant time delay of up to 2 s and then using a variable delay from 50 ms to 1 s. A contact between slave robot and remote environment was included in the test. The system is shown to be stable and the transparency is improved compared to the basic wave variable scheme. However the position error between master and slave robots was observed to be in the order of centimetres.

In 2006, Lee and Spong proposed a new control framework in [10], which utilized the simple proportional-derivative (PD) control for bilateral teleoperation of a pair of multi-degree-of-freedom nonlinear robotic systems. In contrast to the scattering theory, this framework passifies the combination of the communication and control blocks altogether. It can exploit the benefits of position feedback while keeping the system passive. This method was also used for constant time delay, and the parameters of the PD control system were required to be optimized.

The control scheme developed by Lee and Spong used passivity concept, the Lyapunov-Krasovskii technique, and Parsevals identity with any finite constant time delays. They introduced several control parameters in order to achieve the master-slave coordination, force reflection in both ways, and the passivity of the system. For the experiments, four scenarios were assumed. When the time delay was 3 s, the human operator could perceive the force reflection from the object. So the stability and transparency of the teleoperation system could be guaranteed. However, it can be noticed in the experiments for this control scheme that some errors occurred in both the force reflection and position coordination, which were due to the friction used in the algorithm.

A combination of wave variables and predictors, along with the addition of drift control was proposed in 2006 [3]. It claims stability due to the use of wave-variable transfer over the transmission channel and also due to the assumption stating that slave robot dissipates enough energy. It switches between two different predictors depending on the type of remote environment encountered. The semi-adaptive predictor is superior in rigid contact environment while the full-adaptive predictor is better suited for slow varying environment. As shown by the experiments, the transparency is much better with this combined control method than the individual ones.

The varying time delay used in the experiment was between 0.4 s to 0.5 s. Additionally, the researchers were modelling UDP transmission protocol with 10% chance of data loss and system showed to be robust towards this transmission model. In general, the tracking performance was quite satisfactory even though not extremely precise, with the exception of the position tracking during free-space which proves to be quite accurate despite small oscillations observed in the slave robot. One of its weaknesses is its assumptions for guaranteeing passivity under time-varying delays. It is presumed that the slave robot is capable of dissipating enough energy. In reality, this may or may not be a good assumption. This is one of the few methods tested under realistic UDP environment and the system showed robustness towards this transmission model.

The biggest advantage of the wave variable based methods is that they guarantee stability and are completely robust to delays. Also, the wave commands control the output energy in terms of strength and direction only, so it can converge to either motion or force. Hence, the controller can be the same for when manipulator is in free space and when it is in contact with the environment. Additionally, those methods don't require any knowledge of the system since stability is evaluated based on input and output only. The main disadvantage of this control method is the assumption of human operator and environment being both passive, which may not be always be true.

Passivity is a sufficient condition for system stability. Therefore the passivity based control schemes reviewed in this paper could all guarantee the stability of the bilateral teleoperation system. When the system is tested under constant time delay, the scattering theory plays an important role in the scheme design. While when the delay of the system is varying, some extra variables, control parameters, or special communication channels are needed to improve the stability and the tracking performance of the system.

3.2 Prediction Based Schemes

Just around 2000, the concept of Smith Controller, which was first proposed to deal with delays in chemical plants in 1957, has been applied to bilateral teleoperation systems. In this method, the main idea is to replicate the response of the slave and the environment on the master side by building a local model. Hence this predictive local model is a representation of the remote object. Once established, the model permits to completely replicate the response of the remote object, taking out the negative effects of time delay. This kind of control can be robust against time delay of above 1 s, which is a greatly desired feature especially for space teleoperation. However it requires a very good model of the environment in order to reach good transparency. Along with the stability of the overall system, those are the two main challenges that researchers have been trying to overcome. The authors of this paper think that the presence of the model makes the relationship between transparency and stability to be no longer a clear trade-off. Because assuming that the model is a perfect representation of the remote site, the presence of the model guarantees the transparency of the system. Therefore stability remains as the only concern.

In 2003, Huang and Lewis proposed a control scheme based on this concept. In [9], they presented a two-channel position-position control architecture and used a neural network estimator on the slave side to estimate its dynamics. The neural network estimator is an adaptive system that is used to model complex relationships between inputs and outputs. It is widely employed in identification and control of nonlinear systems. This estimator builds up the model on the slave side by linearizing the nonlinear dynamics using local nonlinear compensation method and then sends this information back to the master side. The master side consists of a linear Smith predictor and it takes the information sent from the model in the slave side. At this point, all the nonlinear dynamics of the slave have been approximated to linear by the model. Since the model is transferring the information to the Smith predictor, the predictor is considered to give an accurate representation of the remote environment. Under the assumption that the slave dynamics is well modelled, this control scheme gives satisfactory results. However this can be justified in case of slowly varying or unchanging dynamics only since the predictor model is fixed; this is an important limitation of the system in some applications.

For this control scheme, the stability of the closed-loop system was proven using Lyapunov's method. A few important assumptions to ensure stability include the bounds on the desired states and on the output of the system. Also it assumes that the remote site is well represented by its linearized model. If this was not the case, system may be destabilized. This kind of control is suitable for systems with constant time-delays and gives satisfactory performance for free-motion tracking. However it suffers from the problem of poor transparency at contact, which is a common drawback for position-position control schemes. The results based on this kind of control scheme are obtained using a constant delay of 0.5 s.

Around the same time, Fite, Goldfarb and Rubio proposed a predictive control scheme using force-position architecture [6]. They suggested the implementation of an adaptive controller at the master side in order to adjust the linear estimate on the master-side itself. In this case, the slave information fed back through the communication channel and is then constantly updating the model (predictor) on the master side. This gives the control scheme an asset of being adjustable to time varying environment. However, once again the predictor used here is still linear and it is important to note that not all the environment can be modeled perfectly with a linear model.

This scheme was not rigorously proven stable. It relies on the original idea of Smith predictor stating that if the model is a good representation of the remote site, system is stable. Because of the fact that an adaptive controller is used, this assumption is considered as good. Through the experiments, the transparency of the system is shown to be robust to changing environment dynamics for stiffness ranging from $75\,\text{N·m}^{-1}$ to $1275\,\text{N·m}^{-1}$. The type of delay discussed by the researchers is assumed to be constant and the conducted experiments were using a value of $0.5\,\text{s}$ for each one-way time delay.

In 2006, Smith and Hashtrudi-Zaad presented an improved prediction-based model in [20] using neural network estimators on both master and slave sides, making both of them adaptive. It still uses force-position architecture as the one mentioned in previous paragraph. It builds the model on-line on the slave side using delayed information from the master as input and the corresponding force response from the slave as output, thus making the model free from time delay effects. This model is then transmitted back to the master side and continuously updated. This control scheme permits a real-time mapping of dynamic and nonlinear environment, but it needs a small learning period to build up the model.

In this case, the stability was also not analyzed theoretically. Also from the experiments conducted by the two researchers, even though the stability is shown to be much better compared to non-predictive methods, a noticeable position tracking error is still present. The constant time delay used in the experiments was $0.5\,\text{s}$.

More recently in 2009, Yoshida and Namerikawa proposed another predictive control scheme in [22] and tried to solve the issues in the previous control schemes. Their idea is based on a 3 channel architecture where the delayed position and velocity information of the master are sent to the slave side, and the estimated environment parameters are sent back to the master for the predictor. It guarantees the stability of the system and also ensures that the position tracking error converges to zero while the force tracking error stays small.

For this type of control scheme discussed, the asymptotic stability was proven using Lyapunov method with the finding of a storage function. This is under the assumption that the parameter estimation errors and the master states are bounded. The experimental results show very small tracking errors once steady state is reached, however this system requires a potentially large convergence time in the beginning ($12\,\text{s}$ in the case presented) in order to ensure small to non-existent tracking error. The delay used by the authors was constant and had the value of $1\,\text{s}$.

It can be seen that the predictive control schemes' main asset is being able to accommodate relatively large time delay. Their weakness however remains in the way of finding a good model of the remote site. The use of adaptive controllers and estimators seem to help modeling a non-linear and changing environment and show a much better outcome than the simple linearization method.

3.3 Sliding-Mode Control Schemes

Very different from the passivity based control and the prediction based control, the sliding mode control is a nonlinear system control method which determines a sliding surface and forces the trajectories of the dynamic system to approach it, reach it, and stay on it. Several researchers have tried to use sliding mode based methods in order to ameliorate the bilateral teleoperation system.

In 1999, a modified sliding-mode controller for the bilateral teleoperation was suggested by Park and Cho [14]. In the literature, the sliding-mode controller is known to be robust to model parameter variations, hence has great applications for systems dealing with time-varying delay. In this case, position-force architecture is used. An impedance controller was implemented on the master side while two types of controllers were used on the slave side. In free space, a modified first order sliding-mode controller is employed and upon contact, a local compliance controller is applied. The nonlinear control gain of the modified sliding-mode controller was modified to be independent of the time delay, which ameliorated the transparency and stability of the system.

For this scheme, stability was not mathematically proven. Transparency seems to be acceptable according to the experimental results. The control scheme deals with varying time delay, and the forward delay from master to slave is considered to be different from the feedback delay from slave to master. The round-trip varying time delay used by Park and Cho in their simulations had a maximum value of as high as 4 seconds, hence showing a lot of potential for teleopearations with varying time-delay. A TCP network protocol is used, therefore no data was lost and no packet ordering was necessary.

Along the same line, an observer-based sliding mode control was proposed a few years later where the impedance control is used for the master and a second-order sliding-mode control for the slave [7]. Even though it is a second order sliding mode control, the value of acceleration is not measured with sensor but is obtained with an observer. Using the same principle, the sensor measurement of velocity is also not required.

The resulting system was developed for constant unknown time delay. In free space, the output position, velocity and acceleration tend to the desired ones, however upon contact there exists a trade-off between force and position tracking. Moreover, the system does not achieve a perfect transparency, it only gives a good idea about the remote environment.

3.4 Other Control Schemes

A few other control schemes for bilateral teleoperation also appeared in the literature including an adaptive controller by Zhu and Salcudean [23], event-based controller [17], gain-scheduled compensation [21], etc. Those will not be looked at in details in this paper due to space limitation.

4 Remarks about Future Trends

Time delayed bilateral teleoperation is a fast progressing field of research. Based on its current status, a few remarks are made about its future trends.

It can be seen that an emerging interest has been shown from constant time delay to variable time delay, which models the Internet as the communication channel. In fact, due to its high availability and low cost, it has a great potential and has been a research interest since the late 90s. However in order to implement it, further research where time delay is implemented along with information loss and packet ordering (in case of UDP) needs to be done.

The time delays that were examined in the literature are mainly either constant and small (for physically linked master and slave robots) or varying up to a few seconds (for Internet-based applications). However there is also another type of delay which occurs in the case of space application. When the human operator is on Earth manipulating a robot in the space, the time delay can be much longer, even up to the range minutes. This application would mainly be useful for exploration missions on the Moon or Mars. More tests will need to be done if bilateral teleoperation is to be implemented in such applications.

Also, time-delayed bilateral teleoperation involving multiple human operators (many to one control), multiple slave manipulators (one to many), or both (many to many) would be a very useful control situation to be studied. Cooperation between operators is definitely needed in many contexts.

Moreover, scaled bilateral teleoperation system represents another challenge. In the medical applications, it is sometimes desirable to scale down and perform teleoperation on a micro and even nano scale. In other applications, it may be necessary to scale up, for example to multiply the human operators applied force.

5 Conclusion

Bilateral teleoperation has potentials in many research and industrial areas. As of now, the problem of time delay is one of the difficulties preventing this application to be implemented. A number of control schemes specific to this kind of manipulation were presented in this paper. They were divided into different categories and their pros and cons discussed. The comparison for the control schemes was also done, based on transparency level, stability requirement, delay type, and delay range. They demonstrate the main trends in the field of bilateral teleoperation and prove its development and its maturing steps.

References

1. Anderson, R., Spong, M.: Bilateral control of teleoperators with time delay. IEEE Transactions on Automatic Control 34(5), 494–501 (1989)
2. Arcara, P., Melchiorri, C.: Control schemes for teleoperation with time delay: a comparative study. Robotics and Autonomous Systems 38(1), 49–64 (2002)

3. Ching, H., Book, W.: Internet-based bilateral teleoperation based on wave variable with adaptive predictor and direct drift control. Transactions of the ASME, Journal of Dynamic Systems, Measurement and Control 128, 86–93 (2006)
4. Ferre, M., Buss, M., Aracil, R., Melchiorri, C., Balaguer, C.: Advances in Telerobotics. Springer, Heidelberg (2007)
5. Ferrell, W.: Delayed force feedback. Hum. Factors 8(5), 449–455 (1966)
6. Fite, K., Goldfarb, M., Rubio, A.: Transparent telemanipulation in the presence of time delay. In: Proceedings of the IEEE/ASME International Conference on Advanced Intelligent Mechatronics, vol. 1, pp. 254–259 (2003)
7. Garcia-Valdovinos, L., Parra-Vega, V., Arteaga, M.: Observer-based higher-order sliding mode impedance control of bilateral teleoperation under constant unknown time delay. In: IEEE/RSJ International Conference on Intelligent Robots and Systems, pp. 1692–1699 (2006)
8. Hokayem, P., Spong, M.: Bilateral teleoperation: An historical survey. Automatica 42, 2035–2057 (2006)
9. Huang, J., Lewis, F.: Neural-network predictive control for nonlinear dynamic systems with time-delay. IEEE Trans. on Neural Networks 14(2), 377–389 (2003)
10. Lee, D., Spong, M.: Passive bilateral teleoperation with constant time delay. In: Proceeding of the 2006 IEEE International Conference on Robotics and Automation, vol. 22(2), pp. 2902–2907 (2006)
11. Mulder, M.: Stability in haptic teleoperation literature review (2006)
12. Niemeyer, G., Slotine, J.: Stable adaptive teleoperation. IEEE Journal of Oceanic Engineering 16(1), 152–162 (1991)
13. Niemeyer, G., Slotine, J.: Telemanipulation with time delays. International Journal of Robotics Research 23(9), 873–890 (2004)
14. Park, J., Cho, H.: Sliding-mode control of bilateral teleoperation systems with force-reflection on the internet. In: Proceedings of the IEEE/RSJ International Conference on Intelligent Robots and System, vol. 2, pp. 1187–1192 (2000)
15. Rodriguez-Seda, E.: Comparative experimental study of control schemes for bilateral teleoperation systems. Master's thesis, Univ. Illinois (2007)
16. Rodriguez-Seda, E., Lee, D., Spong, M.: Experimental comparison study of control architectures for bilateral teleoperators. IEEE Transactions on Robotics 25(6), 1304–1318 (2009)
17. Sano, A., Fujimoto, H., Tanaka, M.: Gain-scheduled compensation for time delay of bilateral teleoperation systems. In: Proceedings of 1998 IEEE International Conference on Robotics and Automation, vol. 3, pp. 1916–1923 (1998)
18. Sheridan, T.: Space teleoperation through time delay: review and prognosis. IEEE Transactions on Robotics and Automation 9(5), 592–606 (1993)
19. Sheridan, T.: Teleoperation, telerobotics and telepresence: A progress report. Control Eng. Pract. 3(3), 205–214 (1995)
20. Smith, A., Hashtrudi-Zaad, K.: Smith predictor type control architectures for time delayed teleoperation. The International Journal of Robotics Research 25(8), 797–818 (2006)
21. Xi, N., Tarn, T.: Stability analysis of non-time referenced internet-based telerobotic systems. Robotics and Autonomous Systems 32(2-3), 173–178 (2000)
22. Yoshida, K., Namerikawa, T.: Stability and tracking properties in predictive control with adaptation for bilateral teleoperation. In: Proceedings of the American Control Conference, pp. 1323–1328 (2009)
23. Zhu, W., Salcudean, S.: Stability guaranteed teleoperation: an adaptive motion/force control approach. IEEE Transactions on Automatic Control 45(11), 1951–1969 (2000)

Bilateral Teleoperation System with Time Varying Communication Delay: Stability and Convergence

S. Islam[1,2], P.X. Liu[3], and A. El Saddik[1]

[1] School of Information Technology and Engineering (SITE),
University of Ottawa, 800 King Edward Road, Ottawa, Canada, K1N 6N5
[2] School of Engineering and Applied Sciences,
Harvard University, 29 Oxford Street, Cambridge, MA 02138 USA
[3] Department of Systems and Computer Engineering,
Carleton University, Ottawa, Ontario, Canada, K1S 5B6
sislam@sce.carleton.ca, sislam@seas.harvard.edu

Abstract. The trajectory tracking control problem of internet-based bilateral nonlinear teleoperators with the presence of the symmetric and unsymmetrical time varying communication delay is addressed in this paper. The design comprises proportional derivative (PD) terms with nonlinear adaptive control terms in order to cope with parametric uncertainty of the master and slave robot dynamics. The master-slave teleoperators are coupled by velocity and delayed position signals. The Lyapunov-Krasovskii-like functional is employed to ensure asymptotic stability of the master-slave closed-loop teleoperator systems under time varying communication delay. The stability condition allows the designer to estimate the control gains in order to achieve desired tracking property of the position and velocity signals for the master and slave systems.

Keywords: Teleoperation, Lyapunov-Krasovskii Function, Adaptive Control.

1 Introduction

Bilateral telemanipulation control system technology has many different applications ranging from space and deep-sea exploration to hazardous material handling in nuclear plants [11, 12, 13, 15, 19]. More recently, internet-based bilateral medical teleoperation, such as, robot assisted telesurgery, telemedicine and telediagnosis are attracted considerable amount of interest in medical professionals [1]. In bilateral teleoperation systems, local-master and remote-slave robot manipulators are connected through communication channel. The control objective in teleoperation is to ensure that the slave reproduces the master's command accurately. This means that the position and velocity of the local-master systems are equal to the position and velocity of the remote-slave systems and the master exactly feels the slave interaction force with environment. The human then feels the same interaction force that slave feels. However, any data transmission over the communication network is always associated with time delay. In the presence of communication delay, even it is very small, the entire teleoperation systems may become unstable, if the controller is not designed appropriately. As a result, the desired stability and transparency of teleoperation system may not be achieved in the presence of delay. To deal with the problems associated with time delay in bilateral teleoperation applications, many techniques have been reported in the literature. Results in this area can be found in [2, 3, 4, 5, 6, 7, 8, 14, 16, 20, and references therein]. Majority of these reported results are assumed that the communication time delay between master and slave platform is constant. Recently, authors in [17, 18] proposed teleoperation systems with time varying communication delay. These results assumed that the communication delay from local-master to remote-slave and remote-slave to local-master systems are symmetric. However, time delay in internet-based communication may be irregular and asymmetrical nature which may further complicate the existing teleoperation control system design to achieve desired tracking and transparency. Recently, we have proposed model based bilateral teleoperation control system for unsymmetrical time varying communication delay [10]. The design is based on using strict linear matrix inequality (LMI) condition.

In this paper, we introduce novel control design for the trajectory tracking problem for internet-based teleoperator systems. The proposed design combined velocity and delayed position signals with nonlinear adaptive control terms for the master-slave systems. The communication delay in the internet network is considered to be symmetric and asymmetric time varying nature. In our first design, we introduce symmetric time varying communication delay based teleoperation control systems. The teleoperators under unsymmetrical time varying communication delay is then derived in second design. In both case, we use Lyapunov-Krasovskii-like functional to obtain controller gains in order to ensure asymptotic tracking property of the coupled master-slave teleoperator systems.

The rest of the paper is organized as follows: Section 2 describes the dynamical model of the master-slave teleoperation systems. In section 3 and section 4, we introduce control algorithms for master-slave teleoperator systems. These designs are based on using symmetric and unsymmetrical time varying communication delay between local-master and remote-slave telerobot. A detailed convergence analysis for both symmetric and unsymmetrical time delay case is given to show the boundedness of the position and velocity signals for the closed-loop teleoperator systems. Finally, section 5 concludes the paper.

2 System Dynamics and Objective

Let us first consider the equation of motion for n-links local-master and remote-slave telerobot manipulators

$$M_m(q_m)\ddot{q}_m + C(q_m,\dot{q}_m)\dot{q}_m + G(q_m) = \tau_m + F_h \tag{1}$$

$$M_s(q_s)\ddot{q}_s + C(q_s,\dot{q}_s)\dot{q}_s + G(q_s) = \tau_s - F_e \tag{2}$$

where, m and s represents the master and slave systems, respectively, $\ddot{q}_m \in \Re^n$, $\dot{q}_m \in \Re^n$, $q_m \in \Re^n$ and $\ddot{q}_s \in \Re^n$, $\dot{q}_s \in \Re^n$, $q_s \in \Re^n$ are the joint acceleration, velocity and position, $M_m(q_m) \in \Re^{n \times n}$ and $M_s(q_s) \in \Re^{n \times n}$ are the symmetric and uniformly positive definite inertia matrix, $C_m(q_m, \dot{q}_m)\dot{q}_m \in \Re^n$ and $C_s(q_s, \dot{q}_s)\dot{q}_s \in \Re^n$ are the coriolis and centrifugal loading vector, $G_m(q_m) \in \Re^n$ and $G_s(q_s) \in \Re^n$ are the gravitational loading vector, $\tau_m \in \Re^n$ and $\tau_s \in \Re^n$ are the applied control input vector, and $F_h \in \Re^n$ and $F_e \in \Re^n$ denote the operational torque applied to the local-master by human operator and environmental torque vector by remote-slave robot, respectively.

The presence of the time varying communication delay between local-master and remote-slave platform is the main problem to achieve desired stability and transparency in bilateral teleoperation control system design. In fact, the communication time delay between master and slave always exits as they are separated by a distance. If the distance between master and slave become large, then a large communication time delay appears in the system. Such a large communication time delay can destabilize and degrade the closed-loop performance. On the other hand, the presence of asymmetric nature of the time varying communication delay may further deteriorate system stability and transparency of the entire system.

In this work, we design adaptive control algorithms for master-slave teleoperators system under both symmetric and unsymmetrical time varying communication delay. The system (1) and (2) can be represented in the following state space form

$$\dot{x}_m = A_{ma} x_m + B_{m1}(F_h + \tau_m) + B_{m1} Y_m(x_m)\theta_m \quad (3)$$
$$\dot{x}_s = A_{sa} x_s + B_{s1}(-F_e + \tau_s) + B_{m1} Y_s(x_s)\theta_s \quad (4)$$

where $x_m = \begin{bmatrix} x_{1m} \\ x_{2m} \end{bmatrix}$, $x_s = \begin{bmatrix} x_{1s} \\ x_{2s} \end{bmatrix}$, $x_{1m} = q_m$, $x_{2m} = \dot{q}_m$, $x_{1s} = q_s$, $x_{2s} = \dot{q}_s$, $A_{ma} = \begin{bmatrix} 0_{n \times n} & I_{n \times n} \\ 0_{n \times n} & 0_{n \times n} \end{bmatrix}$, $A_{sa} = \begin{bmatrix} 0_{n \times n} & I_{n \times n} \\ 0_{n \times n} & 0_{n \times n} \end{bmatrix}$, $Y_s(x_s)\hat{\theta}_s = \hat{C}_s(x_{1s}, x_{2s})x_{2s} + \hat{G}_s(x_{1s})$, $Y_m(x_m)\hat{\theta}_m = \hat{C}_m(x_{1m}, x_{2m})x_{2m} + \hat{G}_m(x_{1m})$, $K_{Pm} \in \Re^{n \times n}$, $B_{m1} = \begin{bmatrix} 0_{n \times n} \\ M_m^{-1} \end{bmatrix}$ and $B_{s1} = \begin{bmatrix} 0_{n \times n} \\ M_s^{-1} \end{bmatrix}$. We now consider that the mass matrix $M(q_m)$, $M(q_s)$, $M^{-1}(q_m)$ and $M(^{-1}q_s)$ are symmetric, bounded and positive definite satisfy the inequalities as $\|M(q_m)\| \leq \beta_m$, $\|M(q_s)\| \leq \beta_s$, $\|M^{-1}(q_m)\| \leq \beta_{mi}$ and $\|M(^{-1}q_s)\| \leq \beta_{si}$ with $\beta_{mi} > 0$, $\beta_{si} > 0$, $\beta_m > 0$ and $\beta_s > 0$ [2, 3, 4, 9, 10, many others]. We also assume that the human operator force applied to the end effectors of the master system, F_h, and the remote environmental force applied to the end effectors of the slave system, F_e, can be modeled as passive spring-damper systems as

$$F_h = -k_{m1} q_m - k_{m2} \dot{q}_m, \quad F_e = k_{s1} q_s + k_{s2} \dot{q}_s$$

where $k_{m1} > 0$, $k_{m2} > 0$, $k_{s1} > 0$ and $k_{s2} > 0$. We then use the well-known force to velocity passive maps for human operator and remote environment forces as

$$\int_0^t x_m^T(\eta) B_{m1} F_h(\eta) d\eta \geq -\gamma_m$$
$$\int_0^t -x_s^T(\eta) B_{s1} F_e(\eta) d\eta \geq -\gamma_s \quad (5)$$

with $\gamma_m > 0$ and $\gamma_s > 0$.

3 Teleoperation Control Design Under Symmetric Time Varying Communication Delay

Let us first consider that the internet communication delays between local-master and remote-slave platform are symmetric time varying as $T_{dm}(t) = T_{ds}(t) = T_d(t)$. $T_{dm}(t)$ and $T_{ds}(t)$ represents the time delay from master to slave and slave to master, respectively. Then, design the control law where the master-slave system is coupled by delayed master-slave position signals as follows

$$\tau_m(x_m, \theta_m) = Y_m(x_m)\hat{\theta}_m - K_{Pm}(x_{1m} - x_{1s}(t - T_d(t))) - K_{Dm}x_{2m} \quad (6)$$
$$\tau_s(x_s, \theta_s) = Y_s(x_s)\hat{\theta}_s - K_{Ps}(x_{1s} - x_{1m}(t - T_d(t))) - K_{Ds}x_{2s} \quad (7)$$

with

$$\dot{\hat{\theta}}_m = -\gamma_m Y^T(x_m) B_{m1}^T x_m \quad (8)$$
$$\dot{\hat{\theta}}_s = -\gamma_s Y^T(x_s) B_{s1}^T x_s \quad (9)$$

where $\gamma_m > 0$, $\gamma_s > 0$, $K_{Pm} > 0$, $K_{Dm} > 0$, $K_{Ps} > 0$ and $K_{Ds} > 0$ are the positive definite control gains matrix for the master-slave systems. Then, the closed-loop system can be represented in the following state space form

$$\dot{x}_m = A_m x_m + A_{sm} x_s (t - T_d(t)) + B_{m1} F_h + B_{m1} Y_m(x_m)\tilde{\theta}_m \quad (10)$$
$$\dot{x}_s = A_s x_s + A_{ms} x_m (t - T_d(t)) - B_{s1} F_e + B_{s1} Y_s(x_s)\tilde{\theta}_s \quad (11)$$

where $A_m = \begin{bmatrix} 0_{n \times n} & I_{n \times n} \\ -M_m^{-1} K_{Pm} & -M_m^{-1} K_{Dm} \end{bmatrix}$, $A_s = \begin{bmatrix} 0_{n \times n} & I_{n \times n} \\ -M_s^{-1} K_{Ps} & -M_s^{-1} K_{Ds} \end{bmatrix}$, $A_{ms} = \begin{bmatrix} 0_{n \times n} & 0_{n \times n} \\ M_s^{-1} K_{Ps} & 0_{n \times n} \end{bmatrix}$ and $A_{sm} = \begin{bmatrix} 0_{n \times n} & 0_{n \times n} \\ M_m^{-1} K_{Pm} & 0_{n \times n} \end{bmatrix}$.

3.1 Stability Analysis

To establish stability condition for algorithm (6)-(9) under symmetrical time varying communication delay case, we consider the following positive definite functional

$$V = V_1 + V_2 + V_3 + V_4 + V_5 \quad (12)$$

where
$$V_1 = x_m^T x_m + x_s^T x_s \tag{13}$$

$$V_2 = \zeta_{dm} \int_{(t-T_d(t))}^{t} x_m^T(\eta) x_m(\eta) d\eta + \zeta_{ds} \int_{(t-T_d(t))}^{t} x_s^T(\eta) x_s(\eta) d\eta \tag{14}$$

$$V_3 = \int_{-\alpha_d}^{0} \int_{(t+\xi)}^{t} x_m^T(\eta) x_m(\eta) d\eta d\xi + \int_{-\alpha_d}^{0} \int_{(t+\xi)}^{t} x_s^T(\eta) x_s d(\eta) d\eta d\xi \tag{15}$$

$$V_4 = -2 \int_{0}^{t} x_m^T(\eta) B_{m1} F_h(\eta) d\eta + 2 \int_{0}^{t} x_s^T(\eta) B_{s1} F_e(\eta) d\eta + 2\gamma_m + 2\gamma_s \tag{16}$$

$$V_5 = \tilde{\theta}_m^T \Gamma_m^{-1} \tilde{\theta}_m + \tilde{\theta}_s^T \Gamma_s^{-1} \tilde{\theta}_s \tag{17}$$

where $|T_d(t)| \le \alpha_d$, $\zeta_{dm} = \frac{1}{1-|\dot{T}_d(t)|}$, $\zeta_{ds} = \frac{1}{1-|\dot{T}_d(t)|}$ with $\dot{T}_d(t) \le \mu_d < 1$ and α_d and μ_d are positive constants. Taking derivative along the trajectory (10), (11) leads to

$$\dot{V}_1 = 2x_m^T \left[A_m x_m + A_{sm} x_m(t - T_d(t)) + B_{m1} F_h + B_{m1} Y_m(x_m) \tilde{\theta}_m \right] +$$
$$2x_s^T \left[A_s x_s + A_{ms} x_s(t - T_d(t)) - B_{s1} F_e + B_{s1} Y_s(x_s) \tilde{\theta}_s \right] \tag{18}$$

$$\dot{V}_2 \le \zeta_{dm} x_m^T x_m + \zeta_{dm}(1-\mu_d) x_m^T(t-T_d(t)) x_m(t-T_d(t)) + \zeta_{ds} x_s^T x_s + \zeta_{ds}(1-\mu_d)$$
$$x_s^T(t-T_d(t)) x_s(t-T_d(t)) \tag{19}$$

$$\dot{V}_3 = \alpha_d x_m^T(t) x_m(t) - \int_{(t-T_d(t))}^{t} x_m^T(\eta) x_m(\eta) d\eta + \alpha_d x_s^T(t) x_s(t) - \int_{(t-T_d(t))}^{t} x_s^T(\eta) \tag{20}$$
$$x_s(\eta) d\eta$$

$$\dot{V}_4 \le -2\beta_m x_m^T F_h + 2\beta_s x_s^T F_e \tag{21}$$

$$\dot{V}_5 = 2\tilde{\theta}_m^T \left[\Gamma_m^{-1} \dot{\tilde{\theta}}_m + Y_m^T(x_m) B_{m1}^T x_m \right] + 2\tilde{\theta}_s^T \left[\Gamma_s^{-1} \dot{\tilde{\theta}}_s + Y_s^T(x_s) B_{s1}^T x_s \right] \tag{22}$$

Using (18)-(22) along with the following inequality

$$2x_m^T A_{sm} x_m(t-T_d(t)) \le x_m^T A_{sm} A_{sm}^T x_m + x_m^T(t-T_d(t)) x_m(t-T_d(t))$$
$$2x_s^T A_{ms} x_s(t-T_d(t)) \le x_s^T A_{ms} A_{ms}^T x_s + x_s^T(t-T_d(t)) x_s(t-T_d(t)) \tag{23}$$

$\dot{V}$ becomes

$$\dot{V} \le x_m^T \Omega_m x_m + x_s^T \Omega_s x_s - \int_{(t-T_d(t))}^{t} x_s^T(\eta) x_s(\eta) d\eta - \int_{(t-T_d(t))}^{t} x_m^T(\eta) x_m(\eta) d\eta \tag{24}$$

where $\Omega_m = A_{Gm} + 2A_m + \psi I$, $\Omega_s = A_{Gs} + 2A_s + \psi I$, $\psi = (\zeta_d + \alpha_d)$, $A_{Gm} = A_{sm} A_{sm}^T$, $A_{Gs} = A_{ms} A_{ms}^T$, $\alpha_{dm} = \alpha_{ds} = \alpha_d$ and $\zeta_{dm} = \zeta_{ds} = \zeta_d$. Now, to ensure $\Re(\lambda(\Omega_m)) < 0$ and $\Re(\lambda(\Omega_s)) < 0$, the controller gains K_{Pm}, K_{Dm}, K_{Ps} and K_{Ds} are required to satisfy the following inequality as

$$K_{Pm} > \left(\frac{\psi^2 \|M_m\|(\varepsilon_m - 1)}{4} \right), K_{Dm} > \left(\frac{\|M_m\| \varepsilon_m \psi}{2} + \frac{K_{Pm}^2 \|M_m^{-1}\|}{2} \right)$$
$$K_{Ps} > \left(\frac{\psi^2 \|M_s\|(\varepsilon_s - 1)}{4} \right), K_{Ds} > \left(\frac{\|M_s\| \varepsilon_s \psi}{2} + \frac{K_{Ps}^2 \|M_s^{-1}\|}{2} \right) \tag{25}$$

with $\varepsilon_m \geq 2$ and $\varepsilon_s \geq 2$. The parameters ε_m and ε_m may play an important role to improve overall teleoperation system. Then, using (25), $\dot{V}$ becomes

$$\dot{V} \leq -\lambda_{min}(\Omega_m)\|x_m\|^2 - \lambda_{min}(\Omega_s)\|x_s\|^2 - \int_{(t-T_d(t))}^{t} x_s^T(\eta)x_s(\eta)d\eta$$
$$- \int_{(t-T_d(t))}^{t} x_m^T(\eta)x_m(\eta)d\eta \qquad (26)$$

where $\lambda_{min}(\Omega_m)$ and $\lambda_{min}(\Omega_s)$ are the smallest eigen values of the matrix Ω_m and Ω_s. Taking integral from zero to time T, we have

$$V(T) - V(0) \leq -\int_0^T \lambda_{min}(\Omega_m)\|x_m\|^2 dt - \int_0^T \lambda_{min}(\Omega_s)\|x_s\|^2 dt - \qquad (27)$$
$$\int_0^T \int_{(t-T_d(t))}^{t} x_s^T(\eta)x_s(\eta)d\eta dt - \int_0^T \int_{(t-T_d(t))}^{t} x_m^T(\eta)x_m(\eta)d\eta dt$$

Now, using Schwartz inequality and the bound on the time varying delay for the last two terms, one has

$$V(T) - V(0) \leq -\int_0^T \lambda_{min}(\Omega_m)\|x_m\|^2 dt - \int_0^T \lambda_{min}(\Omega_s)\|x_s\|^2 dt - \qquad (28)$$
$$\int_0^T \alpha_d \int_{(t-T_d(t))}^{t} \|x_s\|^2 d\eta dt - \int_0^T \alpha_d \int_{(t-T_d(t))}^{t} \|x_m\|^2 d\eta dt$$

We now do simple calculus on the last two terms to obtain the bound on the position and velocity of the master-slave teleoperator signals as

$$V(T) - V(0) \leq -\lambda_{min}(\Omega_m)\|x_m\|_2^2 - \lambda_{min}(\Omega_s)\|x_s\|_2^2 + \alpha_d^2\|x_s\|_2^2 + \alpha_d^2\|x_m\|_2^2 \qquad (29)$$

where $\|.\|_2$ represents the $\mathcal{L}_2$ norm of the signals over the given operation time T. Equation (29) can be further simplified as

$$V(T) - V(0) \leq -\chi_m\|x_m\|_2^2 - \chi_s\|x_s\|_2^2 \qquad (30)$$

where $\chi_m = \lambda_{min}(\Omega_m) - \alpha_d^2$ and $\chi_s = \lambda_{min}(\Omega_s) - \alpha_d^2$. So, if the final time $T \to \infty$, then we have $\int_0^\infty \dot{V} dt \leq 0$. Then, we can conclude that $(q_m, q_s, \dot{q}_m, \dot{q}_s) \in \mathcal{L}_\infty$ and $(q_m, q_s, \dot{q}_m, \dot{q}_s) \in \mathcal{L}_2$. This implies that $lim_{t\to\infty}\dot{q}_m = lim_{t\to\infty}\dot{q}_s = lim_{t\to\infty}q_m = lim_{t\to\infty}q_s = 0$. Based on our above analysis, we now state our main results in the following Theorem 1.

Theorem 1: *Consider the coupled master-slave closed loop teleoperator system (10), (11) along with the passive inequality (5). Then, for the given μ_d, there exists K_{Pm}, K_{Ps}, K_{Dm} and K_{Ds} derived in (25) with $\gamma_m > 0$ and $\gamma_s > 0$ such that the position and velocity of the master-slave systems are asymptotically stable.*

4 Teleoperation Control System Under Asymmetrical Time Varying Communication Delay

We now consider that the internet communication delays between local-master and remote-slave platforms are not symmetric time varying as $T_{dm}(t) \neq T_{ds}(t)$. Then, we propose the following master-slave teleoperator control systems as

$$\tau_m(x_m, \theta_m) = Y_m(x_m)\hat{\theta}_m - K_{Pm}(x_{1m} - x_{1s}(t - T_{dm}(t))) - K_{Dm}x_{2m} \quad (31)$$

$$\tau_s(x_s, \theta_s) = Y_s(x_s)\hat{\theta}_s - K_{Ps}(x_{1s} - x_{1m}(t - T_{ds}(t))) - K_{Ds}x_{2s} \quad (32)$$

with the parameter adaptation laws $\dot{\hat{\theta}}_m$ and $\dot{\hat{\theta}}_s$ given in (8), (9). The closed-loop system then can be represented in the following state space form

$$\dot{x}_m = A_m x_m + A_{sm} x_s(t - T_{dm}(t)) + B_{m1} F_h + B_{m1} Y_m(x_m)\tilde{\theta}_m \quad (33)$$

$$\dot{x}_s = A_s x_s + A_{ms} x_m(t - T_{ds}(t)) - B_{s1} F_e + B_{s1} Y_s(x_s)\tilde{\theta}_s \quad (34)$$

4.1 Stability Analysis

We now derive stability condition for the coupled master-slave closed-loop system (33), (34) under unsymmetric time varying communication delay case. For this purpose, we consider the following positive definite functional

$$V = V_1 + V_2 + V_3 + V_4 + V_5 \quad (35)$$

where

$$V_1 = x_m^T x_m + x_s^T x_s \quad (36)$$

$$V_2 = \zeta_{dm} \int_{(t-T_{dm}(t))}^{t} x_m^T(\eta) x_m(\eta) d\eta + \zeta_{ds} \int_{(t-T_{ds}(t))}^{t} x_s^T(\eta) x_s(\eta) d\eta \quad (37)$$

$$V_3 = \int_{-\alpha_{dm}}^{0} \int_{(t+\xi)}^{t} x_m^T(\eta) x_m(\eta) d\eta d\xi + \int_{-\alpha_{ds}}^{0} \int_{(t+\xi)}^{t} x_s^T(\eta) x_s d(\eta) d\eta d\xi \quad (38)$$

$$V_4 = -2 \int_{0}^{t} x_m^T(\eta) B_{m1} F_h(\eta) d\eta + 2 \int_{0}^{t} x_s^T(\eta) B_{s1} F_e(\eta) d\eta + 2\gamma_m + 2\gamma_s \quad (39)$$

$$V_5 = \tilde{\theta}_m^T \Gamma_m^{-1} \tilde{\theta}_m + \tilde{\theta}_s^T \Gamma_s^{-1} \tilde{\theta}_s \quad (40)$$

where $|T_{dm}(t)| \leq \alpha_{dm}$, $|T_{ds}(t)| \leq \alpha_{ds}$, $\zeta_{dm} = \frac{1}{1-|\dot{T}_{dm}(t)|}$, $\zeta_{ds} = \frac{1}{1-|\dot{T}_{ds}(t)|}$ with $\dot{T}_{dm}(t) \leq \mu_{dm} < 1$, $\dot{T}_{ds}(t) \leq \mu_{ds} < 1$ and α_{dm} and μ_{ds} are positive constants. We now take the derivative (35) along the closed-loop tracking trajectory (33), (34). Then, we have

$$\dot{V}_1 = 2x_m^T \left[A_m x_m + A_{sm} x_m(t - T_{dm}(t)) + B_{m1} F_h + B_{m1} Y_m(x_m)\tilde{\theta}_m \right] +$$

$$2x_s^T \left[A_s x_s + A_{ms} x_s(t - T_{ds}(t)) - B_{s1} F_e + B_{s1} Y_s(x_s)\tilde{\theta}_s \right] \quad (41)$$

$$\dot{V}_2 \leq \zeta_{dm} x_m^T x_m + \zeta_{dm}(1 - \mu_{dm}) x_m^T(t - T_{dm}(t)) x_m(t - T_{dm}(t)) + \zeta_{ds} x_s^T x_s + \zeta_{ds}$$

$$(1-\mu_{ds})\,x_s^T(t-T_{ds}(t))x_s(t-T_{ds}(t)) \tag{42}$$

$$\dot{V}_3 = \alpha_{dm}x_m^Tx_m - \int_{(t-T_{dm}(t))}^{t} x_m^T(\eta)x_m(\eta)d\eta - \int_{(t-T_{ds}(t))}^{t} x_s^T(\eta)x_s(\eta)d\eta$$
$$+\alpha_{ds}x_s^Tx_s \tag{43}$$

$$\dot{V}_4 \le 2\beta_m x_m^T F_h - 2\beta_s x_s^T F_e \tag{44}$$

$$\dot{V}_5 = 2\tilde{\theta}_m^T\left[\Gamma_m^{-1}\dot{\tilde{\theta}}_m + Y_m^T(x_m)B_{m1}^T x_m\right] + 2\tilde{\theta}_s^T\left[\Gamma_s^{-1}\dot{\tilde{\theta}}_s + Y_s^T(x_s)B_{s1}^T x_s\right] \tag{45}$$

Using the inequality (23), $\dot{V}$ can be written as

$$\dot{V} \le x_m^T\Omega_{dm}x_m - \int_{(t-T_{dm}(t))}^{t} x_s^T(\eta)x_s(\eta)d\eta - \int_{(t-T_{ds}(t))}^{t} x_m^T(\eta)x_m(\eta)d\eta$$
$$+x_s^T\Omega_{ds}x_s \tag{46}$$

$\Omega_{dm} = A_{Gm} + 2A_m + \psi_m I$, $\Omega_{ds} = A_{Gs} + 2A_s + \psi_s I$, $\psi_m = (\zeta_{dm}+\alpha_{dm})$, $\psi_s = (\zeta_{ds}+\alpha_{ds})$, $A_{Gm} = A_{sm}A_{sm}^T$ and $A_{Gs} = A_{ms}A_{ms}^T$. For $\Re(\lambda(\Omega_{dm})) < 0$ and $\Re(\lambda(\Omega_{ds})) < 0$, the controller gains K_{Pm}, K_{Ps}, K_{Dm} and K_{Ds} required to satisfy the following inequality as

$$K_{Pm} > \left(\frac{(\zeta_{dm}+\alpha_{dm})^2\|M_m\|(\varepsilon_m-1)}{4}\right),\; K_{Ps} > \left(\frac{(\zeta_{ds}+\alpha_{ds})^2\|M_s\|(\varepsilon_s-1)}{4}\right)$$

$$K_{Dm} > \left[\frac{\|M_m\|\varepsilon_m(\zeta_{dm}+\alpha_{dm})}{2} + \frac{K_{Pm}^2\|M_m^{-1}\|}{2}\right],$$

$$K_{ds} > \left[\frac{\|M_s\|\varepsilon_s(\zeta_{ds}+\alpha_{ds})}{2} + \frac{K_{Ps}^2\|M_s^{-1}\|}{2}\right] \tag{47}$$

with $\varepsilon_m \ge 2$ and $\varepsilon_s \ge 2$. Using (47), $\dot{V}$ can be written as

$$\dot{V} \le -\lambda_{min}(\Omega_{dm})\|x_m\|^2 - \int_{(t-T_{ds}(t))}^{t} x_s^T(\eta)x_s(\eta)d\eta - \int_{(t-T_{dm}(t))}^{t} x_m^T(\eta)x_m(\eta)d\eta$$
$$-\lambda_{min}(\Omega_{ds})\|x_s\|^2 \tag{48}$$

where $\lambda_{min}(\Omega_{dm})$ and $\lambda_{min}(\Omega_{ds})$ are the smallest eigen values of the matrix Ω_{dm} and Ω_{ds}. Let us now take the integral from zero to time T, we have

$$V(T) - V(0) \le -\int_0^T \lambda_{min}(\Omega_{dm})\|x_m\|^2 dt - \int_0^T \lambda_{min}(\Omega_{ds})\|x_s\|^2 dt - \tag{49}$$
$$\int_0^T\int_{(t-T_{ds}(t))}^{t} x_s^T(\eta)x_s(\eta)d\eta dt - \int_0^T\int_{(t-T_{dm}(t))}^{t} x_m^T(\eta)x_m(\eta)d\eta dt$$

Applying Schwartz inequality for the last two terms and the bound on the time varying delay, equation (49) becomes

$$V(T) - V(0) \le -\int_0^T \lambda_{min}(\Omega_{dm})\|x_m\|^2 dt - \int_0^T \lambda_{min}(\Omega_{ds})\|x_s\|^2 dt + \tag{50}$$
$$\int_0^T \alpha_{ds}\int_{(t-T_{ds}(t))}^{t} \|x_s\|^2 d\eta dt + \int_0^T \alpha_{dm}\int_{(t-T_{dm}(t))}^{t} \|x_m\|^2 d\eta dt$$

After some manipulations on the last two terms, the bound on the position and velocity of the master-slave teleoperator signals can be written as

$$V(T) - V(0) \leq -\chi_{dm}\|x_m\|_2^2 - \chi_{ds}\|x_s\|_2^2 \tag{51}$$

where $\chi_{dm} = \lambda_{min}(\Omega_{dm}) - \alpha_{dm}^2$ and $\chi_{ds} = \lambda_{min}(\Omega_{ds}) - \alpha_{ds}^2$. If $T \to \infty$, then we have $\int_0^\infty \dot{V} dt \leq 0$. Then, we can conclude that $(q_m, q_s, \dot{q}_m, \dot{q}_s) \in \mathcal{L}_\infty$ and $(q_m, q_s, \dot{q}_m, \dot{q}_s) \in \mathcal{L}_2$. This implies that $\lim_{t\to\infty}\dot{q}_m = \lim_{t\to\infty}\dot{q}_s = \lim_{t\to\infty}q_m = \lim_{t\to\infty}q_s = 0$. Then, we can state our main results in the following Theorem 2.

Theorem 2: *Consider the coupled master-slave closed-loop teleoperator system (33), (34) with the passive inequality (5). Then, for the given μ_{dm} and μ_{ds}, there exists K_{Pm}, K_{Ps}, K_{Dm} and K_{Ds} as defined in (47) with $\gamma_m > 0$ and $\gamma_s > 0$ such that the position and velocity of the master-slave systems are asymptotically stable.*

Remark 1: Adaptation mechanism for master-slave teleoperator given by (8)-(9) may cause discontinuous property in the parameter estimates even after parameter converges to the actual one. To ensure that the parameter estimate remain bounded on the given set $\theta_{(m,s)} \in \Omega_{(m,s)}$, $\hat{\theta}_{(m,s)}$ can be adjusted with the smooth parameter projection scheme as

$$\dot{\hat{\theta}}_{(m,s)i} = [Proj(\hat{\theta}_{(m,s)}, \Phi_{(m,s)})]_i$$

for $\theta_{(m,s)i} \in \Omega_{(m,s)} = \{\theta_{(m,s)} \mid a_{(m,s)i} \leq \theta_{(m,s)i} \leq b_{(m,s)i}\}, 1 \leq i \leq p\}$ where, $\Phi_{(m,s)i}$ is the i-th element of the column vector $-\gamma_{(m,s)} Y^T (x_{(m,s)}) B_{(m1,s1)}^T x_{(m,s)}$ and $\delta_{(m,s)} > 0$ is chosen such that $\Omega_{(m,s)} \subset \Omega_{(m\delta,s\delta)}$ with $\Omega_{(m\delta,s\delta)} = \{\theta_{(m,s)} \mid a_{(m,s)i} - \delta_{(m,s)} \leq \theta_{(m,s)i} \leq b_{(m,s)i} + \delta_{(m,s)}\}, 1 \leq i \leq p\}$.

5 Conclusion

We have presented novel adaptive teleoperation systems for internet-based bilateral nonlinear teleoperation systems under time varying communication delay. In this design, the local-remote sites were coupled by delayed master-slave position signals. The Lyapunov-Krasovskii-like positive definite functional has been used to explore asymptotic control stability condition for both symmetric and unsymmetrical time varying communication delay case.

Acknowledgment

The work presented here is partially supported by the Natural Science and Engineering Research Council (NSERC) of Canada Postdoctoral Fellowship grant, Canada Research Chairs Program grant in Interactive Network, Computer and Teleoperation and Canada Research Chairs Program in Ambient Interactive Media and Communications.

References

1. Marescaus, J., et al.: Translantic robot-assisted telesurgery. Nature 413, 379–380 (2001)
2. Polushin, I.G.: Force reflecting teleoperation over wide-area networks, Ph.D. Thesis, Carleton University (2009)
3. Polushin, I.G., Rhinelander, J.P., Liu, P.X., Lung, C.-H.: A Scheme for virtual reality enhanced bilateral teleoperation with time delay. In: 2008 IEEE Instrumentation and Measurement Technology Conference, Victoria, BC, May 2008, pp. 1819–1822 (2008)
4. Lozano, R., Chopra, N., Spong, M.W.: Convergence analysis of bilateral teleoperation with constant human input. In: Proc. Amer. Control Conf., New York, pp. 1443–1448 (2007)
5. Polushin, I.G., Liu, P.X., Lung, C.-H.: A control scheme for stable force-reflecting teleoperation over IP networks. IEEE Trans. Syst., Man, Cybern. B, Cybern. 36(4), 930–939 (2006)
6. Wang, Z., Ho, D.W.C., Liu, X.: Robust filtering under randomly varying sensor delay with variance constraints. IEEE Trans. Circuits Syst. II, Exp. Briefs 51(6), 320–326 (2004)
7. Polushin, I., Liu, P.X., Lung, C.-H.: A force-reflection algorithm for improved transparency in bilateral teleoperation with communication delay. IEEE/ASME Trans. Mechatronics 12(3), 361–374 (2007)
8. Polushin, G., Liu, P.X., Lung, C.-H.: Projection-based force reflection algorithm for stable bilateral teleoperation over networks. IEEE Trans. Instrum. Meas. 57(9), 1854–1865 (2008)
9. Islam, S., Liu, P.X.: PD output feedback controller for industrial robotic manipulators. IEEE/ASME Transaction on Mechatronics (2009) (in Press)
10. Hua, C., Liu, P.X.: Convergence analysis of teleoperation systems with unsymmetric time-varying delays. IEEE Transaction on Circuits and Systems-II 56(3), 240–244 (2009)
11. Niemeyer, G., Slotine, J.-J.E.: Telemanipulation with time delays. International Journal of Robotic Research 23(9), 873–890 (2004)
12. Arcara, P., Melchiorri, C.: Control schemes for teleoperation with time delay: A comparative study. Robot. Auton. Syst. 38(1), 49–64 (2002)
13. Anderson, R.J., Spong, M.W.: Asymptotic stability of force reflecting teleoperators with time delay. International Journal of Robotic Research 11(2), 135–149 (1992)
14. Lozano, R., Chopra, N., Spong, M.W.: Convergence analysis of bilateral teleoperation with constant human input. In: Proc. American Control Conference, New York, pp. 1443–1448 (2007)
15. Hokayem, P.F., Spong, M.W.: Bilateral teleoperation: An historical survey. Automatica 42(12), 2035–2057 (2006)
16. Kosuge, K., Murayama, H., Takeo, K.: Bilateral feedback control of telemanipulators via Computer Network. In: Proc. of the IEEE/RSJ international Conference on Intelligent Robots and Systems, pp. 1380–1385 (1996)
17. Chopra, N., Spong, M.W., Hirche, S., Buss, M.: Bilateral teleoperation over the Internet: the time varying delay problem. In: Proc. of the American Control Conference, pp. 155–160 (2003)

18. Pan, Y.-J., Canudas-de-Wit, C., Sename, O.: A new predictive approach for bilateral teleoperation with applications to drive-by- wire systems. IEEE Trans. on Robotics 22(6), 1146–1162 (2006)
19. Anderson, R.J., Spong, M.W.: Bilateral control of teleoperators with time Delay. IEEE Trans. on Automatic Control 34(5), 494–501 (1989)
20. Park, J.H., Cho, H.C.: Sliding mode control of bilateral teleoperation systems with force-reflection on the internet. In: Proc. of the 2000 IEEE/RSJ international conference on intelligent robots and systems, pp. 1187–1192 (2000)

Online Incremental Learning of Inverse Dynamics Incorporating Prior Knowledge

Joseph Sun de la Cruz[1,2], Dana Kulić[1,2,*], and William Owen[2]

[1] Department of Electrical and Computer Engineering
[2] Department of Mechanical and Mechatronics Engineering
University of Waterloo
Waterloo, ON, Canada
{jsundela,dkulic,bowen}@uwaterloo.ca

Abstract. Recent approaches to model-based manipulator control involve data-driven learning of the inverse dynamics relationship of a manipulator, eliminating the need for any knowledge of the system model. Ideally, such algorithms should be able to process large amounts of data in an online and incremental manner, thus allowing the system to adapt to changes in its model structure or parameters. Locally Weighted Projection Regression (LWPR) and other non-parametric regression techniques have been applied to learn manipulator inverse dynamics. However, a common issue amongst these learning algorithms is that the system is unable to generalize well outside of regions where it has been trained. Furthermore, learning commences entirely from 'scratch,' making no use of any a-priori knowledge which may be available. In this paper, an online, incremental learning algorithm incorporating prior knowledge is proposed. Prior knowledge is incorporated into the LWPR framework by initializing the local linear models with a first order approximation of the available prior information. It is shown that the proposed approach allows the system to operate well even without any initial training data, and further improves performance with additional online training.

Keywords: Robotics, Learning, Control.

1 Introduction

Control strategies that are based on the knowledge of the dynamic model of the robot manipulator, known as model-based controllers, can present numerous advantages such as increased performance during high-speed movements, reduced energy consumption, improved tracking accuracy and the possibility of compliance [1]. However, this performance is highly dependant upon the accurate representation of the robot's dynamics, which includes precise knowledge of the inertial parameters of link mass, centre of mass and moments of inertia, and friction parameters [2]. In practice, obtaining such a model is a challenging task

---

* This research is funded in part by the Natural Sciences and Engineering Research Council of Canada

which involves modeling physical processes that are not well understood or difficult to model, such as friction [3] and backlash. Thus, assumptions concerning these effects are often made to simplify the modeling process, leading to inaccuracies in the model. Furthermore, uncertainties in the physical parameters of a system may be introduced from discrepancies between the manufacturer data and the actual system [4]. Changes to operating conditions can also cause the structure of the system model to change, thus resulting in degraded performance.

Traditionally, adaptive control strategies have been used to cope with parameter uncertainty [2], [5]. This allows parameters of the dynamic model such as end effector load, inertia and friction to be estimated online [2]. However, such methods are still reliant upon adequate knowledge of the structure of the dynamic model and are thus susceptible to modeling errors and changes in the model structure. Furthermore, accurate estimation of the model parameters requires sufficiently rich data sets [2],[6], and even when available, may result in physically inconsistent parameters meaning that constraints must be applied to the identified parameters [2],[7].

Newer approaches to manipulator control involve data-driven learning of the inverse dynamics relationship of a manipulator, thus eliminating the need for any a-priori knowledge of the system model. Unlike the adaptive control strategy, these approaches do not assume an underlying structure but rather attempt to infer the optimal structure to describe the observed data. Thus, it is possible to encode nonlinearities whose structure may not be well-known. Solutions to this form of supervised learning approach can be broadly categorized into two types [8]. Global methods such as Gaussian Process Regression (GPR) [9] and Support Vector Regression (SVR) [10], and local methods such as Locally Weighted Projection Regression (LWPR). Recent studies comparing these learning methods [11] show that while SVR and GPR can potentially yield higher accuracy than LWPR, their computational cost is still prohibitive for online incremental learning. Furthermore, a major issue of online learning approaches such as LWPR is the failure to generate appropriate control signals away from the trained region of the workspace [12].

Central to the learning approaches introduced thus far is the need for large amounts of relevant training data in order to yield good prediction performance. These algorithms most often assume that there is no prior knowledge of the system dynamics, and thus learning commences entirely from 'scratch', making no use of the rigid body dynamics (RBD) model from the well-established field of analytical robotics [1], which provide a global characterization of the dynamics. Recent research [13] has been done to incorporate the full RBD model into the GPR algorithm to improve its performance in terms of generalization and prediction accuracy in the context of real-time robot control. However, in [13], the high computational requirements of GPR still prohibit incremental online updates from being made, which is a highly desirable feature of any learning algorithm for robot control [14].

In many cases, partial information about the robot dynamics may be available, such as for example, the link masses, which may be helpful in bootstrapping a learning model and accelerating the rate of learning. This paper proposes a novel approach for on-line, incremental learning for model-based control, capable of incorporating full or partial a-priori model knowledge. This is done by using first order approximations of the RBD equation to initialize the local models of the LWPR technique. The developed algorithm can be applied both when full knowledge of the RBD model is available, and also when only partial information (e.g. only the gravity loading vector) is known.

The remainder of this paper is organized as follows. In section 2, the dynamic model of a typical rigid body manipulator and model based control strategies are reviewed. Section 3 briefly overviews the previously developed LWPR algorithm for learning control. Section 4 presents the proposed online, incremental algorithm for incorporating full or partial a-priori knowledge into LWPR. Section 5 presents the simulations and results. Finally, in Section 6, the conclusions and next steps are discussed.

2 Overview of Model-Based Control

The rigid body dynamic equation (RBD) of a manipulator characterizes the relationship between its motion (position, velocity and acceleration) and the joint torques that cause these motions [1]:

$$\mathbf{M(q)\ddot{q} + C(q,\dot{q}) + G(q) = \tau} \tag{1}$$

where $\mathbf{q}$ is the nx1 vector of generalized coordinates consisting of the n joint angles for an n-degree of freedom (DOF) manipulator, $\mathbf{M(q)}$ is the nxn inertia matrix, $\mathbf{C(q,\dot{q})}$ is the nx1 Coriolis and centripetal force vector, $\mathbf{G(q)}$ is the nx1 gravity loading vector and $\boldsymbol{\tau}$ is the nx1 torque vector.

Equation (1) does not include additional torque components caused by friction, backlash, actuator dynamics and contact with the environment. If accounted for, these components are modeled as additional terms in (1).

Model-based controllers are a broad class of controllers which apply the joint space dynamic equation (1) to cancel the nonlinear and coupling effects of the manipulator. A common example of this is the computed torque approach [2],[1] in which the control signal $\mathbf{u}$ is composed of the computed torque signal, $\mathbf{u_{CT}}$, which is set to the torque determined directly from the inverse of the dynamic equations (1). This term globally linearizes and decouples the system, and thus a linear controller can be applied for the feedback term, $\mathbf{u_{FB}}$, which stabilizes the system and provides disturbance rejection. Typically a Proportional-Derivative (PD) feedback scheme is used for this term.

Desirable performance of the computed torque approach is contingent upon the assumption that the values of the parameters in the dynamic model (1) match the actual parameters of the physical system. Otherwise, imperfect cancelation of the nonlinearities and coupling in (1) occurs. Hence, the resulting system is not fully linearized and decoupled and thus higher feedback gains are necessary to achieve good performance.

3 Learning Inverse Dynamics

Learning inverse dynamics involves the construction of a model directly from measured data. The problem of learning the inverse dynamics relationship in the joint space can be described as the map from joint positions, velocities and accelerations to torques

$$(\mathbf{q}, \dot{\mathbf{q}}, \ddot{\mathbf{q}}_\mathbf{d}) \mapsto \boldsymbol{\tau} \qquad (2)$$

where $\boldsymbol{\tau}$ is the nx1 torque vector, and $\mathbf{q}$ is the nx1 vector of generalized coordinates. This means that the mapping has an input dimensionality of $3n$ and an output dimensionality of n.

3.1 Locally Weighted Projection Regression

Locally Weighted Projection Regression (LWPR) approximates the nonlinear inverse dynamics equation (1) with a set of piecewise local linear models based on the training data that the algorithm receives. Formally stated, this approach assumes a standard regression model of the form

$$\mathbf{y} = f(\mathbf{X}) + \epsilon \qquad (3)$$

where $\mathbf{X}$ is the input vector, $\mathbf{y}$ the output vector, and ϵ a zero-mean random noise term. For a single output dimension of $\mathbf{y}$, denoted by y_i, given a data point $\mathbf{X_c}$, and a subset of data close to $\mathbf{X_c}$, with the appropriately chosen measure of closeness, a linear model can be fit to the subset of data such that

$$y_i = \boldsymbol{\beta}^T \mathbf{X} + \epsilon \qquad (4)$$

where $\boldsymbol{\beta}$ is the set of parameters of the hyperplane that describe y_i. The region of validity, termed the receptive field (RF) [8] is given by

$$w_{ik} = \exp(-\frac{1}{2}(\mathbf{X} - \mathbf{X_c})^\mathbf{T}\mathbf{D_{ik}}(\mathbf{X} - \mathbf{X_c})) \qquad (5)$$

where w_{ik} determines the weight of the kth local linear model of the ith output dimension (i.e. the ik^{th} local linear model), $\mathbf{X_c}$ is the centre of the kth linear model, $\mathbf{D_{ik}}$ corresponds to a positive semidefinite distance parameter which determines the size and shape of the ik^{th} receptive field. Given a query point $\mathbf{X}$, LWPR calculates a predicted output

$$\hat{y}_i(X) = \sum_{k=1}^{K} w_{ik}\hat{y}_{ik} / \sum_{k=1}^{K} w_{ik} \qquad (6)$$

where K is the number of linear models, $\hat{y}_{ik}$ is the prediction of the ik^{th} local linear model given by (4) which is weighed by w_{ik} associated with its receptive field. Thus, the prediction $\hat{y}_i(\mathbf{X})$ is the weighted sum of all the predictions of the local models, where the models having receptive fields centered closest to the query point are most significant to the prediction. This prediction is repeated i times for each dimension of the output vector $\mathbf{y}$.

Determining the set of parameters $\boldsymbol{\beta}$ of the hyperplane is done via regression, but can be a time consuming task in the presence of high-dimensional input data, which is a characteristic of large numbers of DOF in robotic systems. To reduce the computational effort, LWPR assumes that the data can be characterized by local low-dimensional distributions, and attempts to reduce the dimensionality of the input space $\mathbf{X}$ using Partial Least Squares regression (PLS). PLS fits linear models using a set of univariate regressions along selected projections in input space which are chosen according to the correlation between input and output data [15].

The distance parameter, $\mathbf{D}$, (5) can be learned for each local model through stochastic gradient descent given by

$$\mathbf{d}^{n+1} = \mathbf{d}^n - a\frac{\partial J_{cost}}{\partial \mathbf{d}} \qquad (7)$$

where a is the learning rate for gradient descent, $\mathbf{D} = \mathbf{d}^\mathrm{T}\mathbf{d}$ and J_{cost} is a penalized leave-one-out cross-validation cost function which addresses the issue of over-fitting of the data [16]. The number of receptive fields is also automatically adapted [8]. Receptive fields are created if for a given training data point, no existing receptive field possesses a weight w_i (5) that is greater than a threshold value of w_{gen}, which is a tunable parameter. The closer w_{gen} is set to one, the more overlap there will be between local models. Conversely, if two local models produce a weight greater than a threshold w_{prune}, the model whose receptive field is smaller is pruned.

Because training points are not explicitly stored but rather encoded into simple local linear models, the computational complexity of LWPR remains low, and is not affected by the number of training points. Assuming that the number of projection directions found by the PLS algorithm remains small and bounded, the complexity of the basic algorithm varies linearly with input dimension [8].

4 Incorporating A-Priori Knowledge into LWPR

Despite its ability to perform online, incremental updates due to its low computational cost [8],[11], as a result of the local learning approach of LWPR, performance deteriorates quickly as the system moves outside of the region of state space it has been trained in [11],[12]. In order to improve the generalization performance of LWPR, a-priori knowledge from the RBD model (1) is incorporated into the LWPR algorithm as a method of initializing the system. This is done by initializing the receptive fields in the LWPR model with a first order approximation of the available RBD model.

$$\boldsymbol{\beta} \leftarrow \left.\frac{\partial \tau}{\partial x}\right|_{x=x_*} \qquad (8)$$

where x_* is a new query point for which no previous training data has been observed and $\tau(x_*)$ is the known or partially known dynamics. For a query point x_* and known RBD equation $\tau(x_*)$, the algorithm shown in Figure 1 for initializing the LWPR model is added to the standard prediction procedure of LWPR:

> **if** there is no existing RF centered near x_* **then**
>   Compute (β) according to (8)
>   Initialize a new RF centered at x_* with hyperparameters β
> **else**
>   Compute prediction with standard LWPR procedure (6)
> **end if**

Fig. 1. LWPR Initialization Pseudocode

The determination of whether an existing RF is centered near x_* is made in the same way as determining whether a new RF should be added, as described in Section 3.1, i.e., if there is no existing RF that produces a weight (5) greater than w_{gen} given x_*, the initialization procedure is applied.

By evaluating the partial derivative of the full (1) or partially known RBD equation at the query point, a first-order linear approximation of the system dynamics is obtained, and is used to initialize the model at that point. The size of the receptive field is initially set as a tunable parameter, and is eventually learned optimally through gradient descent [8].

5 Simulations

The proposed approach is evaluated in simulation on a 6-DOF Puma 560 robot using the Robotics Toolbox (RTB) [17]. The open-source LWPR [8] was modified to incorporate full or partial initialization with the RBD model. In order to closely simulate the nonlinearities present in a physical robot, the effects of Coloumb and viscous friction were accounted for in simulation by the following model:

$$\tau_f = C_f sign(q) + V_f \dot{q} \qquad (9)$$

where τ_f is the torque due to Coulomb and viscous friction, C_f is the Coulomb friction constant, and V_f is the viscous friction constant. The friction constants were obtained from the defaults for the Puma 560 in the RTB.

Furthermore, to simulate the effects imprecise knowledge of the inertial parameters of the robot, a 10% percent error in the inertial parameters of the a-priori knowledge was introduced.

The LWPR algorithm is applied to learn the joint-space mapping in (2). Full a-priori knowledge from the RBD model in (1), as well as partial knowledge from the gravity loading vector $\mathbf{G}(\mathbf{q})$, is used to initialize each algorithm. Standard computed torque control is also implemented with the same RBD model used to initialize the learning algorithms. Although LWPR incorporates many algorithms which enable the system to automatically adjust its parameters for optimal performance, initial values of these parameters can significantly impact the convergence rate. The initial value for the distance metric $\mathbf{D}$ (5) dictates how large a receptive field is upon initialization. Too small a value of $\mathbf{D}$ (corresponding to large receptive fields) tends to delay convergence while a larger value of $\mathbf{D}$ results in overfitting of the data [8]. The initial value of the learning rate α(7) determines the rate at which prediction error converges. Too high a value leads to instability and too low a value leads to a slow convergence. These parameters were generally tuned through a trial-and-error process which involved monitoring the error of the predicted values during the training phase.

Tracking performance of the controllers is evaluated on a 'star-like' asterisk pattern [7] as seen in Figure 2. This trajectory is obtained by first moving in a straight line outwards from a centre point, then retracing back inwards to the centre, repeating this pattern in eight different directions in a sequential manner. This pattern produces high components of acceleration, and thus requires model-based control for tracking accuracy. Two speeds are tested, slow and fast, with the maximum velocities set to 0.5 and 2.5 m/s respectively, where the fast speed corresponds to roughly 80% of the robot's maximum velocity. At low speeds, it is expected that the gravity term $\mathbf{G}(\mathbf{q})$ (1) will be dominant. The high-speed trajectory is used to ensure that the Coriolis and centripetal components, $\mathbf{C}(\mathbf{q},\dot{\mathbf{q}})$, (1) of the dynamics are excited to a sufficient extent so that $\mathbf{G}(\mathbf{q})$ is no longer dominant. A numerical inverse kinematics algorithm [18] is used to convert the workspace trajectories into the joint space.

Figure 2 depicts the task space tracking performance of the computed torque (CT) controller using a model with 10% error in the inertial and friction parameters of the system tracking the slow asterisk trajectory. It is evident that the imprecise knowledge of the RBD model is responsible for poor tracking results. The same RBD model is used to initialize the LWPR model, and the resulting controller is trained online while tracking the slow asterisk trajectory. Table 1 shows the joint space tracking performance after one cycle of the trajectory

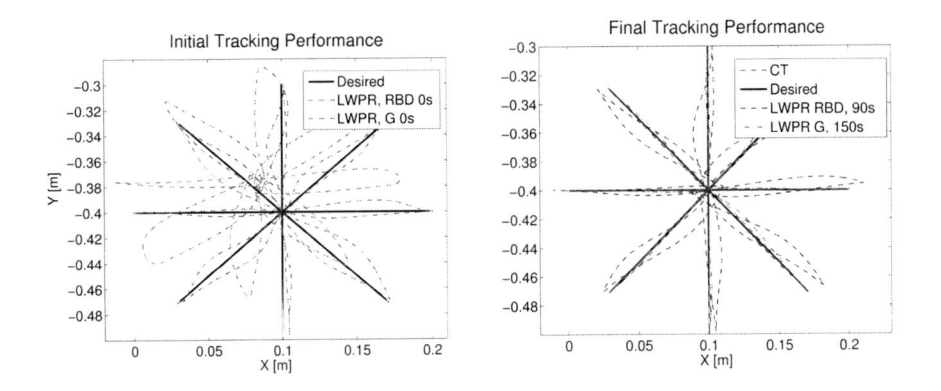

Fig. 2. Task Space tracking results for the slow trajectory

(approximately 11 seconds). Because the LWPR model is initialized with a set of linear approximations of the RBD equation, the initial performance of LWPR is not as good as the CT method. Table 1 shows the joint space tracking after 90 seconds of training. Due to the high computational efficiency of LWPR, incremental updates are made at a high rate of 400 Hz. As time progresses, LWPR is able to collect more training data and eventually outperforms the CT method by learning the nonlinear behaviour of Coloumb and viscous friction and compensating for the initial inaccurate knowledge of the RBD equation. The model learned on the slow trajectory is then applied to tracking the fast trajectory. The performance suffers only slightly compared to the initial performance on the slow trajectory, as the system was initialized with a full, but slightly inaccurate RBD model. As seen in Table 1, further training for 45 seconds mitigates this issue and results in tracking performance that is nearly as good as compared to that of the slow trajectory. Performance of the CT controller was very similar for both slow and fast trajectories, and is also shown for a comparison.

Table 2 illustrates the joint space tracking performance of the LWPR model when initialized with only the gravity loading vector of the RBD equation while tracking the slow asterisk trajectory. The same table also illustrates the performance after 150 seconds of training. Similarly to the case of full knowledge of the RBD, with sufficient training, LWPR is able to compensate for friction and clearly outperforms the CT controller. However, since the system was initialized with only partial knowledge of the RBD equation, it has taken longer to achieve the same tracking performance in the case of full RBD knowledge. The learned model is once again applied to track the fast trajectory. Compared to the case of full RBD initialization, the performance with only gravity initialization on the faster trajectory is much worse, as seen numerically in Table 2. Performance of the CT controller was very similar for both slow and fast trajectories, and is also shown for a comparison.

Table 1. RMS tracking error with full knowledge (deg)

Joint	1	2	3	4	5	6	Avg
Slow Trajectory, Initial	1.25	2.26	1.74	0.65	0.75	0.80	1.24
Slow Trajectory, 90s	0.75	0.94	0.82	0.25	0.38	0.45	0.60
Fast Trajectory, Initial	0.95	1.20	1.05	0.36	0.50	0.65	0.79
Fast Trajectory, 45s	0.82	1.00	0.93	0.35	0.45	0.52	0.68
CT Controller	1.10	2.05	0.95	0.55	0.65	0.62	0.98

Because the learning of the Coriolis and centripetal term, $\mathbf{C}(\mathbf{q},\dot{\mathbf{q}})$, is localized about a specific joint position and velocity, the learned model from the slow trajectory is no longer fully applicable to the fast trajectory. Furthermore, since the system was only initialized with the gravity loading vector, the system performs poorly, and requires significantly more training time to perform well again. This is illustrated in Table 2.

Without the use of a-priori knowledge, learning algorithms were typically initialized with large training data sets obtained through motor babbling [8],

Table 2. RMS tracking error with partial knowledge (deg)

	1	2	3	4	5	6	Avg
Slow Trajectory, Initial	2.57	4.78	3.62	0.95	0.80	0.75	2.25
Slow Trajectory, 150s	0.80	0.95	0.80	0.22	0.41	0.44	0.91
Fast Trajectory, Initial	2.95	5.27	4.16	1.25	1.00	0.94	2.60
Fast Trajectory, 60s	0.86	1.10	0.91	0.32	0.48	0.51	0.70
CT Controller	1.10	2.05	0.95	0.55	0.65	0.62	0.98

[19], [11] in order to achieve decent tracking performance after initialization. By incorporating a-priori knowledge of the RBD equation, whether partial or full, the proposed systems are able to perform reasonably well from the start, even without undergoing such an initialization procedure.

6 Conclusions and Future Work

In this paper, an incremental, online learning method is proposed for approximating the inverse dynamics equation while incorporating full or partial knowledge of the RBD equation. The LWPR algorithm was chosen due to its computational efficiency which allows incremental, online learning to occur. In order to improve the generalization performance of the algorithm, prior knowledge was incorporated into the LWPR framework by initializing its local models with a first-order approximation of the RBD equation. In cases where the full RBD is not known due to lack of knowledge of the inertial parameters, partial information (such as the gravity vector) can still be encoded into the model.

When this occurs, more training time is required to achieve similar performance as with full RBD initialization. In either case, the generalization performance of LWPR is greatly improved, as the model is able to yield acceptable tracking results without having seen any relevant training data beforehand. It is also shown that this approach is able to compensate for the nonlinear effects of friction, as well as the initial inaccuracy in the inertial parameters. Future work will involve more a more comprehensive evaluation of the proposed methods, including experimental results on physical robots. Secondly, other machine learning algorithms for function approximation will be investigated, such as Gaussian Process Learning [9].

References

1. Sciavicco, L., Scicliano, B.: Modelling and Control of Robot Manipulators, 2nd edn. Springer, Heidelberg (2000)
2. Craig, J., Hsu, P., Sastry, S.: Adaptive control of mechanical manipulators. In: Proc. 1986 IEEE Int Conf. on Robotics and Auto., April 1986, vol. 3, pp. 190–195 (1986)
3. Armstrong-Hélouvry, B., Dupont, P., de Wit, C.C.: A survey of models, analysis tools and compensation methods for the control of machines with friction. Automatica 30(7), 1083–1138 (1994)

4. Ayusawa, K., Venture, G., Nakamura, Y.: Identification of humanoid robots dynamics using floating-base motion dynamics. In: IEEE/RSJ Int. Conf. on Intelligent Robots and Systems, IROS 2008, pp. 2854–2859 (September 2008)
5. Ortega, R., Spong, M.W.: Adaptive motion control of rigid robots: a tutorial. In: Proc of the 27th IEEE Conf on Decision and Control, vol. 2, pp. 1575–1584 (December 1988)
6. Khosla, P.: Categorization of parameters in the dynamic robot model. IEEE Transactions on Robotics and Automation 5(3), 261–268 (1989)
7. Nakanishi, J., Cory, R., Mistry, M., Peters, J., Schaal, S.: Operational Space Control: A Theoretical and Empirical Comparison. The Int Journal of Robotics Research 27(6), 737–757 (2008)
8. Vijayakumar, S., D'souza, A., Schaal, S.: Incremental online learning in high dimensions. Neural Comput. 17(12), 2602–2634 (2005)
9. Rasmussen, C., Williams, C.: Gaussian Processes for Machine Learning. MIT Press, Cambridge (2006)
10. Smola, A.J.: A tutorial on support vector regression. Statistics and Computing 14, 199–222 (2004)
11. Nguyen-Tuong, D., Peters, J., Seeger, M., Schölkopf, B., Verleysen, M.: Learning inverse dynamics: A comparison (2008), http://edoc.mpg.de/420029
12. Sun de la Cruz, J., Kulić, D., Owen, W.: Learning inverse dynamics for redundant manipulator control. In: 2010 International Conf. on Autonomous and Intelligent Systems (AIS), pp. 1–6 (June 2010)
13. Nguyen-Tuong, D., Peters, J.: Using model knowledge for learning inverse dynamics. In: IEEE Int. Conf. on Robotics and Automation, pp. 2677–2682 (2010)
14. Nguyen-Tuong, D., Scholkopf, B., Peters, J.: Sparse online model learning for robot control with support vector regression. In: IEEE/RSJ Int. Conf. on Intelligent Robots and Systems, IROS 2009, pp. 3121–3126 (October 2009)
15. Schaal, S., Atkeson, C.G., Vijayakumar, S.: scalable techniques from nonparameteric statistics for real-time robot learning. Applied Intelligence 16(1), 49–60 (2002)
16. Schaal, S., Atkeson, C.G.: Constructive incremental learning from only local information. Neural Computation 10(8), 2047–2084 (1998)
17. Corke, P.I.: A robotics toolbox for MATLAB. IEEE Robotics and Automation Magazine 3(1), 24–32 (1996)
18. Chiaverini, S., Sciavicco, L., Siciliano, B.: Control of robotic systems through singularities. Lecture Notes in Control and Information Sciences, vol. 162, pp. 285–295. Springer, Heidelberg (1991)
19. Peters, J., Schaal, S.: Learning to Control in Operational Space. The Int. Journal of Robotics Research 27(2), 197–212 (2008)

Experimental Comparison of Model-Based and Model-Free Output Feedback Control System for Robot Manipulators

S. Islam[1,2,*], P.X. Liu[3], and A. El Saddik[1]

[1] School of Information Technology and Engineering (SITE),
University of Ottawa, 800 King Edward Road, Ottawa, Canada, K1N 6N5
[2] School of Engineering and Applied Sciences,
Harvard University, 29 Oxford Street, Cambridge, MA 02138 USA
[3] Department of Systems and Computer Engineering,
Carleton University, Ottawa, Ontario, Canada, K1S 5B6
sislam@sce.carleton.ca, sislam@seas.harvard.edu

Abstract. In this paper, we design and implement model-free linear and model-based nonlinear adaptive output feedback method for robot manipulators. The model-free design uses only proportional and derivative (PD) error terms for trajectory tracking control of nonlinear robot manipulators. The design is very simple in the sense that it does not require *a priori* knowledge of the system dynamics. The model-based output feedback method combines PD controller terms with nonlinear adaptive term to cope with uncertain parametric uncertainty. The unknown velocity signals for two output feedback method are generated by the output of the model-free linear observer. Using asymptotic analysis, tracking error bounds for both output feedback design are shown to be bounded and their bounds can be made closed to the bound obtained with state feedback design by using small value of observer design parameters. Finally, we experimentally compare both method on a 3-DOF Phantom^{TM} robot manipulator.

Keywords: Nonlinear Adaptive Control; Robotics; Observer; Output Feedback; Phantom^{TM} Robot.

1 Introduction

Over the last decades, different types of output feedback methods have been reported in the literature for uncertain nonlinear mechanical systems [2, 6, 10, 12, 15, 16, 19, 21, and many others]. The unmeasured velocity signals in these methods are obtained by the output of either linear or nonlinear observer. These designs can be grouped into two categories as model-based and model-free observer-controller algorithms. The model based design comprises PD term with nonlinear adaptive controller to deal with structured and/or unstructured parametric uncertainty. The control performance under model based adaptive output feedback design either linear observer or nonlinear observer, relies on *a priori*

---

* Corresponding Author.

knowledge of the system dynamics. As a result, if the modeling errors uncertainty becomes large value, then the model based design either for state or output feedback will exhibit poor transient tracking performance. To improve the tracking response, the design demands high observer-controller gains in order to minimize the modeling error uncertainty. The requirement of high observer-controller gains makes the model based output feedback design even more complex in realtime application as high observer speed may cause high-frequency chattering and infinitely fast switching control phenomenon during transient state recovery. Indeed, most of today's advanced commercial/practical robot system uses only PD or proportional-Integral-derivative (PID) controllers [7, and references therein]. These control designs do not require *a priori* known system dynamics which makes the design popular in industrial robot control applications. The practical problem is, however, that advanced industrial systems do not provide velocity sensors due to the constraints of weight and cost [3, 18]. In order to obtain velocity signals, one requires to differentiate the position measurements obtained from encoders or resolvers which often leads to high-frequency control chattering phenomenon due to the derivative action of the noisy joint position signals [18]. As a consequence, the performance of the PD and PID controller is limited as, in practice, control chattering activity increase with the increase of the values of the controller gains. To deal with this open practical problem, authors in [4] introduced model-free PD-type observer-controller mechanism for robot manipulators. It has been proven that the observer-controller design parameters can be tuned to ensure desired control objective. In this paper, we experimentally compare model-based and model-free output feedback method on 3-DOF Phantom^{TM} robot system. To the best of our knowledge, there has not been any result reported in the literature where model-based nonlinear adaptive output feedback control method experimentally compared with the model-free output feedback design. Our prime focus in these experiments is to experimentally compare the tracking performance of these two designs with respect to different operating conditions. In the first design, we consider linear PD-type output feedback control method for trajectory tracking of nonlinear robotic systems. The second design combines PD controller terms with nonlinear adaptive term to cope with uncertain parametric uncertainty. The design has two steps. In the first step, we design state feedback controller by assuming that position-velocity signals are available for feedback. We obtain the bound on the tracking error signals for the closed-loop error model via using Lyapunov function. It has been shown the tracking error bound can be made arbitrarily small by increasing the minimal eigenvalue of the control gains. In the second step, we use the linear observer to estimate unknown velocity signals to develop output feedback control law. With the help of asymptotic analysis of the singularly perturbed closed-loop model, it is proven that the observer and tracking error variables are bounded and their bounds can be made closed to a small neighborhood of the origin that obtained under (position-velocity) state feedback design. These arguments are experimentally compared for both output feedback design on a 3-DOF Phantom^{TM} haptic device.

The rest of the paper is organized as follows: Section 2 derives model-based and model-free output-feedback control algorithms. The error bound analysis is also given briefly in section 2. The two methods are implemented and evaluated on 3-DOF Phantom™ robot systems in section 3. Section 4 concludes the paper.

2 Observer-Based Output Feedback Control Methods: Model-Free and Model-Based Design

Let us first consider the equation of motion for an n-link robot manipulators [8, 11, 14, 15, 16, 17, 20] is given by

$$M(q)\ddot{q} + C(q,\dot{q})\dot{q} + G(q) = \tau \tag{1}$$

where $q \in \Re^n$ is the joint position vector, $\dot{q} \in \Re^n$ is the joint velocity vector, $\ddot{q} \in \Re^n$ is the joint acceleration vector, $\tau \in \Re^n$ is the input torque, $M(q) \in \Re^{n \times n}$ is the symmetric and uniformly positive definite inertia matrix, $C(q,\dot{q})\dot{q} \in \Re^n$ is the coriolis and centrifugal loading vector, and $G(q) \in \Re^n$ is the gravitational loading vector. We now represent the robot model (1) in the error state space form as follows

$$\dot{e}_1 = e_2, \dot{e}_2 = \phi_1(e) + \phi_2(e_1)\tau - \ddot{q}_d \tag{2}$$

where $e = [e_1^T, e_2^T]^T$, $\phi_1(e) = -M(e_1 + q_d)^{-1}[C(e_1 + q_d, e_2 + \dot{q}_d)(e_2 + \dot{q}_d) + G(e_1 + q_d)]$, $\phi_2(e_1) = M(e_1 + q_d)^{-1}$, $e_1 = (q - q_d)$, $e_2 = (\dot{q} - \dot{q}_d)$, and $q_d \in \Re^n$ is the desired tracking task for the system to follow. The reference trajectory $q_d(t)$, its first and second derivatives $\dot{q}_d$ and $\ddot{q}_d$ are bounded as $Q_d = [q_d, \dot{q}_d, \ddot{q}_d]^T \in \Omega_d \subset \Re^{3n}$ with compact set Ω_d. Let us now derive the control law under the position and velocity based PD feedback controller given by

$$\tau(e_1, e_2) = -K\lambda e_1 - Ke_2 \tag{3}$$

with $\lambda > 0$ and $K \in \Re^{n \times n}$. The filtered version of this PD controller usually used in advanced industrial robot control algorithm, see for example [18]. Then, we can show that, for any given initial error states $e(0)$, there exists a controller gains K such that the tracking error signals are ultimately bounded by a bound that can be made arbitrarily small closed to origin by increasing the minimal eigenvalue of the control gains K. We now design PD-based output feedback controller for the system (1) as

$$\tau(e_1, \hat{e}_2) = -K(\hat{e}_2 + \lambda e_1) \tag{4}$$

with $\hat{e}_2$ designed as

$$\dot{\hat{e}}_1 = \hat{e}_2 + \frac{H_1}{\epsilon}\tilde{e}_1, \dot{\hat{e}}_2 = \frac{H_2}{\epsilon^2}\tilde{e}_1$$

In view of the controller and observer structure, we can notice that the design is independent of the system dynamics and parameters. Then, the control input with linear observer can be written as $\tau(e_1, \hat{e}_2) = -K\lambda e_1 - Ke_2 + K\tilde{e}_2$. Substitute this input $\tau(e_1, \hat{e}_2)$ to formulate the closed-loop tracking error model as $\dot{e}_1 = e_2$ and $\dot{e}_2 = M^{-1}[-C(e_1 + q_d, e_2 + \dot{q}_d)e_2 - C(e_1 + q_d, e_2 + \dot{q}_d)\dot{q}_d - K\lambda e_1 - Ke_2 + K\tilde{e}_2] - \ddot{q}_d$. We then construct observer error dynamics as

$$\dot{\tilde{e}} = A_0 \tilde{e} + B f_o(Q_d, e, \tilde{e}) \tag{5}$$

where $f_o(Q_d, e, \tilde{e}) = [M^{-1}(-C(e_1 + q_d, e_2 + \dot{q}_d)e_2 - C(e_1 + q_d, e_2 + \dot{q}_d)\dot{q}_d - K\lambda e_1 - Ke_2 + K\tilde{e}_2) - \ddot{q}_d]$, $A_0 = \begin{bmatrix} -H_1 & I_{n\times n} \\ -H_2 & 0_{n\times n} \end{bmatrix}$ and $B = \begin{bmatrix} 0_{n\times n} \\ I_{n\times n} \end{bmatrix}$. Now, we represent the observer-controller closed-loop error dynamics in singularly perturbed form, where the observer errors transform in newly defined variables η as

$$\dot{e} = f_s(Q_d, e, \epsilon, \eta) \tag{6}$$
$$\epsilon \dot{\eta} = A_0 \eta + B\epsilon f_o(Q_d, e, \epsilon, \eta) \tag{7}$$

where $f_s(Q_d, e, \epsilon, \eta) = [e_2; M^{-1}(-C(e_1 + q_d, e_2 + \dot{q}_d)e_2 - C(e_1 + q_d, e_2 + \dot{q}_d)\dot{q}_d - K\lambda e_1 - Ke_2 + K\eta_2) - \ddot{q}_d]$ and $f_o(Q_d, e, \epsilon, \eta) = M^{-1}(-C(e_1 + q_d, e_2 + \dot{q}_d)e_2 - C(e_1 + q_d, e_2 + \dot{q}_d)\dot{q}_d - K\lambda e_1 - Ke_2 + K\eta_2) - \ddot{q}_d$ with $\epsilon\eta_1 = \tilde{e}_1$, $\eta_2 = \tilde{e}_2$ and a small positive parameter ϵ. The functions $f_s(Q_d, e, \epsilon, \eta)$ and $f_o(Q_d, e, \epsilon, \eta)$ represent the effects of the nonlinear dynamics of the manipulator and the effects due to the observer errors, respectively,. The model (6), (7) is in a standard nonlinear singularly perturbed form. Then, for any given compact set $(e(0), \hat{e}(0)) \in \Omega_b$, it can be shown that there exists observer-controller gains ϵ_1^* and K such that, for every $0 < \epsilon < \epsilon_1^*$, the state variables (e, η) of the closed loop system are bounded by a bound that can be made arbitrarily small closed to the origin obtained under PD based state feedback design. This implies that the tracking error bound under PD based state feedback design can be recovered by using PD-output feedback design by using small value of ϵ [4]. We now derive model-based nonlinear adaptive output feedback controller for robot manipulators. We first consider adaptive state feedback control algorithm [16 and references therein]

$$\tau(e, Q_d, \hat{\theta}) = Y(e, \dot{q}_r, \dot{q}_d, \ddot{q}_d)\hat{\theta} - K_P e_1 - K_D e_2, \dot{\hat{\theta}} = -\Gamma Y^T(e, \dot{q}_r, \dot{q}_d, \ddot{q}_d)S \tag{8}$$

where $Y(e, \dot{q}_r, \dot{q}_d, \ddot{q}_d)\hat{\theta} = \hat{M}(q)\ddot{q}_d + \hat{C}(q, \dot{q}_r)\dot{q}_d + \hat{G}(q)$, $K_P \in \Re^{n\times n}$, $K_D \in \Re^{n\times n}$, $S = e_2 + \lambda e_1$, $\dot{q}_r = (\dot{q}_2 - \lambda e_1)$, $\lambda = \frac{\lambda_0}{1+\|e_1\|}$, $\lambda_0 > 0$ and $\hat{M}, \hat{C}(.)$ and $\hat{G}(.)$ define the estimates of the $M(.), C(.)$ and $G(.)$, respectively,. We now replace the velocity signals in (8) to develop an adaptive output feedback controller as

$$\tau(\hat{e}, Q_d, \hat{\theta}) = Y(\hat{e}, \dot{\hat{q}}_r, \dot{q}_d, \ddot{q}_d)\hat{\theta} - K_P e_1 - K_D \hat{e}_2, \dot{\hat{\theta}} = -\Gamma Y^T(\hat{e}, \dot{\hat{q}}_r, \dot{q}_d, \ddot{q}_d)\hat{S} \tag{9}$$

where $Y(\hat{e}, \dot{\hat{q}}_r, \dot{q}_d, \ddot{q}_d)\hat{\theta} = \hat{M}(q)\ddot{q}_d + \hat{C}(q, \dot{\hat{q}}_r)\dot{q}_d + \hat{G}(q)$, $K_P \in \Re^{n\times n}$, $K_D \in \Re^{n\times n}$, $\hat{S} = \hat{e}_2 + \lambda e_1$, $\dot{\hat{q}}_r = (\dot{\hat{q}}_2 - \lambda e_1)$, $\lambda = \frac{\lambda_0}{1+\|e_1\|}$, $\lambda_0 > 0$ and $\hat{M}, \hat{C}(.)$ and $\hat{G}(.)$ define the estimates of the $M(.), C(.)$ and $G(.)$, respectively,. The unknown velocity signals in (9) replaced by linear observer as

$$\dot{\hat{e}}_1 = \hat{e}_2 + \frac{H_1}{\epsilon}\tilde{e}_1, \dot{\hat{e}}_2 = \frac{H_2}{\epsilon^2}\tilde{e}_1 \tag{10}$$

Then, for any given compact set $(e(0), \hat{e}(0)) \in \Omega_b$, there exists observer-controller gains ϵ_1^*, $H_1 \in \Re^{n \times n}$, $H_2 \in \Re^{n \times n}$, $K_P \in \Re^{n \times n}$ and $K_D \in \Re^{n \times n}$ such that for every $0 < \epsilon < \epsilon_1^*$ the state variables (e, η) of the closed loop system are bounded by a bound that can be made arbitrarily small closed to the origin obtained under state feedback based adaptive design. This implies that the tracking error bound obtained under adaptive state feedback design can be recovered by using adaptive output feedback design by assuming small value of observer design constant ϵ.

3 Experimental Comparison: Model-Based and Model-free Approach

In this section, we show design and implementation process of the model-free and model-based design that introduced in previous section on real robotic systems. For this purpose, we use Phantom™ Premimum 1.5A robotic manipulator provided by SensAble Technologies Inc. [3]. The system is equipped with standard gimbal end-effector. Each joint is attached with encoder for joint position measurement. The device does not have velocity sensors to measure the joint position signals. Such robotic systems are used in bilateral teleoperation for biomedical engineering application [3]. To design and implement nonlinear adaptive controller, one requires to model the system dynamics of Phantom™ robot manipulator. The motion dynamics of this system and its physical parameters can be found in [3]. The implementation block diagram for our experimental evaluation (replace the observer by first order derivative for state feedback case) is pictured in Fig. 1. Let us first apply state feedback (position-velocity) based PD control design (3) to track a desired tracking trajectory. The reference trajectory was a square wave with a period of 8 seconds and an amplitude of ±0.5 radians was pre-filtered with a critically damped 2nd-order linear filter using a bandwidth of $\omega_n = 2.0$ rad/sec. As the experimental system only provides joint position measurement, the velocity signals are estimated via differentiating the joint position signals. To obtain the desired control signals for the system to follow the desired path, we use the coefficients of the PD controller as: $\lambda = 15$ and $K = diag(800, 1200, 1200)$. With these design parameters, we then apply PD-based state feedback controller on the given Phantom™ robot systems. The tested results are given in Fig. 2. To examine the robustness property of this design, let us increase the modeling error uncertainty when the manipulator is executing the given desired task. To do that, we first operate the plant under the same parameters that used in previous experiment. Then, at approximate 6 sec., we add 0.86 kg mass externally to the joint 2 in order to change operating dynamics. The experimental results are pictured in Fig. 3. We now apply model-free PD-output feedback control (4) with the linear observer (10) on Phantom™ system. Our objective is to track the same reference trajectory as used in our previous design for algorithm (3). For comparison purpose, we consider the same controller design parameters that applied in state feedback based design. The observer design constants are chosen as $H_1 = 15I_{3\times3}$,

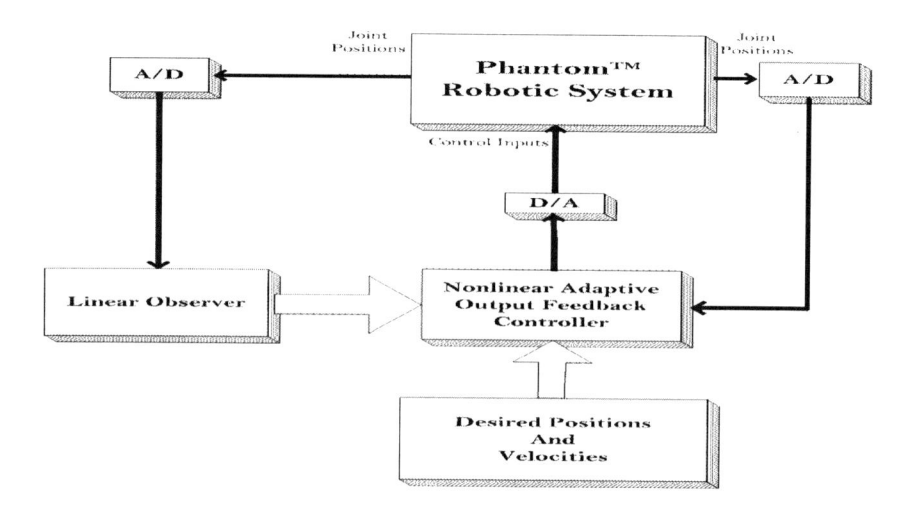

Fig. 1. Implementation block diagram for real-time operation of nonlinear adaptive output feedback design with Computer/Simulink/Real-time Workshop/SensAble Technologies

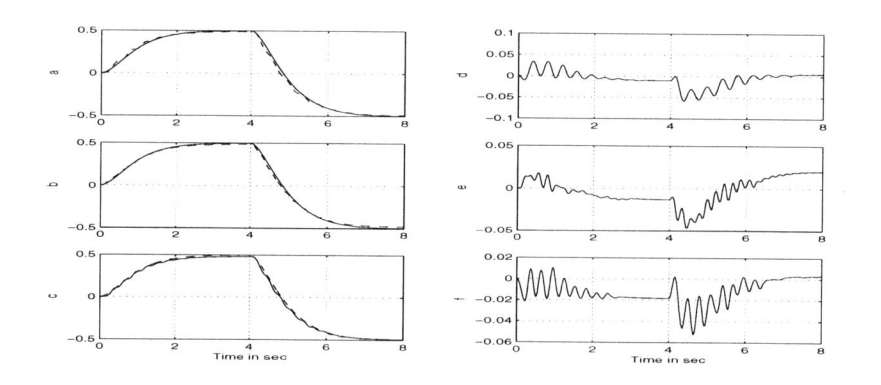

Fig. 2. Left column: a, b & c) The desired (solid-line) and output tracking (dash-line) for joints 1, 2 and 3, Right column: d, e & f) The tracking errors for joints 1, 2 and 3 under PD state feedback design (3)

$H_2 = 5I_{3\times 3}$ and small value of ϵ as $\epsilon = 0.01$. With these design constants, we then apply algorithm (4), (10) on the given haptic robot device. The experimental results are given in Fig. 4. From these results, we can notice that the model-free PD-output feedback design can achieve the desired tracking objectives as the tracking error converges closed to zero. We also observe from our results that the control efforts (not shown here for space limitations) under state feedback based design exhibits control chattering phenomenon due to the derivative action of the position signals. Let us investigate the robust-

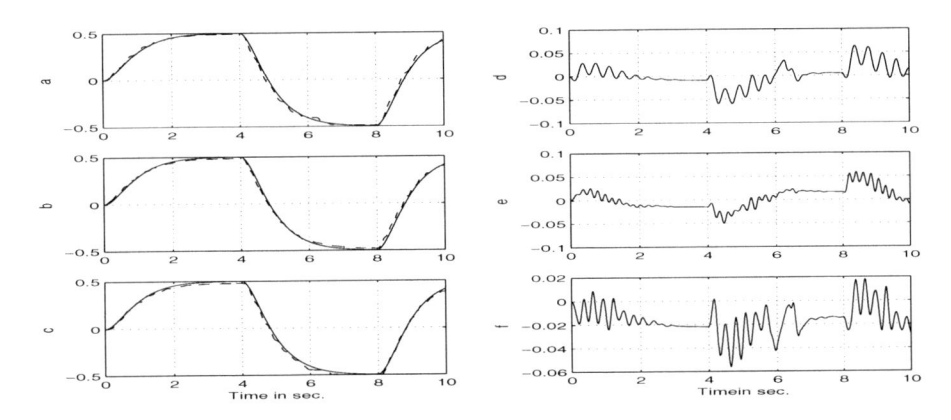

Fig. 3. Left column: a, b & c) The desired (solid-line) and output tracking (dash-line) for joints 1, 2 and 3, Right column: d, e & f) The tracking errors for joints 1, 2 and 3 under PD state feedback design (3) under dynamic model changes

ness property of algorithm (4), (10) with respect to dynamical model parameter changes. In this test, we add approximately 0.86 kg mass to the joint 2 in order to change operating dynamics at 6 sec.. The obtained experimental results are given in Fig. 5. The observer design constants are considered for this experiment as $H_1 = 15I_{3\times 3}$, $H_2 = 5I_{3\times 3}$ and $\epsilon = 0.01$. In view of these experimental results, we can see that the control system is insensitive to the dynamical model parameter changes as the tracking errors slightly increases at the time of the dynamics changes. This is simply because controller design is independent of system dynamics.

We now compare the tracking performance of model-free design with the model-based nonlinear adaptive feedback control design. We first consider the adaptive state feedback control as derived in algorithm (8). The unknown velocity signals in (8) are obtained via differentiating the position signals that measured by encoders attached with the joint of the system. The value of PD controller design parameters are chosen as $\lambda_0 = 2$, $K_P = diag(7500, 1200, 1200)$ and $K_D = diag(500, 800, 800)$. The learning gains Γ are chosen to achieve faster parameter learning via using trial and error search technique as $\Gamma = diag(350, 350, 350, 350, 350, 350, 350, 350)$. The tested results are pictured in Fig. 6. The transient tracking error under adaptive state feedback design is relatively larger than the tracking errors obtained under PD based state feedback. To reduce the tracking errors further, we may further increase the controller gains so as to improve the tracking performance. The main practical problem is that the control system may go unbounded due to the presence of the control chattering phenomenon as a result of the large control efforts. To depict such situation, we present the tested results in Fig. 7. In this experiment, the learning gains increased from $\Gamma = diag(350, 350, 350, 350, 350, 350, 350, 350)$ to $\Gamma = diag(700, 700, 700, 700, 700, 700, 700, 700)$, but we keep all other control design constants similar to our last experiment. As we can see from Fig.7 that we

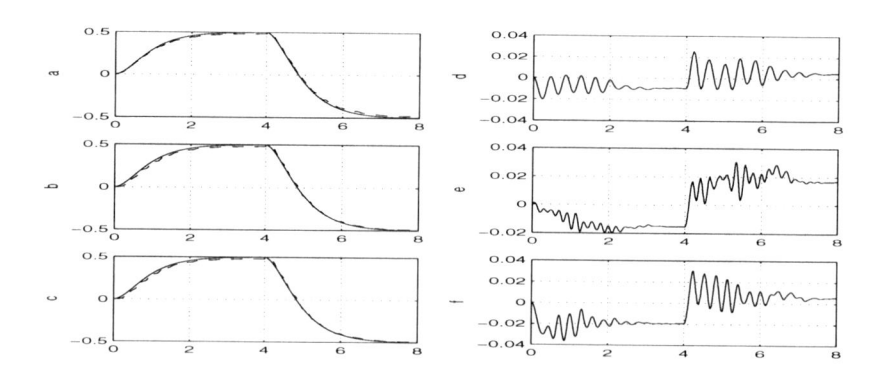

Fig. 4. Left column: a, b & c) The desired (solid-line) and output tracking (dash-line) for joints 1, 2 and 3, Right column: d, e & f) The tracking errors for joints 1, 2 and 3 with PD output feedback design using with $\epsilon = 0.01$

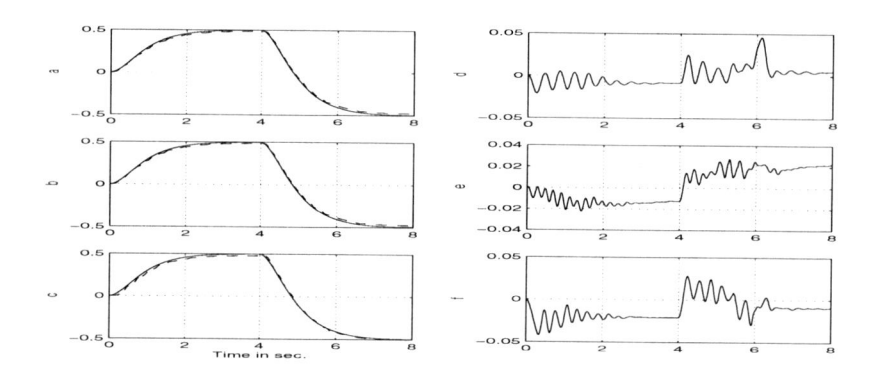

Fig. 5. Left column: a, b & c) The desired (solid-line) and output tracking (dash-line) for joints 1, 2 and 3, Right column: d, e & f) The tracking errors for joints 1, 2 and 3 with (4), (10) under parameter changes with $\epsilon = 0.1$

stopped the experiment at about 1.56 sec. due to control chattering phenomenon causing very large control inputs. Let us now replace the velocity signals in (8) to develop an adaptive output feedback controller (9). Then, we employ the same controller design parameters that used in adaptive state feedback based design as $\lambda_0 = 2$, $K_P = diag(7500, 1200, 1200)$, $K_D = diag(500, 800, 800)$ and $\Gamma = diag(350, 350, 350, 350, 350, 350, 350, 350$). Note that we do not use learning gains $\Gamma = diag(700, 700, 700, 700, 700, 700, 700, 700)$ as they generates excessive control chattering activity. The observer design parameters are chosen as $H_1 = 15I_{3\times3}$, $H_2 = 5I_{3\times3}$ and $\epsilon = 0.03$. The experimental results with nonlinear adaptive output feedback design (9), (10) are given in Fig. 8. To reduce the tracking errors, one requires to increase observer-controller gains. However, we have noticed in our earlier experimental evaluation for state feedback

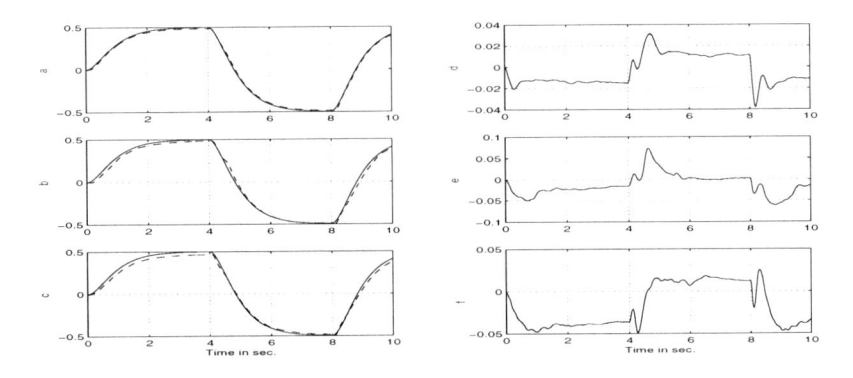

Fig. 6. Left column: a, b & c) The desired (solid-line) and output tracking (dash-line) for joints 1, 2 and 3, Right column: d, e & f) The tracking errors for joints 1, 2 and 3 with algorithm (8)

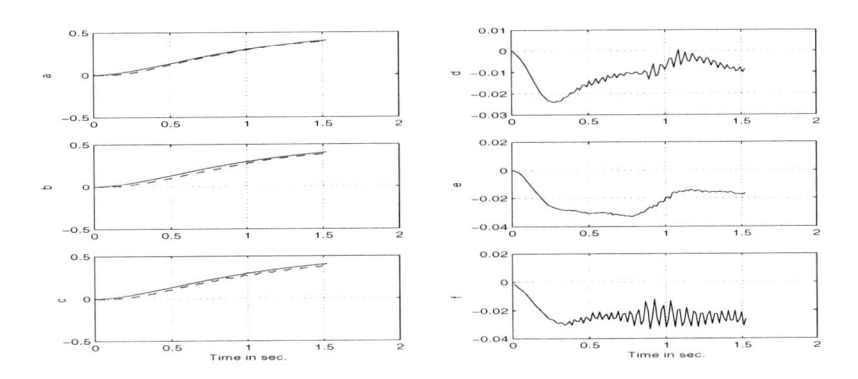

Fig. 7. Left Column: a, b & c) The desired (solid-line) and output tracking (dash-line) for joints 1, 2 and 3, Right Column: d, e & f) The tracking errors for joints 1, 2 and 3 with algorithm (8)

based case that the designer cannot increase the controller gains due to control chattering activity. The requirement of high speed observer makes the nonlinear adaptive output feedback design even more difficult in real-time application as high observer speed generates high-frequency control chattering and infinitely fast switching control phenomenon. To illustrate that, let us increase the observer speed via decreasing the value of ϵ from $\epsilon = 0.03$ to $\epsilon = 0.02$, but we use all other design parameters same as our last test. The experimental results are shown in Fig. 9. From this Fig., we can clearly see that the control efforts under nonlinear adaptive output feedback becomes very large values before the output feedback control system goes unbounded.

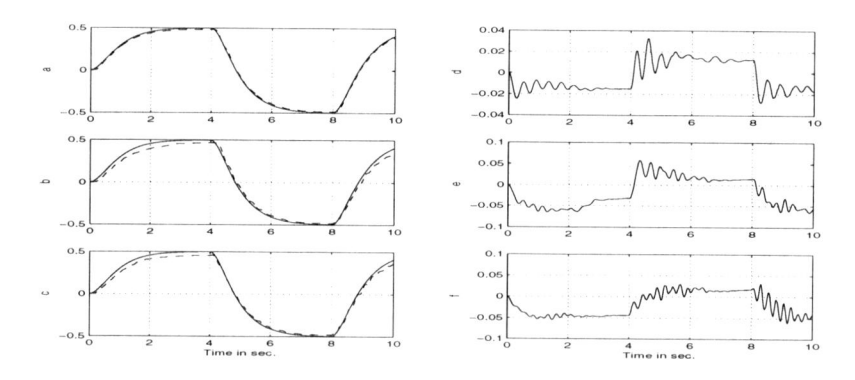

Fig. 8. Left column: a, b & c) The desired (solid-line) and output tracking (dash-line) for joints 1, 2 and 3, Right column: d, e & f) The tracking errors for joints 1, 2 and 3 with nonlinear adaptive output design (9), (10) with $\epsilon = 0.03$

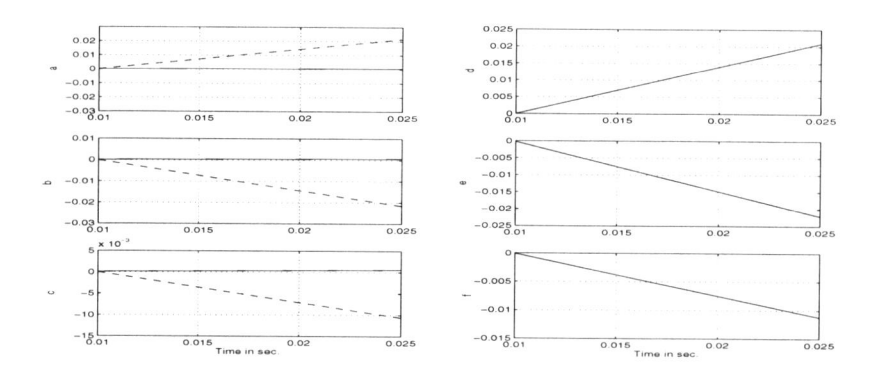

Fig. 9. Left column: a, b & c) The desired (solid-line) and output tracking (dash-line) for joints 1, 2 and 3, Right column: d, e & f) The tracking errors for joints 1, 2 and 3 with nonlinear adaptive output design (9), (10) with $\epsilon = 0.02$

4 Discussion and Conclusion

In this paper, first time, we have experimentally compared model-free and model-based output feedback method for a certain class of nonlinear mechanical systems. It has been observed from our extensive experimental studies that, in the face of the modeling errors uncertainty, the model-based nonlinear adaptive control demanded high observer and learning control gains in order to achieve good tracking performance. However, the observer and learning control gains were limited as they generated very large control effort saturating the control input resulting unbounded control system. In comparison, we were able to increase the observer-controller gains in PD-type output feedback method as the design was an independent of the model dynamics. The proposed PD-type output feedback design strategy can be directly applied to practical/industrial robotic systems that are operating under the filtered version of the PD controller via replacing the velocity signals with the output of the observer.

Acknowledgment

This work is partially supported by the Natural Science and Engineering Research Council (NSERC) of Canada Postdoctoral Fellowship and Canada Research Chairs Program Grant.

References

1. Hua, C., Liu, P.X.: Convergence analysis of teleoperation systems with unsymmetric time-varying delays. IEEE Transaction on Circuits and Systems-II 56(3), 240–244 (2009)
2. Mnasri, C., Gasmi, M.: Robust output feedback full-order sliding mode control for uncertain MIMO systems. In: IEEE International Symposium on Industrial Electronics, pp. 1144–1149 (2008)
3. Polouchine, I.G.: Force-reflecting teleoperation over wide-area networks, Ph. D. Thesis, Carleton University, Ottawa, Canada (2009)
4. Islam, S., Liu, P.X.: PD output feedback for industrial robot manipulators. To appear in IEEE/ASME Transactions on Mechatronics (2010)
5. Pomet, J.-B., Praly, L.: Adaptive nonlinear regulation estimation from the Lyapunov equation. IEEE Transaction on Automatic Control 37, 729–740 (1992)
6. Erlic, M., Lu, W.S.: A reduced-order adaptive velocity observer for manipulator control. IEEE Transaction on Robotics and Automation 11, 293–303 (1995)
7. Takegaki, M., Arimoto, S.: A new feedback method for dynamic control of manipulators. Transaction on ASME Journal of Dynamic Systems, Measurement, and Control 103, 119–125 (1981)
8. Spong, M.W., Vidyasagar, M.: Robot dynamics and control. John Wiley & Sons, New York
9. Rocco, P.: Stability of PID control for industrial robot arms. IEEE Transactions on Robotics and Automation 12(4) (1996)
10. Pagilla, P.R., Tomizuka, M.: An adaptive output feedback controller for robot arms: stability and experiments. Automatica 37 (2001)
11. Meng, M.Q.-H., Lu, W.S.: A unified approach to stable adaptive force/position control of robot manipulators. In: Proceeding of the American Control Conference (1994)
12. Matinez, R., Alvarez, J.: Hybrid sliding-mode-based control of under actuated system with dry friction. IEEE Transaction on Industrial Electronics 55(11) (2008)
13. Arimoto, S., Miazaki, F.: Stability and robustness of PID feedback control for robot manipulators of sensory capability. In: Robotics Research; First International Symposium, pp. 783–799. MIT Press, Cambridge (1984)
14. Islam, S., Liu, P.X.: Robust tracking using hybrid control system. In: Proceeding of the 2009 International Conference on Systems, Man, and Cybernetics (SMC), San Antonio, TX, USA, October 11-14, pp. 4571–4576 (2009)

15. Islam, S., Liu, P.X.: Robust control for robot manipulators by using only join position measurements. In: Proceeding of 2009 IEEE International Conference on Systems, Man, and Cybernetics (SMC), October 11-14, pp. 4113–4118 (2009)
16. Berghuis, H., Ortega, R., Nijmeijer, H.: A robust adaptive robot controller. IEEE Transactions on Robotics and Automation 9, 825–830 (1993)
17. Islam, S., Liu, P.X.: Adaptive sliding mode control for robotic systems using multiple parameter models. In: 2009 IEEE/ASME International Conference on Advanced Intelligent Mechatronics (AIM), Singapore, July 14-17, pp. 1775–1780 (2009)
18. Islam, S.: Experimental comparison of some classical and adaptive iterative learning control on a 5-DOF robot manipulators, Masters Thesis, Lakehead University, Thunder Bay, Canada (2004)
19. Chen, W., Saif, M.: Output feedback controller design for a class of MIMO nonlinear systems using high-Order sliding-mode differentiators with application to a laboratory 3-D Crane. IEEE Transaction on Industrial Electronics 55(11) (2008)
20. Lu, W.S., Meng, M.Q.-H.: Regressor formulation of robot dynamics: computation and applications. IEEE Transactions on Robotics Automation 9(3), 323–333 (1993)
21. Lu, Y.-S.: Sliding-mode disturbance observer with switching-gain adaptation and its application to optical disk drives. IEEE Transaction on Industrial Electronics 56(9), 3743–3750 (2009)

P-Map: An Intuitive Plot to Visualize, Understand, and Compare Variable-Gain PI Controllers

Dongrui Wu

Industrial Artificial Intelligence Laboratory
GE Global Research, Niskayuna, NY 12309 USA
wud@ge.com

Abstract. This paper introduces P-map, an intuitive plot to visualize, understand and compare variable-gain PI controllers. The idea is to represent the difference between a target controller and a baseline controller (or a second controller for comparison) by an equivalent proportional controller. By visualizing the value of the proportional gain and examining how it changes with inputs, it is possible to qualitatively understand the characteristics of a complex variable-gain PI controller. Examples demonstrate that P-map gives more useful information than control surface for variable-gain PI controllers. So, it can be used to supplement or even replace the control surface.

Keywords: Adaptive control, control surface, fuzzy logic control, gain-scheduling control, interval type-2 fuzzy logic control, P-map, PI control, variable-gain controller.

1 Introduction

Proportional-integral-derivative (PID) control [2] is by far the most widely used control algorithm in practice, e.g., according to a survey [7] of over 11000 controllers in the refining, chemicals and pulp and paper industries, 97% of regulatory controllers are PID controllers. Among them, most are proportional-integral (PI) controllers. The input-output relation for an ideal PI controller with constant gains is

$$u(t) = k_p e(t) + k_i \int_0^t e(\tau)d\tau \qquad (1)$$

where the control signal is the sum of the proportional and integral terms. The controller parameters are the constant proportional gain k_p and the constant integral gain k_i. For simplicity in the following we will use $e_i(t)$ to denote $\int_0^t e(\tau)d\tau$.

The properties of constant-gain PID controllers have been extensively studied in the literature [2, 4]. However, in practice the parameters of the system being controlled may be slowly time-varying or uncertain. For example, as an aircraft flies, its mass slowly decreases as a result of fuel consumption. In these cases, a variable-gain controller, whose gains are different at different operational conditions, may be preferred over a constant-gain controller. This is the motivation of gain scheduling control [1, 8, 11], or more generally, adaptive control [3, 18]. The input-output relation for a variable-gain PI controller may be expressed as

$$u(t) = k_p(e(t), e_i(t)) \cdot e(t) + k_i(e(t), e_i(t)) \cdot e_i(t) \qquad (2)$$

In this paper we assume the variable gains k_p and k_i are functions of $e(t)$ and $e_i(t)$ only. However, the P-map method can also be applied when k_p and k_i involve other variables.

Fuzzy logic systems [5, 9, 12, 14, 16, 19], especially the TSK model [15, 17, 19, 22, 23], are frequently used to implement PI controllers. Except in very special cases [15], a fuzzy logic controller (FLC) implements a variable-gain PI control law, where the gains $k_p(e(t), e_i(t))$ and $k_i(e(t), e_i(t))$ change with the inputs. Frequently people use optimization algorithms to tune the parameters of the FLCs [6, 13, 22, 23]. Though the resulting FLCs usually have good performance, it is difficult to understand their characteristics and hence how the performance is achieved.

Traditionally people use control surface to visualize the relationship between $u(t)$ and its inputs, $e(t)$ and $e_i(t)$. An exemplar control surface is shown in the bottom left part of Fig. 1. However, the control surface only gives the user some rough ideas about the controller, e.g., complexity, continuity, smoothness, monotonicity, etc. It is difficult to estimate the control performance from only the control surface. On the other hand, the control systems literature [2, 4] has qualitatively documented the relationship between control performance and PI gains. So, it would be easier to estimate the control performance of a variable-gain PI controller if we can visualize the change of the PI gains with respect to the inputs. This paper proposes such a visualization method called P-map and illustrates how to use it to understand the characteristics of an FLC and to compare the difference between type-1 (T1) and interval type-2 (IT2) FLCs. Examples show that P-map can be used to supplement or even replace the control surface.

The rest of this paper is organized as follows: Section 2 presents the theory on P-map. Section 3 provides illustrative examples on how to use P-map to understand the characteristics of an FLC and to compare the difference between two FLCs. Finally, Section 4 draws conclusions.

2 P-Map: Theory

The effects of changing a proportional gain or an integral gain individually in a PI or PID controller is well-known in the literature [2, 4]. Generally, a larger k_p typically means faster response and shorter rising time, but an excessively large k_p may also lead to larger overshoot, longer settling time, and potential oscillations and process instability. A larger k_i implies steady state errors are eliminated more quickly; however, there may be larger overshoot.

However, if k_p and k_i are varied simultaneously, then their effects are intercorrelated and it is difficult to predict the control performance. So, if we can represent a variable-gain PI controller by a proportional controller only, then it is much easier to understand its characteristics. This is the essential idea of P-map, which is a method to understand the characteristics of a variable-gain PI controller or to compare the difference of two PI controllers using an equivalent proportional gain k_p.

To construct the P-map, we first need to identify a baseline controller for comparison. Next we introduce two methods to obtain the baseline controller, depending on the context of application.

2.1 P-Map for Understanding a Variable-Gain PI Controller

To use P-map for understanding a variable-gain PI controller $u(t)$, we need to construct a baseline constant-gain PI controller

$$u_0(t) = k_{p,0} e(t) + k_{i,0} e_i(t) \qquad (3)$$

from $u(t)$. There may be different ways to construct $k_{p,0}$ and $k_{i,0}$. In this paper we use a least squares approach to fit a plane for $u(t)$: we first obtain n samples $\{e(t), e_i(t), u(t)\}$ from the controller $u(t)$, and then use the least squares method to fit a linear model

$$u(t) = k_{p,0} e(t) + k_{i,0} e_i(t) + \epsilon(e(t), e_i(t)) \qquad (4)$$

Clearly,

$$u(t) = u_0(t) + \epsilon(e(t), e_i(t)) \qquad (5)$$

where $\epsilon(e(t), e_i(t))$ is the difference (residual) between the variable-gain PI controller and the baseline constant-gain PI controller. $\epsilon(e(t), e_i(t))$ can be visualized by the difference between the control surfaces of $u(t)$ and $u_0(t)$, as shown in the bottom left part of Fig. 2.

P-map is used to better visualize and understand the effect of $\epsilon(e(t), e_i(t))$. As the relationship between the proportional gain and the performance of a PI controller is well-known, we want to represent $\epsilon(e(t), e_i(t))$ by a variable-gain proportional controller, i.e.,

$$\epsilon(e(t), e_i(t)) = k'_p(e(t), e_i(t)) \cdot e(t) \qquad (6)$$

In other words, the proportional gain in the P-map is computed as

$$k'_p(e(t), e_i(t)) = \frac{\epsilon(e(t), e_i(t))}{e(t)} \qquad (7)$$

P-map is then used to visualize how $k'_p(e(t), e_i(t))$ changes with $e(t)$ and $e_i(t)$ and hence to understand the characteristics of the variable-gain PI controller. Two illustrative examples are shown in Section 3.

2.2 P-Map for Comparing Two PI Controllers

To use P-map for comparing a variable-gain PI controller $u_2(t)$ against another PI controller $u_1(t)$, $u_1(t)$ is used as the baseline controller. We then compute the difference between $u_2(t)$ and $u_1(t)$, i.e.,

$$\epsilon(e(t), e_i(t)) = u_2(t) - u_1(t) \qquad (8)$$

$\epsilon(e(t), e_i(t))$ can be visualized by the difference between the control surfaces of $u_2(t)$ and $u_1(t)$, as shown in the left part of Fig. 3. However, P-map is a better visualization method, as demonstrated in the next section. The proportional gain in the P-map is again computed by (7). An example will be shown in Section 3.

2.3 Discussions

Observe from (7) that the equivalent gain $k'_p(e(t), e_i(t))$ cannot be computed when $e(t)$ equals 0, and also there may be numerical instabilities when $|e(t)|$ is very close to 0. This is a limitation of the P-map. However, it only concerns with a very small area in the P-map and does not affect the global understanding of the characteristics of a controller.

In this section we have introduced P-map for variable-gain PI controllers. However, P-map can also be applied to variable-gain proportional and PD controllers. For variable-gain proportional controllers, the P-map becomes a curve instead of a plane. P-map can also be computed for PID controllers; however, for visualization we need to plot the P-map in 4-dimensional space, which is difficult. But this is not a unique limitation for the P-map because we have difficulty plotting the control surface of a PID controller in 4-dimensional space too. If a good visualization method for the control surface of a PID controller can be developed, then it can be applied to the P-map, too, and the advantages of the P-map over the control surface can again be demonstrated.

In practice we may implement a PI controller as [10, 22, 23]

$$\dot{u}(t) = k_p(\dot{e}(t), e(t)) \cdot \dot{e}(t) + k_i(\dot{e}(t), e(t)) \cdot e(t) \qquad (9)$$

where $\dot{u}(t)$ is the change in control signal, and $\dot{e}(t)$ is the change in feedback error. For this implementation we compute the proportional gain in P-map as

$$k'_p(\dot{e}(t), e(t)) = \frac{\epsilon(\dot{e}(t), e(t))}{\dot{e}(t)} \qquad (10)$$

where $\epsilon(\dot{e}(t), e(t))$ is the difference between the control surfaces of a variable-gain PI controller and a baseline controller.

3 P-Map: Examples

In this section we show how P-map can be used for understanding the characteristics of FLCs and also comparing difference FLCs. First, the realization of the FLCs is introduced.

3.1 Configurations of the FLCs

In this section fuzzy PI controllers are realized in the form of (9). A typical rulebase may include rules in the form of

$$\text{IF } \dot{e}(t) \text{ is } \dot{E}_n \text{ and } e(t) \text{ is } E_n, \text{Then } \dot{u}(t) \text{ is } y_n \tag{11}$$

where $\dot{E}_n$ and E_n are fuzzy sets, and y_n is a crisp number.

3.2 Example 1: Constant-Gain T1 Fuzzy PI Controller

A T1 FLC with rules in (11) implements the constant-gain PI controller in (9) if [15]:

1. Triangular T1 FSs are used for input membership function (MFs) $\dot{E}_i$ and E_i, and they are constructed in such a way that for any input the firing levels of all MFs add to 1; and,
2. The consequents of the rules are crisp numbers defined as

$$y_n = k_p \dot{e}_n + k_i e_n \tag{12}$$

where $\dot{e}_n$ and e_n are apexes of the antecedent T1 MFs.

An example of such a T1 FLC is shown in the first row of Fig. 1. The apexes of both $\dot{E}_1$ and E_1 are at -1, and the apexes of both $\dot{E}_2$ and E_2 are at 1. The four rules are:

$$\text{IF } \dot{e}(t) \text{ is } \dot{E}_1 \text{ and } e(t) \text{ is } E_1, \text{Then } \dot{u}(t) \text{ is } -k_p - k_i$$
$$\text{IF } \dot{e}(t) \text{ is } \dot{E}_1 \text{ and } e(t) \text{ is } E_2, \text{Then } \dot{u}(t) \text{ is } -k_p + k_i$$
$$\text{IF } \dot{e}(t) \text{ is } \dot{E}_2 \text{ and } e(t) \text{ is } E_1, \text{Then } \dot{u}(t) \text{ is } k_p - k_i$$
$$\text{IF } \dot{e}(t) \text{ is } \dot{E}_2 \text{ and } e(t) \text{ is } E_2, \text{Then } \dot{u}(t) \text{ is } k_p + k_i$$

$k_p = 2.086$ and $k_i = 0.2063$ are used in this example, and the rulebase is given in Table 1.

Table 1. Rulebase and consequents of the T1 FLC shown in Fig. 1.

$\dot{e}(t) \diagdown e(t)$	E_1	E_2
$\dot{E}_1$	-2.2923	-1.8797
$\dot{E}_2$	1.8797	2.2923

The control surface of the constant gain T1 fuzzy PI controller is shown in the bottom left part of Fig. 1. Observe that it is perfectly linear, as suggested by the theoretical results in [15]. The corresponding P-map is shown in the bottom

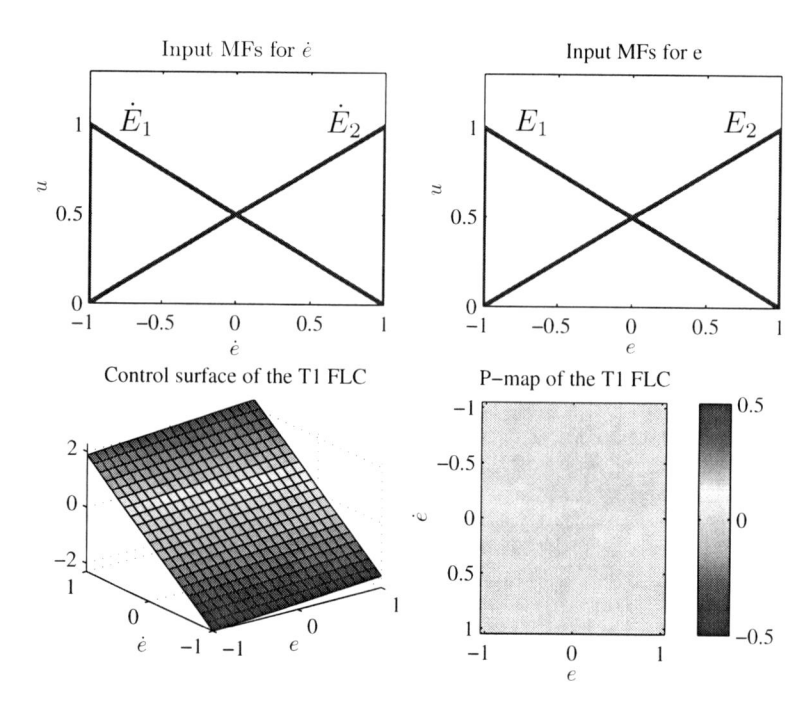

Fig. 1. MFs, control surface, and P-map for the constant-gain T1 FLC in Example 1.

right part of Fig. 1. The least squares method found the baseline constant-gain PI controller to be $\dot{u}_0(t) = 2.086\dot{e}(t) + 0.2063e(t)$, identical to the T1 FLC. As a result, $k'_p(\dot{e}(t), e(t)) = 0$ for $\forall \dot{e}(t)$ and $\forall e(t)$, as indicated in the P-map. So, both the control surface and the P-map show that the T1 fuzzy PI controller is linear.

In summary, the control surface and the P-map are equally good at showing the characteristics of linear PI controllers.

3.3 Example 2: Variable-Gain IT2 Fuzzy PI Controller

In this example, a variable-gain IT2 fuzzy PI controller is obtained from the constant-gain T1 fuzzy PI controller in Example 1 by introducing footprint of uncertainties (FOUs) [14, 22] to the T1 MFs, as shown in the first row of Fig. 2. For simplicity symmetrical FOUs are used, and the four IT2 FSs have the same amount of FOU. The corresponding rulebase is shown in Table 2. Observe that it is essentially the same as the rulebase in Table 1, except that IT2 FSs are used to replace the T1 FSs in Example 1.

Observe from the bottom left part of Fig. 2 that the control surface of the variable-gain IT2 FLC is nonlinear and non-monotonic. However, it is difficult to infer the performance of the IT2 FLC from it. To construct the P-map, the baseline PI controller was found to be $\dot{u}_0(t) = 1.2001\dot{e}(t) + 0.1602e(t)$, and the

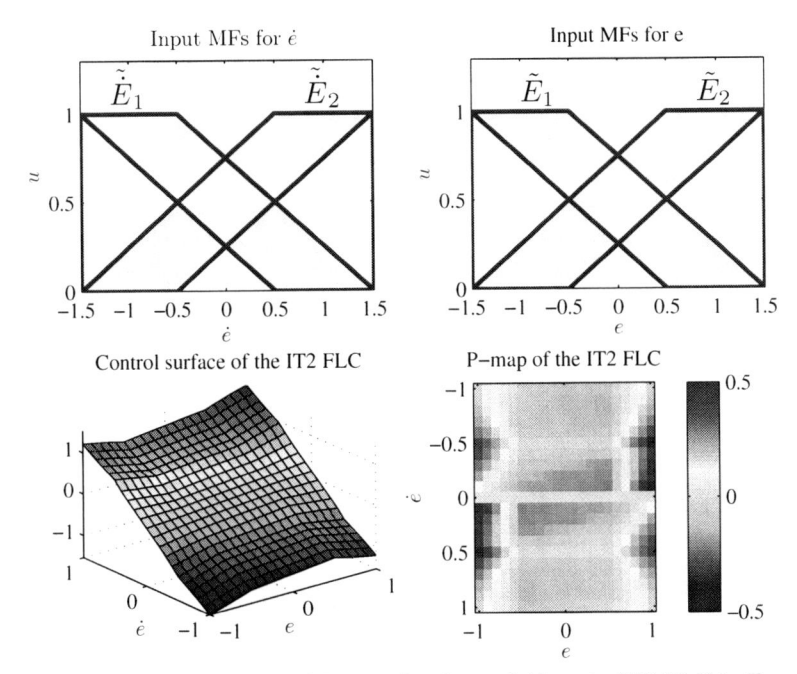

Fig. 2. MFs, control surface, and P-map for the variable-gain IT2 FLC in Example 2

corresponding P-map is shown in the bottom right part of Fig. 2. Notice that both k_p and k_i in the baseline PI controller are smaller than those in Example 1, which confirmed the finding in [24], i.e., generally introducing FOUs to a T1 FLC results a slower IT2 FLC. Observe also from the P-map that:

1. The IT2 fuzzy PI controller is nonlinear, as the proportional gain in the P-map is not a constant. However, unlike the control surface, it is difficult to observe from the P-map whether the control law is monotonic. But this is not important as the monotonicity of a controller is not used very often in control practice.
2. When $|\dot{e}(t)| \to 0$ and $|e(t)| \to 0$, $k'_p(\dot{e}(t), e(t)) < 0$, i.e., compared with the baseline linear controller $\dot{u}_0(t) = 1.2001\dot{e}(t) + 0.1602e(t)$, the IT2 FLC has a smaller proportional gain near the steady-state; so, it implements a gentler control law near the steady-state, which helps eliminate oscillations. This is difficult to be observed from the control surface.
3. When $|e(t)|$ becomes large, $k'_p(\dot{e}(t), e(t))$ increases. So, compared with the baseline linear controller $\dot{u}_0(t) = 1.2001\dot{e}(t) + 0.1602e(t)$, the IT2 FLC has a larger proportional gain when the error is large. Consequently, it implements a faster control law when the current plant state is far away from the set-point, which helps reduce rising time. Again, this is difficult to be observed from the control surface.

In summary, in this example the P-map gives us more useful information about the characteristics of the IT2 FLC than the control surface.

Table 2. Rulebase and consequents of the IT2 FLC shown in Fig. 2.

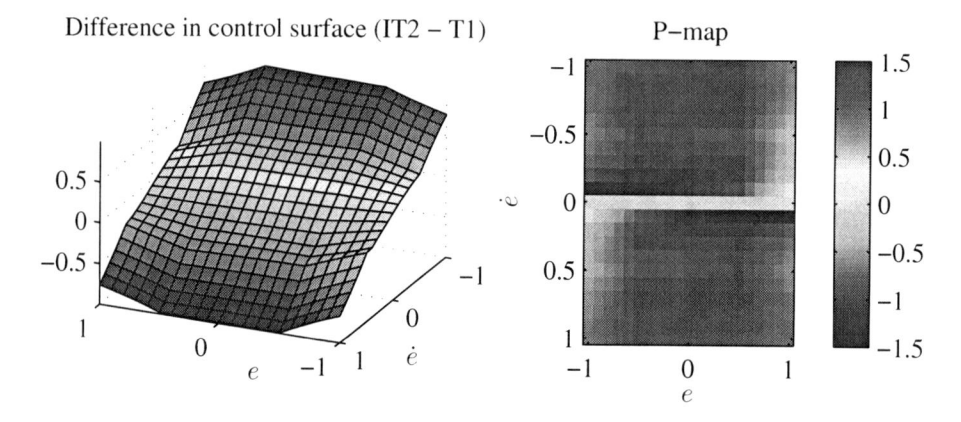

$\dot{e}(t)$ \ $e(t)$	$\tilde{E}_1$	$\tilde{E}_2$
$\tilde{E}_1$	-2.2923	-1.8797
$\tilde{E}_2$	1.8797	2.2923

Fig. 3. Difference between the two control surfaces in Examples 2 and 1, and the corresponding P-map

3.4 Example 3: Comparing Two Fuzzy PI controllers

T1 and IT2 FLCs are fundamentally different. In [21] we showed that an IT2 FLC can output control signals that cannot be implemented by a T1 FLC. In [20] we further showed that an IT2 FLC cannot be implemented by traditionally T1 FLCs. In [24] we derived the closed-form equivalent PI gains of the IT2 FLC introduced in Example 2 for the input domain near the steady-state ($|e(t)| \to 0$ and $|\dot{e}(t)| \to 0$), and we observed that the equivalent PI gains are smaller than those in the baseline T1 FLC in Example 1; however, it is difficult to perform similar analysis for input domains far away from the steady-state.

In this example we show how P-map can be used to quantitatively compare the two FLCs in Examples 1 and 2 in the entire input domain. First we show the difference between the IT2 FLC control surface and the T1 FLC control surface in the left part of Fig. 3. We can observe that the difference is nonlinear and non-monotonic, but it is difficult to discover other useful information. However, from the P-map in the right part of Fig. 3, we can observe that[1]:

---

[1] Note that the following observations are specific for the T1 FLC introduced in Example 1 and the IT2 FLC introduced in Example 2. They may not be generic for all T1 and IT2 FLCs. However, for each new pair of T1 and IT2 FLCs, we can always use the P-map to discover their specific differences.

1. The difference is nonlinear, as the proportional gain in the P-map is not a constant.
2. Most part of the P-map is smaller than zero; so, generally the IT2 FLC introduced in Example 2 has a smaller proportional gain than the T1 FLC introduced in Example 1. This may result in less oscillation, but also slower response.
3. When $|e(t)|$ gets larger, the value in P-map approaches 0, i.e., the difference between the IT2 FLC and the T1 FLC becomes smaller.

In summary, in this example the P-map gives us more useful information about the difference between the two FLCs than the control surface.

4 Conclusions

In this paper we have introduced P-map, which is an intuitive plot to visualize, understand and compare variable-gain PI controllers. The idea is to represent the difference between a target PI controller and a baseline PI controller (or a second PI controller for comparison) using an equivalent proportional controller. By visualizing the value of the proportional gain and examining how it changes with inputs, we can qualitatively understand the characteristics of a complex variable-gain PI controller. As demonstrated by three examples, P-map gives more useful information than control surface for variable-gain PI controllers. So, it can be used to supplement or even replace control surface.

References

1. Apkarian, P., Adams, R.J.: Advanced gain-scheduling techniques for uncertain systems. IEEE Trans. on Control Systems Technology 6(1), 21–32 (1998)
2. Astrom, K.J., Murray, R.M.: Feedback Systems. Princeton University Press, Princeton (2008)
3. Astrom, K.J., Wittenmark, B.: Adaptive Control, 2nd edn. Addison-Wesley Longman Publishing Co., Inc., Boston (1994)
4. Astrom, K., Hagglund, T.: Advanced PID Control. ISA Press (2006)
5. Bonivento, C., Rovatti, R., Fantuzzi, C.: Fuzzy Logic Control: Advances in Methodology. In: Proc. Int'l Summer School, Ferrara, Italy (16-20, 1998)
6. Cazarez-Castro, N.R., Aguilar, L.T., Castillo, O.: Fuzzy logic control with genetic membership function parameters optimization for the output regulation of a servomechanism with nonlinear backlash. Journal of Expert Systems with Applications 37(6), 4368–4378 (2010)
7. Desborough, L., Miller, R.: Increasing customer value of industrial control performance monitoringhoneywells experience. In: Proc. 6th Int. Conf. on Chemical Process Control, Tucson, AZ, pp. 172–192 (2001)
8. Driankov, D.: Fuzzy gain scheduling. In: Bonivento, C., Fantuzzi, C., Rovatti, R. (eds.) Fuzzy Logic Control: Advances in Methodology, pp. 111–148. World Scientific, Singapore (1998)
9. Driankov, J., Hellendoorn, H., Reinfrant, M.: An Introduction to Fuzzy Control. Prentice-Hall, NY (1993)

10. Du, X., Ying, H.: Derivation and analysis of the analytical structures of the interval type-2 fuzzy-PI and PD controllers. IEEE Trans. on Fuzzy Systems 18(4), 802–814 (2010)
11. Filev, D.P.: Gain scheduling based control of a class of TSK systems. In: Farinwata, S.S., Filev, D., Langari, R. (eds.) Fuzzy Control: Synthesis and Analysis, ch. 16, pp. 321–334. John Wiley & Sons, Chichester (2000)
12. Lee, C.: Fuzzy logic in control systems: Fuzzy logic controller — Part II. IEEE Trans. on Systems, Man, and Cybernetics 20(2), 419–435 (1990)
13. Martinez-Marroquin, R., Castillo, O., Aguilar, L.T.: Optimization of interval type-2 fuzzy logic controllers for a perturbed autonomous wheeled mobile robot using genetic algorithms. Information Sciences 179(13), 2158–2174 (2009)
14. Mendel, J.M.: Uncertain Rule-Based Fuzzy Logic Systems: Introduction and New Directions. Prentice-Hall, Upper Saddle River (2001)
15. Mizumoto, M.: Realization of PID controls by fuzzy control methods. Fuzzy Sets and Systems 70, 171–182 (1995)
16. Passino, K.M., Yurkovich, S.: Fuzzy Control. Addison Wesley Longman, Inc., Menlo Park (1998)
17. Wang, L.X.: Stable adaptive fuzzy control of nonlinear systems. IEEE Trans. On Fuzzy Systems 1(2), 146–155 (1993)
18. Wang, L.X.: Adaptive Fuzzy Systems and Control: Design and Stability Analysis. Prentice Hall, Englewood Cliffs (1994)
19. Wang, L.X.: A Course in Fuzzy Systems and Control. Prentice Hall, Upper Saddle River (1997)
20. Wu, D.: An interval type-2 fuzzy logic system cannot be implemented by traditional type-1 fuzzy logic systems. In: Proc. World Conference on Soft Computing, San Francisco, CA (May 2011)
21. Wu, D., Tan, W.W.: Type-2 FLS modeling capability analysis. In: Proc. IEEE Int'l Conf. on Fuzzy Systems, Reno, NV, pp. 242–247 (May 2005)
22. Wu, D., Tan, W.W.: Genetic learning and performance evaluation of type-2 fuzzy logic controllers. Engineering Applications of Artificial Intelligence 19(8), 829–841 (2006)
23. Wu, D., Tan, W.W.: A simplified type-2 fuzzy controller for real-time control. ISA Transactions 15(4), 503–516 (2006)
24. Wu, D., Tan, W.W.: Interval type-2 fuzzy PI controllers:Why they are more robust. In: Proc. IEEE Int'l. Conf. on Granular Computing, San Jose, CA, pp. 802–807 (August 2010)

Sufficient Conditions for Global Synchronization of Continuous Piecewise Affine Systems

Hanéne Mkaouar[1,2] and Olfa Boubaker[1,2]

[1] National Institute of Applied Sciences and Technology
[2] University of Carthage
INSAT, Centre Urbain Nord BP. 676 – 1080 Tunis Cedex, Tunisia
olfa.boubaker@insat.rnu.tn

Abstract. This paper focuses on master slave synchronization problem of piecewise affine (PWA) systems of continuous type. The synchronization problem is solved as a global asymptotic stability problem of the error dynamics between the master and the slave systems via a state feedback control law. Using a Lyapunov approach and the so-called S-procedure, new sufficient conditions are proposed. The synchronization criteria are formulated as Linear Matrix Inequalities (LMIs) which are efficiently solved via common software. Two engineering systems are finally used to illustrate the efficiency of the proposed approach: A non smooth mechanical system and a non linear electrical circuit. Simulation results prove that the synchronization problem is achieved with success applying the proposed strategy.

Keywords: Synchronization, Continuous Piecewise Affine Systems, Lyapunov Theory, Linear Matrix Inequalities.

1 Introduction

In the last decade, synchronization has received considerable interest in engineering problems [1]. Synchronization is a complex phenomenon that occurs when two or more nonlinear oscillators are coupled. Different kinds of synchronization exist. They are summarized in [2]. From the control point of view, the controlled synchronization is the most interesting. There exist two cases of controlled synchronization: mutual synchronization, see e.g. [3], and master-slave synchronization, see e.g. [4]. Many results on controlled synchronization exist where the synchronization problem is considered in the scope of the regulator problem [5, 6] or in the observer design problem [7].

PWA systems are nonlinear processes described by dynamical systems exhibiting switching between a bank of linear dynamics [8]. Many systems in engineering field are described by such models [9, 10]. Switching rules of such systems can result from piecewise linear approximations of nonlinear dynamics or result from piecewise-linear characteristics of some components of the process [11]. Two categories of PWA systems exist, as reported in [12]: continuous PWA systems and discontinuous PWA systems.

Many references were solved the analysis and synthesis problems of continuous PWA systems [9, 10, 11, 13, 14, 15] using a Lyapunov approach. However, only few research results exist on the ground of the controlled synchronization problem. The lack of results in this field is due to the complexity of developments introduced by Lyapunov strategies when applied in the scope of PWA systems. The switching nature of the vector-field seriously complicates the translation of the controlled synchronization problem of PWA systems into some stabilization problem. The few results found in this framework are reported in [16] for continuous PWA systems and in [17] for discontinuous PWA systems. For the two cases, the authors were solved the synchronization problem of PWA systems using the convergence concept of nonlinear systems [12, 18]. The design of the synchronizing controllers is based, in these cases, on the idea of rendering the closed loop slave system convergent by means of feedback and guaranteeing that the closed-loop slave system has a bounded solution corresponding to zero synchronization error.

In this paper, we prove that the difficult problem of master slave synchronization of PWA systems can be solved by conventional methods based Lyapunov approach using the S-procedure. New sufficient conditions will be then provided. Application of the proposed strategy is very practical since it designed via LMI tools.

This paper is structured as follows: the problem of master slave synchronization of PWA systems and the assumptions under which the problem is solved are formulated in section 2. The control law designed under a Lyapunov approach is proposed in section 3. In section 4, two examples from the mechanical and electrical engineering fields are used to illustrate the efficiency of the proposed approach.

2 Problem Formulation

Consider the class of continuous piecewise affine systems described for all $j \in Q = \{1,\ldots,N\}$ and $x \in \Lambda_j$ by:

$$\dot{x} = A_j x + Bu + b_j . \tag{1}$$

where $x \in \Re^n$ and $u \in \Re^m$ are the state and the input vectors, respectively. $A_j \in \Re^{n \times n}$ is a state matrix, $B \in \Re^{n \times m}$ is the input matrix and $b_j \in \Re^n$ is a constant vector. Λ_j is a partition of the state-space into polyhedral cells defined by the following polytopic description:

$$\Lambda_j = \{x | H_j^T x + h_j \leq 0\} . \tag{2}$$

where $H_j \in \Re^{n \times r_j}$ and $h_j \in \Re^{r_j \times 1}$.

Let (1) representing the dynamics of the master system. Assume, for $z \in \Lambda_i$ and $i \in Q = \{1,\ldots,N\}$, that the slave system is described by the following continuous piecewise affine system:

$$\dot{z} = A_i z + b_i + Bv \ . \tag{3}$$

where $z \in \Re^n$ and $v \in \Re^m$ are the state and the input vectors, respectively. $A_i \in \Re^{n \times n}$ is a state matrix, $B \in \Re^{n \times m}$ is the input matrix and $b_i \in \Re^n$ is a constant vector.

If we define $e = z - x$ as an error vector between the dynamics of the slave and those of the master respectively, the synchronization error dynamics is described by:

$$\dot{e} = A_i e + Bw + A_{ij} x + b_{ij} \quad i, j \in Q \ . \tag{4}$$

where: $w = v - u$, $A_{ij} = A_i - A_j$, $b_{ij} = b_i - b_j$.

The objective of the synchronization problem is to design a control law w for the error system (4) based on the error state vector e such that $z \to x$ as $t \to \infty$.

To solve the synchronization problem, let us adopt the following assumptions:

A.1. The right-hand side of the equation (1) and (3) are Lipchitz.

A.2. The state vector x of the master system and the state vector z of the slave system are measured.

A.3. A unique state feedback control law, $w = Ke$, is applied in closed loop where $K \in \Re^{m \times n}$ is some constant gain matrix to be designed such that $z \to x$ as $t \to \infty$.

Note that the right-hand side of the error dynamics (4) is Lipchitz since the right-hand sides of the equations (1) and (3) are Lipchitz. Applying the state feedback control law $w = Ke$ to the error dynamics (4) we obtain:

$$\dot{e} = (A_i + BK)e + A_{ij} x + b_{ij} \ . \tag{5}$$

3 Controller Synthesis

In order to synchronize the dynamics between the master and the slave PWA systems it is necessary to prove global stability of the error dynamics (5) via the control law $w = Ke$. Impose then to the system (5) to have a unique Lyapunov function defined by:

$$V(e) = e^T Pe \ . \tag{6}$$

Where P is a constant positive and symmetric definite matrix. The master-slave synchronization error dynamics (5) are then globally asymptotically stable if $V(e)$ satisfies the following three conditions [19]:

$$V(e) = 0 \ . \tag{7}$$

$$V(e) > 0 \text{ for } e \neq 0. \tag{8}$$

$$\dot{V}(e) < 0 \text{ for } e \neq 0. \tag{9}$$

Furthermore, let $\alpha_1 > 0$ be the desired decay rate for this Lyapunov function such that:

$$\dot{V}(e) \leq -\alpha_1 V(e). \tag{10}$$

The condition (10) satisfies the third Lyapunov condition (9) and furthermore makes $\dot{V}(e)$ more negative. Note that for any small positive constants $0 < \alpha_2 \ll 1$, $0 < \alpha_3 \ll 1$, we can write:

$$\dot{V}(e) + \alpha_1 V(e) - \alpha_2 x^T x - \alpha_3 \leq 0. \tag{11}$$

Theorem 1

For a given decay $\alpha_1 > 0$ and for all pairs of indices $i, j \in Q$, if there exist constant symmetric positive definite matrix $S \in \Re^{n \times n}$, constant matrix $R \in \Re^{m \times n}$, diagonal negative definite matrices $E_{ij} \in \Re^{r_i \times r_i}$ and $F_{ij} \in \Re^{r_j \times r_j}$ and strictly negative constants β_{ij} and ξ_{ij}, such that the following LMIs:

$$\begin{bmatrix} \xi_{ij} & \xi_{ij}|h_i|^T & \xi_{ij}|h_j|^T \\ * & \frac{1}{2}E_{ij} & 0 \\ * & * & \frac{1}{2}F_{ij} \end{bmatrix} < 0, \tag{12}$$

$$\begin{bmatrix} \Delta_{1,i} & A_{ij} & SH_i & 0 & \xi_{ij}b_{ij}|h_i|^T - \frac{1}{2}SH_iM_i & \xi_{ij}b_{ij}|h_j|^T \\ * & \beta_{ij}I & H_i & H_j & -\frac{1}{2}H_iM_i & -\frac{1}{2}H_jM_j \\ * & * & 2E_{ij} & 0 & 0 & 0 \\ * & * & * & 2F_{ij} & 0 & 0 \\ * & * & * & * & \frac{1}{2}E_{ij} - \xi_{ij}|h_i||h_i|^T & -\xi_{ij}|h_i||h_j|^T \\ * & * & * & * & * & \frac{1}{2}F_{ij} - \xi_{ij}|h_j||h_j|^T \end{bmatrix} < 0. \tag{13}$$

$$\Delta_{1,i} = A_i S + S A_i^T + BR + R^T B^T + \alpha_1 S - \xi_{ij} b_{ij} b_{ij}^T$$

are satisfied, then the master-slave synchronization error system (5) is globally asymptotically stable for the control law:

$$w = Ke. \tag{14}$$

where:

$$K = RS^{-1}. \tag{15}$$

In (12), (13) and in the following, * denotes a symmetric bloc.

Proof Theorem 1: Using the synchronization error dynamics (5) and the Lyapunov function (6) we can write for all pairs of indices $i, j \in Q$:

$$\dot{V}(e) + \alpha_1 V(e) = e^T \left[(A_i + BK)^T P + P(A_i + BK) + \alpha_1 P \right] e \\ + \left(x^T A_{ij}^T Pe + e^T PA_{ij} x \right) + \left(b_{ij}^T Pe + ^T Pb_{ij} \right). \quad (16)$$

Substituting (16) in (11) we can write:

$$W^T F_0 W \leq 0. \quad (17)$$

where :

$$W = [e \ x \ 1]^T$$

$$F_0 = \begin{bmatrix} (A_i + BK)^T P + P(A_i + BK) + \alpha_1 P & PA_{ij} & Pb_{ij} \\ * & -\alpha_2 I & 0 \\ * & * & -\alpha_3 \end{bmatrix}$$

$I \in \Re^{n \times n}$ is the identity matrix.
Relying on polytopic expression (2) of polyhedral cell and for all column vectors with positive elements $\delta_i \in \Re^{r_i \times 1}$, $\gamma_j \in \Re^{r_j \times 1}$ and all small positive constants satisfying $0 < \varepsilon_j \ll 1$ and $0 < \sigma_j \ll 1$, we can write for all pairs of indices $i, j \in Q$, that [20]:

$$W^T F_1 W \leq 0. \quad (18)$$

$$W^T F_2 W \leq 0. \quad (19)$$

where:

$$F_1 = \begin{bmatrix} 0 & 0 & H_i \delta_i \\ 0 & 0 & H_i \delta_i \\ \delta_i^T H_i^T & \delta_i^T H_i^T & 2\delta_i^T h_i \end{bmatrix} \quad F_2 = \begin{bmatrix} 0 & 0 & 0 \\ 0 & 2\varepsilon_j I & H_j \gamma_j \\ 0 & \gamma_j^T H_j^T & 2\sigma_j + 2\gamma_j^T h_j \end{bmatrix}$$

If there exist $\tau_{1,ij} \geq 0, \tau_{2,ij} \geq 0$ and using the well known S-procedure [21], the inequalities (17), (18) and (19) could be written, for all pairs of indices $i, j \in Q$, as:

$$F_0 - \tau_{1,ij} F_1 - \tau_{2,ij} F_2 \leq 0 \quad (20)$$

Let now, for all pairs of indices $i, j \in Q$,:

$$\begin{cases} \beta_{1,ij} = -\tau_{1,ij} \delta_i \in \Re^{r_i \times 1} \\ \beta_{2,ij} = -\tau_{2,ij} \gamma_j \in \Re^{r_j \times 1} \\ \beta_{3,ij} = -\alpha_2 - 2\tau_{2,ij} \varepsilon_j \in \Re \\ \beta_{4,ij} = -\alpha_3 - 2\tau_{2,ij} \sigma_j \in \Re \end{cases}$$

such that:
$$\beta_{1,ij}(k) \leq 0, \quad \beta_{2,ij}(k) \leq 0, \quad \beta_{3ij} \leq 0 \text{ and } \beta_{4,ij} \leq 0$$
$$k=1,\cdots,r_i \quad k=1,\cdots,r_j$$

For all pairs of indices $i, j \in Q$, the inequality (20) could be then rewritten as:

$$\begin{bmatrix} \Delta_{2,i} & PA_{ij} & Pb_{ij} + H_i \beta_{1,ij} \\ * & \beta_{3,ij} I & H_i \beta_{1,ij} + H_j \beta_{2,ij} \\ * & * & \beta_{4,ij} + 2\beta_{1,ij}^T h_i + 2\beta_{2,ij}^T h_j \end{bmatrix} \leq 0 \quad (21)$$

$$\Delta_{2,i} = (A_i + BK)^T P + P(A_i + BK) + \alpha_1 P$$

Let now:
$$\begin{cases} \beta_{1,ij} = E_{ij}^{-1} |h_i| \\ \beta_{2,ij} = F_{ij}^{-1} |h_j| \end{cases}$$

and
$$\begin{cases} h_i = M_i |h_i| \\ h_j = M_j |h_j| \end{cases}$$

where $E_{ij} \in \Re^{r_i \times r_i}$ and $F_{ij} \in \Re^{r_j \times r_j}$ are two diagonal negative definite matrices, $|h_i| \in \Re^{r_i \times 1}$ and $|h_j| \in \Re^{r_j \times 1}$ are two column vectors defined such that: $|h_q|(k) = |h_q(k)|$ $q = i, j$ and $M_i \in \Re^{r_i \times r_i}$ and $M_j \in \Re^{r_j \times r_j}$ are two diagonal matri-
$k=1,\cdots,r_q \quad k=1,\cdots,r$
ces defined as follows:

if $h_q(k) \geq 0$ then $M_q(k,k) = 1$, $q = i, j$.
$k=1,\ldots,r_q$

if $h_q(k) < 0$ then $M_q(k,k) = -1$, $q = i, j$.
$k=1,\ldots,r_q$

Using the Schur complement [21], the Bilinear Matrix Inequality (21) is satisfied, for all pairs of indices $i, j \in Q$, if the two following conditions are verified:

$$\beta_{4,ij} + 2|h_i|^T M_i E_{ij}^{-1} |h_i| + 2|h_j|^T M_j F_{ij}^{-1} |h_j| < 0. \quad (22)$$

$$\Lambda_1 - \Lambda_2 \Omega^{-1} \Lambda_2^T \leq 0 \quad (23)$$

where:
$$\Lambda_1 = \begin{bmatrix} (A_i + BK)^T P + P(A_i + BK) + \alpha_1 P & PA_{ij} \\ * & \beta_{3,ij} I \end{bmatrix}$$

$$\Lambda_2 = \begin{bmatrix} Pb_{ij} + H_i E_{ij}^{-1}|h_i| \\ H_i E_{ij}^{-1}|h_i| + H_j F_{ij}^{-1}|h_j| \end{bmatrix}$$

and

$$\Omega = \beta_{4,ij} + 2|h_i|^T M_i E_{ij}^{-1}|h_i| + 2|h_j|^T M_j F_{ij}^{-1}|h_j|$$

Let now, for all pairs of indices $i, j \in Q$:

$$\begin{cases} \beta_{4,ij} = \xi_{ij}^{-1} \\ \Psi = [|h_i| \quad |h_j|]^T \end{cases}$$

we can write that:

$$\Omega = \xi_{ij}^{-1} + \Psi^T \left(\frac{1}{2} \begin{bmatrix} M_i E_{ij} & 0 \\ 0 & M_j F_{ij} \end{bmatrix} \right)^{-1} \Psi < 0, \quad (24)$$

Multiplying (24) by ξ_{ij}^2, assuming that $E_{ij} \leq M_i E_{ij} \leq -E_{ij}$ and $F_{ij} \leq M_j F_{ij} \leq -F_{ij}$ and using the Schur complement, the LMI (12) is obtained. Let now using the matrix inversion Lemma [22], this gives, for all pairs of indices $i, j \in Q$, from (24):

$$\Omega^{-1} = \xi_{ij} - \xi_{ij}^2 \Psi^T \left(\xi_{ij} \Psi \Psi^T + \frac{1}{2} \begin{bmatrix} M_i E_{ij} & 0 \\ 0 & M_j F_{ij} \end{bmatrix} \right)^{-1} \Psi \quad (25)$$

Let now, for all pairs of indices $i, j \in Q$, $\beta_{3,ij} = \beta_{ij}$ and $S = P^{-1}$, left and right multiplying expression (23) by the diagonal matrix $(diag \, [S \quad I])$ gives:

$$\Lambda_3 - \Lambda_4 \Omega^{-1} \Lambda_4^T \leq 0 \quad (26)$$

where:

$$\Lambda_3 = \begin{bmatrix} (A_i S + BR) + (A_i S + BR)^T + \alpha_1 S & A_{ij} \\ * & \beta_{ij} I \end{bmatrix}$$

$$\Lambda_4 = \begin{bmatrix} b_{ij} + SH_i E_{ij}^{-1}|h_i| \\ H_i E_{ij}^{-1}|h_i| + H_j F_{ij}^{-1}|h_j| \end{bmatrix}$$

Substituting (25) in (26), assuming that for all pairs of indices $i, j \in Q$, that $E_{ij} \leq M_i E_{ij} \leq -E_{ij}$ and $F_{ij} \leq M_j F_{ij} \leq -F_{ij}$ and using the Schur complement, the LMI (13) is obtained via some transformations.

4 Application to Mechanical and Electrical Systems

4.1 A Non-smooth Mechanical System

Consider the problem of master slave synchronization of the two non-smooth mechanical systems shown by Fig.1. Each system is constituted by two masses m_1 and m_2, three springs characterized by their given spring stiffnesses k_1, k_2 and k_3 and two dampers characterized by their given damping coefficients d_1 and d_2 [23]. The positions of masses m_1 and m_2 are donated by y_1 and y_2, respectively. Their respective velocities are denoted by $\dot{y}_1$ and $\dot{y}_2$. We assume that y_1, y_2 $\dot{y}_1$ and $\dot{y}_2$ are measured. The mass m_1 is actuated by the input u.

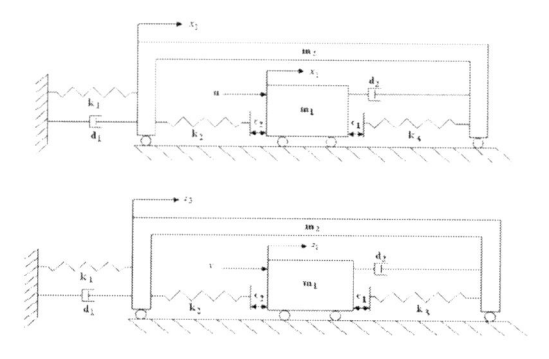

Fig. 1. Non smooth master slave mechanical system

Mass m_1 experiences a possibly asymmetric backlash with backlash gaps c_1, $c_1 > 0$, before hitting the one-sided springs. The dynamics of the system can be described by:

$$\begin{cases} \ddot{y}_1 = \dfrac{1}{m_1}\left[-k_2 g(x) - d_2(\dot{y}_1 - \dot{y}_2) + u\right] \\ \ddot{y}_2 = \dfrac{1}{m_2}\left[k_2 g(x) - d_2(\dot{y}_2 - \dot{y}_1) - k_1 y_2 - d_1 \dot{y}_2\right] \end{cases} \quad (27)$$

where $g(x)$ is a piecewise linear function depending on the positions of masses m_1 and m_2 given by:

$$\begin{cases} g(x) = x_1 - x_3 + c_2 & \text{if } x \in \Lambda_1 \\ g(x) = 0 & \text{if } x \in \Lambda_2 \\ g(x) = x_1 - x_3 - c_1 & \text{if } x \in \Lambda_3 \end{cases}$$

where Λ_1, Λ_2 and Λ_3 are the three polyhedral cells described by:

$$\begin{cases} \Lambda_1 = \{x | x_3 \geq x_1 + c_2\} \\ \Lambda_2 = \{x | x_3 < x_1 + c_2 \cap x_3 > x_1 - c_1\} \\ \Lambda_3 = \{x | x_3 \leq x_1 - c_1\} \end{cases}$$

Let, for $j = \{1,2,3\}$, the state vector for the slave system given by:

$$x = [x_1 \quad x_2 \quad x_3 \quad x_4]^T = [y_1 \quad \dot{y}_1 \quad y_2 \quad \dot{y}_2]^T$$

The dynamics of the system (27) could be written in the form (1) where:

$$A_1 = \begin{bmatrix} 0 & 1 & 0 & 0 \\ -\frac{k_2}{m_1} & -\frac{d_2}{m_1} & \frac{k_2}{m_1} & \frac{d_2}{m_1} \\ 0 & 0 & 0 & 1 \\ \frac{k_2}{m_2} & \frac{d_2}{m_2} & -\frac{k_2+k_1}{m_2} & -\frac{d_2+d_1}{m_2} \end{bmatrix} \quad A_2 = \begin{bmatrix} 0 & 1 & 0 & 0 \\ 0 & -\frac{d_2}{m_1} & 0 & \frac{d_2}{m_1} \\ 0 & 0 & 0 & 1 \\ 0 & \frac{d_2}{m_2} & -\frac{k_1}{m_2} & -\frac{d_2+d_1}{m_2} \end{bmatrix}$$

$$A_3 = \begin{bmatrix} 0 & 1 & 0 & 0 \\ -\frac{k_3}{m_1} & -\frac{d_2}{m_1} & \frac{k_3}{m_1} & \frac{d_2}{m_1} \\ 0 & 0 & 0 & 1 \\ \frac{k_3}{m_2} & \frac{d_2}{m_2} & -\frac{k_3+k_1}{m_2} & -\frac{d_2+d_1}{m_2} \end{bmatrix} \quad B = \begin{bmatrix} 0 \\ 1 \\ \frac{1}{m_1} \\ 0 \\ 0 \end{bmatrix} \quad b_1 = \begin{bmatrix} 0 \\ -\frac{k_2 c_2}{m_1} \\ 0 \\ \frac{k_2 c_2}{m_2} \end{bmatrix} \quad b_2 = \begin{bmatrix} 0 \\ 0 \\ 0 \\ 0 \end{bmatrix} \quad b_3 = \begin{bmatrix} 0 \\ \frac{k_3 c_1}{m_1} \\ 0 \\ -\frac{k_3 c_1}{m_2} \end{bmatrix}$$

System parameters for the master and the slave are fixed as follows: $m_1 = m_2 = 1$, $k_1 = 1$, $k_2 = 0.4$, $k_3 = 0.4$, $d_1 = 0.2$, $d_2 = 0.1$ $c_1 = 0.1$, $c_2 = 0.05$.

A sinusoidal input $u = A_d \sin(\omega t)$ where $A_d = 0.15$ and $\omega = 2\pi$ were applied to the master system. Using LMI Toolbox of MatLab, the LMIs (12) and (13) are solved and the state feedback gain (15) is then deduced:

$$K = [-3.3602 \quad -1.3929 \quad 1.9483 \quad 0.9660]$$

Simulation results are obtained for master and slave initial conditions fixed to $x_0 = [0.25 \quad 0 \quad 0 \quad 0]^T$ and $z_0 = [-0.25 \quad 0 \quad 0 \quad 0]^T$, respectively. Position and velocity synchronization is displayed by Fig. 2.

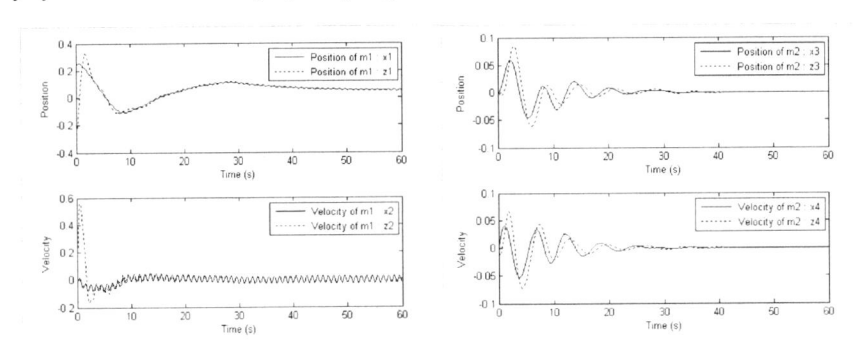

Fig. 2. Position and velocity synchronization of the mechanical system

4.2 A Non Linear Electrical Circuit

Let's now consider the master slave synchronization problem of two identical electrical circuits. The non linear electrical circuit considered [10] is shown by Fig.3.

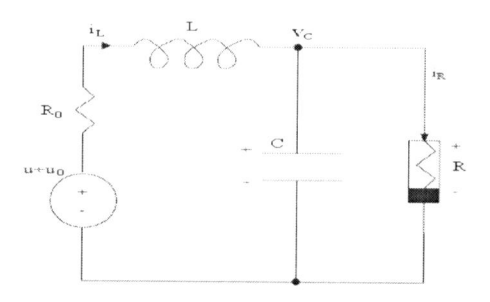

Fig. 3. Non linear electrical circuit

Based on the Kirchhoff law, the electrical circuit with its non linear resistor R could be modeled by the following dynamical system:

$$\begin{bmatrix} \dot{x}_1 \\ \dot{x}_2 \end{bmatrix} = \begin{bmatrix} -\dfrac{R_0}{L} & -\dfrac{1}{L} \\ \dfrac{1}{C} & 0 \end{bmatrix} \begin{bmatrix} x_1 \\ x_2 \end{bmatrix} + \begin{bmatrix} \dfrac{u_0}{L} \\ -\dfrac{1}{C} g(x_2) \end{bmatrix} + \begin{bmatrix} \dfrac{1}{L} \\ 0 \end{bmatrix} u \qquad (28)$$

where $x = [x_1 \ x_2]^T = [i_L \ v_C]$ is the state vector, i_L is the current in the inductance L, v_C is the voltage at the end of the capacitor C. The function $g(x_2)$ is a piecewise linear function describing the current in the non linear resistor R and described by:

$$\begin{cases} g(x_2) = 5x_2 & \text{if } x \in \Lambda_1 \\ g(x_2) = -2x_2 + 1.4 & \text{if } x \in \Lambda_2 \\ g(x) = 4x_2 - 2.2 & \text{if } x \in \Lambda_3 \end{cases}$$

where Λ_1, Λ_2 and Λ_3 are the three polyhedral cells described by:

$$\begin{cases} \Lambda_1 = \{x/ -2 \le x_2 \le 0.2\} \\ \Lambda_2 = \{x/ \ 0.2 \le x_2 \le 0.6\} \\ \Lambda_3 = \{x/ \ 0.6 \le x_2 \le 2\} \end{cases}$$

For simulation proposes, system parameters are fixed as follows: $L = 5.10^{12} H$, $R_0 = 1500\Omega$, $C = 2.10^{-12} F$, $u_0 = 1.2V$, time is expressed in 10^{10} seconds, the inductor current in milliamps and the capacitor voltage in Volts. Furthermore we consider for each polyhydric cell the following equilibrium points [11]: $x_{eq,1} = [0.14 \ 0.71]^T$, $x_{eq,2} = [0.45 \ 0.5]^T$, $x_{eq,3} = [0.64 \ 0.37]^T$ and we apply, for the master system, a sinusoidal input $u = A_d \sin(\omega t)$ where $A_d = 15$ and $\omega = 0.5\pi$. Solving the LMIs (12) and (13) gives the state feedback gain $K = [1.228 \quad -3.6442]^T$. Initial conditions for the master and the slave systems are fixed to $x_0 = [1 \quad 0.5]^T$ and $z_0 = [0.1 \quad 0]^T$, respectively. The synchronization of the state variables is displayed by Fig.4. The switching dynamics of the state variables between polyhedral cells are displayed by Fig.5 for the slave system.

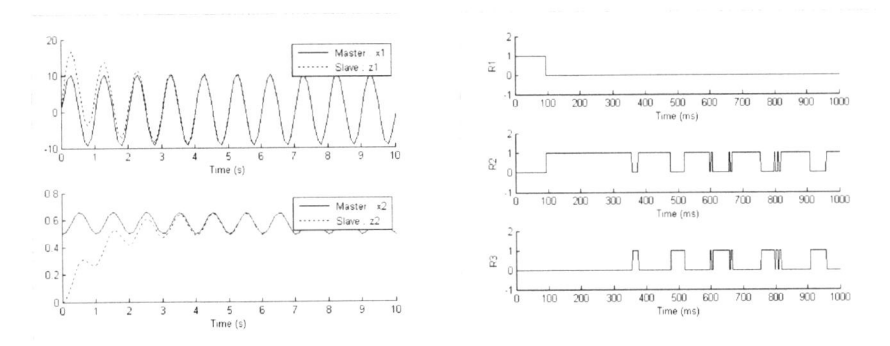

Fig. 4. Synchronization of the electrical systems

Fig. 5. Dynamics of the slave electrical system

4.3 Comparative Study

The synchronization problem for the system (27) is achieved using the strategy proposed in [16] which gives the gain vector K =[-1.6168 -1.1158 0.2928 -0.0680]. The comparison of error synchronization between our approach and those given by [16] is shown by Fig.6. It is clear that the two approaches give comparable results.

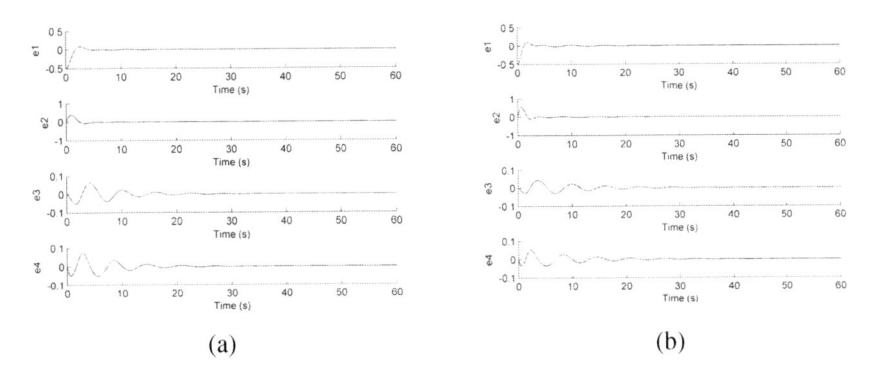

Fig. 6. Error synchronization: (a) solving LMI conditions (12) and (13), (b) solving LMI conditions proposed in [16]

5 Conclusion

In this paper, we have proved, despite the complexities introduced by the switching dynamics of PWA systems, that the master slave synchronization problem can be solved by translating the synchronization problem into a stabilization problem. The global asymptotic stability of error dynamics is achieved using a Lyapunov approach and the so-called S-procedure. The new sufficient conditions are formulated via Linear Matrix Inequalities. The efficiency of the proposed approach was proved via simulation results using two engineering problems. Simulation results were finally compared to a previous synchronization approach using the convergence concept of nonlinear systems. The two approaches give almost comparable system dynamics, however, as a purely mathematical problem, the new conditions proposed in this paper could be considered as tighter stability criteria.

References

1. Nijmeijer, H., Rodriguez-Angeles, A.: Synchronization of mechanical systems. World Scientific Publishing Co. Pte. Ltd, Singapore (2003)
2. Blekhman, I., Fradkov, A., Nijmeijer, H., Pogromsky, A.Y.: On self synchronization and controlled synchronization. Systems & Control Letters 31, 299–305 (1997)
3. Rodriguez-Angeles, A., Nijmeijer, H.: Mutual synchronization of robots via estimated state-feedback: a cooperative approach. IEEE Transactions on Control Systems Technology 12, 542–554 (2004)
4. Curran, P.F., Suykens, J.A.K., Chua, L.O.: Absolute stability theory and master–slave synchronization. International Journal of Bifurcation and Chaos 7, 2891–2896 (1997)
5. Suykens, J.A.K., Curran, P.F., Chua, L.O.: Master–slave synchronization using dynamic output feedback. International Journal of Bifurcation and Chaos 7, 671–679 (1997)
6. Suykens, J.A.K., Curran, P.F., Chua, L.O.: Robust synthesis for master-slave synchronization of Lur'e Systems. IEEE Transactions on Circuits and Systems I: Fundamental Theory and Applications 46, 841–850 (1999)
7. Nijmeijer, H., Mareels, I.M.Y.: An observer looks at synchronization. IEEE Transactions on Circuits and Systems I: Fundamental Theory and Applications 44, 882–890 (1997)
8. Bemporad, A., Morari, M.: Control of systems integrating logic, dynamics, and constraints. Automatica 35, 407–427 (1999)
9. Rodrigues, L., Boyd, S.: Piecewise-affine state feedback for piecewise-affine slab systems using convex optimization. Systems and Control Letters 54, 835–853 (2005)
10. Rodrigues, L., How, J.: Observer-based control of piecewise-affine systems. International Journal of Control 76, 459–477 (2003)
11. Hassibi, A., Boyd, S.P.: Quadratic stabilization and control of piecewise-linear systems. In: American Control Conference, pp. 3659–3664. IEEE Press, New York (1998)
12. Pavlov, A., Pogromsky, A., Van de Wouw, N., Nijmeijer, H.: On convergence properties of piecewise affine systems. International Journal of Control 80, 1233–1247 (2007)
13. Johansson, M., Rantzer, A.: Computation of piecewise quadratic Lyapunov functions for hybrid systems. IEEE Transactions on Automatic Control 43, 555–559 (1998)
14. Rodrigues, L., How, J.: Synthesis of piecewise-affine controllers for stabilization of nonlinear systems. In: 42nd IEEE Conference on Decision and Control, pp. 2071–2076. IEEE Press, New York (2003)
15. Rodrigues, L.: Stability of sampled-data piecewise-affine systems under state feedback. Automatica 43, 1249–1256 (2007)

16. Van de Wouw, N., Pavlov, A., Nijmeijer, H.: Controlled synchronization of continuous PWA systems. In: Pettersen, K.Y., Gravdahl, J.T., Nijmeijer, H. (eds.) LNCS, vol. 336, pp. 271–289. Springer, Heidelberg (2006)
17. Van de Wouw, N., Pavlov, A.: Tracking and synchronization for a class of PWA systems. Automatica 44, 2909–2915 (2008)
18. Pavlov, A., Van de Wouw, N., Nijmeijer, H.: Convergent dynamics, a tribute to B.P. Demidovich. Systems and Control Letters 52, 257–261 (2004)
19. Slotine, J.J.E., Li, W.: Applied nonlinear control. Prentice-Hall, Englewood Cliffs (1991)
20. Boubaker, O.: Master-slave synchronization for PWA systems. In: 3rd IEEE International Conference on Signals, Circuits and Systems. IEEE Press, New York (2009)
21. Boyd, S.P., Ghaoui, L.E., Feron, E., Balakrishnan, V.: Linear matrix inequalities in system and control theory. Studies in Applied Mathematics. SIAM, Philadelphia (1994)
22. Kailath, T.: Linear systems. Prentice-Hall, Englewood Cliffs (1989)
23. Van de Wouw, N., Pavlov, A., Pettersen, K.Y., Nijmeijer, H.: Output tracking control of PWA systems. In: 45th IEEE Conference on Decision and Control, pp. 2637–2642. IEEE Press, New York (2006)

Question Type Classification Using a Part-of-Speech Hierarchy

Richard Khoury

Department of Software Engineering, Lakehead University,
955 Oliver Road, Thunder Bay, Ontario, Canada, P7A 5E1
Richard.Khoury@lakeheadu.ca

Abstract. Question type (or answer type) classification is the task of determining the correct type of the answer expected to a given query. This is often done by defining or discovering syntactic patterns that represent the structure of typical queries of each type, and classify a given query according to which pattern they satisfy. In this paper, we combine the idea of using informer spans as patterns with our own part-of-speech hierarchy in order to propose both a new approach to pattern-based question type classification and a new way of discovering the informers to be used as patterns. We show experimentally that using our part-of-speech hierarchy greatly improves type classification results, and allows our system to learn valid new informers.

Keywords: Natural Language Processing, Question Type Classification, Answer Type Classification, Part-of-Speech, Syntactic Pattern, Informer Span.

1 Introduction

Question type (or answer type) classification is the Natural Language Processing (NLP) task of determining the correct type of the answer expected to a given query, such as whether the answer should be a person, a place, a date, and so on. This classification task is of crucial importance in question-answering (QA) systems. Indeed, correctly pinpointing the expected answer type of a question allows a QA system to use type-specific answer retrieval algorithms and to reject possible answers of the wrong type [1]. In fact, it has been shown that questions that have been classified into the correct type are answered correctly twice as often as misclassified questions [1]. Over the years, many varied approaches to question type classification have been proposed. A large proportion of systems, including many of those entered in the TREC QA competition, simply try to accomplish this task in one of two ways, either by detecting keywords in the query [1] or by relying on the wh-term (who, what, where, when, why, which, whom, whose, how) at the beginning of the query for disambiguation. However, both approaches are too simplistic and can be easily misled. Keywords alone are not enough to differentiate between question types: for example, the query "who was the French emperor defeated at Waterloo" and "when was a French emperor defeated at Waterloo" will have the same keywords after stopword removal, but belong to different types. And wh-terms can take different meanings based on the context. For example, "who was Napoleon" and "who defeated Napoleon" both begin with *who* but only the second is asking for a person while the first is asking for a historical definition, and "what French emperor was defeated at Waterloo" is a *who* query phrased as a *what* query. In fact, research has shown that rephrasing a question to use a different wh-term is the single most common way that humans use to paraphrase queries [2]. In light of this, syntactic-pattern-based methods have become popular in question type classification systems [3]. These methods use or discover syntactic patterns that represent the structure of typical queries of each type, and classify a given query according to which pattern they satisfy. For example, a system could discover, by comparing example queries, that those featuring the pattern "...in which year..." should be classified in the *time* type [3].

In this paper, we combine an idea from Krishnan *et al.* of using informer spans as question type classification patterns [4] with our own part-of-speech hierarchy [5] in order to propose both a new approach to pattern-based question type classification and a new way of discovering the informers to be used as patterns. We show experimentally that using our part-of-speech hierarchy together with a few simple informer spans greatly improves type classification results, and moreover that the hierarchy can be used to learn new valid informers from example queries.

The rest of this paper is organized as follows. In Section 2, we review a representative sample of work done on the task of question type classification in order to clearly illustrate the nature of our contribution. The theoretical frameworks of our classification system, of our learning algorithm and of our part-of-speech hierarchy are all presented in Section 3. Our ideas have all been implemented and tested, and experimental results are presented and discussed in Section 4. Finally, we offer some concluding remarks in Section 5.

2 Background

Question type classification is an integral task in most QA systems developed today. Consequently, there are considerable variations in the nature of the classification systems, in the set of question types recognized, and in the nature and size of the knowledge base underlying the classification systems.

A purely Bayesian solution to the type classification problem was proposed in [6]. The authors used a straightforward Naïve Bayes classifier, and computed the probability of a query belonging to a question type given the prior probability of that question type multiplied by the conditional probability of the query's features (i.e. the words left after stemming and stopword removal) given that type. They used six question types for their classification, namely time, location, human, number, object, and description. Their system achieves an average precision of 59% over these six types; however the authors do not discuss how large the table of conditional probabilities has to be to account for a reasonable number of features in six categories.

A team from the University of Concordia developed a simple keyword-based question type classification system as part of their QA system for the TREC-2007 competition [1]. Their system classifies queries into seven types, namely date, location, person, organization, number, website, and other, by recognizing keywords in the query. The authors do not give the size of the lexicon they manually built over the years for that purpose, but in [1] they report adding 50 new words into it. Their research does show that a keyword-matching approach can work quite well given a large and detailed enough lexicon: they report that their system achieves 93% accuracy. However, a smaller and coarser lexicon is not necessarily useful, and can even confuse a system, as indicated by Tomuro in his study on the impact of semantic information in question type classification [7]. Tomuro built two question type classifiers, one using a decision tree and one using a k-nearest-neighbour algorithm, and trained two instances of each classifier. The first instance used only a closed lexicon of about 100 words frequently found in queries. This lexicon is thus composed mostly of domain-independent, non-content, closed-class words and includes wh-terms. The second instance of both classifiers uses the same lexicon and also adds in the categorization (i.e. WordNet hypernym) of query words that are not part of the lexicon. The classifiers were trained to recognize 12 question types, namely time, location, entity, definition, reference, reason, procedure, manner, degree, atrans, interval, and yes-no questions. His results show that taking the extra keywords into account and adding the categorization information does not affect the results in a statistically significant way. Moreover, he found that the extra knowledge added can mislead the classifier when a query's syntax affects its semantic meaning. For example, the query "what does Hanukah mean?" is clearly a definition-type question, but the keyword "Hanukah" is a hyponym of "time period" in WordNet, and thus misleads his system into classifying it as a time-type question [7]. Another WordNet-based question type classifier proposed in [8] faced the same difficulties, and overcame them by creating a two-step classification scheme. It begins by defining a set of simple question type syntactic patterns, such as "what {is|are} <phrase>?" for definition-type queries. Their system attempts to match queries to these patterns, and then goes to WordNet hypernym searches for queries that are not recognized by them. The results in [7] and [8] illustrate how syntactic patterns can be more powerful than simple keyword recognition for question type classification, as they offer a more complete picture of the query.

Zhang and Nunamaker [9] built a pattern-based question type classifier for their video indexing and retrieval system. They defined nine question types in their system, namely time, location, person, organization, number, object, reason, definition, and undefined. Their system classifies each user's query according to a set of simple patterns that combine both wh-term and the categories of keywords found in the query. To illustrate, a sample pattern given in [9] is "if (question starts with 'what' + person) then (answer type is person)". Unfortunately, the authors do not give a complete list of patterns nor the total number of patterns in their system. The authors of [4] take the idea of type classification patterns one step in a different direction and propose that the patterns could consist simply of a short string of contiguous words found in the query, which may or may not include wh-terms, and which they called the *informer span* (or simply *informer*) of the query. Their work shows clearly that a support vector machine classifier using hand-made informers yields better results than question bigrams, and that informers discovered automatically work almost as well as hand-made ones.

We can point out a common thread in these methods, namely their reliance on large knowledge bases. The keyword-based system of [1] requires a massive lexicon of keywords likely to be observed in queries of each type, and the Bayesian system of [6] further needs to define the conditional probability of each type given each keyword, while the categorization systems of [7] and [8] use the WordNet lexicon to recognize classes of keywords. However, keyword-based question type classification is inherently limited and misleading, since the syntactic structure of the query can change the semantic importance and meaning of its keywords [2], [7]. The systems proposed in [9] and [4] compensate for this problem by developing syntactic patterns instead of keyword lists, but they still suffer from the need to develop massive lists of patterns from which to find the one that exactly matches the query. In fact, Krishman *et al.* point out in their review that such systems are built on hundreds of unpublished patterns [4].

3 Methodology

In this paper, we develop a new system for question type classification based on a part-of-speech hierarchy that we present below. As will become evident, using the part-of-speech hierarchy makes it possible to get good classification results using only a handful of simple informers. This stands in stark contrast to the other systems reviewed in Section 2, which need large sets of patterns, lexicons, or probability tables to work well.

The question types we use are "person" (who), "date" (when), "physical object" (what), "location" (where), "numeric value" (how many/how much), and "description" (how/why). We manually define a set of 14 simple informers to represent these question types; a simple enough task given that they each have clear wh-terms. We design these informers to be two words long each so that none would have a length advantage over the others. There are two informers per question type, except for the description type which has two different wh-terms and therefore four informers. The informers are listed in Figure 1.

Who is	What was	How did
Who was	Where is	How does
When is	Where was	Why did
When was	How many	Why does
What is	How much	

Fig. 1. 14 basic informer spans

3.1 Part-of-Speech Hierarchy

A part-of-speech (POS) is commonly defined as a linguistic category of lexical items that share some common syntactic or morphological characteristics. However, it is telling that, despite the concept of a POS being thousands of years old, grammarians and linguists still cannot agree on what exactly the shared characteristics are. This question has very real practical consequences: depending on where the line is drawn between common characteristics and distinguishing differences, one can end up with anywhere from the eight parts-of-speech defined in English textbooks to the 198 parts-of-speech of the London-Lund Corpus of Spoken English [10].

Our solution to this problem is to organize lexical items not into a single set of parts-of-speech but into a part-of-speech hierarchy. The lower levels of this hierarchy feature a greater number of finely-differentiated parts-of-speech, starting with the Penn Treebank POS tags at the lowest level, while the higher levels contain fewer and more general parts-of-speech, and the topmost level is a single all-language-encompassing "universe" part-of-speech. Another innovation has been the inclusion in our hierarchy of several "blank" parts-of-speech, to represent and distinguish between the absences of different types of words. In total, our hierarchy contains 165 parts of speech organized in six levels. We originally developed it in the context of a keyword-extraction project; a detailed description of the hierarchy can be found in the paper describing that project [5].

We can define the semantic importance of different lexical items by assigning weights on the connections between POS in the hierarchy. This allows us to specialize the hierarchy for use in different NLP tasks. In our earlier work on keyword extraction [5], weight was given to verbs and nouns – the typical parts-of-speech of the keywords we were looking for. For question type classification, however, verbs and nouns are not the most semantically important words to take into account. Indeed, Tomuro [7] has shown that question type classification relies mostly on closed-class non-content words. The semantic weight in our hierarchy was thus shifted to the subtrees corresponding to popular query adverbs and pronouns, including wh-terms, and calibrated using queries from the 2007 TREC QA track [11] as examples. The value of each POS in our hierarchy is then computed on the basis of its semantic weight and of the number of descendents it has. The resulting hierarchy, with the value of each POS, is presented in Figure 2.

We showed in [5] how using a part-of-speech hierarchy makes it possible to mathematically define several linguistic comparison operations. It is possible to compute the similarity between two words or POS simply as a function of their distance in the hierarchy. This computation takes several factors into account, including the number of levels in the hierarchy that need to be traversed on the path from one POS to the other and the value of each intermediate POS visited. Likewise, we can find a general place-holder POS to represent two words simply by finding the lowest common ancestor of both words in the hierarchy, and the similarity of the place-holder compared to the original two words it represents is again a function of the distance in the hierarchy between each word and the place-holder POS. By extension, we can measure the similarity between two sentences by pairing their words together by similarity; the fact that our hierarchy includes "blank" parts-of-speech means that the two sentences do not need to be of the same length to be compared. And finally, we can merge two sentences by pairing their words together by similarity and replacing each pair by its lowest common ancestor in the hierarchy.

It is now straightforward to see how our part-of-speech hierarchy can be used for the task of question type classification. Given a list of informers representing different question types, we can use the hierarchy to compute the similarity between a user-specified query and each informer. The query is then classified to the same type as the informer it is most similar to.

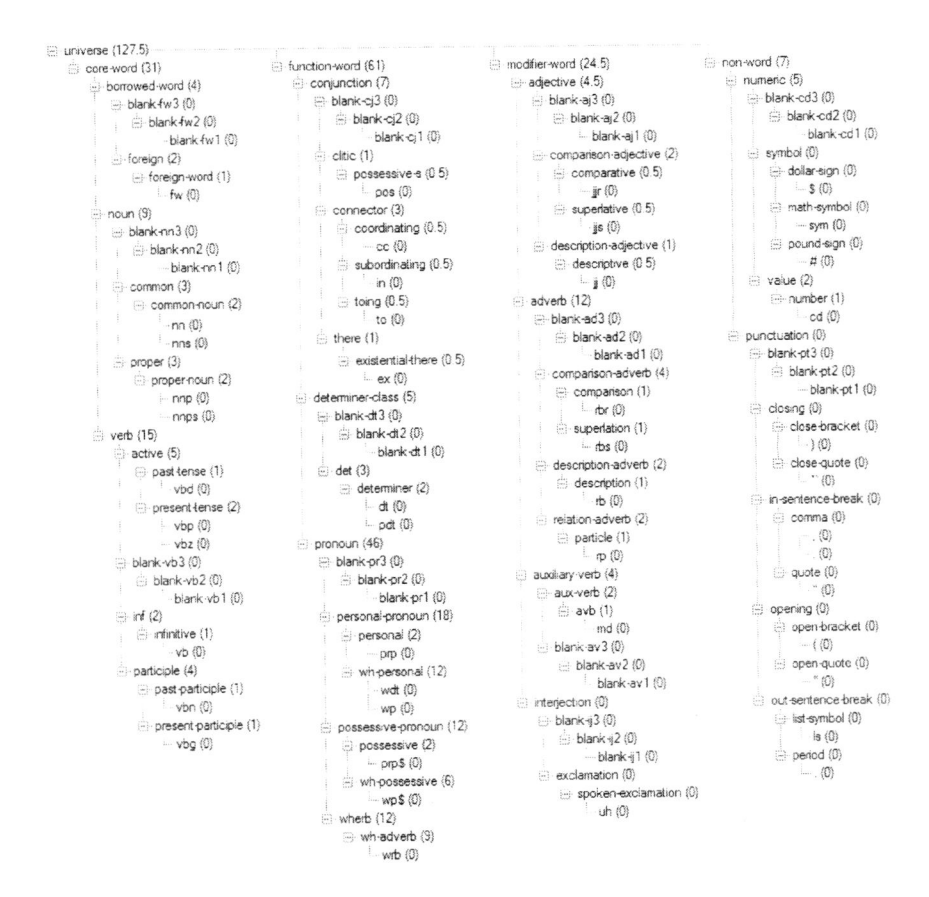

```
universe (127.5)
  core-word (31)
    borrowed-word (4)
      blank-fw3 (0)
        blank-fw2 (0)
          blank-fw1 (0)
      foreign (2)
        foreign-word (1)
          fw (0)
    noun (9)
      blank-nn3 (0)
        blank-nn2 (0)
          blank-nn1 (0)
      common (3)
        common-noun (2)
          nn (0)
          nns (0)
      proper (3)
        proper-noun (2)
          nnp (0)
          nnps (0)
    verb (15)
      active (5)
        past-tense (1)
          vbd (0)
        present-tense (2)
          vbp (0)
          vbz (0)
      blank-vb3 (0)
        blank-vb2 (0)
          blank-vb1 (0)
      inf (2)
        infinitive (1)
          vb (0)
      participle (4)
        past-participle (1)
          vbn (0)
        present-participle (1)
          vbg (0)
  function-word (61)
    conjunction (7)
      blank-cj3 (0)
        blank-cj2 (0)
          blank-cj1 (0)
      clitic (1)
        possessive-s (0.5)
          pos (0)
      connector (3)
        coordinating (0.5)
          cc (0)
        subordinating (0.5)
          in (0)
        toing (0.5)
          to (0)
      there (1)
        existential-there (0.5)
          ex (0)
    determiner-class (5)
      blank-dt3 (0)
        blank-dt2 (0)
          blank-dt1 (0)
      det (3)
        determiner (2)
          dt (0)
          pdt (0)
    pronoun (46)
      blank-pr3 (0)
        blank-pr2 (0)
          blank-pr1 (0)
      personal-pronoun (18)
        personal (2)
          prp (0)
        wh-personal (12)
          wdt (0)
          wp (0)
      possessive-pronoun (12)
        possessive (2)
          prp$ (0)
        wh-possessive (6)
          wp$ (0)
      wherb (12)
        wh-adverb (3)
          wrb (0)
  modifier-word (24.5)
    adjective (4.5)
      blank-aj3 (0)
        blank-aj2 (0)
          blank-aj1 (0)
      comparison-adjective (2)
        comparative (0.5)
          jjr (0)
        superlative (0.5)
          jjs (0)
      description-adjective (1)
        descriptive (0.5)
          jj (0)
    adverb (12)
      blank-ad3 (0)
        blank-ad2 (0)
          blank-ad1 (0)
      comparison-adverb (4)
        comparison (1)
          rbr (0)
        superlation (1)
          rbs (0)
      description-adverb (2)
        description (1)
          rb (0)
      relation-adverb (2)
        particle (1)
          rp (0)
    auxiliary-verb (4)
      aux-verb (1)
        avb (1)
          md (0)
      blank-av3 (0)
        blank-av2 (0)
          blank-av1 (0)
    interjection (1)
      blank-ij3 (0)
        blank-ij2 (0)
          blank-ij1 (0)
      exclamation (0)
        spoken-exclamation (0)
          uh (0)
  non-word (7)
    numeric (5)
      blank-cd3 (0)
        blank-cd2 (0)
          blank-cd1 (0)
      symbol (0)
        dollar-sign (0)
          $ (0)
        math-symbol (0)
          sym (0)
        pound-sign (0)
          # (0)
      value (2)
        number (1)
          cd (0)
    punctuation (0)
      blank-pt3 (0)
        blank-pt2 (0)
          blank-pt1 (0)
      closing (0)
        close-bracket (0)
          ) (0)
        close-quote (0)
          " (0)
      in-sentence-break (0)
        comma (0)
          , (0)
        quote (0)
          " (0)
      opening (0)
        open-bracket (0)
          ( (0)
        open-quote (0)
          " (0)
      out-sentence-break (0)
        list-symbol (0)
          ls (0)
        period (0)
          . (0)
```

Fig. 2. The POS hierarchy (POS values are in brackets)

3.2 Informer-Learning Algorithm

In addition to being used to compare queries and informers together, our part-of-speech hierarchy can be used as the core of an informer-learning algorithm. The intuition behind the learning algorithm is that, when merging two queries together, irrelevant words will be replaced by high-level POS while informer words common to both queries (if there are any) will remain words or low-level POS in the merged query. We can then extract the informer from the merged query by deleting POS of a level higher than a set threshold and keeping only contiguous words and POS below that threshold.

The informer-learning algorithm we use is summarized in Figure 3. It begins with a set of training queries classified in their correct types and a list of informers representing each type such as the list already proposed in Figure 1. It then divides the set of training queries into two subsets, one containing queries that can be correctly classified by the current informers and the other containing queries that cannot. In Figure 3, we call these sets C-queries and I-queries, for Correctly-classified queries and Incorrectly-classified queries respectively. The learning algorithm then merges together pairs of incorrectly-classified queries of the same type to generate new informers. Good informers, that can be used to correctly classify some of the incorrectly-classified queries without leading to misclassification of the already correctly-classified queries, are added to the list of informers, and the queries they correctly classify are moved to C-queries, the set of correctly-classified queries. The enriched list of informers is the final result of the learning algorithm.

```
1. Input: list of informers, training queries
2. For each training query
   3. Classify using informers
   4. If classified correctly, add to C-queries
   5. Else, add to I-queries
6. For each pair of misclassified queries of same type
   7. Merge and generate informer
   8. If informer correctly classifies some of I-queries and does not misclassify any of C-queries, add to the list of informers
   9. Move newly-correctly-classified I-queries to C-queries
10. Return the list of informers
```

Fig. 3. Structure of the learning algorithm

4 Experimental Results

For our experiments, we built a test corpus of queries using the 459 queries from the 2006 TREC QA track [12]. We tagged the words of the queries with their parts-of-speech using the standard Brill tagger, and we manually classified the queries into their correct question types.

Our first experiment is meant to study the classification results obtained by using our 14 basic informers with and without our part-of-speech hierarchy, to show the impact of the hierarchy. Classification using the hierarchy and the basic informers is done as described in Section 3.1. The results without using the hierarchy are meant to be a benchmark. They will show how well a system can perform the classification task by only recognizing the informers in the queries. An initial check shows that the 14 informers are a resource of limited usefulness: they only appear in 43% of our test queries, and in about 12% of these cases they are used in the wrong question type because of the paraphrasing phenomenon studied in [2]. The informers give no useful information to help classify the remaining 57% of queries they do not appear in. This gives insight into the reason why other systems must rely on hundreds of patterns [4].

Next, we used the learning algorithm to expand the list of informers. The set of training queries we used is the query list from the 2007 TREC QA track [11], which we tagged and classified into correct types in the same way as we did for the 2006 TREC queries. The algorithm learned four new informers, which we present in Table 1. In the informers in that table, actual words are written plainly while POS from our hierarchy are written in square brackets. The words corresponding to the informers in the sample queries are marked in bold.

In each case, we computed the precision and recall of the classification of queries into each of our six question types, using the standard equations given in (1) and (2). We then computed the average precision and recall over all six types, and the average F-measure using equation (3). The results of these three experiments are given in Table 2.

$$\text{Precision} = \frac{\text{True Positive}}{\text{True Positive} + \text{False Positive}} \quad (1)$$

$$\text{Recall} = \frac{\text{True Positive}}{\text{True Positive} + \text{False Negative}} \quad (2)$$

$$F - \text{Measure} = \frac{2 \times \text{Precision} \times \text{Recall}}{\text{Precision} + \text{Recall}} \quad (3)$$

Table 1. New informers learned

Informer	Question type	Sample query
[nn] [common-noun]	person (who)	**Name members** of the group.
[wh-personal] [present-tense]	location (where)	**Which** college did she go to?
what does	description (how/why)	**What does** LPGA stand for?
[nn] is [dt] [nn] ?	description (how/why)	What kind of **animal is an agouti?**

Table 2. Question Type Classification Results

#	Experiment	Precision	Recall	F-Measure
1	Without hierarchy	43%	41%	42%
2	With hierarchy	85%	56%	68%
3	With learning	74%	65%	69%

4.1 Discussion

From the results presented in Table 2, it can clearly be seen that our system (line #2) does much better than the benchmark (line #1). It achieves nearly twice the precision and yields a 26% increase in F-measure. The reason for this improvement, and the only difference between our system and the benchmark, is the use of the part-of-speech hierarchy in addition to the 14 basic informers. The benchmark system can only exactly match queries to the informers; if there is no such match or if several informers appear in a query, the system is clueless. On the other hand, our system compares queries and informers together using the POS hierarchy as described in Section 3.1, and computes the similarity of each pair. In other words, it associates a query to its most similar informer, rather than look for an exact match. This ability to handle similar but inexact matches is clearly an important advantage.

In our third experiment, the learning algorithm discovers four new informers. They represent different syntaxes of queries that were not accounted for in our initial 14 informers, as illustrated in Table 1. In particular, the first informer is learned to handle an error in the tagging: the Brill tagger mistakenly identified the word "name" in these queries as a noun instead of a verb. The next two informers in Table 1 are learned to handle the wh-term parts of varying styles of queries; the second one in particular is a general inexact match in queries made possible by our hierarchy. The fourth informer represents a longer sentence structure often found in description-type queries.

The classification results using our hierarchy and the informer list including the four learned informers (line #3) show an important improvement compared to the benchmark (line #1). However, the advantage is less clear when compared to the 14 basic informers alone (line #2). The new classification shows a worse precision but a better recall, leaving the F-measure nearly unchanged. It is worth noting that using the basic informers alone leads to an important 30% difference between precision and recall, while that gap is reduced to 10% when the extra four informers are added in, giving our system a more balanced classification performance.

Although both experiments using the POS hierarchy outperform the benchmark, there is still clearly room for improvement. Errors in our system are misclassifications caused by an informer from a wrong question type being more similar to a query than any of the informers from its own type. The main bad informer is "what was", one of our 14 basic informers, which is alone responsible for 75% of misclassifications in each of the two experiments with our system. However, that informer cannot be simply eliminated, as it is also responsible for a lot of correct classifications; indeed, more than a third of the queries it classifies are done so correctly. Rather, the solution is for the system to learn new informers that will be more similar to the misclassified queries than "what was" and will classify them correctly. The fact that our current learning algorithm does not discover these informers might be due to our strict learning criterion of only saving informers that do not misclassify any queries from the C-queries subset. A more permissive criterion, for example of accepting informers that correctly classify more of the I-queries subset than they misclassify the C-queries subset, could lead to learning a better informer list. Changes to the learning algorithm in order to discover better informers will be studied in future research.

5 Conclusion

In this paper, we present a new method to learn and apply informers for the task of syntactic-pattern-based question type classification. What sets our method apart is the use of our part-of-speech hierarchy, which makes it possible to compute mathematically the distance between the informers and the queries, and to associate queries with their most similar informers rather than look for exact keyword matches. In fact, experimental results show that using the hierarchy for this task yielded an average 26% improvement in the F-measure of type classification. Another advantage of our method is that it obtains good results using only 14 simple informers, while traditional methods use hundreds of keywords, patterns or probabilities. Future work will focus on further improving the classification by refining the informer-learning algorithm to discover new and better informers, which will account for and rectify some of the mistakes the informers in the current system cause.

References

1. Razmara, M., Fee, A., Kosseim, L.: Concordia University at the TREC 2007 QA track. In: Proceedings of the Sixteenth Text REtrieval Conference (TREC 2007), Gaithersburg, USA (2007)
2. Tomuro, N.: Interrogative reformulation patterns and acquisition of question paraphrases. In: Proceedings of the Second International Workshop on Paraphrasing, Sapporo, Japan, vol. 16, pp. 33–40 (2003)
3. Sung, C.-L., Day, M.-Y., Yen, H.-C., Hsu, W.-L.: A template alignment algorithm for question classification. In: IEEE International Conference on Intelligence and Security Informatics (ISI 2008), pp. 197–199 (2008)
4. Krishnan, V., Das, S., Chakrabarti, S.: Enchanced Answer Type Inference from Questions using Sequential Models. In: Proceedings of Human Language Technology Conference / Conference on Empirical Methods in Natural Language Processing (HLT/EMNLP 2005), Vancouver, Canada, pp. 315–322 (2005)
5. Khoury, R., Karray, F., Kamel, M.: Keyword extraction rules based on a part-of-speech hierarchy. International Journal of Advanced Media and Communication 2(2), 138–153 (2008)
6. Liang, Z., Lang, Z., Jia-Jun, C.: Structure analysis and computation-based Chinese question classification. In: Sixth International Conference on Advanced Language Processing and Web Information Technology (ALPIT 2007), Luoyang, China, pp. 39–44 (2007)
7. Tomuro, N.: Question terminology and representation of question type classification. In: Second International Workshop on Computational Terminology, vol. 14 (2002)
8. Harabagiu, S., Moldovan, D., Pasca, M., Mihalcea, R., Surdeanu, M., Bunescu, R., Girju, R., Rus, V., Morarescu, P.: Falcon: Boosting knowledge for answer engines. In: Proceedings of the 9th Text REtrieval Conference (TREC-9), Gaithersburg, USA, pp. 479–488 (2000)
9. Zhang, D., Nunamaker, J.F.: A Natural language approach to content-based video indexing and retrieval for interactive e-learning. IEEE Transactions on Multimedia 6(3), 450–458 (2004)
10. Marcus, M., Santorini, B., Marcinkiewicz, M.A.: Building a large annotated corpus of English: the Penn Treebank. Computational Linguistics 19(2), 313–330 (1993)
11. Dang, H.T., Kelly, D., Lin, J.: Overview of the TREC 2007 Question Answering Track. In: Proceedings of the Sixteenth Text REtrieval Conference (TREC 2007), Gaithersburg, USA (2007)
12. Dang, H.T., Lin, J., Kelly, D.: Overview of the TREC 2006 Question Answering Track. In: Proceedings of the Fifteenth Text REtrieval Conference (TREC 2006), Gaithersburg, USA (2006)

Exploring Wikipedia's Category Graph for Query Classification

Milad Alemzadeh[1], Richard Khoury[2], and Fakhri Karray[1]

[1] Department of Electrical and Computer Engineering, University of Waterloo,
200 University Avenue West, Waterloo, Ontario, Canada, N2L 3G1
{malemzad,karray}@uwaterloo.ca
[2] Department of Software Engineering, Lakehead University, 955 Oliver Road,
Thunder Bay, Ontario, Canada, P7A 5E1
richard.khoury@lakeheadu.ca

Abstract. Wikipedia's category graph is a network of 400,000 interconnected category labels, and can be a powerful resource for many classification tasks. However, its size and the lack of order can make it difficult to navigate. In this paper, we present a new algorithm to efficiently explore this graph and discover accurate classification labels. We implement our algorithm as the core of a query classification system and demonstrate its reliability using the KDD CUP 2005 competition as a benchmark.

Keywords: Natural Language Processing, Query Classification, Category Labeling, Wikipedia.

1 Introduction

Query classification is the task of Natural Language Processing (NLP) whose goal is to identify the category label, in a predefined set, that best represents the domain of a question being asked. An accurate query classification system would be beneficial in many practical systems, including search engines and question-answering systems. But while similar categorization tasks are found in several branches of NLP, the challenge of query classification is accentuated by the fact that a typical query is only between one and four words long [1], [2], rather than the hundreds or thousands of words one can get from an average text document. Such a limited number of keywords makes it difficult to select the correct category label, and moreover it makes the selection very sensitive to "noise words", or words unrelated to the query that the user entered for some reason such as because they didn't remember a correct name or technical term to query for. A second challenge of query classification comes from the fact that, while document libraries and databases can be specialized to a single domain, the users of query systems expect to be able to ask queries about any domain at all [1].

In this paper, we build upon our previous work on query labeling using the Wikipedia category graph [3]. We have already shown that Wikipedia offers a set of nearly 400,000 interconnected categories which can be used for query classification. Moreover, since these categories cover most domains of human knowledge at varying degrees of granularity, it is easy for system designers to identify a subset of them as "target categories" they wish to use as classification goals, rather than deal with the full set of 400,000 categories. This paper now presents a new algorithm to explore the graph of categories, to efficiently discover the best target category to classify a query into.

The rest of the paper is organized as follows. Section 2 presents overviews of the literature in the field of query classification with a special focus on the use of Wikipedia for that task. We present in detail our exploration and classification algorithm Section 3, then we move on in Section 4 to describe and analyze the experimental results we obtained with our system. Finally, we give some concluding remarks in Section 5.

2 Background

Query classification is the task of NLP that focuses on inferring the domain information surrounding user-written queries, and on assigning each query to the category label that best represents its domain in a predefined set of labels. Given the ubiquity of search engines and question-handling systems today, the challenge of query classification has been receiving a growing amount of attention. Notably, it was the topic of the ACM's annual KDD CUP competition in 2005 [4], where 37 systems competed to classify a set of 800,000 real web queries into a set of 67 categories designed to cover most topics found on the internet. The winning system was designed to classify a query by comparing its word vector to that of each website in a set pre-classified in the Google directory. The query was assigned the category of the most similar website, and the directory's set of categories was mapped to the KDD CUP's set [2]. This system was later improved by introducing a bridging classifier and an intermediate-level category taxonomy [5].

There are a lot of other active research groups working in query classification. They all follow the basic pattern of mapping a query into an external knowledge source to classify it. There exist a great variety of such systems, using for example ontologies [6], web query logs [7], and Wikipedia [8], [9].

In fact, exploiting Wikipedia as a knowledge source has become commonplace in scientific research. Several hundreds of journal and conference papers have been published using this tool since its creation in 2001. However, to the best of our knowledge, aside from our previous work mentioned in Section 1, there have been only two query classification systems designed based on Wikipedia.

The first of these two systems was proposed by Hu et al. [8]. Their work assumes that there is a set of seed concepts that their query classification should be trained to recognize. They thus target the articles and categories relevant to these concepts, and construct a graph of Wikipedia domains by following the links in these articles using a Markov random walk algorithm. Each step from one concept to the next on the graph is assigned a transition probability, and these probabilities are then used to compute the likelihood of each domain. Once the knowledge base has been build in this way, a new user query can be classified simply by using its keywords to retrieve a list of relevant Wikipedia domains, and sorting them by likelihood. Unfortunately, their system remained small-scale and limited to only three basic domains, namely "travel", "personal name" and "job". It is not a general-domain classifier such as the one we aim to create.

The second query classification system was designed by one of our co-authors in [9]. It follows Wikipedia's encyclopedia structure to classify queries step-by-step, using the query's words to select titles, then selecting articles based on these titles, then categories from the articles. At each step, the weights of the selected elements are computed based on the relevant elements in the previous step: a title's weight depends on the words that selected it, an article's weight on the titles', and a category's weight on the articles'. Unlike [8], this system was a general classifier that could handle queries from any domain, and its performance would have ranked near the top of the KDD CUP 2005 competition.

3 Methodology

Wikipedia's category graph is a massive set of almost 400,000 category labels, describing every domain of knowledge and ranging from the very precise, such as "fictional secret agent and spies", to the very general, such as "information". The categories are connected by hypernym relationships, with a child category having an "is-a" relationship to its parents. However, the graph is not strictly hierarchic: there exist shortcuts in the connections (i.e. starting from one child category and going up two different paths of different lengths to reach the same parent category) as well as loops (i.e. starting from one child category and going up a path to reach the same child category again).

The fact that the set of category labels covers practically every domain at every level of precision makes it easy for a system designer to identify a subset of categories to be used as "target categories" for a classification system. The query classifier we propose in this paper is designed to explore the graph of categories from any starting point until it reaches the nearest such target categories. The pseudocode of our new algorithm is shown in Figure 1.

3.1 Building the Category Graph

The list of categories in Wikipedia and the connections between categories can easily be extracted from the database dump made freely available by the Wikimedia Foundation. For this project, we used the version available from September 2008.

However, our graph includes one extra piece of information in addition to the categories, namely the article titles. In Wikipedia, each article is an encyclopedic entry on a given topic which is classified in a set of categories, and which is pointed to by a number of titles: a single main title, some redirect titles (for common alternative names, including foreign translations and typos) and some disambiguation titles (for ambiguous names that may refer to it). For example, the article for the United States is under the title "United States", as well as the redirect titles "USA", "United States of America" and "United Staets", and the disambiguation title "America". Our graph maps the titles directly to the categories of the articles, and then discards the articles. After this processing, we find that our category graph features 5,453,808 titles and 390,807 categories [3].

```
Define:    CategoryGraph,
           TargetCategories (a subset of CategoryGraph),
           Classification (classification results),
           CassificationSize (number of classification
           results allowed per query)
Input:     User query
0. Classification ← {}
1. TitleList ← the most relevant Wikipedia titles to
   the user query
2. CatList ← the categories relating to TitleList
3. Do for 20 iterations:
      4. NewClassification ← subset of CatList that are
         in TargetCategories
      5. If COUNT(Classification + NewClassification)
         <= CassificationSize
            6. Classification ← Classification +
               NewClassification
      7. If COUNT(Classification + NewClassification) >
         CassificationSize AND COUNT(Classification) > 0
         8. Break from loop
      9. If COUNT(Classification + NewClassification) >
         CassificationSize AND COUNT(Classification) = 0
            10. Classification ← Select CassificationSize
                elements from NewClassification
            11. Break from loop
      12. CatList ← unvisited parent categories di-
          rectly connected to CatList
13. Return Classification
```

Fig. 1. Structure of our classification algorithm

3.2 Starting the Search

The first step of our algorithm is to map the user's query to an initial set of categories from which the exploration of the graph will begin. This is accomplished by going through the titles included in the graph. The query is stripped of stopwords to keep only keywords; the system then generates the exhaustive list of titles that feature at least one of these keywords, and expands the exhaustive list of categories pointed to by these titles. Next, the algorithm considers each keyword/title/category triplet where it is the case that the keyword is in the title and the title points to the category, and assigns each one a weight that is a function of how many query keywords are featured in the title with a penalty for title keywords not featured in the query. The exact formula to compute the weight W_t of keywords in title t is given in equation (1). In this formula, N_k is the total number of query keywords featured in the title, C_k is the character count of the keywords featured in the title, and C_t is the total number of characters in the title. The rationale for using character counts in this formula is to shift some density weight to titles that match longer keywords in the query. The assumption is that, given that the user typically only provides less than four keywords in the query, having one much longer keyword in the set could mean that this one keyword is more important. Consequently, we give a higher weight to keywords in a title featuring the longer query keywords and missing the shorter ones, as opposed to a title featuring the shorter query keywords and missing the longer ones.

$$W_t = \frac{N_k}{1 + \frac{|C_k - C_t|}{C_k}} \qquad (1)$$

The weight of a keyword given a category is then defined as the maximum value that keyword takes in all titles that point to that category. Finally, the density value of each category is computed as the sum of the weights of all query keywords given that category. This process will generate a long list of categories, featuring some categories pointed to by high-weight words and summing to a high density score, and a lot of categories pointed to by only lower-weight words and having a lower score. The list is trimmed by discarding all categories having a score less than half that of the highest-density category. This trimmed set of categories is the initial set the exploration algorithm will proceed from. It corresponds to "CatList" at step 2 of our pseudocode in Figure 1. Through practical experiments, we found that this set typically contains approximately 28 categories.

3.3 Exploration Algorithm

Once the initial list of categories is available, the search algorithm explores the category graph step by step. At each step, the algorithm compares the set of newly-visited categories to the list of target categories defined as acceptable classification labels and adds any targets discovered to the list of classification results. It then generates the next generation of unvisited categories directly connected to the current set as parent and repeats the process. The exploration can thus be seen as radiating through the graph from each initial category. This process corresponds to steps 4 and 12 of the pseudocode algorithm in Figure 1.

There are two basic termination conditions for the exploration algorithm. The first is when a predefined maximum number of classification results have been discovered. This maximum could for example be 1, if the user wants a unique classification for each query, while it was set at 5 in the KDD CUP 2005 competition rules. However, since the exploration algorithm can discover several target categories in a single iteration, it is possible to overshoot this maximum. The algorithm has two possible behaviors defined in that case. First, if some results have already been discovered, then the new categories are all rejected. For example, if the algorithm has already discovered four target categories to a given query out of a maximum of five and two more categories are discovered in the next iteration, both new categories are rejected and only four results are returned. The second behavior is for the special case where no target categories have been discovered yet and more than the allowed maximum are discovered at once. In that case, the algorithm simply selects randomly the maximum allowed number of results from the set. For example, if the algorithm discovers six target categories at once in an iteration, five of them will be kept at random and returned as the classification result.

The second termination condition for the algorithm is reaching a maximum of 20 iterations. The rationale for this is that, at each iteration, both the set of categories visited and the set of newly-generated categories expand. The limit of 20 iterations thus reflects a practical consideration, to prevent the size of the search from growing without constraint. But moreover, after 20 steps, we find that the algorithm has explored too far from the initial categories for the targets encountered to still be relevant. For comparison, in our experiments, the exploration algorithm discovered the maximum number of target categories in only 3 iterations on average, and never reached the 20 iterations limit. This limit thus also allows the algorithm to cut off the exploration of a region of the graph that is very far removed from target categories and will not generate relevant results.

4 Experimental Results

In order to test our system, we submitted it to the same challenge as the KDD CUP 2005 competition [4]. The 37 solutions entered in that competition were evaluated by classifying a set of 800 queries into up to 5 categories from a predefined set of 67 target categories c_i, and comparing the results to the classification done by three human labelers. The 800 test queries were meaningful English queries selected randomly from MSN search logs, unedited and including the users' typos and mistakes. The solutions were ranked based on overall precision and overall F1 value, as computed by Equations (2-6). The competition's Performance Award was given to the system with the top overall F1 value, and the Precision Award was given to the system with the top overall precision value within the top 10 systems evaluated on overall F1 value. Note that participants had the option to enter their system for precision ranking but not F1 ranking or vice-versa rather than both precision and F1 ranking, and several participants chose to use that option. Consequently, the top 10 systems on F1 value ranked for precision are not the same as the top 10 systems ranked for F1 value, and there are some N/A values in the results in Table 1.

$$\text{Precision} = \frac{\sum_i \text{Number of queries correctly labeled as } c_i}{\sum_i \text{Number of queries labeled as } c_i} \quad (2)$$

$$\text{Recall} = \frac{\sum_i \text{Number of queries correctly labeled as } c_i}{\sum_i \text{Number of queries belonging to } c_i} \quad (3)$$

$$F1 = \frac{2 \times \text{Precision} \times \text{Recall}}{\text{Precision} + \text{Recall}} \quad (4)$$

$$\text{Overall Precision} = \frac{1}{3} \sum_{j=1}^{3} \text{Precision against labeler } j \quad (5)$$

$$\text{Overall F1} = \frac{1}{3}\sum_{j=1}^{3} \text{F1 against labeler } j \qquad (6)$$

In order for our system to compare to the KDD CUP competition results, we need to use the same set of category labels. As we mentioned in Section 3, the size and level of detail of Wikipedia's category graph makes it possible to identify categories to map any set of labels to. In our case, we identified 84 target categories in Wikipedia corresponding to the 67 KDD CUP category set.

With the mapping done, we classified the 800 test queries with our system and evaluated the results on overall precision and F1 following the KDD CUP guidelines. Our results are presented in Table 1 along with the KDD CUP mean and median, the best system on precision, the best system on F1, and the worst system overall as reported in [4]. As can be seen from that table, our system performs well above the competition average, and in fact ranks in the top-10 of the competition.

Table 1. Classification results

System	F1 Rank	Precision Rank	Overall Precision	Overall F1
Best F1	1	N/A	0.4141	0.4444
Best Precision	N/A	1	0.4237	0.4261
Our System	10	7	0.3081	0.3005
Mean	18	13	0.2545	0.2353
Median	19	15	0.2446	0.2327
Worst	37	37	0.0509	0.0603

It is interesting to consider not only the final classification result, but also the performance of our exploration algorithm. To do this, we studied how frequently each of the termination conditions explained in Section 3.3 was reached. We can summarize from Section 3.3 that there are five distinct ways the algorithm can terminate. The first is "no initial list", which is to say that the initial keyword-to-category mapping failed to generate any categories for our initial set and the exploration cannot begin. If there is an initial set of categories generated and the exploration begins, then there are still four ways it can terminate. The first is "failure", if it reaches the cutoff value of 20 iterations without encountering a single target category. The second termination condition is "exploration limit", if the algorithm reaches the cutoff value of 20 iterations but did discover some target categories along the way. These categories are returned as the classification results. The third termination is the "overshoot", if the algorithm discovers more than the maximum number of results in a single iteration and must select results randomly. And the final termination condition is "category limit", which is when the algorithm has already found some categories and discovers more categories that bring it to or above the set maximum; in the case it goes above the maximum the newly-discovered categories are discarded. In each case, we obtained the number of query searches that ended in that condition, the average number of iterations it took the algorithm to reach that condition, the average number of categories found (which can be greater than the maximum allowed when more categories are found in the last iteration) and the average number of target categories returned. These results are presented in Table 2.

Table 2. Exploration performance

Termination	Number of queries	Average number of iterations	Average number of target categories found	Average number of target categories returned
No initial list	52	0	0	0
Failure	0	20	0	N/A
Exploration limit	0	20	0	N/A
Overshoot	28	2.4	7.0	5
Category limit	720	3.3	7.8	3.3

As can be seen from table 2, two of the five termination conditions we identified never occur at all. They are the two undesirable conditions where the exploration strays 20 iterations away from the initial categories. This result indicates that our exploration algorithm never does diverge into wrong directions or miss the target categories, nor does it end up exploring in regions without target categories. However, there is still one undesirable condition that does occur, namely that of the algorithm selecting no initial categories to begin the search from. This occurs when no titles featuring query words can be found; typically because the query consists only of unusual terms and abbreviations. For example, one query consisting only of the abbreviation "AATFCU" failed for this reason. Fortunately, this does not happen frequently: only 6.5% of queries in our test set terminated for this reason. The most common termination conditions, accounting for 93.5% of query searches, are when the exploration successfully discovers the maximum number of target categories, either in several iterations or all in one, with the former case being much more common than the latter. In both cases, we can see that the system discovers these categories quickly, in less than 4 iterations on average. This demonstrates the success and efficiency of our exploration algorithm.

5 Conclusion

In this paper, we presented a novel algorithm to explore the Wikipedia category graph and discover the target categories nearest to a set of initial categories. To demonstrate its efficiency, we used the exploration algorithm as the core of a query classification system, and showed that its classification results compare favorably to those of the KDD CUP 2005 competition: our system would have ranked 7th on precision in that competition, with an increase of 6.4% compared to the competition median, and 10th on F1 with a 6.9% increase compared to the median. By using Wikipedia, our system gained the ability to classify queries into a set of almost 400,000 categories covering most of human knowledge and which can easily be mapped to a simpler application-specific set of categories when needed. But the core of our contribution remains the novel exploration algorithm, which can efficiently navigate the graph of 400,000 interconnected categories and discover the target categories to classify the query into in 3.3 iterations on average. Future work will focus on further refining the exploration algorithm to limit the number of categories generated at each iteration step by selecting the most promising directions to explore, as well as on developing ways to handle the 6.5% of queries that remain unclassified with our system.

References

1. Jansen, M.B.J., Spink, A., Saracevic, T.: Real life, real users, and real needs: a study and analysis of user queries on the web. Information Processing and Management 36(2), 207–227 (2000)
2. Shen, D., Pan, R., Sun, J.-T., Pan, J.J., Wu, K., Yin, J., Yang, Q.: Q2C@UST: our winning solution to query classification in KDDCUP 2005. ACM SIGKDD Explorations Newsletter 7(2), 100–110 (2005)
3. Alemzadeh, M., Karray, F.: An efficient method for tagging a query with category labels using Wikipedia towards enhancing search engine results. In: 2010 IEEE/WIC/ACM International Conference on Web Intelligence and Intelligent Agent Technology, Toronto, Canada, pp. 192–195 (2010)
4. Li, Y., Zheng, Z., Dai, H.: KDD CUP-2005 report: Facing a great challenge. ACM SIGKDD Explorations Newsletter 7(2), 91–99 (2005)
5. Shen, D., Sun, J., Yang, Q., Chen, Z.: Building bridges for web query classification. In: Proceedings of SIGIR 2006, pp. 131–138 (2006)
6. Fu, J., Xu, J., Jia, K.: Domain ontology based automatic question answering. In: International Conference on Computer Engineering and Technology (ICCET 2008), vol. 2, pp. 346–349 (2009)
7. Beitzel, S.M., Jensen, E.C., Lewis, D.D., Chowdhury, A., Frieder, O.: Automatic classification of web queries using very large unlabeled query logs. ACM Transactions on Information Systems 25(2), article 9 (2007)
8. Hu, J., Wang, G., Lochovsky, F., Sun, J.-T., Chen, Z.: Understanding user's query intent with Wikipedia. In: Proceedings of the 18th International Conference on World Wide Web, Spain, pp. 471–480 (2009)
9. Khoury, R.: Using Encyclopaedic Knowledge for Query Classification. In: Proceedings of the 2010 International Conference on Artificial Intelligence (ICAI 2010), Las Vegas, USA, vol. 2, pp. 857–862 (2010)

Combination of Error Detection Techniques in Automatic Speech Transcription

Kacem Abida[1], Wafa Abida[2], and Fakhri Karray[1]

[1] Electrical and Computer Engineering Dept.,
University of Waterloo, Ontario, Canada
{mkabida,karray}@uwaterloo.ca
http://www.uwaterloo.ca

[2] Voice Enabling Systems Technology Inc.,
Waterloo, Ontario, Canada
wafa@vestec.com
http://www.vestec.com

Abstract. Speech recognition technology has been around for several decades now, and a considerable amount of applications have been developed around this technology. However, the current state of the art of speech recognition systems still generate errors in the recognizer's output. Techniques to automatically detect and even correct speech transcription errors have emerged. Due to the complexity of the problem, these error detection approaches have failed to ensure both a high recall and a precision ratio. The goal of this paper is to present an approach that combines several error detection techniques to ensure a better classification rate. Experimental results have proven that such an approach can indeed improve on the current state of the art of automatic error detection in speech transcription.

Keywords: Combination, Error Detection, Point Wise Mutual Information, Latent Semantic Indexing, Recall, Precision.

1 Introduction

The advancement in the signal processing field along with the availability of powerful computing devices have led to the achievement of descent performance in the speech transcription field. However this performance could only be obtained under restrictive conditions, such as broadcast speech data, noise-free environment, etc. Making speech recognition effective under all conditions is the ultimate goal, which can be achieved if we are able to minimize or even eliminate and/or correct most of the recognition errors. However, detecting errors in a speech transcription is not an easy task. In fact, there are several types of errors, namely deletion, substitutions, stop word errors, etc which can be manifested at the lexical and syntactic levels. This makes it difficult to a single generic error detector to capture all these types of errors at different levels. This observation is clearly noticeable on the current state of the art of the error detection techniques. Each technique could only guarantee descent performance for a subset of these errors. To tackle this problem, recently few researchers started investigating ways to combine these error classifiers to cover a broader range of error types and to improve on the current performance of the existing techniques. This work lies in this same direction, and proves that it's beneficial to combine several detectors in terms of the precision and the recall rates. The paper starts by presenting some background material in section 2. Section 3 describes the combination algorithm, followed by a case study in section 4 using two commonly known error detection techniques. Experimental results and analysis are presented in section 5, followed by a conclusion in section 6.

2 Background Material

In large vocabulary speech transcription, recognition errors can be caused by a variety of factors[1]. The most common of these causes are: missing terms from the language model, hypothesis pruning during the Viterbi decoding procedure, noisy environment, etc. Since there is no foreseeable solution to counter effect all of these factors, researchers investigated ways to automatically detect possible recognition errors and eliminate them or even attempt to correct them. In this research work, we are interested in recognizer-independent (generic) approaches for error detection. In this type of approaches, the speech decoder is a black box. Throughout the literature, we could distinguish two main directions under the recognizer-independent approaches: probabilistic and non-probabilistic-based techniques. The most common non-probabilistic technique is pattern matching. The idea is to identify and collect error patterns that occur in the speech transcription. This technique suffers from lot of issues. First, pattern matching is unable to cope with unseen patterns. Second, for a broad domain, it's quite difficult to collect all possible error patterns due to the language variation. Finally, pattern matching is susceptible to false positives in cases where correct words occur in a known error context[1]. According to [2], the speech recognition errors tend to occur in regular patterns rather than at random. This lead us to consider probabilistic type of techniques to capture these error patterns. In fact, if we are given a large enough training data from a certain domain, the collected frequencies can be very well extended beyond the training corpus to cover the whole domain. The most commonly used techniques in this regard are Latent Semantic Indexing (LSI)[3] and Point-wise Mutual Information[4] (PMI)-based error detection. LSI is an information retrieval tool that relies on the terms co-occurrences to identify the degree of similarity between them. It represents the terms and documents in a reduced dimensionality, without loosing much of the relationship information among them. PMI computes similarity scores between terms using word frequencies from a given corpus. LSI has been first applied to the field of error spotting in [3] by exploiting this semantic analysis. Experiments showed that it's possible to achieve high recalls, but with low precision rate. The same conclusions have been also observed in [4], where PMI measures have been used to identify semantic outliers. Specifically, high precision rates have been achieved with low recall ratios. Few attempts already have been successful in combining error detection techniques to improve both precision and recall ratios[3,1,5]. This current work lies in this same direction, and is mainly targeting error detection techniques that relies on thresholding a confidence score to decide whether or not a given token from the speech transcription is an erroneous or correct output. In the remaining of this research work, and specifically in the experimentation section, we use both the PMI- and LSI-based error detection techniques to illustrate in a case study the combination approach we are presenting.

3 Proposed Approach

The low recall and precision ratios of the current automatic error detection techniques in speech transcription led us to investigate ways to improve both of these ratios. The idea behind this work is to combine different error detection approaches in the hope that the new technique achieves higher recall, respectively precision, without degrading the precision, respectively recall, ratio. Our ultimate goal is to be able to improve both ratios simultaneously upon the combination of the error detection techniques. The implicit assumption in this thinking, is that the error detection techniques have to have different performance in terms of these two ratios. In other words, when one approach achieves high recall and low precision, the other techniques to be combined with, needs to achieve high precision and low recall to ensure improvement in both ratios with the new technique. The logic of our proposed approach is to preserve each technique's advantage or powerful characteristics in the final combination.

This work is generic in the sense that it can be applied to any error detection technique that relies on thresholding a confidence score to determine whether a given word is erroneous or correct. If the confidence score is less than the selected threshold value, then the word is considered as an error, otherwise, it's tagged as a correct token. Figure 1 describes the flow of the error detection combination approach.

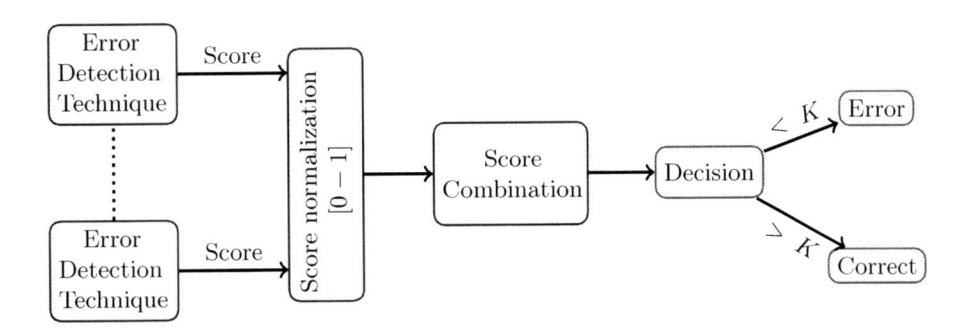

Fig. 1. Error Detection Techniques Combination Procedure

The scale of each error detection technique confidence score is different. Therefore a score normalization stage is needed to standardize all confidence score from the various detection techniques to lie between zero and one. Equation 1 is used to normalize the confidence scores, where X is the score to be normalized, and min, respectively max, is the minimum, respectively maximum, value of the technique's confidence score.

$$X_{scaled} = \frac{X - min}{max - min} \quad (1)$$

Once all the confidence scores have been normalized, a score combination formula is then applied to build a new score. The classification threshold, K, is then applied on this new score to detect erroneous output. Two score combination formulas were used, namely Weighted Average (WA) and Harmonic Mean (HM), as shown in equations 2 and 3, where $Score_i$ refers to the confidence score of the ith error detection technique, N is the total number of combined error detection techniques, and α_i are weighting scales to each technique in such a way $\sum_{i=1}^{N} \alpha_i = 1$

$$Score_{WA} = \sum_{i=1}^{N} \alpha_i Score_i \quad (2)$$

$$Score_{HM} = \frac{N}{\sum_{i=1}^{N} \frac{1}{Score_i}} \quad (3)$$

The weighting factors play an important role in realizing a trade off between various detection techniques to optimize recall and precision ratios. This coefficients needs to be optimized a priori during training. The algorithm 1 summarizes the combination procedure for N different speech transcription error detection techniques.

Algorithm 1. Combination of Error Detection Techniques

1: Compute the score of the word w, $Score_i$, for each technique.
2: Scale the confidence score to the [0,1] interval, using eq.1.
3: Compute the new confidence score, $Score_{comb}$, using eq.2 or eq.3.
4: Tag the word w as an error if $Score_{comb} \leq K$.

In step 4 of algorithm 1, the threshold parameter K is to be optimized through a training stage. It is used to control the error detection rate. The higher K is, the more aggressive the error filtering, and vice versa. If K is quite low, more erroneous words slip past the combined error detector.

4 Case Study: PMI- and LSI-Based Error Detection Combination

We illustrate the proposed combination approach on two widely used error detection techniques. Both techniques suffer from low recall and precision ratios. A brief description of both techniques now follows.

4.1 PMI-Based Error Detection

A few items need to be defined before describing the error-detection procedure in detail. The neighborhood $N(w)$ of a word w is the set of context tokens around w that appear before and after it. This concept is defined within a window of tokens. The Point-wise Mutual Information (PMI)-based semantic similarity is a measure of how similar and how close in meaning w_i and w_j are. In a nutshell, the PMI score in equation 4 below is defined as the probability of seeing both words (w_i and w_j) together, divided by the probability of observing each of these words separately.

$$PMI(w_i, w_j) = log\left(\frac{P(w_i, w_j)}{P(w_i).P(w_j)}\right) \quad (4)$$

Given a large textual corpus with size N tokens, the probabilities introduced in 4 can be computed using $P(w_i) = \frac{c(w_i)}{N}$, and $P(w_i, w_j) = \frac{c(w_i, w_j)}{N}$, where $c(w_i)$ and $c(w_i, w_j)$ are the frequency counts collected from the corpus. The process of detecting an error using the PMI-based technique[4] is described in algorithm 2.

Algorithm 2. PMI-based Error Detection

1: Identify the neighborhood $N(w)$.
2: Compute PMI scores $PMI(w_i, w_j)$ for all pairs of words $w_i \neq w_j$ in the neighborhood $N(w)$, including w.
3: Compute Semantic Coherence $SC(w_i)$ for every word w_i in the neighborhood $N(w)$, by aggregating the $PMI(w_i, w_j)$ scores of w_i with all $w_j \neq w_i$.
4: Define SC_{avg} to be the average of all the semantic coherence measures $SC(w_i)$ in $N(w_i)$
5: Tag the word w as an error if $SC(w) \leq K.SC_{avg}$.

According to [6], the bigger the window size the better for the first step. In step 3 of the algorithm, the semantic coherence has been computed using the Maximum aggregation variant: $SC(w_i) = \max_{i \neq j} PMI(w_i, w_j)$, since it yielded better results in [6]. The filtering parameter K is used to control the error detection rate.

4.2 LSI-Based Error Detection

LSI determines the similarity between terms by analyzing their co-occurrences within several documents. The LSI procedure mines for features that highlight the similarities between words. These features are obtained by applying a dimensionality reduction technique to high dimensional word-feature matrix. Given a large textual corpus, a term-document matrix is built where rows stand for words, and columns stand for documents. The value in the cell (i, j), $w_{i,j}$, holds the weighted frequency of occurrence of word i in the document j. These weights are a combination of local and global weighting schemes, as shown in eq.5.

$$w_{i,j} = localweight(i,j) * globalweight(i) \qquad (5)$$

In this research work, we have used the combination of entropy (as global weighting) and the logarithm term frequency (as local weighting) as shown by eq.6 and eq.7, where $freq(i,j)$ denotes the frequency of occurrence of term i in document j, N is the total number of documents constituting the corpus[7].

$$localweight(i,j) = log(1 + freq(i,j)) \qquad (6)$$
$$globalweight(i) = 1 - entropy(i) \qquad (7)$$
$$= 1 + \frac{\sum_{j=1}^{N} P(i,j) * log(P(i,j))}{log(N)}$$

Singular Values Decomposition (SVD) is applied on the large term-document matrix. Since the term-document matrix is very sparse and its rank is much more lower than its actual dimensionality, it is safe to represent the terms and documents vectors in a much lower dimensional space with little loss of information. The most commonly used measure between vectors is the cosine metric. The cosine similarity is computed using eq.8, which is the angle between the two vectors u and v.

$$cosine(u,v) = \frac{<u,v>}{||u|| * ||v||} \qquad (8)$$

Now that we have collected the most significant word features (through the SVD decomposition), and selected a metric to measure the similarity between any two words (through the Cosine similarity measure), all what is left is to compute the semantic similarity score of a given word in an utterance of length M. Two different aggregations have been used: the mean semantic scoring (MSS) and the mean rank of the semantic scores (MR)[3]. MSS and MR scores are computed as shown in eq.9 and eq.10 respectively.

$$MSS_i = \frac{1}{M} \sum_{j=1}^{M} cosine(w_i, w_j) \qquad (9)$$

$$MR_i = \frac{1}{M} \sum_{j=1}^{M} RANK(cosine(w_i, w_j)) \qquad (10)$$

The $cosine(w_i, w_j)$ is computed using eq.8. The rank of the semantic score shown in eq.10 is computed as follows. First, the set of semantic scores L_i is computed. L_i is the set of cosine scores between the word w_i and all the remaining words w_j in the corpus. The MR_i score is then the mean of the rank of each $cosine(w_i, w_j)$ score in the set of L_i. The LSI-based error detection works as follows: given a recognizer's transcription output, a word is tagged as erroneous if and only if its MSS (respectively MR) is below a threshold K. The error filtering procedure applied on a word w_i in a given transcription output, is detailed in algorithm 3.

Algorithm 3. LSI-based Error Detection

1: Compute Cosine scores between w_i and all other words in the transcription output.

2: Compute MSS_i score (eq.9) or MR_i score (eq.10).
3: Tag the word w_i as an error if $MSS_i \leq K$ or $MR_i \leq K$.

The threshold K is to be optimized through a training stage. It is used to control the error detection rate.

4.3 Combination of LSI and PMI

Now that we have gone through the details of each of the error detection technique, the next step is to apply the combination procedure detailed in algorithm 1. Given a word from a transcription output, we first apply both PMI- and LSI-based error detection in order to obtain the confidence score, from step 3 of algorithm 2 and step 2 of algorithm 3. Once we have these two scores, we apply step 2 of the combination algorithm 1. This step aims at scaling both scores to lie within the $[0, 1]$ range. Using equations 2 or 3, we combine both scores into one single confidence value, on which we apply a thresholding to tag the given word. If the new score is below the threshold, the word is tagged as an error, otherwise it's a correct output. Obviously, this procedure is generic and scalable since it can be applied to any number of error detection techniques. The weighted average score combination method is particularly adequate when the confidence scores from various detection techniques are separately scattered within the $[0, 1]$ interval. The weighting coefficient acts therefore as a catalyst to boost or weaken the scores from different techniques. This will play an important role in improving the precision and/or recall ratios. The weighting coefficients are to be optimized a priori through a training phase. The following section presents experimental results of the binary combination of both PMI- and LSI-based error detection methods.

5 Experimental Assessment

5.1 Experimental Framework

The experiments were conducted using the CMU Sphinx 4 decoder. The HUB4 [8] testing framework was used. Transcriptions of this HUB4 corpus, have been

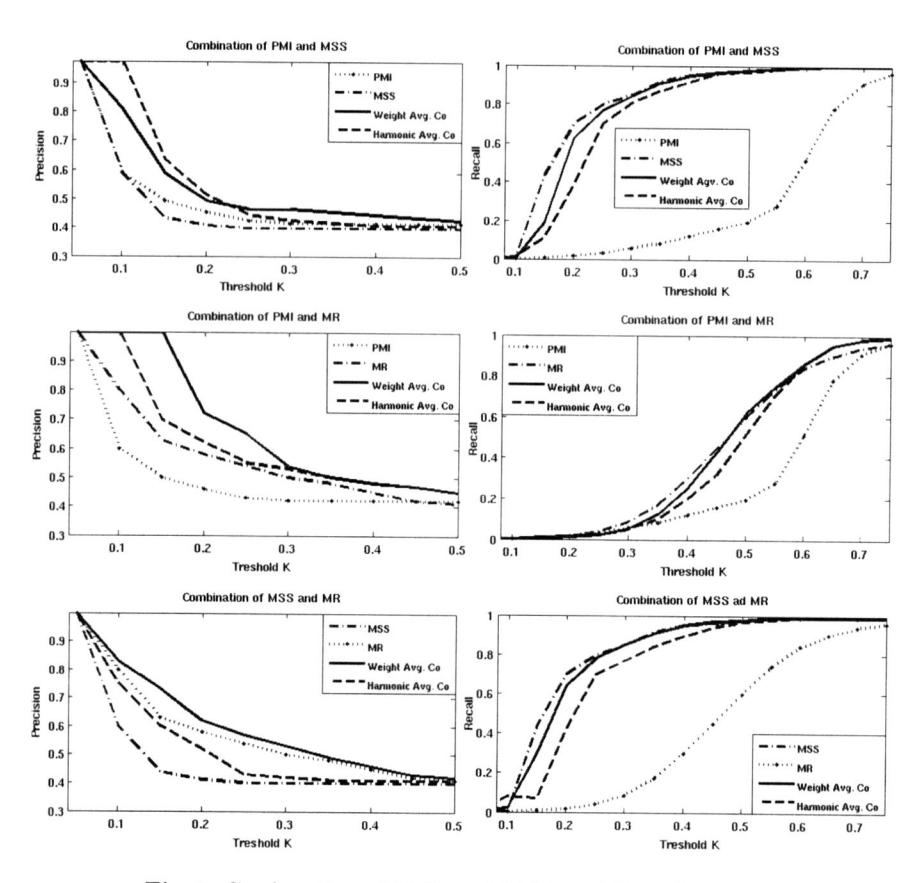

Fig. 2. Combination of PMI- and LSI-based Error Detectors

used to train the language model. For the acoustic model, the already trained HUB4 model provided on the Sphinx download site has been used. The transcriptions output from the Sphinx recognizer has been used to create the testing framework for the error detection combination procedure. In order to assess the performance of the error detection algorithms, the errors in each transcription were automatically flagged, and then correct words were selected at random, to obtain a testing dataset consisting of both erroneous and correctly recognized words. 20% of the testing dataset was used to optimize the different parameters (the filtering threshold K, the window size, the combination weighting factors).

The PMI-based error detection technique requires uni-gram and bi-gram frequency counts, which have been taken from Google Inc.'s trillion-token corpus[9]. It is a dataset of n-gram counts which were generated from approximately one trillion word tokens of text. For the LSI-based error detector, the term-document matrix has been built from the latest Wikipedia XML dump. A total of 3.2M documents, and 100K unique terms have been identified.

In order to assess the performance of the error detection module, metrics from the machine learning theory were used. For a two-class problem (Error/Not Error), we can define the following quantities: precision and recall. The precision can be defined as $Precision = \frac{TP}{TP+FP}$, and the recall is defined by $Recall = \frac{TP}{TP+FN}$, where TP, FP, TN, and FN represents true positives, false positives, true negatives, and false negatives respectively.

5.2 Experimental Results

In this case study, our goal is to prove that the combination procedure we're presenting is able to boost the precision ratio without degrading the recall. Figure 5 summarizes our findings. We have in total three various detectors, one PMI-based detector, one LSI detector using the MR aggregation scheme and another LSI-based detector using the MSS aggregation scheme. Each row refers to one possible combination of the three classifiers. The left plot in each row is for precision, whereas the right one is for recall. At every threshold point K, the precision and recall values are reported from the exact same setting. We avoided to put both plots in the same figure for the sake of clarity. In each figure, we plot the precision or recall curve for each of the error detector as well as for both combination schemes, namely the weighted average and the harmonic mean.

Notice here that when the pruning threshold increases, the precision rate falls whereas the recall goes up. This is due to the fact that when K is high, the error filtering is more aggressive and therefore more and more words are tagged as erroneous, especially the actual erroneous tokens, which lead to the increase of the recall. However, when more and more words are classified as errors, the number of correct words mistakenly being tagged as erroneous increases as well, which lead to a decrease of the precision ratio.

A good improvement in the precision rate has been achieved for all the three possible combination of the error detectors. This improvement has also led to an increase in the recall. These are already very promising results since we could, drastically improve the precision without affecting the recall. Actually, there has been even a considerable improvement in the recall as well. The precision rate has been relatively boosted by up to 0.52 in some cases, and the recall improved by up to 0.6 compared to the lowest performing error detector. These finding are valuable in the sense that it shows it's possible to compensate for the weakness of some classifiers with the strength of other detectors via this simple combination schema.

Experiments also show that the weighted average combination scheme of the confidence scores yield higher precision than the harmonic average. This is actually expected. In fact, the weighting coefficients act as catalyst to create a trade off between the different error detection techniques confidence scores. However, since the harmonic mean doesn't have such a scaling factors, we observe less improvement for the precision.

6 Conclusion

In this research work, an approach to combine error detection techniques applied to speech transcription has been presented. The proposed method is generic and scalable, in the sense it can be applied to any error detection technique which relies on thresholding a confidence score for the decision, as well as it can be applied to any number of techniques. A case study has been provided, to combine PMI- and LSI-based error detection approaches. Experiments have shown that combination of both these techniques yields to higher precision and recall rates. This proves that it's possible to compensate for the low precision and recall rates of the current error detection techniques through their combination. Future work will investigate more novel ways to combination, ultimately using machine learning tools, such as neural networks or support vector machines, to achieve a more robust and intelligent combination of the different detection techniques.

References

1. Shi, Y.: An investigation of linguistic information for speech recognition error detection, Ph.D. thesis, University of Maryland (2008)
2. Patwardhan, S., Pedersen, T.: Using wordnet-based context vectors to estimate the semantic relatedness of concepts. In: EACL 2006 Workshop Making Sense of Sense - Bringing Computational Linguistics and Psycholinguistics Together, pp. 1–8 (April 2006)
3. Cox, S., Dasmahapatra, S.: High-level approaches to confidence estimation in speech recognition. IEEE Transactions on Speech and Audio Processing 10(7), 460–471 (2002)
4. Inkpen, D., Desilets, A.: Semantic similarity for detecting recognition errors in automatic speech transcripts. In: HLT/EMNLP, pp. 49–56 (2005)
5. Voll, K.D.: A Methodology of error detection: Improving speech recognition in radiology, Ph.D. thesis, Simon Fraser University (2006)
6. Abida, K., Karray, F., Abida, W.: cROVER: Improving ROVER using automatic error detection. In: IEEE International Conference on Acoustics, Speech, and Signal Processing, ICASSP (May 2011)
7. Abida, K., Karray, F., Abida, W.: mproving rover using latent semantic indexing-based error detection. In: IEEE International Conference on Multimedia and Expo, ICME (July 2011)
8. Fiscus, J., et al.: 1997 english broadcast news speech (HUB4). LDC, Philadelphia (1998)
9. Brants, T., Franz, A.: Web 1T 5-gram version 1. LDC, Philadelphia (2006)

Developing a Secure Distributed OSGi Cloud Computing Infrastructure for Sharing Health Records

Sabah Mohammed, Daniel Servos, and Jinan Fiaidhi

Department of Computer Science, Lakehead University,
Thunder Bay, ON P7B 5E1, Canada
`{sabah.mohammed,dservos,jfiaidhi}@lakeheadu.ca`

Abstract. Cloud Computing has become an emerging computing paradigm which brings new opportunities and challenges to overcome. While the cloud provides seemingly limitless scalability and an alternative to expensive data center infrastructure, it raises new issues in regards to security and privacy as processing and storage tasks are handed over to third parties. This article outlines a Distributed OSGi (DOSGi) architecture for sharing electronic health records utilizing public and private clouds which overcomes some of the security issues inherent in cloud systems. This system, called HCX (Health Cloud eXchange), allows for health records and related healthcare services to be dynamically discovered and interactively used by client programs accessing services within a federated cloud. A basic prototype is presented as proof of concept along with a description of the steps and processes involved in setting up the underlying security services. Several improvements have been added to HCX including a Role-Based Single-Sign-On (RBSSO).

Keywords: Cloud Computing, Distributed OSGi, Cloud Security, Electronic Healthcare Records.

1 Introduction

"Cloud Computing" has become a popular buzzword in the web applications and services industry. Cloud computing provides a large pool of easily usable and accessible virtualized resources [1]. These resources can be dynamically reconfigured to adjust to a variable load, allowing for optimum resource utilization. Recent cloud offerings from vendors such as Amazon, IBM, Google, Oracle, Microsoft, RedHat, etc., have created various public cloud computing services where organizations are no longer required to own, maintain or create the infrastructure or applications that power their business and online presence. Cloud computing provides the potential for those in the healthcare industry, including patients, physicians, healthcare workers and administrators, to gain immediate access to a wide range of healthcare resources, applications and tools. For hospitals, physician practices and emergency medical service providers, the lowered initial investment and the elimination of IT costs offered by cloud computing can help overcome the financial barriers blocking the wide adoption of EHR systems [2]. Along with potential cost savings and scalable infrastructure, cloud computing brings with it new security and privacy issues that need to be addressed. When utilizing a cloud based platform, potentially sensitive information must be transmitted to and stored on a cloud provider's infrastructure. It is left to the cloud provider to properly secure their hardware infrastructure and isolate customer's processing and storage tasks. This transfer of trust may be acceptable in most cloud use cases, however, in industries that must comply with data privacy laws such as PIPEDA [3], HIPA [4] and HIPAA [5], allowing sensitive information to be processed or stored on a public cloud may not be directly feasible. This article continues work on the Health Cloud Exchange (HCX) [6] system to provide security measures for protecting shared health records (EHRs) over the cloud. In particular, this article presents a lightweight Roll Based Single-Sign-On (RBSSO) authentication service for authenticating users securely on our HCX cloud infrastructure. Our RBSSO prototype extends the Apache CXF reference Distributed OSGi (DOSGi) implementation to operate securely on a cloud shared by potentially untrustworthy cloud users and providers.

2 Sharing EHRs over the Cloud

Sharing EHRs over the Internet countenances two major obstacles to break the barrier of the isolated digital silos and to enable interoperability: (1) availability of unified EHR specification and (2) the availability of a suitable centralized and secure collaboration infrastructure. Both constraints require the availability of measures and standards.

2.1 EHR Specification

Related to the EHR specifications, there are several notable EHR standards for storing, processing and transmitting patient healthcare records such as HL7 CDA [7], Continuity of Care Record (CCR) [8] and HL7 Continuity of Care Document (CCD) [9]. The CCR and CCD standards have gained wide usage and popularity among healthcare communities because of the support provided by visible IT vendors' products including Google Health[1], and Microsoft's HealthVault[2]. The CCR and CCD standards represent an XML based patient health summary which contains various sections such as patient demographics, insurance information, diagnosis, medications, allergies and care plan in a platform independent format that may be read by any XML parser. For this reason we have chosen to adopt the CCR and CCD standards as an intermediate format between EHR systems in our HCX design.

2.2 Cloud Challenges

Cloud based computing infrastructure offers organizations a pay-per-use based virtual infrastructure solution made dynamically scalable by the ability to spawn and destroy virtual machine instances on demand. This allows virtual servers to be created as needed to meet current demand and then scaled back when strain on the system is reduced, lowering resource usage and cost. This scalability provides an ideal infrastructure for deploying a large scale but affordable system that connects end users, providers and

---

[1] www.google.com/health/
[2] www.healthvault.com

isolated EHR systems. Adopting certain cloud computing technologies suitable for sharing sensitive information such as EHRs remains one of the challenges that vendors and researchers try to answer. Some of these security and privacy issues include:

- **Confidentiality:** Protecting cloud based storage and network transmissions from possible unwanted access by cloud providers or data leakage to other cloud users.
- **Auditability:** Maintaining logs of users' actions with the system and ensuring that no part of the system has been tampered with or compromised.
- **Security:** Preventing user credentials, which may be used for multiple services on and off the cloud from being obtained by untrusted parties including the cloud provider or other cloud users?
- **Legal:** Complying with data privacy laws that may be in effect in given geographical regions (eg. PIPEDA, HIPA, HIPAA, etc).

While many solutions for these issues exist for traditional systems, public cloud infrastructure removes control of the physical infrastructure that makes it possible to ensure a cloud provider properly secures their services and is not performing any potentially malicious activities. It may seem unlikely that large public cloud operators would intentionally violate their user's privacy, but external factors in some regions (such as legal pressure from local governments, e.g. USA PATRIOT Act[3]) may force discloser of sensitive information. Hardware based solutions, such as Trusted Platform Module (TPM)[4], that would normally provide protection for a remote system are difficult to implement in cloud environments due to instances being created on a number of physical servers that share the same hardware and lack of support from major cloud providers. Additionally, cloud computing has several challenges related to taking full advantage of the scalability gained from cloud infrastructure that limit potential solutions including:

- **Bottlenecks**: The cloud may provide seemingly limitless scalability for virtual server resources and storage, but any connections to systems outside of the cloud or lacking the same scalability quickly become a new bottleneck for the system. For example if multiple machine instances are spawned to meet an increase in demand but all connect to the same database or authentication backend provided by the same server, a bottleneck will be formed that will limit the scalability of the whole system.
- **Distributed Design**: While cloud computing is distinct from traditional distributed computing, many of the same concepts apply and must be considered in the design of a cloud application or platform. Cloud applications must be built to offer their services from multiple machine instances distributed in the same cloud rather than a traditional single server to client architecture.
- **Volatile Storage**: Most cloud infrastructure solutions (such as Amazon's EC2) do not provide persistent storage by default to their machine instances. Applications built upon such infrastructures need to take into account this static nature in their design and use additional services or solutions (such as Amazon's S3 or EBS) for permanent storage.
- **Dynamic IPs**: In most cases when cloud instances are launched, a public IP address is dynamically assigned. While this may be selected from a list of static IP addresses, autonomous cloud systems are often used which automatically create and destroy instances each obtaining an unused address when initialized. This can create issues for traditional systems that expect static or unchanging addresses for servers.

---

[3] http://en.wikipedia.org/wiki/USA_PATRIOT_Act
[4] http://en.wikipedia.org/wiki/Trusted_Platform_Module

3 Designing a Secure Cloud for Sharing EHRs

This section introduces our proposed architecture for securely sharing CCR/CCD patient records over a public or private cloud. The Health Cloud Exchange (HCX) system provides a dynamic and scalable cloud platform solution built on DOSGi for discovering, and providing cloud based health record services. The Role-Based Single-Sign-On (RBSSO) system provides a means of authenticating users from various organizations with these services that keeps users' credentials off the public cloud and isolated while maintaining the systems scalability.

3.1 Health Cloud Exchange (HCX)

Our proposed HCX design provides three primitive DOSGi service interfaces: EHRServices, AuditLog and EHRClient. Any number of services implementing these interfaces may run within the same cloud and are registered upon execution with a central service registry for dynamic service discovery. Multiple instances of the same service may be run simultaneously, balancing the load between them. A Service Controller which resides in a trusted network outside of the public cloud is used to start and stop machine instances on the cloud that provide HCX services based on the current level of demand as well as load instances with their initial data. These interactions can be seen in Figure 1.

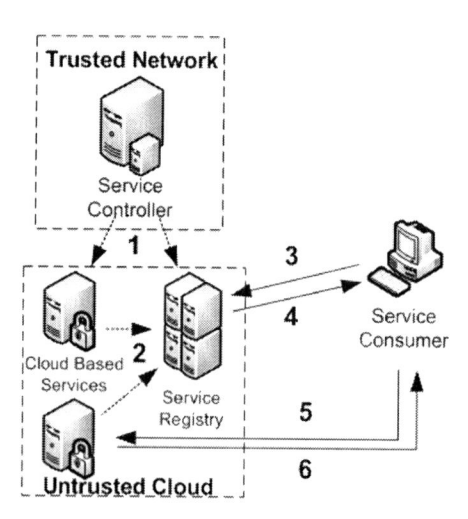

Fig. 1. An Overview of the HCX Architecture
   1. Service controller starts and initializes machine instances for HCX services and the service Registry. 2. HCX servicers register them self with the service registry. 3-4. Service consumer queries service registry for a listing of available HCX services. 5-6. Service consumer sends a request to a HCX service to view, or update an EHR and receives an appropriate response. *Dotted lines* indicate interactions transparent to the service consumer.

EHRServices

EHRServices are dedicated to sharing CCR and CCD formatted health records with consumers. Consumers query EHRServices for either a listing of available health records for which they have access to or a specific record using the shared EHRService interface implemented by all HCX services that share records. The EHRService interface is implemented by three main derivative services called RecordStore, EHRAdapter, and EHRBridge. RecordStores are databases of health records stored on the cloud in a relational database or other cloud storage (eg. Amazon's S3 or EBS). EHRAdapters are interfaces to existing isolated EHR systems located on the same cloud and EHRBridges are interface to external EHR systems operating outside the cloud. This allows for loose coupling between the consumers and EHRServices as consumer need only know about the standard EHRService interface and the location of the service registry to use any EHRService that becomes available on the cloud, including bridges to other systems outside of the cloud. This interaction can be seen in Figure 2.

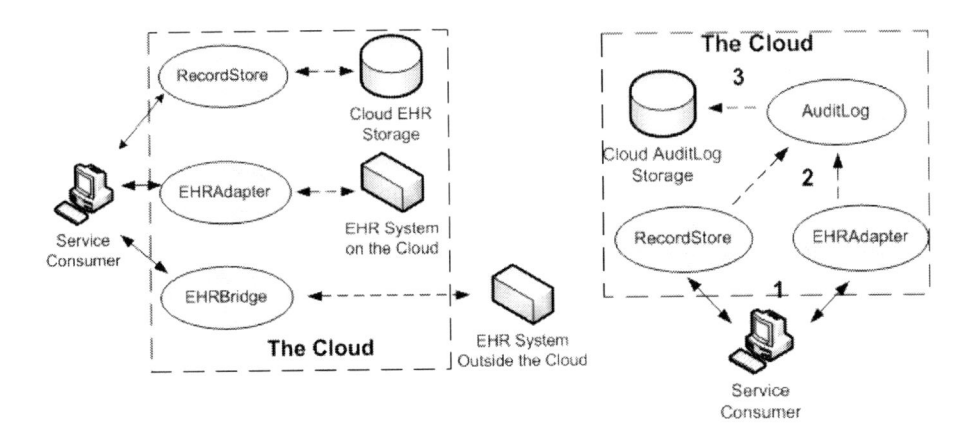

Fig. 2. Service consumer and EHRService interactions. *Dotted arrows* are interactions transparent to the consumer, for which all three services offer the same interface but return records from different locations.

Fig. 3. AuditLog interactions. 1. Client/Service consumer makes a request on a service. 2. The service sends the RequestToken, AuditToken and request summary to the AuditLog service. 3. The AuditLog service stores a log entry for the request on cloud storage.

AuditLog

The AuditLog service is used to keep and store an uneditable audit log of all actions that have been performed on the EHRs and services, including views, changes and removal of records. These logs keep a permanent record of user's actions that can be used as evidence in case an abuse of a user's credentials occurs. The AuditLog service is called directly from EHRServices (as well as any other HCX service that may require a detailed audit trail) and is not accessible to normal users directly. The interface of the AuditLog has a single operation which takes the user's AuthToken, RequestToken (see section 3.2), and a summary of the users request upon the service. If the AuthToken and RequestToken are valid, the AuditLog service adds an entry to its audit log which is stored on persistent cloud storage. Figure 3 shows the interaction between an EHRService and an AuditLog service.

EHRClient

The EHRClient service is a cloud application and web interface which allows users of the system to view and update records through their browser rather than through an application implementing an EHRService consumer. The EHRClient contains both a service which controls the web interface and an EHRClient consumer which connects to other HCX record services. User authentication is still performed through the RBSSO system described in section 3.2.

3.2 Role-Based Single-Sign-On (RBSSO)

To secure HCX from access by undesired users, a roll based authentication service is required. Due to the distributed nature of the HCX architecture, traditional authentication methods are not appropriate as they would require duplication of authentication mechanism and user databases or the creation of a single point failure resulting in a system bottleneck. Additionally, users will likely need to make a request on multiple HCX services during a single session. Forcing users to provide credentials when accessing each would be an unreasonable burden, as well as inefficient if each request needed to be authenticated with a separate service. Finally it may be desired to not

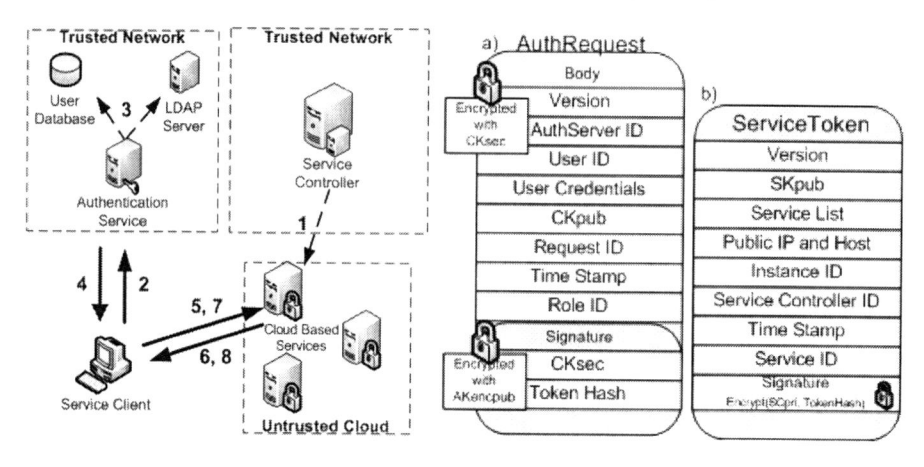

Fig. 4. RBSSO Protocol

Fig. 5. (a) AuthRequest and (b) ServiceToken

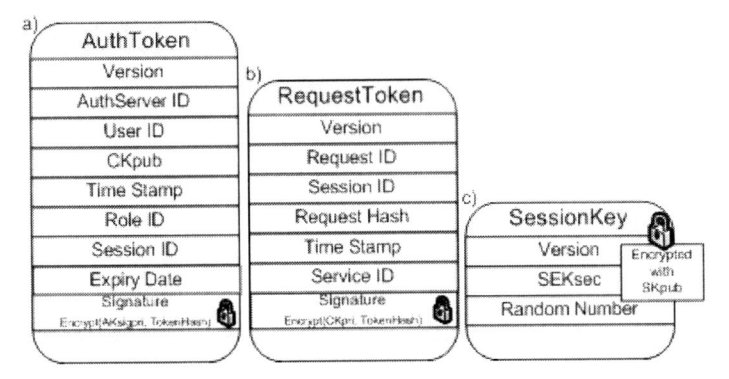

Fig. 6. (a) AuthToken, (b) RequestToken, (c) SessionKey

store a user's credentials and information directly on the cloud. A possible solution to these problems is to develop a single-sign authentication service, in which users first authenticate with a trusted party to receive an authentication token that enables access to services that trust the same party. Several technologies currently exist which enable single-sign on capabilities, such as Kerberos [10], SAML [11], and X.509 [12]. However, the nature of the cloud and architecture of HCX make traditional solutions complicated as new machine instances are spawned and destroyed automatically based on demand and have no persistent memory to store public/private key pairs or certificates. Additionally, authentication servers become a scalability bottleneck when run outside the cloud (which may be necessary if no trusted party exists in a cloud environment). This makes any security service that have a large number of requests between services and authentication servers unreasonable or even impossible if the server is made available only to a private network to which the client is part. To solve these problems and provide user rolls which are lacking in most existing single-sign solutions, we have developed a lightweight roll based single-sign in protocol called RBSSO (Roll Based Single-Sign On) for cloud based services which may not have direct access to authentication servers. RBSSO is loosely based on X.509 single-sign on and aims to minimize the number of request on an authentication server, support a large number of authentication methods, supports sessions spanning multiple services and be relatively easy to implement and understand.

The Protocol

Figure 4 displays the interactions involved in the RBSSO protocol. Each client is assumed to be provided with the public signing (AKsigpub) and encryption (AKencpub) keys for their organizations authentication server, as well as the public signing key for the service controller (SCKsigpub). Authentication servers contain or access an organization's user credentials and are located on their private trusted networks which need only be accessible to their clients. Service controllers initialize the virtual machine instances that offer HCX services and are located off the pubic cloud. The protocol for RBSSO follows the proceeding steps, as shown in Figure 4:

1. The service controller initializes machine instances with a ServiceToken (Figure 5b), a list of HCX services the instance will provide, a list of trusted authentication servers and their set of public keys, a list of globally black listed users and the instances private key, SKpri.

2. The HCX consumer authenticates with their organizations authentication server by generating a secret key CKsec and an AuthRequest (Figure 5a). The AuthRequest, containing the user's credentials, roll they wish to activate and a public client key from the client public/private key pair created when the client program is initialized, is then transmitted to the authentication server.
3. The authentication server decrypts the AuthRequest using AKencpri and CKsec, validates the user's credentials, and checks that the time stamp and request id are acceptable. Credentials may be validated against a local database of user credentials or existing authentication infrastructure on the same trusted network (eg. LDAP).
4. Once the user is validated, the authentication server issues and signs an AuthToken (Figure 6a) with AKsigpri for the client's session with the HCX services. This transmission is encrypted with CKsec to protect the user's privacy (i.e. so the user may not be identified by outside observers).
5. Before the service consumer makes a normal request upon a service it first requests the service's ServiceToken from the instance on which it resides. The service consumer then validates the service controller's signature using SCKpub and ensures that the service is listed in the service listing and is connecting from the stated IP or hostname.
6. The consumer may now authenticate and make a request upon any HCX service on the instance by generating the secret session key SEKsec and using it to encrypt its AuthToken, the request and a newly generated RequestToken (Figure 6b) together. SEKsec is appended with a delimiter and random number and encrypted with SKpub (obtained from the ServiceToken) (Figure 6c). The ciphertexts are appended and transmitted to the service.
7. The service decrypts SEKsec using SKpri and decrypts the request, RequestToken and AuthToken using SEKsec. The service then proceeds to validate the signatures contained in AuthToken and RequestToken using AKsigpub and CKpub (from the AuthToken) and validate the fields they contain (time stamp has not expired, etc). If valid SEKsec and the AuthToken are temporarily stored for future requests with the instance until the session expires.
8. If the user has a roll active which allows the request to be performed on the service, the service complies with the request and provides the appropriate response. All further communications between the consumer and service for the length of the session will be encrypted using SEKsec. Subsequent requests on any service on the instance need only to provide a RequestToken and the content of the request encrypted with SEKsec until the session expires.

4 Implementation Details

4.1 HCX

Two major open source projects were used in our implementation to provide private cloud as a service infrastructure and service discovery: Eucalyptus[5] and the Apache CXF DOSGi[6] respectively. Additionally, Amazon's EC2 and S3[7] cloud infrastructure

---

[5] http://www.eucalyptus.com/
[6] http://cxf.apache.org/
[7] http://aws.amazon.com/

services were used for public cloud storage and computing. Compared to other private cloud frameworks such as Nimbus[8] and abiCloud[9] Eucalyptus was chosen for its stronger community support, detailed documentation and benefit of coming prepackaged in the Ubuntu Enterprise Cloud[10] Linux distribution. Apache CXF DOSGi was chosen as it is the reference implementation for remote OSGi services and discovery. The distributed nature of DOSGi allows for a loose coupling between services and consumers through the uses of a service discovery mechanism for finding the location and type of services currently being offered in a given grouping. This is accomplished through the use of an Apache ZooKeeper[11] based cluster, in which a central service registry enables simple and scalable service look up and discovery while keeping the advantages of a distributed system (e.g. not rely on a single point of failure). Service consumers are notified of new services becoming available or going offline (a common occurrence in a cloud based setting) and are able to automatically use or discontinue use of a given service. The hardware infrastructure of the private cloud consisted of 15 identical IBM xSeries rack mounted servers connected to each other via a 1000 Mbit/s switch. Of the 15 servers, 14 were designated as Eucalyptus Node controllers which ran the Xen based virtual machine instances, while the Cloud Controller, Walrus (S3 Storage), Storage Controller and Cluster Controller services were installed on the remaining server to provide scheduling, S3 based bucket storage, EBS based storage and a front end for the cloud.

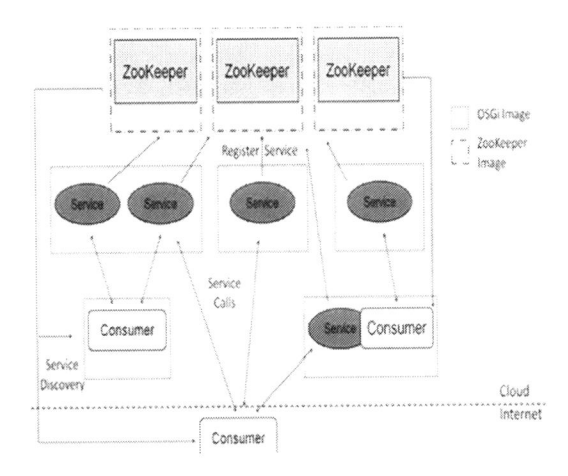

Fig. 7. Xen Images (OSGi and ZooKeeper) to Support DOSGi on the Cloud

Adapting DOSGi for use on the cloud required the creation of two Xen based machine images. An image was required to host a standard OSGi implementation (such as Eclipse Equinox, Apache Felix or Knopflerfish) upon which the Apache CXF DOSGi, HCX bundles would be run to provide HCX's services. A second image was

---

[8] http://www.nimbusproject.org/
[9] http://abicloud.org
[10] http://www.ubuntu.com/business/cloud/overview
[11] http://hadoop.apache.org/zookeeper/

required to host ZooKeeper servers for service registry and discovery. As demand increases on a particular service additional OSGi machine instances may be run to load balance request between multiple instances of that service. As demand increases on the service registry, more ZooKeeper machine instances may be run to add additional ZooKeeper servers to the cluster. This set-up is shown in figure 7.

4.2 RBSSO

To evaluate the performance of the RBSSO protocol a prototype of the Authentication Server and Client were created using standard Java TCP sockets. The protocol was expanded to include length bytes to make processing the message easier. 128bit AES encryption was used for the symmetric encryption of the AuthRequest body and AuthToken body. 3072bit RSA encryption was used for the asymmetric encryption of the AuthRequest tail and the signature on the AuthToken. SHA-256 was used for generating hashes for the AuthRequest.

Two controls, SSL and Kerberos (a popular SSO system), were used to compare the performance of the protocol against. For the first control an Authentication Server was created that replaced the encryption of the body and signature of the AuthRequest with an SSL connection (the tail containing CKsec and the token hash were removed from the SSL implementation). Secondly the RBSSO protocol was also compared against the performance of a Java based Kerberos client and the MIT Kerberos 5[12] implementation which retrieved a ticket granting ticket and a service ticket (somewhat equivalent to an AuthToken in RBSSO). The performance of all three protocols (measured in average time per request) was measured on both a private isolated local area network and over a noisier internet connection. Each protocol was tested with 10,000 authentication requests for each network in sequential runs of 1000 requests. The results on these tests are shown in Figures 8, 9 and 10.

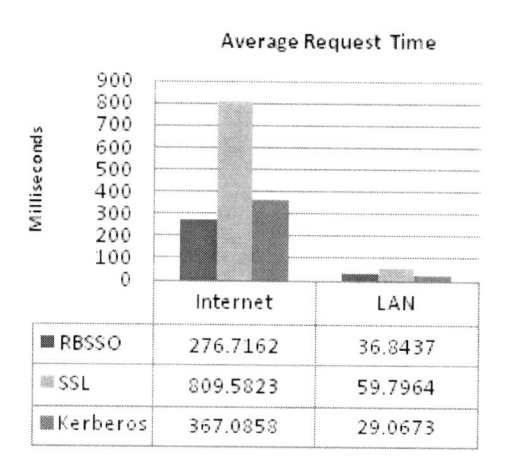

Average Request Time

	Internet	LAN
RBSSO	276.7162	36.8437
SSL	809.5823	59.7964
Kerberos	367.0858	29.0673

Fig. 8. Average time (in milliseconds) requried to complete and verify an authencation request using each protocol. Based on 10,000 requests.

---

[12] http://web.mit.edu/kerberos/

Average Request Time Per Run
LAN

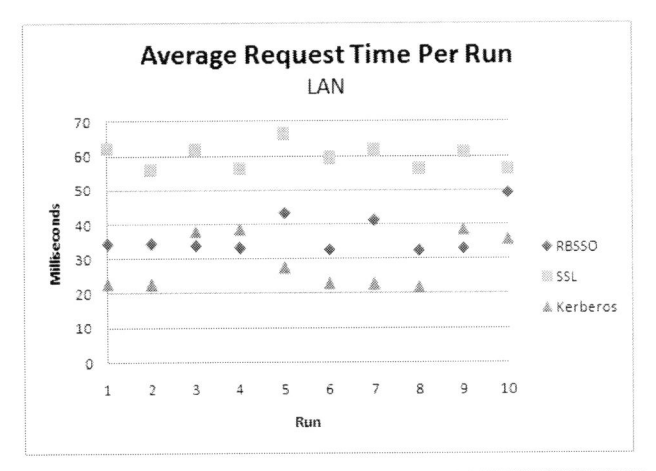

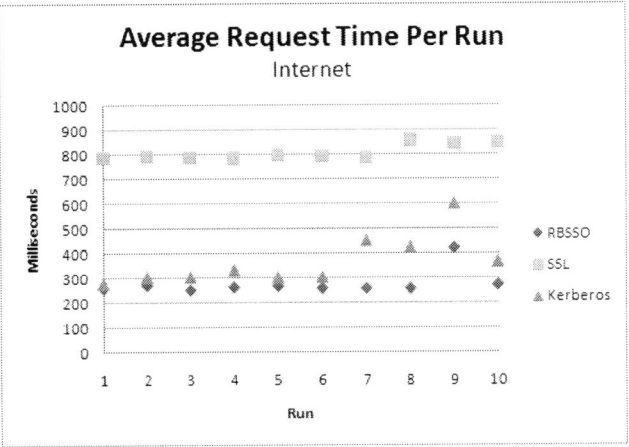

Average Request Time Per Run
Internet

Fig. 9: Average time requried to complete and verify an authencation request over the internet and LAN. Based on 1000 requests per run.

The RBSSO protocol performed approximately 38% faster on average than the SSL implementation on the local area network and 66% faster over the internet connection. This is likely a result of the decreased number of request involved the RBSSO protocol (no handshake is required and only a single request in made containing the AuthRequest) and explains the difference between the local and internet connections (the cost per request being higher on the connection with increased latency). Similarly RBSSO performed 25% faster than Kerberos over an internet connection but performed 21% slower over a local area connection. This is also likely a result of the number of requests, Kerberos requiring a connection to both to a Kerberos authentication server and a ticket granting server before it can make a request on a service.

5 Conclusions

The HCX system described in this paper provides a distributed, modular and scalable system for sharing health records over the cloud. This system is made secure through the extension of the RBSSO protocol also presented in this paper. We showed how to build and integrate a composite application using the Apache CXF DOSGi open source framework for sharing CCR/CCD EHR records, and how the distributed roll based single-sign on can be accomplished using the presented RBSSO protocol. The developed HCX prototype comprising of composite modules (distributed across the cloud) can be integrated and function as a single unit. HCX allows adaptors and bridges to be created for existing EHR systems and repositories allowing records to be exchanged through a standard interface and CCR/CCD record format. This is accomplished by building DOSGi based services and consumers made scalable through the cloud. The RBSSO protocol allows users to sign in once with their home organization and transparently have a session open with all HCX services for which the user's rolls give them access. There are several tasks left to our future research including providing higher information and data privacy for cloud storage thus blocking access from potentially untrustworthy cloud providers, providing in-depth details of the role based components of RBSSO, protecting cloud machine instances from tampering, and fully evaluating the performance of RBSSO in realistic cloud settings.

References

[1] Vaquero, L., et al.: A break in the clouds: towards a cloud definition. ACM SIGCOMM Computer Communication Review 39, 50–55 (2008)
[2] Urowitz, S., et al.: Is Canada ready for patient accessible electronic health records? A national scan. BMC Medical Informatics and Decision Making 8(1), 33 (2008)
[3] PIPEDA Personal Information Protection and Electronic Documents Act (2000), http://laws.justice.gc.ca/en/P-8.6/index.html
[4] Ontario Statutes and Regulations, Personal Health Information Protection Act, S.O. 2004 Ch. 3 Schedule A (2004)
[5] 104th United States Congress, Health Insurance Portability and Accountability Act (HIPAA), P.L.104-191, August 21 (1996)
[6] Mohammed, S., Servos, D., Fiaidhi, J.: HCX: A Distributed OSGi Based Web Interaction System for Sharing Health Records in the Cloud. In: IEEE/WIC/ACM International Conference on Web Intelligence and Intelligent Agent Technology, Toronto, Canada, August 31-September 3 (2010)
[7] Health Level Seven International. HL7 v3.0, http://www.hl7.org
[8] ASTM Subcommittee: E31.25, ASTM E2369 - 05e1 Standard Specification for Continuity of Care Record (CCR), ASTM Book of Standards, vol. 14.01 (2005)
[9] Care Management and Health Records Domain Technical Committee, HITSP/C32: HITSP Summary Documents Using HL7 Continuity of Care Document (CCD) Component, Healthcare Information Technology Standards Panel, Version 2.5 (2009)
[10] Neuman, B.C., et al.: Kerberos: An authentication service for computer networks. IEEE Communications Magazine 32, 33–38 (1994)
[11] OASIS Open, Assertions and Protocols for the OASIS Security Assertion Markup Language (SAML) V2.0 – Errata Composite (December 2009)
[12] Housley, R., et al.: Internet X.509 Public Key Infrastructure Certificate and Certificate Revocation List (CRL) Profile. RFC3280 (2002)

Extreme Learning Machine with Adaptive Growth of Hidden Nodes and Incremental Updating of Output Weights

Rui Zhang[1,2], Yuan Lan[1], Guang-Bin Huang[1,*], and Yeng Chai Soh[1]

[1] School of Electrical and Electronic Engineering,
Nanyang Technological University, Nanyang Avenue, Singapore 639798
{zh0010ui,lany0001,egbhuang,eycsoh}@ntu.edu.sg
[2] Department of Mathematics, Northwest University, Xi'an, China 710069
rzhang@nwu.edu.cn

Abstract. The extreme learning machines (ELMs) have been proposed for generalized single-hidden-layer feedforward networks (SLFNs) which need not be neuron alike and perform well in both regression and classification applications. An active topic in ELMs is how to automatically determine network architectures for given applications. In this paper, we propose an extreme learning machine with adaptive growth of hidden nodes and incremental updating of output weights by an error-minimization-based method (AIE-ELM). AIE-ELM grows the randomly generated hidden nodes in an adaptive way in the sense that the existing hidden nodes may be replaced by some newly generated hidden nodes with better performance rather than always keeping those existing ones in other incremental ELMs. The output weights are updated incrementally in the same way of error minimized ELM (EM-ELM). Simulation results demonstrate and verify that our new approach can achieve a more compact network architecture than EM-ELM with better generalization performance.

1 Introduction

The capabilities of single-hidden-layer feedforward networks (SLFNs) to approximate complex nonlinear mappings directly from the input samples have been widely investigated due to their applications in various areas of scientific research and engineering[2,5,6,20]. Different from the conventional neural network theories where the parameters of SLFNs are well adjusted, Huang *et al* [4,7,8,9,10,11,12,13,14,15,16,17,18,19,21] propose extreme learning machines (ELMs) where the hidden nodes of SLFNs need not be tuned. ELMs were originally developed for neuron based SLFNs [8,13,14] and then extended to the "generalized" SLFNs [4,7,9] which need not be neuron alike. The original extreme learning machine (ELM) has been successfully applied in many applications and has been shown to be an extremely fast learning algorithm and having

---

* This work was supported by the grant from Academic Research Fund (AcRF) Tier 1 of Ministry of Education, Singapore, Project No. RG 22/08 (M52040128), and Grant 61075050 of the National Natural Science Foundation, China.

good generalization performance. However, since ELM uses the batch learning scheme, the size of the SLFN (the hidden-node number) needs to be set by users before learning. Therefore, how to choose the optimal hidden-node number of ELM is the main objective of this paper.

In [8], an incremental extreme learning machine (I-ELM) was proposed where the hidden nodes are randomly added one by one. In I-ELM, the output weights of the existing hidden nodes are freezed when a new hidden node is added. The universal approximation capability of I-ELM has also been proved in the same paper. To improve the convergence rate of I-ELM, Huang et al [7] proposed the convex incremental extreme learning machine (CI-ELM), which adopts another incremental method originated from the Barron's convex optimization concept [1]. Unlike I-ELM, CI-ELM allows to properly adjust the output weights of the existing hidden nodes when a new hidden node is added. It has been shown that CI-ELM can achieve faster convergence rate and more compact network architectures than I-ELM since the network output combines the former information and the new information instead of only depending on the new information as in I-ELM. During the learning procedure of both I-ELM and CI-ELM, some newly added hidden nodes only play a very minor role in the final output of the network. Under this situation, the network complexity may be increased. To avoid this issue and to obtain a more compact network architecture, the enhanced incremental extreme learning machine (EI-ELM) was proposed in [9]. At each learning step of EI-ELM, several hidden nodes are randomly generated and only the one leading to the largest residual error reduction will be added to the existing network. On the other hand, EI-ELM also extends I-ELM to "generalized" SLFNs which need not be neuron alike. However, both I-ELM and its extensions add the hidden node one by one continuously until the given maximum hidden-node number is attained, which is obviously ineffective to choose the optimal network architecture. Recently, Feng et al [4] provided a new approach referred to as error minimized extreme learning machine (EM-ELM). EM-ELM allows to add random hidden nodes one by one or group by group (with varying size) and incrementally update the output weights during the network growth. However, in the implementation of all the above referred incremental extreme learning machines, the number of the hidden nodes monotonically increases with the increase of the iterative steps. By doing so, large number of hidden nodes will be obtained eventually if there needs many iterative steps, while some of the hidden nodes may play a very minor role in the network output.

This paper proposes a novel incremental extreme learning machine with adaptive growth of hidden nodes and incremental updating of output weights using error-minimization-based method (AIE-ELM). At each step, we first generate several temporary networks with different numbers of hidden nodes. Among all these possible networks and the network obtained at the previous step, the best one will be selected, which has the least hidden-node number as well as the least error. Then one more random hidden node is added to such selected network, which results in the network for the current step. In AIE-ELM, the hidden nodes are randomly generated and the output weights are updated incrementally by using the error-minimization-based method proposed in [4]. Different from I-ELMs adding the hidden nodes one by one or group by group and freezing those existing nodes, AIE-ELM grows the randomly generated hidden nodes in an adaptive way in the sense that the existing hidden nodes may be replaced by some newly generated hidden nodes with better performance. The empirical study shows that AIE-ELM leads to a more compact network structure than EM-ELM.

The rest of this paper is organized as follows. Section 2 gives a brief review of EM-ELM. The proposed algorithm AIE-ELM is introduced in Section 3 and the performance of AIE-ELM is evaluated in Section 4. Then we conclude the paper in Section 5.

2 Brief on Error Minimized Extreme Learning Machine

In this section, we briefly describe the error minimized extreme learning machine (EM-ELM).

EM-ELM is a simple and efficient approach to determine the hidden-node number in generalized SLFNs. An SLFN with d inputs, L hidden nodes and m linear outputs can be represented in a general form by

$$f_L(\mathbf{x}) = \sum_{i=1}^{L} \beta_i G(\mathbf{a}_i, b_i, \mathbf{x})$$

where $G(\mathbf{a}_i, b_i, \mathbf{x})$ denotes the output of the ith hidden node with the hidden node parameters $(\mathbf{a}_i, b_i) \in R^d \times R$ and $\beta_i \in R^m$ is the weight vector between the ith hidden node and the output nodes. It should be noted that the hidden node output $G(\mathbf{a}_i, b_i, \mathbf{x})$ can be neuron alike, such as additive nodes $G(\mathbf{a}_i, b_i, \mathbf{x}) = g(\mathbf{a}_i \cdot \mathbf{x} + b_i)$, where g is hidden node activation function, and radius basis function (RBF) nodes $G(\mathbf{a}_i, b_i, \mathbf{x}) = \exp(-b_i \parallel \mathbf{x} - \mathbf{a}_i \parallel^2)$[8], or non-neuron alike, such as multiquadratic nodes $G(\mathbf{a}_i, b_i, \mathbf{x}) = (\parallel \mathbf{x} - \mathbf{a}_i \parallel^2 + b_i^2)^{\frac{1}{2}}$.

Given a set of training data $\{(\mathbf{x}_i, \mathbf{t}_i)\}_{i=1}^{N} \subset R^d \times R^m$, if the output of the network is equal to the target, then we have

$$f_L(\mathbf{x}_j) = \sum_{i=1}^{L} \beta_i G(\mathbf{a}_i, b_i, \mathbf{x}_j) = t_j, \; j = 1, ..., N$$

which can also be equivalently expressed in the matrix form

$$\mathbf{H}\beta = \mathbf{T} \qquad (1)$$

where

$$\mathbf{H} = \begin{pmatrix} G(\mathbf{a}_1, b_1, \mathbf{x}_1) & \dots & G(\mathbf{a}_L, b_L, \mathbf{x}_1) \\ G(\mathbf{a}_1, b_1, \mathbf{x}_2) & \dots & G(\mathbf{a}_L, b_L, \mathbf{x}_2) \\ \vdots & & \vdots \\ G(\mathbf{a}_1, b_1, \mathbf{x}_N) & \dots & G(\mathbf{a}_L, b_L, \mathbf{x}_N) \end{pmatrix}_{N \times L}, \; \beta = \begin{pmatrix} \beta_1^T \\ \beta_2^T \\ \vdots \\ \beta_L^T \end{pmatrix}_{L \times m}, \; \mathbf{T} = \begin{pmatrix} t_1^T \\ t_2^T \\ \vdots \\ t_N^T \end{pmatrix}_{N \times m}$$

Here, $\mathbf{H}$ is called the hidden-layer output matrix of the network. In ELM, the hidden-layer output matrix $\mathbf{H}$ is randomly generated. Thus, training SLFNs simply amounts to getting the solution of the linear system (1) with respect to the output weight vector β, specifically

$$\hat{\beta} = \mathbf{H}^\dagger \mathbf{T} \tag{2}$$

where $\mathbf{H}^\dagger$ is the Moore-Penrose generalized inverse of $\mathbf{H}$.

In EM-ELM, given target error $\epsilon > 0$, an SLFN $f_{L_0}(\mathbf{x}) = \sum_{i=1}^{L_0} \beta_i G(\mathbf{a}_i, b_i, \mathbf{x})$ with L_0 hidden nodes is first initialized. Denote by $\mathbf{H}_1$ the hidden-layer output matrix of this network. Then according to (2) the output weight vector can be calculated by $\hat{\beta}_1 = \mathbf{H}_1^\dagger \mathbf{T}$.

If the network output error $E(\mathbf{H}_1) > \epsilon$, then new hidden nodes $\delta L_0 = L_1 - L_0$ are added to the existing SLFN and the new hidden-layer output matrix becomes $\mathbf{H}_2 = [\mathbf{H}_1, \delta \mathbf{H}_1]$. In this case, different from ELM which calculates $\hat{\beta}_2 = \mathbf{H}_2^\dagger \mathbf{T}$ based on the entire new hidden-layer matrix $\mathbf{H}_2^\dagger$ whenever the network architecture is changed, EM-ELM proposes a fast incremental output weights updating method to calculate $\mathbf{H}_2^\dagger$ in the following way:

$$\mathbf{H}_2^\dagger = \begin{bmatrix} \mathbf{U}_1 \\ \mathbf{D}_1 \end{bmatrix} = \begin{bmatrix} ((\mathbf{I} - \mathbf{H}_1 \mathbf{H}_1^\dagger)\delta \mathbf{H}_1)^\dagger \\ \mathbf{H}_1^\dagger - \mathbf{H}_1^\dagger \delta \mathbf{H}_1 \mathbf{D}_1 \end{bmatrix}$$

Similarly, the output weights are updated incrementally as

$$\hat{\beta}^{(k+1)} = \mathbf{H}_{k+1}^\dagger \mathbf{T} = \begin{bmatrix} \mathbf{U}_k \\ \mathbf{D}_k \end{bmatrix} \mathbf{T}$$
$$\mathbf{D}_k = ((\mathbf{I} - \mathbf{H}_k \mathbf{H}_k^\dagger)\delta \mathbf{H}_k)^\dagger$$
$$\mathbf{U}_k = \mathbf{H}_k^\dagger - \mathbf{H}_k^\dagger \delta \mathbf{H}_k \mathbf{D}_k \tag{3}$$

The EM-ELM algorithm can be summarized as follows [4].

EM-ELM Algorithm: Given the maximum number of hidden nodes $L_{\max}$ and the expected learning accuracy ϵ.

Initialization Phase:

1) Randomly generate a small group of hidden nodes $\{(\mathbf{a}_i, b_i)\}_{i=1}^{L_0}$, where L_0 is a small positive integer given by users.
2) Calculate the hidden layer output matrix $\mathbf{H}_1$.
3) Calculate the corresponding output error $E(\mathbf{H}_1) = \|\mathbf{H}_1 \mathbf{H}_1^\dagger \mathbf{T} - \mathbf{T}\|$.
4) Let $j = 0$.

Recursively Growing Phase: while $L_j < L_{\max}$ and $E(\mathbf{H}_j) > \epsilon$.

1) Randomly add δL_j hidden nodes to the existing SLFN. The hidden-node number now becomes $L_{j+1} = L_j + \delta L_j$ and the corresponding hidden layer output matrix becomes $\mathbf{H}_{j+1} = [\mathbf{H}_j, \delta \mathbf{H}_j]$.
2) Update the output weight $\hat{\beta}$ according to (3).
3) $j = j + 1$.
endwhile.

3 The Proposed Algorithm AIE-ELM

In this section, we will introduce the proposed extreme learning machine with adaptive growth of hidden nodes and incremental updating of output weights (AIE-ELM). Let $f_n(\mathbf{x})$ be the network obtained at the nth step, e_n and L_n be the residual error and the hidden-node number of f_n respectively. In the following, we express the network in the matrix form as $f_n(\mathbf{x}) = \sum_{i=1}^{n} \beta_i G(\mathbf{a}_i, b_i, \mathbf{x}) = \boldsymbol{\beta_n} \cdot \mathbf{G_n}(\mathbf{x})$ where $\boldsymbol{\beta_n} = [\beta_1, \beta_2, \cdots, \beta_n]^T$ and $\mathbf{G_n}(\mathbf{x}) = [G(\mathbf{a}_1, b_1, \mathbf{x}), G(\mathbf{a}_2, b_2, \mathbf{x}), \cdots, G(\mathbf{a}_n, b_n, \mathbf{x})]^T$. Here, $\boldsymbol{\beta_n} \cdot \mathbf{G_n}(\mathbf{x})$ denotes the inner product of $\boldsymbol{\beta_n}$ and $\mathbf{G_n}(\mathbf{x})$. The details of the proposed AIE-ELM are presented as follows.

AIE-ELM Algorithm: Given a set of training data $\{(\mathbf{x}_i, t_i)\}_{i=1}^{N}$, the maximum number of hidden nodes $L_{\max}$, and the expected learning accuracy ε.

Phase I — Initialization Phase:
Let $n = 1$,
1) randomly generate the hidden node parameter $(\mathbf{a}_1, b_1) \in R^d \times R$;
2) calculate the hidden node output $\mathbf{G_1}(\mathbf{x}) = [G(\mathbf{a}_1, b_1, \mathbf{x}_1), \ldots, G(\mathbf{a}_1, b_1, \mathbf{x}_N)]^T$;
3) calculate the optimal output weight $\hat{\boldsymbol{\beta}}_1 = [\mathbf{G_1}(\mathbf{x})]^\dagger \mathbf{T}$.

Therefore, we obtain the approximated function f_1 and its corresponding error e_1:

$$f_1(\mathbf{x}) = \hat{\boldsymbol{\beta}}_1 \cdot \mathbf{G_1}(\mathbf{x}) \quad \text{and} \quad e_1 = \| f_1 - f \|.$$

Phase II — Recursively Growing Phase: while $L_n < L_{\max}$ and $e_n > \varepsilon$:
Let $n = 2$.
1) *Building a temporary network* φ_n:
  For $k = 1 : L_{n-1}$,
  a) randomly generate $(\widetilde{\mathbf{a}}_n^k, \widetilde{b}_n^k) \in R^d \times R$;
  b) calculate $\widetilde{\varphi}_n^k(\mathbf{x}) = \sum_{i=1}^{k} \widetilde{\beta}_n^i G(\widetilde{\mathbf{a}}_n^i, \widetilde{b}_n^i, \mathbf{x}) = \widetilde{\boldsymbol{\beta}_k} \cdot \widetilde{\mathbf{G}_k}(\mathbf{x})$ with k hidden nodes and its corresponding error $\widetilde{e}_n^k = \| \widetilde{\varphi}_n^k - f \|$ where $\widetilde{\boldsymbol{\beta}_k}$ is calculated by (3);
  c) compare f_{n-1} with $\widetilde{\varphi}_n^k(\mathbf{x})$: if $\widetilde{e}_n^k \leq e_{n-1}$, then stop; otherwise, let $k = k+1$ and go to step a).
  Denote by φ_n the final obtained network in step 1) of *Phase II*:

$$\varphi_n(\mathbf{x}) = \begin{cases} \widetilde{\varphi}_n^k(\mathbf{x}), & \text{if } k < L_{n-1}; \\ \widetilde{\varphi}_n^{L_{n-1}}(\mathbf{x}), & \text{if } k = L_{n-1} \text{ and } \widetilde{e}_n^{L_{n-1}} < e_{n-1}; \\ f_{n-1}(\mathbf{x}), & \text{if } k = L_{n-1} \text{ and } \widetilde{e}_n^{L_{n-1}} \geq e_{n-1} \end{cases}$$

and by $\mathbf{G_{nb}}(\mathbf{x})$ the hidden-layer output of $\varphi_n(\mathbf{x})$.
2) *Incremental step:* randomly generate $(\mathbf{a}_n, b_n) \in R^d \times R$. Adding a hidden node on φ_n in terms of the parameters $(\mathbf{a}_n, b_n)$ yields a new network with the hidden-layer output $\mathbf{G_n}(\mathbf{x}) = [\mathbf{G_{nb}}(\mathbf{x})^T, G(\mathbf{a}_n, b_n, \mathbf{x})]^T$. Compute the output weight $\hat{\boldsymbol{\beta}}_n$ according to (3).

Therefore, we obtain the approximated function f_n and its corresponding error e_n:

$$f_n(\mathbf{x}) = \hat{\boldsymbol{\beta}}_n \cdot \mathbf{G_n}(\mathbf{x}) \quad \text{and} \quad e_n = \| f_n - f \|.$$

3) *Recursive step:* Let $n = n+1$ and go to step 1) of *phase II*.

As stated above, AIE-ELM includes two phases: the initialization phase and the recursively growing phase.

In the first phase, we initialize a network $f_1(\mathbf{x}) = \hat{\boldsymbol{\beta}}_1 \cdot \mathbf{G}_1(\mathbf{x})$ with one hidden node where the hidden-layer output $\mathbf{G}_1(\mathbf{x})$ are randomly generated and the output weight is calculated by $\hat{\boldsymbol{\beta}}_1 = [\mathbf{G}_1(\mathbf{x})]^\dagger \mathbf{T}$.

In the second phase, we first build up a temporary network φ_n whose hidden-node number may be any one of $1, 2, \cdots, L_{n-1}$ and whose residual error must be less than or equal to e_{n-1}. We start from generating a network $\widetilde{\varphi}_n^1$ with one hidden node and compare it with the network f_{n-1} according to the residual errors $\widetilde{e}_n^1$ and e_{n-1}, and then retain the one whose error is less. If the retained network is $\widetilde{\varphi}_n^1$, then this network is just the temporary one we are looking for at this step. Otherwise, we keep generating a network $\widetilde{\varphi}_n^2$ with two hidden nodes and compare it with f_{n-1}. This procedure continues until one of these networks is better than f_{n-1} or all of these networks with from one hidden node to L_{n-1} hidden nodes have been compared with f_{n-1}. Denote by φ_n the best one we obtained at this step eventually. Then the network f_n at the current step is obtained by adding one more hidden node in φ_n. At each step of this phase, the output weights are updated incrementally according to (3).

Remark 1. Differen from those existing I-ELMs to add the hidden nodes one by one or group by group, in AIE-ELM, the hidden-node number is determined in an adaptive way. At each step, among all the possible networks $\widetilde{\varphi}_n^1(\mathbf{x})$, $\widetilde{\varphi}_n^2(\mathbf{x}), \cdots, \widetilde{\varphi}_n^{L_{n-1}}(\mathbf{x})$ and the network f_{n-1} obtained at the previous step, the one having the least hidden-node number whose error is less than or equal to e_{n-1}, will be selected as the temporary network φ_n. Such selected φ_n can be viewed as the best one since it has the most compact network architecture so far. Meanwhile, in AIE-ELM, the existing hidden nodes may be replaced by some newly generated hidden nodes with better performance rather than always keeping those existing ones in other incremental ELMs. This is the reason why we say that the proposed AI-ELM adaptively determines the network architecture.

4 Experimental Verification

In this section, the proposed algorithm AIE-ELM is compared with EM-ELM on six benchmark regression problems from UCI database [3]. The specification of these benchmark problems are shown in Table 1.

In our experiments, all the inputs (attributes) have been normalized into the range $[-1, 1]$ while the outputs (targets) have been normalized into $[0, 1]$. Neural networks with sigmoid type of additive hidden nodes $G(\mathbf{a}, b, \mathbf{x}) = 1/(1+\exp(-(\mathbf{a} \cdot \mathbf{x}+b)))$ are tested in both EM-ELM and AIE-ELM. The hidden node parameters $\mathbf{a}$ and b are randomly chosen from the range $[-1, 1]$ based on a uniform sampling distribution probability. For each trial of simulations, the dataset of the application was divided into training and testing datasets with the number of samples indicated in Table 1. All the simulations are running in MATLAB 7.4.0 environment and the same PC with Pentium 4 2.66GHZ CPU and 3GB RAM.

Table 1. Specification of Benchmark Datasets

Problems	Attributes	Cases	Training Data	Testing Data
Abalone	8	4177	2000	2177
Boston Housing	13	506	250	256
California Housing	8	20640	8000	12640
Census(House8L)	8	22784	10000	12784
delta ailerons	5	7129	3000	4129
delta elevators	6	9517	4000	5517

For each problem, the average results over 20 trials are obtained for EM-ELM and AIE-ELM. The apparant better results are shown in boldface.

4.1 Comparison of AIE-ELM and EM-ELM with the Same Expected Accuracy

In this subsection, we compare the network structure and the generalization performance obtained by both AIE-ELM and EM-ELM with the same expected accuracy (or the same maximum step). Table 2 shows the network structure and generalization performance of both AIE-ELM and EM-ELM for all six applications. The stopping root mean square error (RMSE) and maximum step for all the six regression cases are presented in the table respectively. From the table, we can observe that as compared to EM-ELM, AIE-ELM always has lower network complexity with less testing RMSE in most cases. In addition, the testing standard deviation (Std dev) of AIE-ELM is always better than or comparable with EM-ELM except the delta aileron case, which shows the stability of AIE-ELM.

Table 2. Performance comparison between AIE-ELM and EM-ELM with the same expected accuracy

Datasets	Stop RMSE	Max Steps	Algorithms	Training Time (s)	Testing RMSE	Testing Dev	#Nodes	NodeDev
Abalone	0.09	200	EM-ELM	0.0055	0.0865	0.0038	6.95	2.481
			AIE-ELM	0.0141	0.086	0.0039	**4.75**	1.2513
Boston Housing	0.12	200	EM-ELM	0.0023	0.1299	0.0103	11.6	2.1619
			AIE-ELM	0.0305	0.1268	0.0097	**9.2**	1.9084
California Housing	0.14	200	EM-ELM	0.1094	0.1402	0.0025	19.85	3.7735
			AIE-ELM	0.7484	0.1398	0.0032	**16.1**	2.1981
Census (House8L)	0.07	200	EM-ELM	0.6281	0.0719	0.0023	44.55	9.6271
			AIE-ELM	8.9531	0.072	0.0026	**36.35**	5.5939
delta ailerons	0.05	200	EM-ELM	0.0063	0.0439	0.0027	4.8	0.8335
			AIE-ELM	0.0156	0.0456	0.0026	**3.65**	0.9333
delta elevators	0.06	200	EM-ELM	0.007	0.0574	0.0019	6.05	1.2344
			AIE-ELM	0.0289	0.0564	0.0024	**5.5**	0.8272

Table 3. Performance comparison between AIE-ELM and EM-ELM with the same step

Dataset	AIE-ELM					EM-ELM			
	Training Time (s)	Testing RMSE	Testing Dev	#Nodes	Node Dev	Training Time (s)	Testing RMSE	Testing Dev	#Nodes
Abalone	13.8617	0.0815	0.0091	**80.5**	8.6176	0.5172	0.0976	0.0442	100
Boston Housing	2.2695	0.1162	0.0297	**84**	6.8364	0.0563	0.121	0.0177	100
California Housing	61.6961	0.132	0.085	**78.5**	6.2197	2.382	0.1319	0.005	100
Census (House8L)	84.2344	0.0681	0.0015	**79.95**	7.2219	3.0094	0.0677	0.0017	100
delta ailerons	20.3625	0.0398	0.0006	**79.45**	5.3849	0.7734	0.04	0.0008	100
delta elevators	28.7828	0.0533	0.0005	**78.9**	6.3403	1.0758	0.0534	0.0005	100

4.2 Comparison of AIE-ELM and EM-ELM with the Same Step

In this subsection, we compare the network structure and the generalization performance of AIE-ELM and EM-ELM with the same step. In EM-ELM, the hidden nodes are added one by one. Table 3 shows the network structure and generalization performance of both AIE-ELM and EM-ELM network under the case where the learning steps are fixed to 100. Seen from Table 3, for each case, the hidden-node number of AIE-ELM are always smaller than that of EM-ELM. Meanwhile, the testing RMSE of AIE-ELM for each case is smaller than or comparable with that of EM-ELM. These facts can also be verified from Fig. 1 and Fig. 2 which show the updating progress of the hidden-node number and the testing error for Abalone case respectively. We can observe that

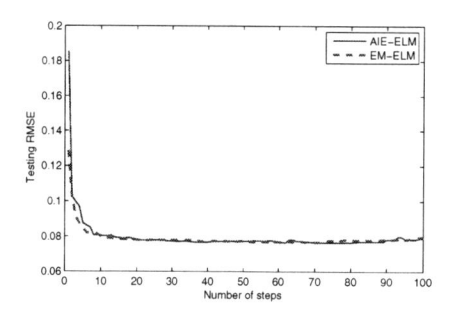

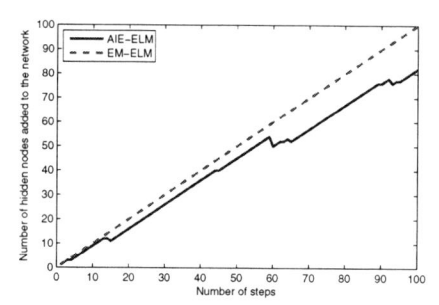

Fig. 1. The testing RMSE updating curves of AIE-ELM and EM-ELM for Abalone case

Fig. 2. The hidden-node number updating progress in 100 steps for Abalone case

the testing RMSE of AIE-ELM is comparable with that of EM-ELM from Fig. 1, and meanwhile, the hidden-node number of AIE-ELM is always smaller than that of EM-ELM from Fig. 2. These facts reveal that to obtain the same residual error, AIE-ELM can achieve lower network complexity than EM-ELM. Similar curves could be found for other applications.

5 Conclusion

In this paper, we have proposed an extreme learning machine with adaptive growth of hidden nodes and incremental updating of output weights using an error-minimization-based method (AIE-ELM) to automatically determine the hidden-node number in generalized SLFNs which need not be neuron alike. In AIE-ELM, the randomly generated hidden nodes grow in an adaptive way in the sense that the existing hidden nodes may be replaced by some newly generated hidden nodes with better generalization performance rather than always keeping those existing ones in other incremental ELMs. With such mechanism of the growth of the networks, the output weights are then updated incrementally based on the method proposed in [4]. The simulation results on sigmoid type of hidden nodes show that the new approach can achieve a more compact network architecture than EM-ELM while retaining the low computation complexity of EM-ELM. The performance of our method on other type of hidden nodes will be reported in the future work due to the limited space in this paper.

References

1. Barron, A.R.: Universal approximation bounds for superpositions of a sigmoidal functions. IEEE Transactions on Information Theory 39, 930–945 (1993)
2. Bishop, C.M.: Neural networks for pattern recognition. Oxford University Press, New York (1995)
3. Black, C.L., Merz, C.J.: UCI repository of machine learning databases, department of information and computer sciences, University of California, Irvine, USA (1998), http://www.ics.uci.edu/~mlearn/mlrepository.html
4. Feng, G., Huang, G.-B., Lin, Q., Gay, R.: Error minimized extreme learning machine with growth of hidden nodes and incremental learning. IEEE Transactions on Neural Networks 20(8), 1352–1357 (2009)
5. Hagan, M.T., Demuth, H.B., Beale, M.H.: Neural network design. PWS Publshing Company, Boston (1996)
6. Haykin, S.: Neural networks: a comprehensive foundation. Macmillan College Publishing Company, New York (1994)
7. Huang, G.-B., Chen, L.: Convex incremental extreme learning machine. Neural Computing 70, 3056–3062 (2007)
8. Huang, G.-B., Chen, L., Siew, C.-K.: Universal approximation using incremental constructive feedforward networks with random hidden nodes. IEEE Transactions on Neural Networks 17(4) (2006)
9. Huang, G.-B., Chen, L.: Enhanced random search based incremental extreme learning machine. Neural Computing 71, 3460–3468 (2008)

10. Huang, G.-B., Liang, N.-Y., Rong, H.-J., Saratchandran, P., Sundararajan, N.: On-line sequential extreme learning machine. In: The IASTED International Conference on Computational Intelligence (2005)
11. Huang, G.-B., Siew, C.-K.: Extreme learning machine: RBF network case. In: Proceedings of 8th International Conference Control, Automation, Robotics Vision, vol. 2, pp. 1029–1036 (2004)
12. Huang, G.-B., Siew, C.-K.: Extreme learning machine with randomly assigned RBF kernels. International Journal of Technology 11(1) (2005)
13. Huang, G.-B., Zhu, Q.-Y., Siew, C.-K.: Extreme learning machine: A new learning scheme of feedforward neural networks. In: Proceedings of International Joint Conference Neural Networks, vol. 2, pp. 985–990 (2004)
14. Huang, G.-B., Zhu, Q.-Y., Siew, C.-K.: Extreme learning machine: theory and applications. Neural Computing 70, 489–501 (2006)
15. Huang, G.-B., Zhu, Q.-Y., Siew, C.-K.: Real-time learning capability of neural networks. IEEE Transactions on Neural Networks 17(4), 863–878 (2006)
16. Huang, G.-B., Zhu, Q.-Y., Zhi, M.-K., Siew, C.-K., Saratchandran, P., Sundararajan, N.: Can threshold networks be trained directly? IEEE Transactions on Circuits and Systems II 53(3), 187–191 (2006)
17. Lan, Y., Soh, Y.C., Huang, G.-B.: Ensemble of online sequential extreme learning machine. Neural Computing 72, 3391–3395 (2009)
18. Li, M.-B., Huang, G.-B., Saratchandran, P., Sundararajan, N.: Fully complex extreme learning machine. Neural Computing 68, 306–314 (2005)
19. Liang, N.-Y., Huang, G.-B., Saratchandran, P., Sundararajan, N.: A fast and accurate online sequential learning algorithm for feedforward networks. IEEE Transactions on Neural Networks 17(6), 1411–1423 (2006)
20. Looney, C.G.: Pattern recognition using neural networks: theory and algorithms for engineers and scientists. Oxford University Press, New York (1997)
21. Rong, H.-J., Huang, G.-B., Sundararajan, N., Saratchandran, P.: Online sequential fuzzy learning machine for function approximation and classfication problem. IEEE Transactions on System, Man and Cybernetics B 39(4), 1067–1072 (2009)

Face Recognition Based on Kernelized Extreme Learning Machine*

Weiwei Zong, Hongming Zhou, Guang-Bin Huang, and Zhiping Lin

School of Electrical and Electronic Engineering, Nanyang Technological University, Nanyang Avenue, Singapore 639798
{zong0003,hmzhou,egbhuang,ezplin}@ntu.edu.sg

Abstract. The original extreme learning machine (ELM), based on least square solutions, is an efficient learning algorithm used in "generalized" single-hidden layer feedforward networks (SLFNs) which need not be neuron alike. Latest development [1] shows that ELM can be implemented with kernels. Kernlized ELM can be seen as a variant of the conventional LS-SVM without the output bias b. In this paper, the performance comparison of LS-SVM and kernelized ELM is conducted over a benchmarking face recognition dataset. Simulation results show that the kernelized ELM outperforms LS-SVM in terms of both recognition prediction accuracy and training speed.

Keywords: Face Recognition, Extreme Learning Machine, Kernelized Extreme Learning Machine, Least Square Support Vector Machine.

1 Introduction

As a popular topic in the area of computer vision and pattern recognition, face recognition has been extensively studied for the past decades [2,3]. For the purpose of fast and accurate face recognition, original data normally need to be mapped into the lower dimensional space. Among various dimensionality reduction methods [3,4,5], two classic approaches are principal component analysis [6] and linear discriminant analysis [7], which have been widely used in pattern recognition tasks.

After mapped into the lower dimensional space, face images with compact representations are fed into a classifier for training, which can generalize to new images later. In the literature of face recognition, classifiers such as Nearest Neighbor [3,6,8] and Support Vector Machine (SVM) [9,10,11] have been mainly used. Generally speaking, it is believed that supervised classifier is superior than unsupervised one because the label information of the training data is utilized in training. Therefore, we will focus on supervised classifier in this paper.

Support Vector Machine [12] and its variants [13,14] have demonstrated good performance on classification tasks. The learning procedure of traditional SVM can be summarized in two steps. First, the training data is mapped into a high dimensional feature space with a nonlinear mapping function. Then a hyperplane needs to be found

---

* This research work was sponsored by the grant from Academic Research Fund (AcRF) Tier 1, Ministry of Education, Singapore, under project No. RG22/08(M52040128).

to bisect the training data in the feature space. The resultant optimization problem of attempting to maximize the marginal distance between the two different classes and minimize the training error is resolved by the standard optimization method.

The computational complexity of the conventional SVM is closely related to the number of training samples, more precisely, at least quadratic since quadratic programming tool is used in finding the solution. Least square support vector machine (LS-SVM) [13] was proposed as an easier and faster implementation of SVM. By reformulating the optimization structure, inequality optimization constraints in original SVM are replaced by equality equations, which can be solved by least square methods instead of quadratic programming.

Extreme learning machine [15,16] was proposed as the training algorithm for "generalized" single-hidden layer feedforward networks (SLFNs) where the hidden nodes can be neuron alike [17] or non-neuron alike [18,19]. The hidden layer nodes are randomly generated without tuning and the output weights are computed as the least square solution. Therefore, ELM is much faster than SVM and also easier to implement.

Huang, et al [20] pointed out that SVM and ELM are actually consistent from optimization point of view; but ELM has milder optimization constraints. They have further studied SVM with ELM kernel but without the bias b in the optimization constraint. The resultant classifier is verified as efficient as conventional SVM but with less sensitive user specified parameters.

Huang, et al [1] recently proposed a kernelized ELM and found that kernelized ELM actually simplifies the implementation of least square SVM (LS-SVM).

As far as we know, there are few works where ELM is considered as the classifier in face recognition applications. In this paper, the kernelized ELM learning framework [1] is proposed as the classifier in face recognition. The performance is compared with LS-SVM over a benchmarking face data set. Note that, both SVM and LS-SVM are essentially binary classifiers. One against all [4] and one against one [21] were mainly used to decompose the multi-class problems into binary subproblems which can be solved by multiple binary SVM classifiers. The coding scheme was proposed for LS-SVM to transform the original problem into binary subproblems. However, multiple binary classifiers are still needed in the implementation. In contrast to SVM and LS-SVM, single ELM classifier is enough to carry out multi-category classification tasks.

2 Dimensionality Reduction Methods: PCA and LDA

Since face images usually contain significant redundancies, generally dimensionality reduction methods are adopted to transform the images to lower dimensional space for an efficient recognition [3]. Among the dimensionality reduction methods, linear mapping has been widely used due to its simplicity and tractability. Principal component analysis (PCA) [6] and linear discriminant analysis (LDA) [7] are two representative approaches.

Given training samples $\mathbf{x} \in R^{n \times d}$ representing n face images with dimension d, a transformation matrix $\mathbf{W} \in R^{d \times k}$ is searched to map $\mathbf{x}$ into the new feature space with dimensionality k by $\mathbf{xW} = \mathbf{y}$ ($k < d$ and $\mathbf{y} \in R^{n \times k}$).

In PCA, the eigen decomposition of the covariance matrix of the data is manipulated. The obtained subspace is spanned by the orthogonal set of eigen vectors which reveal the maximum variance in the data space.

By utilizing the label information, LDA seeks the subspace that gains maximum discrimination power. The mathematical solutions is to compute a generalized eigen equation which maximizes the ratio of the determinants of between-class scatter matrix and within-class scatter matrix. In the feature space, the samples from different classes tend to be farther apart and the samples from the same class tend to be closer.

3 Review of Least Square Support Vector Machine (LS-SVM)

Given a set of training data $(\mathbf{x}_i, t_i)$, $i = 1, \cdots, N$, where $\mathbf{x}_i \in \mathbf{R}^d$ and $t_i \in \{-1, 1\}$, in most cases these training data may not be linearly separable. And thus, ideally speaking, one can map the training data $\mathbf{x}_i$ from the input space to a feature space Z through a nonlinear mapping $\phi(\mathbf{x}) : \mathbf{x}_i \rightarrow \phi(\mathbf{x}_i)$ so that the corresponding mapped data $\phi(\mathbf{x}_i)$ can be separated in the feature space. In this feature space, to separate these data with training errors ξ_i we have:

$$t_i(\mathbf{w} \cdot \phi(\mathbf{x}_i) + b) \geq 1 - \xi_i, \quad i = 1, \cdots, N \tag{1}$$

The essence of SVM [12] is to maximize the separating margin of the two classes in the feature space and to minimize the training error, which is equivalent to:

$$\text{Minimize: } L_{P_{SVM}} = \frac{1}{2}\|\mathbf{w}\|^2 + C \sum_{i=1}^{N} \xi_i$$
$$\text{Subject to: } t_i(\mathbf{w} \cdot \phi(\mathbf{x}_i) + b) \geq 1 - \xi_i, \quad i = 1, \cdots, N \tag{2}$$
$$\xi_i \geq 0, \quad i = 1, \cdots, N$$

where C is a user specified parameter and provides a tradeoff between the distance of the separating margin and the training error.

To train such an SVM is equivalent to solving the following dual optimization problem:

$$\text{Minimize: } L_{D_{SVM}} = \frac{1}{2} \sum_{i=1}^{N} \sum_{j=1}^{N} t_i t_j \alpha_i \alpha_j \phi(\mathbf{x}_i) \cdot \phi(\mathbf{x}_j) - \sum_{i=1}^{N} \alpha_i$$
$$\text{Subject to: } \sum_{i=1}^{N} t_i \alpha_i = 0 \tag{3}$$
$$0 \leq \alpha_i \leq C, i = 1, \cdots, N$$

where each Lagrange multiplier α_i corresponds to a training sample $(\mathbf{x}_i, t_i)$. Vectors $\mathbf{x}_i$ for which $t_i(\mathbf{w} \cdot \phi(\mathbf{x}_i) + b) = 1$ are termed support vectors [12].

Suykens and Vandewalle [13] proposed a least square version to SVM classifier. In the least square support vector machine (LS-SVM), the classification problem is formulated as:

$$\text{Minimize: } L_{P_{LS-SVM}} = \frac{1}{2}\mathbf{w}\cdot\mathbf{w} + C\frac{1}{2}\sum_{i=1}^{N}\xi_i^2 \quad (4)$$

$$\text{Subject to: } t_i(\mathbf{w}\cdot\phi(\mathbf{x}_i)+b) = 1-\xi_i, \ i=1,\cdots,N$$

where ξ_i is the training error.

The inequality constraints (1) and (2) are used in the conventional SVM while equality constraints (4) are used in LS-SVM. Different from SVM which needs to solve quadratic programming, LS-SVM only needs to solve a set of linear equations and thus the least square approach can be implemented. LS-SVM produces good generalization performance and has low computational cost in many applications.

To train such a LS-SVM is equivalent to solving the following dual optimization problem:

$$L_{D_{LS-SVM}} = \frac{1}{2}\mathbf{w}\cdot\mathbf{w} + C\frac{1}{2}\sum_{i=1}^{N}\xi_i^2 - \sum_{i=1}^{N}\alpha_i(t_i(\mathbf{w}\cdot\phi(\mathbf{x}_i)+b)-1+\xi_i) \quad (5)$$

Based on the Karush-Kuhn-Tucker theorem [22], the following solution can be obtained for LS-SVM:

$$\begin{bmatrix} 0 & \mathbf{T}^T \\ \mathbf{T} & \frac{\mathbf{I}}{C}+\boldsymbol{\Omega}_{LS-SVM} \end{bmatrix}\begin{bmatrix} b \\ \boldsymbol{\alpha} \end{bmatrix} = \begin{bmatrix} 0 & \mathbf{T}^T \\ \mathbf{T} & \frac{\mathbf{I}}{C}+\mathbf{Z}\mathbf{Z}^T \end{bmatrix}\begin{bmatrix} b \\ \boldsymbol{\alpha} \end{bmatrix} = \begin{bmatrix} 0 \\ \mathbf{1} \end{bmatrix} \quad (6)$$

where

$$\mathbf{Z} = \begin{bmatrix} t_1\phi(\mathbf{x}_1) \\ \vdots \\ t_N\phi(\mathbf{x}_N) \end{bmatrix} \quad (7)$$

$$\boldsymbol{\Omega}_{LS-SVM} = \mathbf{Z}\mathbf{Z}^T$$

The feature mapping $\phi(\mathbf{x})$ is a row vector[1], $\mathbf{T}=[t_1,t_2,\cdots,t_N]^T$, $\boldsymbol{\alpha}=[\alpha_1,\alpha_2,\cdots,\alpha_N]^T$ and $\mathbf{1}=[1,1,\cdots,1]^T$. In LS-SVM, as $\phi(\mathbf{x})$ is usually unknown, Mercer's condition [23] can be applied to matrix $\boldsymbol{\Omega}_{LS-SVM}$:

$$\boldsymbol{\Omega}_{LS-SVM\ i,j} = t_it_j\phi(\mathbf{x}_i)\cdot\phi(\mathbf{x}_j) = t_it_jK(\mathbf{x}_i,\mathbf{x}_j) \quad (8)$$

The decision function of LS-SVM classifier is: $f(\mathbf{x}) = \text{sign}\left(\sum_{i=1}^{N}\alpha_i t_i K(\mathbf{x},\mathbf{x}_i)+b\right)$.

4 Kernelized Extreme Learning Machine

Extreme learning machine (ELM) [15,16,17] was originally proposed for the single-hidden layer feedforward *neural* networks and was then extended to the generalized single-hidden layer feedforward networks (SLFNs) where the hidden layer need not be

---
[1] Feature mappings $\phi(\mathbf{x})$ and $\mathbf{h}(\mathbf{x})$ are defined as a row vector while the rest vectors as defined as column vectors in this paper.

neuron alike [18,19]. The essence of ELM is that in ELM the hidden layer need not be tuned. The output function of ELM for generalized SLFNs is

$$f_L(\mathbf{x}) = \sum_{i=1}^{L} \beta_i h_i(\mathbf{x}) = \mathbf{h}(\mathbf{x})\beta \qquad (9)$$

where $\beta = [\beta_1, \cdots, \beta_L]^T$ is the vector of the output weights between the L hidden nodes and the output nodes, and $\mathbf{h}(\mathbf{x}) = [h_1(\mathbf{x}), \cdots, h_L(\mathbf{x})]$ is the (row) vector presenting the outputs of the L hidden nodes with respect to the input $\mathbf{x}$. $\mathbf{h}(\mathbf{x})$ actually maps the data from the d-dimensional input space to the L-dimensional hidden layer feature space (*ELM feature space*) H.

The minimal norm least square method was used in the original implementation of ELM [16,15]:

$$\beta = \mathbf{H}^\dagger \mathbf{T} \qquad (10)$$

where $\mathbf{H}^\dagger$ is the *Moore-Penrose generalized inverse* of matrix $\mathbf{H}$ [24,25] and $\mathbf{H}$ is the hidden layer output matrix:

$$\mathbf{H} = \begin{bmatrix} \mathbf{h}(\mathbf{x}_1) \\ \vdots \\ \mathbf{h}(\mathbf{x}_N) \end{bmatrix} \qquad (11)$$

As introduced in [1], one of the methods to calculate Moore-Penrose generalized inverse of a matrix is the orthogonal projection method [25]: $\mathbf{H}^\dagger = \mathbf{H}^T \left(\mathbf{H}\mathbf{H}^T\right)^{-1}$. According to the ridge regression theory [26], one can add a positive value to the diagonal of $\mathbf{H}\mathbf{H}^T$ the resultant solution is more stable and tends to have better generalization performance:

$$\mathbf{f}(\mathbf{x}) = \mathbf{h}\beta = \mathbf{h}(\mathbf{x})\mathbf{H}^T \left(\frac{\mathbf{I}}{C} + \mathbf{H}\mathbf{H}^T\right)^{-1} \mathbf{T} \qquad (12)$$

Different from SVM, feature mapping $\mathbf{h}(\mathbf{x})$ is usually known to users in ELM. However, if a feature mapping $\mathbf{h}(\mathbf{x})$ is unknown to users a kernel matrix for ELM can be defined as follows [1]:

$$\mathbf{\Omega}_{ELM} = \mathbf{H}\mathbf{H}^T : \mathbf{\Omega}_{ELM\,i,j} = h(\mathbf{x}_i) \cdot h(\mathbf{x}_j) = K(\mathbf{x}_i, \mathbf{x}_j) \qquad (13)$$

Thus, the output function of ELM classifier can be written compactly as:

$$\mathbf{f}(\mathbf{x}) = \mathbf{h}(\mathbf{x})\mathbf{H}^T \left(\frac{\mathbf{I}}{C} + \mathbf{H}\mathbf{H}^T\right)^{-1} \mathbf{T}$$

$$= \begin{bmatrix} K(\mathbf{x}, \mathbf{x}_1) \\ \vdots \\ K(\mathbf{x}, \mathbf{x}_N) \end{bmatrix}^T \left(\frac{\mathbf{I}}{C} + \mathbf{\Omega}_{ELM}\right)^{-1} \mathbf{T} \qquad (14)$$

Interestingly, ELM solution (14) is consistent to LS-SVM solution when the bias b is not used in the constraint conditions. Different from other learning algorithms [27], ELM is to minimize the training error as well as the norm of the output weights [15,16]:

$$\text{Minimize: } \|\mathbf{H}\beta - \mathbf{T}\|^2 \text{ and } \|\beta\| \qquad (15)$$

In fact, according to (15) the classification problem for ELM with multi output nodes can be formulated as:

$$\text{Minimize: } L_{P_{ELM}} = \frac{1}{2}\|\boldsymbol{\beta}\|^2 + C\frac{1}{2}\sum_{i=1}^{N}\|\boldsymbol{\xi}_i\|^2 \qquad (16)$$

$$\text{Subject to: } \mathbf{h}(\mathbf{x}_i)\boldsymbol{\beta} = \mathbf{t}_i^T - \boldsymbol{\xi}_i^T, \ i = 1,\cdots,N$$

where $\boldsymbol{\xi}_i = [\xi_{i,1},\cdots,\xi_{i,m}]^T$ is the training error vector of the m output nodes with respect to the training sample $\mathbf{x}_i$. Different from LS-SVM which has the bias b in the constraint conditions (4), ELM has no bias in the constraint conditions resulting in much simpler solutions. As proved in Huang, et al [1], one can apply Mercer's conditions on ELM classification problem (16). And then the same solution as (14) can be obtained:

$$\begin{aligned}\mathbf{f}(\mathbf{x}) &= \mathbf{h}(\mathbf{x})\mathbf{H}^T\left(\frac{\mathbf{I}}{C} + \mathbf{H}\mathbf{H}^T\right)^{-1}\mathbf{T} \\ &= \begin{bmatrix}K(\mathbf{x},\mathbf{x}_1)\\ \vdots \\ K(\mathbf{x},\mathbf{x}_N)\end{bmatrix}^T \left(\frac{\mathbf{I}}{C} + \boldsymbol{\Omega}_{ELM}\right)^{-1}\mathbf{T}\end{aligned} \qquad (17)$$

Comparing (14) and (17), it is found that under ridge regression theory kernelized ELM actually simplifies the implementation of the original LS-SVM. Details of analysis of kernelized ELM and the reasons why the bias b should not be in LS-SVM constraint conditions can be found in Huang, et al [1].

5 Performance Evaluation

In this section, the performances of LS-SVM and ELM are evaluated on the benchmarking face data set YALE [28]. There are 15 different persons with 11 images from each person. The images span a variation on illumination, facial expression, and decorations. Images are in the size of 40×40 pixels. Therefore, the dimension of the original face vectors is 1600, with each entry representing the grey value of the corresponding pixel. Further, we have normalized the entries of each image vector into the range of $[0,1]$. 3, 5 and 7 images are randomly selected from each person for training. The sample images from YALE can be seen in Figure 1. After PCA projection, the dimension of the image is reduced to 45, 75 or 105 when 3, 5 or 7 images per person are used for training, respectively. On the other hand, the dimension of the image vector is reduced to 14 when LDA projection is applied.

The simulations are conducted in MATLAB 7.0.1 environment running on a Core 2 Quad, 2.66G CPU with 2G RAM.

With the detail of parameter settings specified as below, the performances of LS-SVM and ELM are compared in the aspect of testing accuracy, computation complexity, training time and sensitivity of user specified parameters.

Fig. 1. Display of sample images from YALE.

Table 1. Comparison of performance of LS-SVM and ELM on YALE

YALE	LS-SVM (Gaussian Kernel)				ELM (Gaussian Kernel)			
	Test Rate (%)	Test Dev (%)	Train Time (s)	(C, γ)	Test Rate (%)	Test Dev (%)	Train Time (s)	(C, γ)
PCA(3)	57.54	4.73	0.0130	$(2^4, 2^{10})$	**69.25**	3.72	0.0008	$(2^4, 2^{11})$
PCA(5)	68.72	4.32	0.0149	$(2^{10}, 2^{20})$	**80.80**	3.32	0.0008	$(2^5, 2^{10})$
PCA(7)	74.58	4.08	0.0203	$(2^{10}, 2^{16})$	**85.58**	3.35	0.0039	$(2^9, 2^{13})$
LDA(3)	65.25	5.52	0.0130	$(2^{-4}, 2^6)$	**67.83**	4.81	0.0015	$(2^5, 2^4)$
LDA(5)	79.61	3.50	0.0164	$(2^{-4}, 2^4)$	**81.28**	3.65	0.0008	$(2^{-4}, 2^{-1})$
LDA(7)	84.83	5.97	0.0218	$(2^{10}, 2^{12})$	**85.92**	4.31	0.0023	$(2^2, 2^{-7})$

5.1 Parameter Settings

Both LS-SVM and ELM adopt the popular Gaussian kernel function $K(\mathbf{x}, \mathbf{x}_i) = \exp(-\gamma |\mathbf{x} - \mathbf{x}_i|^2)$. The choice of the cost parameter C and the kernel parameter γ is varied in order to achieve the optimal generalization performance. A wide range of C and γ have been tried $\{2^{-24}, 2^{-23}, \cdots, 2^{24}, 2^{25}\}$ for LS-SVM and ELM, respectively.

5.2 Testing Accuracy

The average testing accuracy is obtained based on 20 trials with randomly generated train-test set for each trial. The comparisons of average testing accuracy between LS-SVM and ELM with Gaussian kernels are given in Table 1. Note that PCA (5) indicates the case where 5 images from each person are randomly chosen for training. It can be observed that ELM with Gaussian kernel outperforms LS-SVM in the aspect of recognition accuracy, especially when unsupervised PCA is used to reduce the dimension. In addition, LS-SVM has larger standard deviation caused by 20 random train-test splittings, which implies that ELM with Gaussian kernel tends to be more stable.

5.3 Computational Complexity and Training Time

During simulations, 4 different coding schemes have been tried for LS-SVM to decompose the multi-category problem into multiple binary subproblems. Among the minimum output coding (MOC), one against all (OAA), one against one (OAO) and error correcting output coding (ECOC), LS-SVM with one against one (OAO) coding scheme has shown much better generalization performance than with the other three. With OAO coding scheme, the label is extended to a label vector of dimension $(M - 1) * M/2$, which requires $(M - 1) * M/2$ binary LS-SVM classifiers in training. In the case of YALE, for instance, a total of 105 binary classifiers is involved, which obviously

(a) PCA(5)+LS-SVM

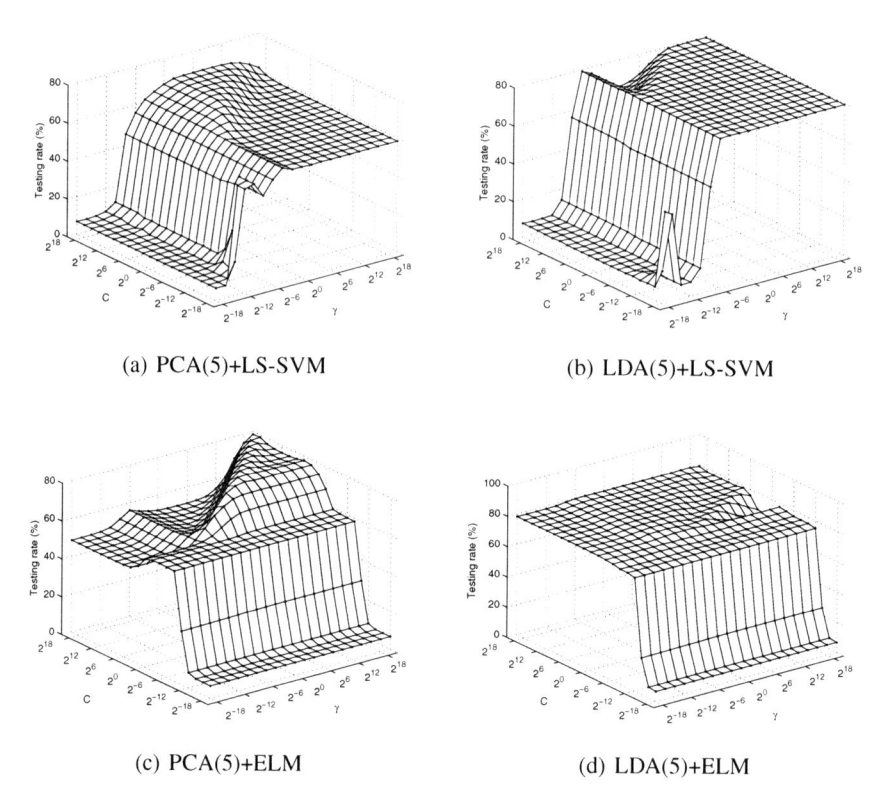

(b) LDA(5)+LS-SVM

(c) PCA(5)+ELM

(d) LDA(5)+ELM

Fig. 2. The Relationship of Testing Accuracy to the parameter selection (C, γ) of LS-SVM and ELM with Gaussian kernels

increases the computation complexity as well as the training time. Different from LS-SVM, a single ELM classifier is able to accomplish multi-category classification tasks. By default, the number of the output nodes is set the same as the number of classes, which is 15 in our simulations. This has explained the advantage of ELM in training time shown in Table 1.

5.4 Sensitivity of User Specified Parameters

Since parameter selection has great impact on the generalization performance of SVM with various kernels, the sensitivity of user specified parameters in both LS-SVM and ELM is studied in this section. As observed in Figure 2, the generalization performances of both LS-SVM and ELM are sensitive to user specified parameters (C, γ), where C is the cost parameter and γ is the kernel parameter. Similar observations can be found in [1] based on benchmarking regression and classification data sets.

6 Conclusions

ELM with kernel functions can be considered as the simplified version of LS-SVM without bias b in the constraints conditions. By comparing the kernelized ELM with

LS-SVM on face recognition applications, it is found that ELM achieves better recognition accuracy with much easier implementation and faster training speed. Reasonably speaking, the kernelized ELM can be effectively applied in many real world biometric problems.

References

1. Huang, G.-B., Zhou, H., Ding, X., Zhang, R.: Extreme learning machine for regression and multi-class classification. Submitted to IEEE Transactions on Pattern Analysis and Machine Intelligence (2010)
2. Lin, S.-H., Kung, S.-Y., Lin, L.-J.: Face recognition/detection by probabilistic decision-based neural network. IEEE Transactions on Neural Networks 8(1), 114–132 (1997)
3. Zhao, W., Chellappa, R., Phillips, P.J., Rosenfeld, A.: Face recognition: A literature survey. ACM Computing Surveys 35(4), 399–458 (2003)
4. Heisele, B., Ho, P., Wu, J., Poggio, T.: Face recognition: component-based versus global approaches. Computer Vision and Image Understanding 91(1-2), 6–21 (2003)
5. Li, Z., Tang, X.: Using support vector machines to enhance the performance of bayesian face recognition. IEEE Transactions on Information Forensics and Security 2(2), 174–180 (2007)
6. Turk, M., Pentland, A.: Eigenfaces for recognition. Journal of Cognitive Neuroscience 3(1), 71–86 (1991)
7. Fukunag, K.: Introduction to statistical pattern recognition. Academic Press, New York (1990)
8. Zhao, W., Krishnaswamy, A., Chellappa, R.: Discriminant analysis of principal components for face recognition. In: Proceedings of the Third IEEE International Conference on Automatic Face and Gesture Recognition, Nara, Japan, April 14-16, pp. 336–341 (1998)
9. Zhang, G.-Y., Peng, S.-Y., Li, H.-M.: Combination of dual-tree complex wavelet and svm for face recognition. In: Proceedings of International Conference on Machine Learning and Cybernetics, Kunming, China, July 12-15, vol. 5, pp. 2815–2819 (2008)
10. Gan, J.-Y., He, S.-B.: Face recognition based on 2DLDA and support vector machine. In: Proceedings of International Conference on Wavelet Analysis and Pattern Recognition (ICWAPR 2009), Baoding, China, July 12-15, pp. 211–214 (2009)
11. Zhao, L., Song, Y., Zhu, Y., Zhang, C., Zheng, Y.: Face recognition based on multi-class svm. In: Proceedings of Chinese Control and Decision Conference (CCDC 09), Guilin, China, June 17-19, p. 5871 (2009)
12. Cortes, C., Vapnik, V.N.: Support vector networks. Machine Learning 20, 273–297 (1995)
13. Suykens, J.A.K., Vandewalle, J.: Least squares support vector machine classifiers. Neural Processing Letters 9(3), 293–300 (1999)
14. Fung, G., Mangasarian, O.L.: Proximal support vector machine classifiers. In: International Conference on Knowledge Discovery and Data Mining, San Francisco, California, USA, pp. 77–86 (2001)
15. Huang, G.-B., Zhu, Q.-Y., Siew, C.-K.: Extreme learning machine: Theory and applications. Neurocomputing 70(1-3), 489 (2006)
16. Huang, G.-B., Zhu, Q.-Y., Siew, C.-K.: Extreme learning machine: A new learning scheme of feedforward neural networks. In: Proceedings of International Joint Conference on Neural Networks (IJCNN 2004), Budapest, Hungary, July 25-29, vol. 2, pp. 985–990 (2004)
17. Huang, G.-B., Chen, L., Siew, C.-K.: Universal approximation using incremental constructive feedforward networks with random hidden nodes. IEEE Transactions on Neural Networks 17(4), 879–892 (2006)

18. Huang, G.-B., Chen, L.: Convex incremental extreme learning machine. Neurocomputing 70, 3056–3062 (2007)
19. Huang, G.-B., Chen, L.: Enhanced random search based incremental extreme learning machine. Neurocomputing 71, 3460–3468 (2008)
20. Huang, G.-B., Ding, X., Zhou, H.: Optimization method based extreme learning machine for classification. Neurocomputing 74, 155–163 (2010)
21. Allwein, E.L., Schapire, R.E., Singer, Y.: Reducing multiclass to binary: a unifying approach for margin classifiers. The Journal of Machine Learning Research 1, 113–141 (2001)
22. Fletcher, R.: Practical Methods of Optimization: Volume 2 Constrained Optimization. Wiley, New York (1981)
23. Haykin, S.: Neural Networks: A comprehensive foundation. Prentice-Hall, New Jersey (1999)
24. Serre, D.: Matrices: Theory and Applications. Springer, Heidelberg (2002)
25. Rao, C.R., Mitra, S.K.: Generalized Inverse of Matrices and its Applications. John Wiley & Sons, Inc., New York (1971)
26. Hoerl, A.E., Kennard, R.W.: Ridge regression: Biased estimation for nonorthogonal problems. Technometrics 12(1), 55–67 (1970)
27. Rumelhart, D.E., Hinton, G.E., Williams, R.J.: Learning representations by back-propagation errors. Nature 323, 533–536 (1986)
28. Belhumeur, P.N., Hespanha, J.P., Kriegman, D.J.: Eigenfaces vs. fisherfaces: Recognition using class specific linear projection. IEEE Transactions on Pattern Analysis and Machine Intelligence 19(7), 711–720 (1997)

Detection and Tracking of Multiple Similar Objects Based on Color-Pattern

Hadi Firouzi and Homayoun Najjaran

Okanagan School of Engineering,
the University of British Columbia,
3333 University Way, Kelowna, BC, V1V1V7, Canada
{hadi.firouzi,h.najjaran}@ubc.ca

Abstract. In this paper an efficient and applicable approach for tracking multiple similar objects in dynamic environments is proposed. Objects are detected based on a specific color pattern i.e. color label. It is assumed that the number of objects is not fixed and they can be occluded by other objects. Considering the detected objects, an efficient algorithm to solve the multi-frame object correspondence problem is presented. The proposed algorithm is divided into two steps; at the first step, previous mismatched correspondences are corrected using the new information (i.e. new detected objects in new image frame), then all tail objects (i.e. objects which are located at the end of a track) are tried to be matched with unmatched objects (either a new object or a previously mismatched object). Apart from the correspondence algorithm, a probabilistic gain function is used to specify the matching weight between objects in consecutive frames. This gain function benefits Student T distribution function for comparing different object feature vectors. The result of the algorithm on real data shows the efficiency and reliability of the proposed method.

Keywords: Object Tracking, Correspondence Algorithm, Color Pattern, Object Detection.

1 Introduction

Object tracking has been considered as one of the most important topics in the field of machine vision and robotics since many years ago. The efficiency and robustness of many real world applications such as surveillance systems [10][11], traffic control [12][13], autonomous vehicles [14], video tracking [15], and medical imaging [16] relies on their object tracking method. Roughly speaking object tracking can be viewed as the problem of locating and corresponding objects in every image frame. However, developing an efficient and robust object correspondence method can be a difficult task especially in dynamic and unstructured environments, in which objects can be occluded by other objects, wrongly detected, missed, or added gradually.

Usually object detection is the first step in all tracking methods. Depending on tracking approaches, detection is performed in every frame or only for identifying the objects at their first appearance. Also objects can be detected either by extracting object features (e.g. color, edge, and texture) in a single frame or by comparing sequential frames, i.e. image change detection. For a complete survey on the latter refer to [9]. In general, Object detection methods can be divided into four common categories; 1) point detectors, 2) background subtraction, 3) segmentation, and 4) supervised classifiers [8].

The second step of a typical object tracking method is to match the detected objects over the image frames. Commonly correspondence methods can be categorized into two wide groups, i.e. deterministic and statistical methods [8]. Most of deterministic methods are designed to optimize a heuristic cost function. This function is usually defined based on spatiotemporal information such as motion model rather than surface features e.g. color, texture, edge. Inearly methods such as [19], [20], [21], and [22] only two frames have been considered to establish correspondences. However, Sethi and Jain [7] considered an extended frame sequence to find correspondence based on the coherence in spatiotemporal properties of consecutive points. They proposed an iterative greedy exchange (GE) algorithm. In self-initialization step, each point is assigned to a trajectory (track) using nearest neighbor criteria, then in an iterative process; points are exchanged between trajectories to maximize the total gain. Rangarajan and Shah [23] proposed a non-iterative algorithm to minimize the proximal uniformity cost function and establish correspondence over multiple frames. They used a gradient based optical flow to initialize the correspondences. Veenman et [24] proposed an approximation of an optimal algorithm named Greedy Optimal Assignment (GOA) tracker for the correspondence problem. Also they present a self-initializing version of GOA tracker that is less sensitive to the tracking parameters e.g. maximum speed constant. It was claimed that for fixed number of points the proposed algorithm is efficient and qualitatively best. Shafique and Shah [1] proposed a non-iterative greedy algorithm for the multi-frame correspondence formulated as a graph theory problem. They used the direction and speed coherence to define the cost function.

On the other hand, statistical correspondence methods try to estimate the object state e.g. position and velocity by taking the measurement and model uncertainties into account. Kalman filter and Particle filter are two efficient and general state estimators which have been used to solve the correspondence problem in case of single object [25], [26]. However, Kalman Filter and Particle filter may fail to converge when multiple objects exist.Several statistical data association methods have been proposed to solve this problem. Joint Probability Data AssociationFiltering[27] (JPDAF) and Multiple Hypothesis Tracking[28] (MHT) are two widely used methods for statistical data association. For A survey on statistical point correspondence, see [29].

In the next section the proposed detection and tracking method are described in details respectively, and then the experimental result on real data are shown. Finally some conclusions will be provided.

2 Proposed Approach

The proposed approach is consisted of two consecutive steps; first objects are detected using their visual features i.e. color pattern, and then, a multi-frame object correspondence algorithm are used to track the detected objects. In the following sections the object detection and tracking algorithms are explained in details.

2.1 Object Detection Based on Color Patterns

In this paper, it is assumed that there is no specific visual feature to discriminate objects; so that a predefined color pattern is used to differentiate objects in a dynamic and unstructured environment. For simplicity and with no loss of generality, a specific color label is attached on the objects. This label is consisted of two regions with equal size - but different colors - opposite to each other. A range in HSV color space [19] is learnt empirically to detect each area. Once an image is received, it is converted to HSV color space, and then; compared with the two color ranges to locate the objects based on their color labels. Indeed two binary images, called silhouette images, are generated based on two different color ranges. These silhouettes are preprocessed to remove noise and a typical connected component analysis method is applied to detect color regions. Two connected components, extracted from different silhouette, create an object if the first component is the best match with the second one and vice versa.

Let each connected component be defined by its bounding rectangle and a correlation function be used to find the best match. The correlation function (1) is a convex combination of two terms called *closeness* and *equality*. The first term decreases the matching weight of the far components, whereas the *equality* term increases the weight of equal size components.

$$\text{Cor}(i,j) = \alpha \times \text{Cl}(i,j) + (1-\alpha) \times \text{Eq}(i,j), \alpha \in [0,1] \qquad (1)$$

where α is a coefficient indicating the impact of each term on the matching weight. The *closeness* term (2) is defined as a normalized Euclidean distance of components' center points.

$$\text{Cl}(i,j) = \frac{\left|\sqrt{(x_i - x_j)^2 + (y_i - y_j)^2} - \frac{\mu_w + \mu_h}{2}\right|}{\sqrt{\frac{\sigma_w^2 + \sigma_h^2}{2}}} \qquad (2)$$

where $(\mu_w, \sigma_w^2), (\mu_h, \sigma_h^2)$ are the mean and variance of the components' width and height respectively, and (x_i, y_i) is the component i center point.

The second term (3), i.e. *equality*, is the Mahalanobis distance of components' sizes.

$$\text{Eq}(i,j) = \sqrt{\frac{(w_i - w_j)^2}{\sigma_w^2} + \frac{(h_i - h_j)^2}{\sigma_h^2}} \qquad (3)$$

where w_i, h_i are the width and height of the component i respectively. Based on the correlation function, the best match for component i is obtained as follows:

$$j_i^* = \arg\max_{j \in [1, C_N]} \text{Cor}(i,j) \qquad (4)$$

where C_N is the number of extracted connected components. Every pair components i and j, can create a new object if and only if $i_j^* = j_i^*$. The schematic flowchart of the proposed object detection algorithm is illustrated in Fig 1.

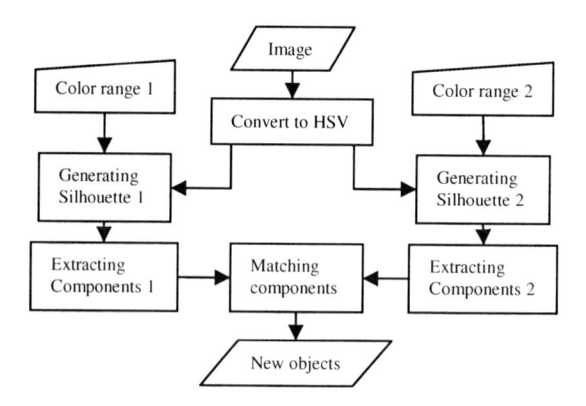

```
                    ┌──────────┐
                    │  Image   │
                    └────┬─────┘
┌───────────┐            ▼             ┌───────────┐
│Color range│     ┌─────────────┐      │Color range│
│     1     │     │Convert to HSV│     │     2     │
└─────┬─────┘     └──────┬──────┘      └─────┬─────┘
      ▼                                       ▼
┌───────────┐                          ┌───────────┐
│Generating │                          │Generating │
│Silhouette1│                          │Silhouette2│
└─────┬─────┘                          └─────┬─────┘
      ▼              ┌──────────┐            ▼
┌───────────┐        │ Matching │       ┌───────────┐
│Extracting │───────▶│components│◀──────│Extracting │
│Components1│        └─────┬────┘       │Components2│
└───────────┘              ▼            └───────────┘
                    ┌──────────┐
                    │New objects│
                    └──────────┘
```

Fig 1. An schematic flowchart of the proposed object detection method

Thus far an applicable and reliable algorithm was proposed to detect objects based on their specific color patterns, i.e. color labels; however, the main contribution of this work is presenting an efficient and real-time method to track similar objects in dynamic environments. In the next section the proposed tracking method will be explained in details.

3 Object Tracking

3.1 Multi-frame Object Correspondence Problem

The proposed tracking is a correspondence method that optimizes a cost function over multiple frames. The cost function is defined based on the object motion and size, however, any other deterministic or statistical based functions that satisfy the optimization constrain can be used. The main contribution of this work is the presentation of an efficient and robust algorithm that optimizes the cost function based on the graph theory. The presented correspondence algorithm is most closely related to Shafique and Shah [1], however, in this paper, a probabilistic scheme is used to define the gain function.

The main goal of a correspondence algorithm is to find a set of tracks. Ideally each track specifies the distinguished information of a unique object in every frame from entry to exit in the scene. In this work, it is assumed that the information available for each object is its position and direction in the image plane.

Let $\{F_{n-c+1}, \ldots, F_c\}$ be the N previous sequential image frames up to the current time c and $O_i = \{o_1^i, o_2^i, \ldots, o_{M_i}^i\}$ be the set of detected objects in image frame F_i. Note that number of objects, i.e. M_i, is not fixed for different image frames. Indeed, in every frame, objects may enter or exit the scene. An object o_j^i can be a representation for a real world object, a combination of multiple real world objects, or a misdetection caused by noise. A track $T_r = \{o_{j_1}^{i_1}, o_{j_2}^{i_2}, \ldots, o_{j_L}^{i_L}\}$ is defined as a sequence of L timely ordered objects in N previous image frames, where $c - N + 1 \leq j_1 < j_2 < \ldots < j_L \leq c$ and $L \leq N$.

In summary, the tracking problem can be reduced to finding a set of tracks $T = \{T_1, T_2, \ldots, T_R\}$ such that $\forall o_j^i \in T_r$, $r \in [1, R]$ is the representation of either of following cases:

1. A unique real world object or occlusion of that object with other objects.
2. A misdetection or noise.

The first condition indicates the element integrity principle [2] and in the second condition it is assumed that no noisy detection is part of any track corresponding to a real world object.

3.2 Multi-frame Object Correspondence Solution

Before solving the correspondence problem, it is necessary to have a brief introduction on the graph theory[3]. Let $D = (V, E)$ be a weighted directed graph without self loops and multiple edges, where V and E are, respectively, the set of vertices and edges of digraph. A directed path P of length K in digraph D is asequence of es$v_1, v_2, \ldots, v_{K+1}$, where $v_i \in V$, and, forevery v_i; $1 \leq i \leq K$, there is a directed edge in E fromv_i to v_{i+1}. Let $G = (V, E)$ be a bipartite graph if all vertices (V) are partitioned into two or more classes such that every edge has its ends in different classes, for more information refer to [3].

In the graph theoretical point of view, the multi-frame object correspondence problem can be viewed as an N-partite directed weighted graph G, see **Fig 2**, such that the partite set V_i corresponds to objects O_{c-N+i} in time instant $(c - N + i)$ where $1 \leq i \leq N$, c is the current time, and N is the number of frames. Also in the graph G, a track T is expressed by a directed path P. In the following sections, track and path are used interchangeability.

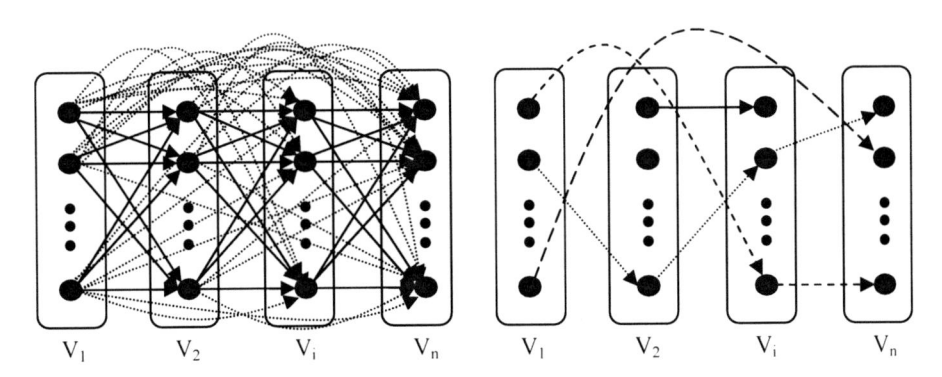

Fig. 2. An N-partite directed graph

Fig. 3. A solution for the multi-frame point correspondence problem

As a result of the graph theoretical formulation, the tracking problem can be considered as finding a set of directed paths (tracks) of length 0 or more within the

N-partite directed graph G, such that all vertices are covered and the total weight is maximized among all such tracks. For each sequential objects, there is an edge $e = v\left(o_{j_b}^{i_a}\right)v\left(o_{j_b}^{i_b}\right)$ which corresponds to a match hypothesis between the object $o_{j_b}^{i_a}$ in frame i_a and the object $o_{j_b}^{i_b}$ in frame i_b, whereas the associated edge weight $\omega(e)$ is estimated based on a cost (gain) function $g\left(o_{j_b}^{i_a}, o_{j_b}^{i_b}\right)$. Thus, by finding the best match for each object, the total matching weight is maximized. A candidate solution for the problem illustrated in Fig 2 is shown in Fig 3.

Although an efficient algorithm for finding the optimal tracks in a 2-partite graph has been proposed [17], the extended solution for a N-frame correspondence problem where N>2 is NP-Hard [4]. In fact, all proposed solutions are exponential, and cannot be applied on a practical application [5], [6]. Nevertheless, it is proved in [1] that in a case of having a directed acyclic graph, a polynomial solution exists if for a track $T_r = \{o_{j_1}^{i_1}, o_{j_2}^{i_2}, ..., o_{j_L}^{i_L}\}$, $\forall a, b$, $1 < a + 1 < b \le L$ the gain function $g\left(o_{j_b}^{i_a}, o_{j_b}^{i_b}\right)$ satisfies the following inequality:

$$g\left(o_{j_b}^{i_a}, o_{j_b}^{i_b}\right) < g\left(o_{j_b}^{i_a}, o_{j_{b+1}}^{i_{a+1}}\right) + g\left(o_{j_{b-1}}^{i_{b-1}}, o_{j_b}^{i_b}\right) \quad (5)$$

Using inequality (5), no shorter track is created when a longer valid track exists, as a result, maximization of the total gain can be guaranteed.

3.3 Proposed Multi-frame Object Correspondence Algorithm

It is assumed in (5) that the gain function is independent of the backward correspondence of object $o_{j_b}^{i_a}$, however it is not always possible, specially, where the gain function is obtained from the object motion or generally object history. For instance gain functions based on smoothness of object motion [7] do not satisfy this condition. In this work the gain function is calculated based on the object motion and size, so that an efficient algorithm is designed to solve the correspondence problem whereas the condition (5) is satisfied. It is assumed that the best matching objects in frames (c-N+1) to (c-1) have been found using the information available up to time instant (c-1), where c is the current time instant and N is the number of frames. Every new frame is either used to extent the previously established tracks using the new detected objects or to correct the previous mismatched correspondences. In fact a new object can create either two different edges i.e. extension or correction edge. Let $O_c = \{o_1^c, o_2^c, ..., o_{M_c}^c\}$ be the detected objects in the current frame F_c and $T = \{T_1, T_2, ..., T_R\}$ be the set of tracks up to time instant (c-1). For a track $T_r = \{o_{j_1}^{i_1}, o_{j_2}^{i_2}, ..., o_{j_L}^{i_L}\}$ where $i_L < c$, an extension edge is created if a new object o_a^c ($1 \le a \le M_c$) is appended to the track such that $T_r = \{o_{j_1}^{i_1}, o_{j_2}^{i_2}, ..., o_{j_L}^{i_L}, o_a^c\}$, and a correction edge is created if a new object o_a^c is added at the middle of the track such that $T_r = \{o_{j_1}^{i_1}, o_{j_2}^{i_2}, ..., o_{j_b}^{i_b}, o_a^c\}$ where $1 \le b < L$. In this case, objects $\{o_{j_{b+1}}^{i_{b+1}}, ..., o_{j_L}^{i_L}\}$, which called orphan objects, are removed from the track and need to be again considered for matching. The proposed correspondence algorithm is divided into two parts; i.e. 1) correction and 2) expansion steps. The aim of the first step is to correct mismatched edges based on new detected objects and in the second step, both orphan and new objects are matched to the existing tracks.

Correction Step
In this step, firstly each new object is connected to all previous objects to make an expansion graph, and then, by finding the best matches for new objects, mismatched edges are corrected.
Considering the inequality (5), the best match for object o_a^c among the objects of the track T_r is obtained as follows:

$$b_r^* = \arg\max_{b \in [1, L_r]} f_r(a, b) := \left(\sum_{k=1}^{b-1} w_k^r + q_{r,b}^a \right) \quad (6)$$

where $o_{j_{b_r^*}}^{i_{b_r^*}}$ is the best matched object in track T_r, $w_k^r = g\left(o_{j_k}^{i_k}, o_{j_{k+1}}^{i_{k+1}}\right)$ is the matching weight between two consecutive objects $o_{j_k}^{i_k}, o_{j_{k+1}}^{i_{k+1}}$ in track T_r, and $q_{r,b}^a = g\left(o_{j_b}^{i_b}, o_a^c\right)$ is the matching weight between object $o_{j_b}^{i_b} \in T_r$ in frame F_{i_b} and object o_a^c in the current frame F_c. As a result, the best global match for object o_a^c is obtained:

$$b^* = \arg\max_{r \in [1, R]} f_r(a, b_r^*) \quad (7)$$

where R is the number of existing tracks and $o_{j_{b^*}}^{i_{b^*}}$ is the best match from previous frames for object o_a^c in the current frame F_c.

For a track T_r, one or more mismatched edges exist if the best matched object $o_{j_{b^*}}^{i_{b^*}}$ is not located at the end of the track, i.e. $b^* < L_r$.

In the correction step, the best matched object is not used to update existing tracks if it is located at the end of a track. Indeed correcting the previous edges may change the best matches; therefore the previously found best matches are not valid anymore. In the next step, i.e. expansion step, all orphan objects and the rest of new objects will be matched to the previous objects.

Expansion Step
For every frame starting from the earliest frame, i.e. frame F_{c-N+1}, all objects located at the end of tracks are connected to all unmatched objects, including orphan and new objects, in ahead frames.
Let U be the set of all unmatched objects, the best expansion match for an ending object $o_{j_L}^{i_L}$ belonging to the track T_r is obtains as follows:

$$o_{j_{k^*}}^{i_{k^*}} = \arg\max_{o_{j_k}^{i_k} \in U} g\left(o_{j_L}^{i_L}, o_{j_k}^{i_k}\right), i_k \in [i_L + 1, c] \quad (8)$$

where c is the current time instant.
The best matched object $o_{j_{k^*}}^{i_{k^*}}$ is removed from the unmatched set U and appended to the matched track T_r, and thereafter the track T_r is $T_r = \{o_{j_1}^{i_1}, o_{j_2}^{i_2}, \ldots, o_{j_L}^{i_L}, o_{j_{k^*}}^{i_{k^*}}\}$.

After matching all ending objects in frame F_i, the same procedure will apply on the next frame F_{i+1}.

Cost Function

It is stated in above section that the main aim of a multi-frame object correspondence algorithm is to find a set of tracks by which the total cost (gain) is minimized (maximized), therefore, a proper cost (gain) function has a significant impact on the final correspondence solution.

In this work a probabilistic scheme is used to define the gain function. Suppose that for each track $T_r = \{o_{j_1}^{i_1}, o_{j_2}^{i_2}, \ldots, o_{j_L}^{i_L}\}$ such that $o_{j_a}^{i_a} \in T_r$ is a feature vector describing the related object (e.g. a combination of object location, direction, size, and/or color) $\hat{o}_{j_a}^{i_a}$ is the predicted vector of the object feature vector change in the next image frame. The multivariate Student T distribution [18] is used to estimate the probabilistic similarity of the predicted object $\hat{o}_{j_a}^{i_a}$ with a candidate observed object o_i^c (e.g. newly detected object).

$$p\left(o_i^c | \hat{o}_{j_a}^{i_a}\right) = \frac{\Gamma\left(\frac{v+1}{2}\right)}{\sqrt{v\pi}\,\Gamma\left(\frac{v}{2}\right)} \left(1 + \frac{\Theta^T \Theta}{v}\right)^{-\left(\frac{v+1}{2}\right)} \tag{9}$$

where v is the number of matched objects to the track T_r up to the object $o_{j_a}^{i_a}$ and θ is a similarity vector defined by comparing the predicted object and the candidate matching object. In this work θ is calculated as follows:

$$\Theta = 1 - \begin{bmatrix} \frac{d \cdot \hat{d}}{\|d\|\|\hat{d}\|} \\ \tau \frac{\|\|d\| - \|\hat{d}\|\|}{\|d\| + \|\hat{d}\|} \end{bmatrix} \tag{10}$$

$$d = o_i^c - o_{j_a}^{i_a}, \quad \hat{d} = \hat{o}_{j_a}^{i_a} - o_{j_a}^{i_a}, \quad \tau = c - i_a$$

where τ is the time interval between the object $o_{j_a}^{i_a}$ and new object o_i^c frames, the first term of Θ specifies the direction change coherence between the predicted and candidate object features, and the second term is a kind of relative distance of the predicted and candidate objects respectively.

4 Results

In this section the proposed method has been used to detect and track multiple similar real objects in an indoor environment. In this experiment, four moving objects specified by a square color label exist in the scene. The labels are the same and consisted of two equal-size red and blue rectangles which can be detected using the object detection method described in section 2.1. Objects have been moved independently by a thread; indeed the velocity and direction are not fixed for the objects. As shown in Fig 4, at first frame, three objects were in the scene, and then at frame 5 another object entered the scene, accordingly at frame 13 two objects crossed each other, finally objects left the environment. A ceiling color-camera has been used to capture the environment. Although only two frames per second have been processed, all objects have been tracked completely and precisely from entry to exit, see Fig 5.

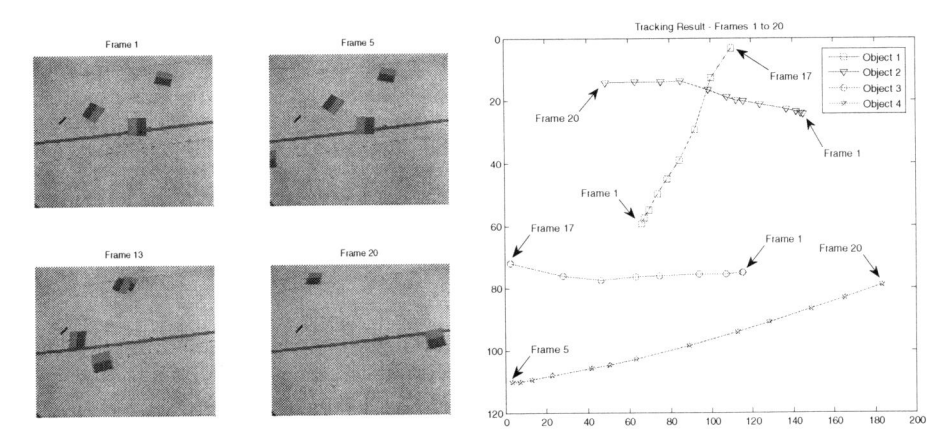

Fig. 4. Tracking multiple real objects in different frames

Fig. 5. Tracking result on real data by the proposed method

Tracking results and object trajectories obtained by the proposed method are illustrated in Fig 5. In this figure, objects positions during the experiment are also specified by different-shape markers.

In this experiment the following conditions have been applied; the image size used by object correspondence method is 200 by 160 pixels.

5 Conclusions

In this work a method to detect and track multiple same appearance objects based on their color patterns i.e. color labels is presented. These color labels are detected based on a HSV color range. A normalized Euclidean and Mahalanobis distances are used to match the detected color areas. This method is robust to noise and suitable for real-time applications.

The proposed object tracking method on the other hand is a heuristic graph-based object correspondence algorithm over multiple frames. Indeed, the proposed correspondence algorithm is able to find the set of tracks that minimize the overall cost in the graph. This graph is made of all possible correspondences between objects over multiple frames. Moreover, a probabilistic cost function based on Student T distribution is used to estimate the matching weight between objects.

The experimental result on real data illustrates the robustness of object detection and tracking method dealing with variable object direction and movements, occlusion and the real-world environment.

References

[1] Shafique, K., Shah, M.: A noniterative greedy algorithm for multiframe point correspondence. IEEE Trans. Pattern Anal. Mach. Intell. 27, 51–65 (2005)
[2] Dawson, M.R.W.: The How and Why of What Went Where in Apparent Motion: Modeling Solutions to the Motion Correspondence Problem. Psychological Rev. 98, 569–603 (1991)

[3] Diestel, R.: Graph Theory, 4th edn. Springer, Heidelberg (2010)
[4] Gary, M.R., Johnson, D.S.: Computers and Intractability. Freeman, NewYork (1979)
[5] Deb, S., Yeddanapudi, M., Pattipati, K., Bar-Shalom, Y.: A Generalized S-D Assignment Algorithm for Multisensor-MultitargetState Estimation. IEEE Trans. Aerospace and Electronic Systems 33(2), 523–538 (1997)
[6] Poore, A.B., Yan, X.: Data Association in Multi-Frame Processing. In: Proc. Second Int'l Conf. Information Fusion, pp. 1037–1044 (1999)
[7] Sethi, I.K., Jain, R.: Finding Trajectories of Feature Points in a Monocular Image Sequence. IEEE Trans. Pattern Analysis andMachine Intelligence 9(1), 56–73 (1987)
[8] Yilmaz, A., Javed, O., Shah, M.: Object tracking: A survey. ACM Computing Surveys 38(4), 45 pages (December 2006), Article 13
[9] Radke, R., Andra, S., Al-Kofahi, O., Roysam, B.: Image Change Detection Algorithms: A Systematic Survey. IEEE Trans. ImageProcessing 14(3), 294–307 (2005)
[10] Javed, O., Shah, M.: Tracking and object classification for automated surveillance. Journal of Computer Vision 50(2), 103–110 (2002)
[11] Huang, K., Wang, L., Tan, T., Maybank, S.: A real-time object detecting and tracking system for outdoor night surveillance. Pattern Recognition 41, 432–444 (2008)
[12] Huang, M.C., Yen, S.H.: A real-time and color-based computer vision for traffic monitoring system. In: IEEE Int. Conf. Multimedia Expo., vol. 3, pp. 2119–2122 (2004)
[13] Tai, J., Tseng, S., Lin, C., Song, K.: Real-time image tracking for automatic traffic monitoring and enforcement applications. Image and Vision Computing 22(6), 485–501 (2004)
[14] Coifman, B., Beymer, D., McLauchlan, P., Malik, J.: A real-time computer vision system for vehicle tracking and traffic surveillance. Transportation Research Part C: Emerging Technologies 6, 271–288 (1998)
[15] Weng, S.-K., Kuo, C.-M., Tu, S.-K.: Video object tracking using adaptive Kalman filter. J. Vis. Commun. Image Represent. 17, 1190–1208 (2006)
[16] Behrens, T., Rohr, K., Stiehl, H.S.: Robust segmentation of tubular structures in 3-D medical images by parametric object detection and tracking. IEEE Transactions on Systems, Man, and Cybernetics, Part B: Cybernetics 33, 554–561 (2003)
[17] Veenman, C.J., Reinders, M.J.T., Backer, E.: Resolving Motion Correspondence for Densely Moving Points. IEEE Trans. Pattern Analysis and Machine Intelligence 23(1), 54–72 (2001)
[18] Venables, W.N., Ripley, B.D.: Modern Applied Statistics with S, 4th edn. Springer, Heidelberg (2002)
[19] Ullman, S.: The Interpretation of Visual Motion. MIT Press, Cambridge, Mass (1979)
[20] Jenkin, M.R.M.: Tracking Three Dimensional Moving Light Displays. In: Proc., Workshop Motion: Representation Contr., pp. 66–70 (1983)
[21] Barnard, S.T., Thompson, W.B.: Disparity Analysis of Images. IEEE Trans. Pattern Analysis and Machine Intelligence 2(4), 333–340 (1980)
[22] Kuhn, H.: The Hungarian method for solving the assignment problem. Naval Research Logistics Quart. 2, 83–97 (1955)
[23] Rangarajan, K., Shah, M.: Establishing Motion Correspondence. Computer Vision, Graphics, and Image Processing 54(1), 56–73 (1991)
[24] Veenman, C.J., Reinders, M.J.T., Backer, E.: Resolving Motion Correspondence for Densely Moving Points. IEEE Trans. Pattern Analysis and Machine Intelligence 23(1), 54–72 (2001)
[25] Rosales, R., Sclaroff, S.: 3d trajectory recovery for tracking multiple objects and trajectory guidedrecog-nition of actions. In: IEEE Conference on Computer Vision and Pattern Recognition (CVPR), pp. 117–123 (1999)

[26] Isard, M., Blake, A.: Condensation - conditional density propagation for visual tracking. Int. J.Comput. Vision 29(1), 5–28 (1998)
[27] Rasmussen, C., Hager, G.: Probabilistic data association methods for tracking complex visualobjects. IEEE Trans. Patt. Analy. Mach. Intell. 23(6), 560–576 (2001)
[28] Hue, C., Cadre, J.L., Prez, P.: Sequential montecarlo methods for multiple targettracking anddata fusion. IEEE Trans. Sign. Process. 50(2), 309–325 (2002)
[29] Cox, I.J.: A review of statistical data association techniques for motion correspondence. Int. J. Comput. Vision 10(1), 53–66 (1993)

Argo Vehicle Simulation of Motion Driven 3D LIDAR Detection and Environment Awareness

Mohammad Azam Javed, Jonathan Spike, Steven Waslander,
William W. Melek, and William Owen

Mechanical & Mechatronics Engineering Department, University of Waterloo, Waterloo
majaved@engmail.uwaterloo.ca, jspike@engmail.uwaterloo.ca,
stevenw@uwaterloo.ca, bowen@mecheng1.uwaterloo.ca,
wmelek@mecheng1.uwaterloo.ca

Abstract. This paper presents a method to detect three dimensional (3D) objects using an occupancy grid based mapping technique. This paper adapts the 3D occupancy grid based mapping technique from the two dimensional (2D) occupancy grid based mapping technique commonly used in Simultaneous Localization and Mapping applications. The 3D occupancy mapping technique uses a 3D inverse measurement model and has been developed for a LIDAR based off-road ground rover vehicle that drives on 3D terrain. The technique is developed and simulated in MATLAB to demonstrate its 3D object detection capabilities. This technique was developed as part of off-road autonomous vehicle research being conducted at the University of Waterloo.

1 Introduction

This paper is part of an ongoing research at the University of Waterloo on a six-wheeled autonomous off-road vehicle (called an 'Argo vehicle'). The focus of the research is to equip various sensors onto the vehicle, with the primary sensory equipment being a Light Detection and Ranging (LIDAR) sensor. Through the use of the LIDAR (or a 3D LIDAR), the autonomous vehicle acquires a spatial map of its environment while simultaneously localizing itself to this model [1]. This problem is known as a SLAM or Simultaneous Localization and Mapping problem.

The 2D version of the algorithm is more popular for use with mobile robotics for 2D mapping and navigation applications. The 2D occupancy grid based mapping algorithm is described as a framework that supports incremental environment discovery, more agile and robust sensor interpretation methods, and composition of information from multiple sensors and over multiple positions of the robot [2]. There are also many systems that rely on global occupancy grid maps for global path planning and navigation ([3],[4]). Occupancy maps were initially developed using sonar sensors [5], laser rangers [6] and stereo vision ([4],[7]). Moreover, they have been built using forward and inverse sensor models [4]. Furthermore, Burgard et al. [8] explored building occupancy grid maps while simultaneously localizing the robot [4]. Lastly, a 3D occupancy mapping algorithm was first developed by Moravec and Martin [9].

The contributions presented in this paper relate to mapping 3D surfaces and objects through the use of a 3D inverse measurement model for occupancy grid based mapping for a ground rover vehicle using a 3D LIDAR model in a MATLAB environment. The 2D occupancy grid based mapping technique has been extended in order to accommodate detection of 3D surfaces and objects.

2 Background

This section reviews some of the basic concepts pertaining to the occupancy grid based mapping Algorithm.

2.1 Grid-Based Mapping

Occupancy grid based mapping is useful for producing consistent maps from noisy measurement data [10]. The algorithm assumes that the robot pose is known [10]. This technique proves its worth in post-processing after the SLAM problem has been solved through another method and the exact pose of the robot is available [10].

The basic premise behind the occupancy grid based mapping technique is to calculate the following probability over the given map data [10]:

$$p(m \mid y_{1:t}, x_{1:t}). \tag{1}$$

where m represents the map, the $y_{1:t}$ term is the set of all measurements up to time t, and $x_{1:t}$ represents the robot's path (the cumulative sequence of its poses) [10]. Quite simply, the occupancy grid framework breaks up its map into equally spaced grid cells [10].

When the robot moves through a map, the probability of a grid cell containing an obstacle is updated depending on the robot's location and sensor measurement model.

A 3D surface is used in the simulation environment for the Argo model to maneuver on. For the purposes of the grid based mapping technique, the 3D surface is defined as a location based global map. The obstacles and 3D terrain contained in the map are defined by a large 50 x 50 x 50 matrix m:

$$m^{ijk} \in [0, 1]. \tag{2}$$

In the instance that an object exists at a location m, a value of 1 is assigned to that grid cell. If no obstacle is encountered in the grid cell, m is assigned a value of 0.

2.2 Algorithm Assumptions

There are a number of key assumptions imposed for the algorithm to function effectively. The key assumptions are that the environment is static, the vehicle state at each time step is known, the sensor measurement model is also known, and the grid cells in the global map are independent (hence the probabilities assigned to one grid cell does not affect other grid cells).

3 3D Occupancy Grid Based Mapping Algorithm

The 3D occupancy grid based mapping algorithm is divided into four parts:

1. Save range distance data for all obstacles in robot's Field of View (FOV)
2. Perform inverse measurement calculation
3. Ensure the local inverse measurement coordinate system is within bounds
4. Perform log-odds ratio on the global coordinate system

3.1 Saving Range Distance Data for Obstacles in FOV

This part of the algorithm returns a 2D matrix which contains the range distance to all of the obstacles at a specific bearing (φ^{LIDAR}) that exists in the x-y plane and a specific elevation (γ^{LIDAR}) that exists in the x-z plane:

$$\varphi^{LIDAR} = [-\varphi_{max}^{LIDAR} \ldots \varphi_{max}^{LIDAR}]. \tag{3}$$

$$\gamma^{LIDAR} = [-\gamma_{max}^{LIDAR} \ldots \gamma_{max}^{LIDAR}]. \tag{4}$$

These bearings and elevations are saved in the vectors illustrated in Equation 3 and Equation 4, and compose the robot's complete FOV. They also serve as a model for the 3D LIDAR's scanning area.

For example, the 3D LIDAR's scanning area in the x-y plane is 180°. This scan angle is subdivided into 5° increments, where each such increment represents a unique bearing angle in the vector. Similarly, the 3D LIDAR's scanning range in the x-z plane is 50°. Once again, this angle is divided into 5° increments, where every increment represents a unique elevation angle in the vector. Moreover, each specific bearing angle in the x-y plane has an associated set of elevation angles contained in the vector. Figure 1 presents the subdivided angles from the two described views.

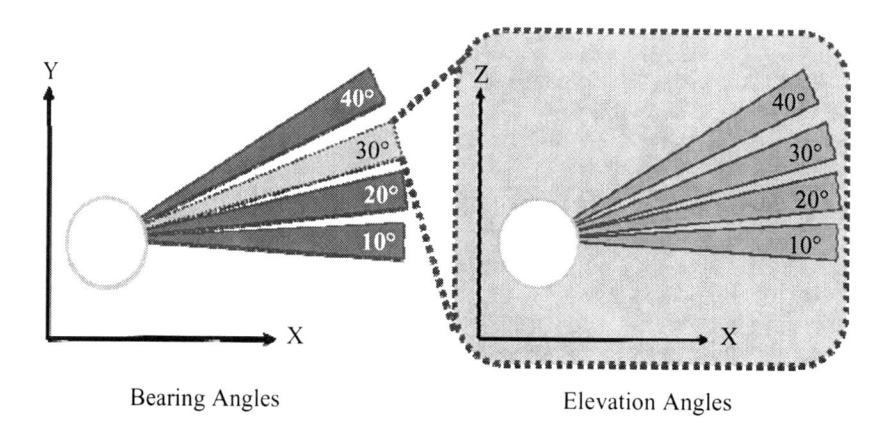

Fig. 1. Illustration of how a set of elevation angles are associated with a specific bearing angle

With the concept of the bearing and elevation angles clarified, a quick step-by-step overview of the range distance function is illustrated as follows:

```
function range_distance (φLIDAR,γLIDAR, rø,γ,Rot. Mtx.,
Trans. Mtx., Global Mtx.)
1. For i = 1 : φLIDAR
    2. For j = 1 : γLIDAR
        3. For j = 1 : rø,γ
            4.  x = rø,γ·cos(φLIDAR)
            5.  y = rø,γ·sin(φLIDAR)
            6.  z = rø,γ· sin (γLIDAR)
            7.  Multiply x,y,z by Rotation Mtx.
            8.  Add x,y,z to TranslationMtx.
            9.  Ensure x,y,z are inside global bounds
            10. If in bounds and obstacle exists at x,y,z
                11. Return rø,γ^LIDAR
        12. End for
    13. End for
14. End for
```

The range distance function presents the basic steps behind obtaining a 2D matrix which contains range distance data, and is represented by the variable $r_{k,j}^{LIDAR}$.

The function is initiated by traversing along a specific bearing angle, through a set of elevation angles that increment from negative to positive. Equation 5 details the sensor range limits for the 3D LIDAR sensor:

$$r_{\emptyset,\gamma} \in [0, r_{max}]. \tag{5}$$

where r_{max} is the sensor's maximum range. Lines 4-6 of the function describe how every grid cell along every scan line is identified in 3D space. Lines 7-8 rotate and translate the grid cell location to match it to the robot's position on the global coordinate system. Line 9 describes how the global boundaries are checked against the global grid cell location. If this grid cell is out of bounds, further looping is discontinued, and iteration through the next elevation angle is initiated. Lines 10-11 show details on returning an obstacle range value in the 2D matrix. Moreover, detection of the surface counts as a detection of an obstacle and the range value associated with it is saved in the $r_{\emptyset,\gamma}^{LIDAR}$ matrix. Equation 6 shows the form of the matrix:

$$r_{\varphi,\gamma}^{LIDAR} = \begin{bmatrix} r_{-\varphi_{max},-\gamma_{max}} & \cdots & r_{-\varphi_{max},\gamma_{max}} \\ & \vdots & \\ r_{\varphi_{max},-\gamma_{max}} & \cdots & r_{\varphi_{max},\gamma_{max}} \end{bmatrix}. \tag{6}$$

The occupancy grid based mapping technique requires the use of the inverse measurement model.

The reason this probability is called inverse is because it reasons from effects to causes – hence, it provides information about the world conditioned on a measurement that is caused by the world [10].

The inverse measurement algorithm iterates through all of the cells in a local coordinate system, identifies the cells that are in the LIDAR's FOV, and assigns a high or low probability to those cells depending on the existence of an obstacle. The use of Bresenham's line algorithm could substantially improve this process – see conclusions for a detailed description regarding this.

3.2 Performing Inverse Measurement Calculation

The 3D inverse measurement algorithm is presented as:

```
function inverse_measurement
```
$(x_{1,t}, x_{1,t}, x_{1,t}, \emptyset^{LIDAR}, \gamma^{LIDAR}, r_{max}, r_{k,j}^{LIDAR}, \alpha, \beta, \zeta)$

1. Let m_x^{ijk}, m_y^{ijk}, m_z^{ijk} be the center of mass of m^{ijk}
2. $r^i = \sqrt{(m_x^{ijk}-x_{1,t})^2 + (m_y^{ijk}-x_{2,t})^2 + (m_z^{ijk}-x_{3,t})^2}$
3. $\emptyset^i = \tan^{-1}\left(\frac{m_y^{ijk}-x_{2,t}}{m_x^{ijk}-x_{1,t}}\right)$
4. $\gamma^i = \tan^{-1}\left(\frac{m_z^{ijk}-x_{3,t}}{m_x^{ijk}-x_{1,t}}\right)$
5. $k = \text{argmin}(\emptyset^i - \emptyset^{LIDAR})$
6. $j = \text{argmin}(\gamma^i - \gamma^{LIDAR})$
7. if $r > \min\left(r_{max}, r_{k,j}^{LIDAR} + \frac{\alpha}{2}\right)$ or $|\emptyset^i - \emptyset_k^{LIDAR}| > \frac{\beta}{2}$ or $|\gamma^i - \gamma_j^{LIDAR}| > \frac{\zeta}{2}$
8.   then no info-cell probability remains 0.5
9. else if $r_{k,j}^{LIDAR} < r_{max}$ and $| r^i - r_{k,j}^{LIDAR} | < \frac{\alpha}{2}$
10.   then high probability-cell probability is at 0.9
11. else if $r^i < r_{k,j}^{LIDAR}$
12.   then low probability-cell probability at 0.1
13. end if

In the inverse measurement function presented, Line 2 captures the range distance from the robot's current position to the marginalized grid cell. Next, Line 3-4 calculate the bearing and elevation angles from the robot to the center of mass of the current grid cell. Lines 5-6 find the closest LIDAR bearing and elevation angles to the angles of the grid cell in an attempt to discern if the grid cell is inside the LIDAR's FOV. Lines 7-8 evaluate whether the grid cell is outside of the LIDAR's FOV – in the case that this is true, the probability of 50% is assigned to the grid cell. Line 9-10 verify whether or not an obstacle is detected on the outer edge of the LIDAR's FOV. If the latter is true, then the grid cell containing that obstacle is assigned a 90% probability of an obstacle existing at that grid cell. This 90% probability value is a hypothetical estimate (can be adjusted as needed) for the confidence that one should have in the LIDAR's readings. Lines 11-12 ensure that the current grid cell is inside the LIDAR's FOV without obstacles. In this case, a 10% probability is assigned to the grid cell for an obstacle to exist in the grid cell – this value can also be changed as needed.

3.3 Ensure Local Inverse Measurements are within Bounds

Once the inverse measurement algorithm traverses through all of the local grid cells, a matrix containing all grid cells with assigned probabilities is returned. This matrix is in the local coordinate system and has to be rotated and translated to match up with the global coordinate system.

Figure 2 illustrates the simple procedure that converts the returned matrix from the local coordinate system to the global one.

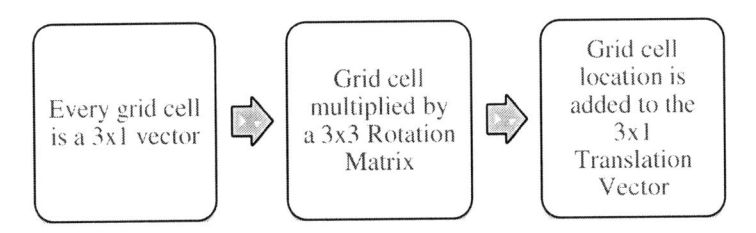

Fig. 2. Illustration of procedure followed to convert the local three dimensional probability matrix into the global format

With the locally updated section converted to a global matrix, all areas that are not updated from the sensor are set to a 50% probability. This indicates that no new information is contained in grid cells not detected by the LIDAR sensor.

3.4 Performing Log-Odds Ratio on the Global Coordinate System

The last step in the 3D occupancy grid based algorithm is to apply the log odds ratio because we want to track the log odds ratio of each cell, not the probability of each cell. The main advantage of using the log odds ratio over probabilities is in dealing with high and low probability discrete states. The truncation errors seen with multiplicative combinations of probabilities are avoided because the update rule for the log odds ratio involves addition.

Once the probability matrix is transformed into the global form, the log odds ratio is applied in the global sense to update the current belief of the global system with the measured readings.

$$[L_{new}] = [L_{old}] + \log\left(\frac{Inv_{msmt_{element}}}{1 - Inv_{msmt_{element}}}\right) - [L_0] . \qquad (7)$$

$$[Prob_{map}] = \frac{e^{\lfloor L \rfloor}{}_{element}}{1-e^{\lfloor L \rfloor}}. \tag{8}$$

Equation 7 and 8 show the calculations that pertain to the applied log odds ratio in the global sense.

Once the log odds calculations are completed and the current belief of the global obstacle map is updated, the probabilities of the cells surrounding the Argo vehicle are updated and plotted. Subsequently, the entire algorithm is then repeated.

4 Simulation Parameters and Environment

To simplify the development of the simulation environment, a simple kinematic model is implemented with the intention of obtaining motion, not necessarily real life motion. The assumed kinematic model used to estimate the skid-steer system behaves much like a two wheeled robotic system (see [10]).

The six-wheeled Argo vehicle is a skid-steer off-road vehicle. A skid steer vehicle can make turns in one spot by throttling one side of the vehicle's wheels forward, and engaging the brakes of the wheels on the other side to complete the turn. Upon investigation of other literature based on the Argo vehicle and other skid-steer systems, we found that the complex and highly dynamic nature of the vehicle mechanism limited the effectiveness of modeling such systems for a simulation more concerned with object detection. These complexities extended into a nonlinear drive system (see [11]). Furthermore, the actual vehicle motion is explored in other articles ([12], [13]), leading to the belief that skid-steer systems have highly dynamic ground integration that extend beyond the scope of the intended simulation configuration.

The 3D surface the Argo model moves on is defined using simple shape functions and is loaded from a file. The simulated surface can be replaced with an alternative one easily and the continuously updates its current perception of the world, replacing it with every iteration that the simulation runs.

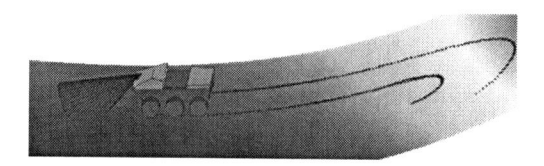

Fig. 3. Argo model driving on a hilled portion of the 3D surface

For simulation purposes, the 3D LIDAR model has a scanning angle that consists of a bearing angle of 180°, and an elevation angle of 50°, with a measuring range that is open to adjustments. Additionally, one grid cell in the simulation environment is equivalent to one meter in reality.

Figure 4 is obtained by expanding sensor's FOV to include the area above the Argo. This is done for illustrative purposes only. In this figure, colored dots indicate detected free space, where magenta indicates 70% or higher confidence in free space, and cyan indicates 90% or higher confidence in free space.

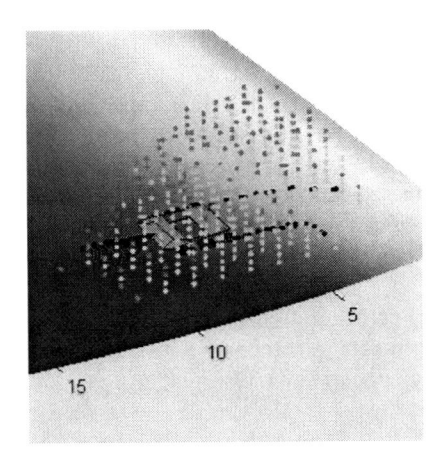

Fig. 4. 3D Occupancy grid based mapping assigning obstacle probabilities while the Argo travels on a 3D surface

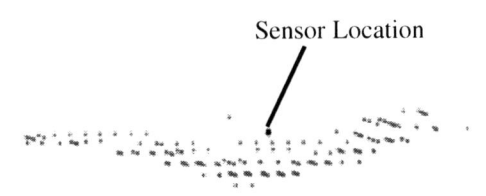

Fig. 5. 3D surface detection using 3D occupancy grid based mapping

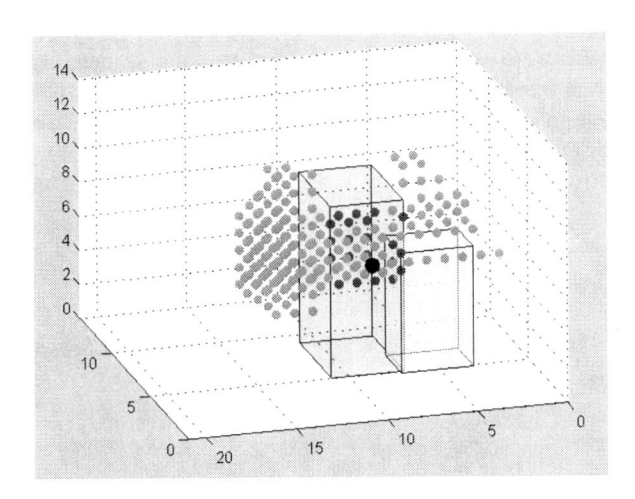

Fig. 6. 3D LIDAR scanning area – the sensor (black dot) detecting two objects

The 3D LIDAR measurements collected by the moving Argo model are setup to be affected by the vehicle's 3D orientation through the use of local and global coordinates. In addition, the simulated 3D LIDAR is capable of detecting the 3D surface as an obstacle (Figure 5), as well as any other external shape located on the 3D surface (Figure 6).

Figure 5 illustrates the detection of the 3D surface when the Argo model is static. The Argo model is not shown but the location of the 3D LIDAR is shown.

Figure 6 illustrates two obstacles being detected by the sensor (black dot). The green dots represent empty grid cells and the red dots represent cells that have been detected as obstacles. For illustration purposes, the sensor has been located in the middle of the local coordinate system with a shorter scanning range (7 meters/7 grid cells), and scanning angle (bearing of 120°, and an elevation of 50°).

5 Conclusion

This paper presents the development of a 3D occupancy grid based mapping technique using a 3D inverse measurement model and discusses the simulated results for the technique. The simulation environment is configured to model the motion of a six-wheeled Argo vehicle equipped with a 3D LIDAR sensor. The model of the Argo uses 3D occupancy mapping to attain an accurate map of 3D objects and surfaces in the vicinity of the vehicle.

The simulation environment is developed in a modular manner and many future extensions can be added to the environment. It is important to note that the use of the Bresenham line algorithm could greatly improve the efficiency of the 3D grid based mapping algorithm developed in this paper, and this is one area for future extensions. Use of the Bresenham line algorithm would prevent the 3D occupancy algorithm from traversing through all voxels in the local map. Since line algorithms like Bresenham's generate a series of discrete pixels by selecting the next pixel from two adjacent neighbors of the current pixel, the 3D grid based mapping technique could be changed to only update the belief of the pixels in contact with or encased by the pixels that are inside the scanning area of the 3D LIDAR [14]. As such, unnecessary computation can be avoided.

Other extensions could include the integration of path planning algorithms as well as localization algorithms.

Acknowledgements

We wish to acknowledge the review and helpful comments of Aseel Al Dallal, and Mohammad Biglarbegian.

References

1. Siciliano, B., Khatib, O.: Springer Handbook of Robotics. Springer, Heidelberg (2008)
2. Elfes, A.: Using Occupancy Grids for Mobile Robot Perception and Navigation. IEEE Computer 6, 46–57 (1989)

3. Kortenkamp, D., Bonasso, R.P., Murphy, R.: AI-Based Mobile Robots: Case Studies of Successful Robot Systems. American Association for Artificial Intelligence, Menlo Park (1998)
4. Thrun, S.: Learning Occupancy Grids with Forward Models. In: Proceedings of the Conference on Intelligent Robots and Systems (2001)
5. Moravec, H.P., Elfes, A.: High Resolution Maps from Wide Angle Sonar. In: Proceedings of IEEE International Conference on Robotics and Automation, pp. 116–121 (1985)
6. Thrun, S., Fox, D., Burgard, W.: A Probabilistic Approach to Concurrent Mapping and Localization for Mobile Robots. Machine Learning 31, 29–53 (1998)
7. Buhmann, J., Burgard, W., Cremers, A.B., Fox, D., Hofmann, T., Schneider, F., Strikos, J., Thrun, S.: The Mobile Robot Rhino. AI Magazine 16 (1995)
8. Burgard, W., Fox, D., Jans, H., Matenar, C., Thrun, S.: In: Proceedings of the International Conference on Machine Learning (1999)
9. Moravec, H.P., Martin, M.C.: Robot Navigation by 3D Spatial Evidence Grids. Internal Report, Carnegie Mellon University (1994)
10. Thrun, S., Burgard, W., Fox, D.: Probabilistic Robots. Massachusetts Institute of Technology Press, Cambridge (2006)
11. Tran, T.H., Ha, Q.P., Grover, R., Scheding, S.: Modeling of an Autonomous Amphibious Vehicle. ARC Centre of Excellence in Autonomous Systems (2006)
12. Yu, W., Chuy, O., Collins, E.G., Hollis, P.: Dynamic Modeling of a Skid-Steered Wheeled Vehicle with Experimental Verification. In: Proceedings of the IEEE/RSJ International Conference on Intelligent Robots and Systems, pp. 4212–4219 (2009)
13. Lucet, E., Grand, C., Salle, D., Bidaud, P.: Dynamic Sliding Mode Control of a Four-Wheel Skid-Steering Vehicle in Presence of Sliding. In: 17th CISM-IFToMM Symposium on Robot Design, Dynamics, and Control (2008)
14. Cohen-Or, D., Kaufman, A.: 3D Line Voxelization and Connectivity Control. IEEE Computer Graphics and Applications 17, 80–87 (1997)

Signal Processing and Pattern Recognition for Eddy Current Sensors, Used for Effective Land-Mine Detection

Hendrik Krüger and Hartmut Ewald

University of Rostock, Institute of General Electrical Engineering,
Justus-von-Liebig-Weg 2, 18059 Rostock, Germany
{hendrik.krueger,hartmut.ewald}@uni-rostock.de

Abstract. Magnetic and inductive sensors are widely used in research and industry for a variety of applications, e.g. for geophysical prospecting, non-destructive testing (NDT) of materials, in the food industries, for distance and proximity sensing, security systems as well as for landmine detection. Metal detectors (MD), based on the eddy current principle, are the most used systems in humanitarian demining. Their main disadvantage is the high false alarm rate, caused by harmless metal objects and "uncooperative" soils with magnetic properties. The sensor signal (i.e. the induced complex coil voltage) is influenced by the object properties (material, shape). This paper describes an object recognition based on multi-parameter MD signals, which are classified by the fuzzy method. The ability to identify mines by their characteristic signature was demonstrated in test lanes for mine detection provided by the University of Rostock (Germany), JRC-Ispra (Italy) and the CTRO-Benkovac (Croatia).

Keywords: Metal detection, fuzzy classification, multivariate data analysis.

1 Introduction

In recent years, there has been an increase in the application of magnetic and inductive sensing in many areas. However, for many applications advanced signal processing and pattern recognition systems are necessary, e.g. in nondestructive testing [1], [2] (determination of the width and depth of a surface crack or welding seam recognition) and in humanitarian landmine detection [3], [4].

Metal detectors, based on the eddy current principle, are widely used in humanitarian demining, but their main disadvantage is the high false alarm rate (up to 1000 per mine) caused by harmless clutter or by "uncooperative" soils.

In most cases, the types of mines in the field are known a-priori. So their influence on the metal detector signal can be measured (without soil) and stored in a signature database. This database is used in our approach to lower the false alarm rate by optimized soil compensation and object classification for continuous-wave metal detectors. Nevertheless, the methods are also transferable to pulse inducting systems.

In general, magnetic and/or inductive sensors generate one, or multi-dimensional, signals while the probe is moving over the test piece or the ground. The absolute value of the sensor signal (e.g. secondary voltage of the coil) is of low interest because the value depends on soil properties, the coil geometry etc. The change of the

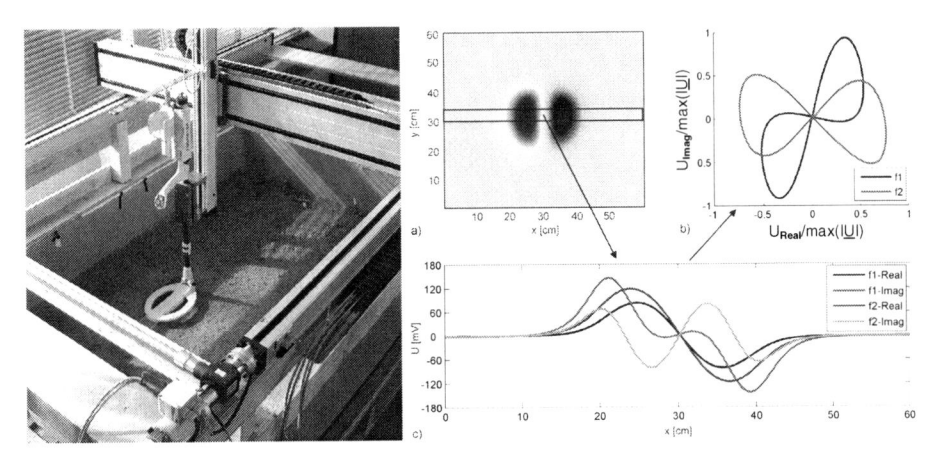

Fig. 1. Mine M2B 5 cm below the sensing coil, acquired with a 3-axis scanner: a) 2D-Scan $U_{Re\_f1}(x,y)$, b) phase-loop plot, c) measured signals $U(x)$ for f_1 = 2,4 kHz and f_2 = 19,2 kHz

voltage due to moving the coil above the suspected area with buried metal objects gives all the information that is needed. Thus in the first stage of any MD operation the secondary voltage has to be compensated or 'balanced'. The second stage has to be feature-extraction, followed by 'adapted' classification.

In general, if we measure the metal detector signals in a 2D plane over the soil we can image each MD Signal (real and imaginary part of the complex coil voltage per excitation frequency) as a 2D intensity distribution (Fig. 1a). Because of the wide aperture function (point spread function) of the sensing coil compared to the size of the metal components of anti-personnel mines, the images are typical for the sensing coil but not for the different targets. Another representation is obtained by plotting the signal components of a scanned line over the maximum of the 2D signature (Fig. 1c). Related to this, we can also plot pairs of signal components (i.e. the real vs. imaginary part of the complex voltage) in an xy-plane to get the so-called phase-loops, which are typical for different objects (Fig. 1b). Reasons for the differences are the size, the electrical and magnetic material properties and also the metal distribution in the mine.

The butterfly shape of the phase loops is always present if we use a differential (double-D) coil. But the phase angle is related to the conductivity and permeability of the metal object.

Today, there are only a few metal detectors on the market (i.e. AKA Vector) which assist the mine searcher with a phase-loop visualization. The main disadvantage is that these signatures have to be interpreted by an experienced user. Furthermore, the shape of the phase-loop changes completely in the presence of "uncooperative" soils with magnetic properties.

2 Compensation of "Uncooperative" Soils

Uncooperative soils are inhomogeneous magnetic permeable or electrical conductive soils, which influences the metal detector like a piece of metal. The false alarm rate increases significantly in presence of such soils.

If the detector provides N linear independent signal components $c_{1..N}$, (i.e. L and R or U_{Real}, U_{Imag} from multiple frequencies ω_i) it is well known from eddy-current testing, that up to N-1 components can be used to suppress N-1 disturbance values by a linear combination of all components c. Adapted from the old concept of Libby [5], in [4] and [6] an enhanced method (based on singular value decomposition) is described, that uses a-priori information about the types of mines in the field (which are known in most cases) to reduce the soil influence regarding the mine-to-soil signal ratio. This amplitude ratio is calculated as the quotient of the signal preserved from the mine divided by the amplitude remaining from the soil after the soil compensation.

The soil compensation of four MD signals $c_{1..4}$ gives us up to three new signals $c'_{1..3}$, which can be plotted in a phase-loop representation (Fig. 2).

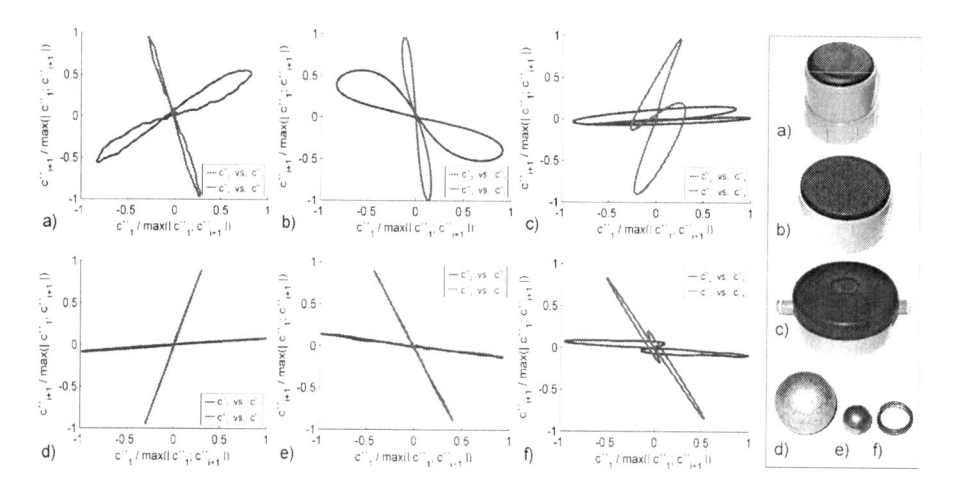

Fig. 2. Soil compensated signatures, ("uncooperative" laterite soil, depth: 5 cm), mine surrogates: a) M1A (low metal content), b) M2B (high metal content), c) M3B (high metal content) und clutter: d) aluminum sphere Ø 28 mm, e) steel ball Ø 5 mm, f) ring of steel Ø 11 mm

The shapes of these signatures are not disturbed by the soil influences however, because of the soil compensation, the new signatures are not comparable with those generated from raw data. Therefore the same soil compensation must be applied to the reference signatures (signature database, mines measured in air without soil influences) before feature extraction (Fig. 3).

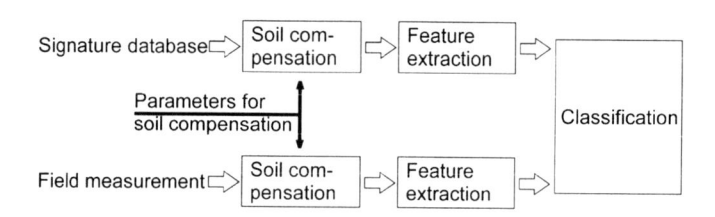

Fig. 3. Strategy for the classification of disturbed measurements using soil compensation and a signature database

3 Feature Extraction and Classification

The methods of feature extraction and the chosen structure of the classification system are related to the problem which has to be solved. All well known methods of pattern recognition such as phase and level detection of 1D data, classical methods of cluster analysis of the multi frequency data sets and Artificial Neural Network (ANN) including Support Vector Machine (SMV) as well as Fuzzy Logic methods (FL, Fuzzy clustering and rule based) can be used. An application of a rule-based fuzzy classifier for pattern recognition in case of one, two and three dimensional magnetic leakage flux sensor and an eddy-current system has been presented [1], [2], [8]. In case of a landmine classification system advanced signal processing methods using time and high spatial resolved metal detector data are necessary: Identification of multi target situation, a feature-preserving soil compensation strategy and a data-based pattern recognition system [4], [6], [7].

From the point of signal processing, 1D-, 2D- and the 3D-signal processing algorithms will be able to extract or 'detect' much more information from the metal detector data.

After the soil compensation has been applied to both the in-air signatures and the field measurement of an unknown object an experienced user can compare signature plots of them with the mine database. However, the real challenge is an automatic object classification. Therefore we have to extract features from the phase-loop representation which are also stable in the case of noisy signatures to include classification of objects near the limit of the sensing range. In the following we use geometrical features, which are:

- the amplitude of the voltage U_{max}.
- the phase values α_i of each phase-loop and
- the normalized loop width b_i defined by the ratio of signature width and length.

While the first feature is strongly related to the distance of sensing coil to the object, the amplitude is used for the depth estimation if the object is classified as a mine. This classification is based manly on the phase and loop width features.

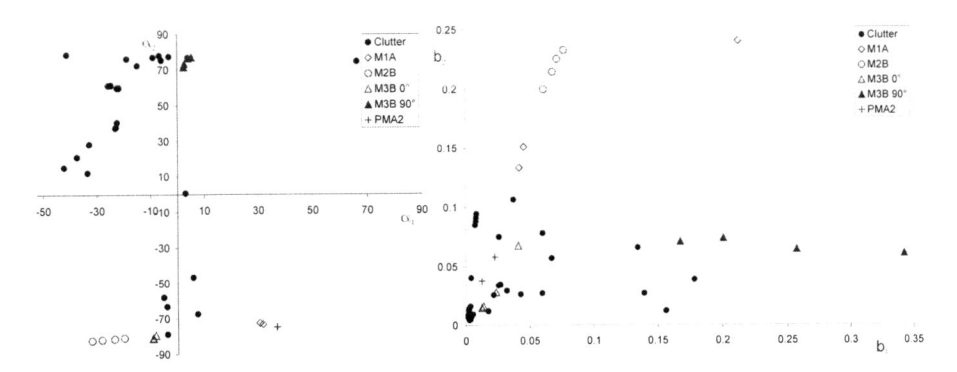

Fig. 4. Feature values after soil compensation: phase (left) and width (right) of 4 mines in 2.5, 5, 10, 15 cm depth and 24 clutter objects. ("uncooperative" laterite soil).

It can be seen from Fig. 4, that the phase values of each mine in different depths are localized, while most clutter objects are outside this area. The loop width features are less concentrated. A large value of b_i is typical for objects with spatial distributed metal components. While a small value b_i (for which the phase-loop is similar to a straight line) indicates a small single piece of metal. But in both cases, b_i increases because of measurement noise for larger depths. Although large b_i values give no useful information, a small b_i value can exclude a mine with multiple metal components very effectively.

Using the fuzzy method, we can introduce such knowledge about the significance of the different features to the fuzzy classifier.

3.1 Signal Adapted Fuzzy Membership Functions

The membership degree of an object to a fuzzy set in dependency of a feature value x is defined by a membership function $\mu(x)$. Unlike the Boolean logic, the membership is defined continuously over the range $0 \leq \mu \leq 1$.

Because the dimension of the feature space and the feature variability depends on the mine type, depth and the soil compensation parameters, an automatic adaptation of the membership functions is needed. Therefore we use membership functions in π - and S-form (Fig. 5 a, b).

The π–form is a good choice for the phase angle and the amplitude features while for the loop width the S-form is preferred because a mine cannot be excluded by a large loop width, which is also typical for a noisy measurement.

To each signature in the database (mine type i and depth j) a membership function for each feature x_k is defined. The form (slope and plateau) of each membership function $\mu_{i,j,k}$ is adapted with help of the signature database. The function is centered on x_0, which is the value of a feature x extracted from a signature stored in the mine database. The plateau Δx is given by the maximum distance of the feature value x_0 to that of the same mine with a depth one step below or above. The slope area is defined as $0.5*\Delta x$. All membership functions of the feature x for one mine at different depths combined show a flat top with $\mu=1$ (Fig. 5c).

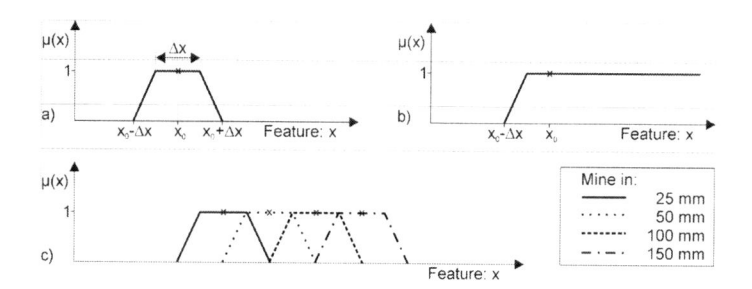

Fig. 5. Membership functions as a) π-form, and b) S-form. c) Superposed membership functions (π -form) for a mine in different depths

3.2 The Fuzzy Classifier

To classify the signature of an unknown object, the membership values $\mu_{i,j,k}(x_k)$ are calculated for the extracted features $x_1,\ldots,x_K$.

Afterwards, all feature-related membership values $\mu_{i,j,1\ldots K}$ of one mine type and depth are combined with the interference value $y_{i,j}$ by a fuzzy AND operation (1).

$$y_{i,j}(\mathbf{x}) = \mu_{i,j,1}(x_1) \wedge \ldots \wedge \mu_{i,j,K}(x_K) = \min\left[\mu_{i,j,1}(x_1),\ldots,\mu_{i,j,K}(x_K)\right]. \quad (1)$$

Whilst the interference values $y_{i,1..J}$ of one mine in all depths $j=1..J$ a fuzzy OR operation is applied.

$$y_i(\mathbf{x}) = y_{i,1}(\mathbf{x}) \vee \ldots \vee y_{i,J}(\mathbf{x}) = \max\left[y_{i,1}(\mathbf{x}),\ldots,y_{i,J}(\mathbf{x})\right]. \quad (2)$$

The interference value y_i gives the fuzzy conclusion if the feature set $\mathbf{x}$ belongs to the i-th mine type. The class of unknown clutter objects is expressed by an exclusion of all mines (3).

$$y_0(\mathbf{x}) = 1 - \max_{i=1..I}\left[y_i(\mathbf{x})\right]. \quad (3)$$

The value y_i ($i=0\ldots I$) is defuzzyfied in order to obtain a crisp decision (mine or not a mine of type i or clutter). Therefore the maximum-method is used on all interference values y_i.

4 Results

To quantify the reduction of the false alarm rate, we took measurements on the test fields for mine detection (University of Rostock and JRC Ispra). Over-all we captured measurement data of surrogate mines (M1A, M2B, M3B and PMA2) and clutter objects in 6 different soils.

4.1 Test Field from the University of Rostock (Germany)

The test field in Rostock consists of three containers with a precise 3-axis scanner for spatial referenced data acquisition (Fig. 1). The containers are filled with a cooperative sandy- and strongly uncooperative laterite soil. A third unfilled container is used to acquire the in-air-signatures of the mines in four depths (25 - 150mm) for the database.

The cooperative soil with a magnetic susceptibility of $\kappa = 30 \cdot 10^{-6}$ was taken from Fuhrberg near Hanover and the uncooperative soil with $\kappa = 43560 \cdot 10^{-6}$ from Vogelsberg-Lich. Both are filled in containers preventing their natural stratification.

Fig. 6 shows the reduction of the false alarm rate (100% means no false alarm) in both lanes related to the count of false alarms indicated by a non classifying detector. As expected the results are better if we use all provided features and furthermore, if we can reduce the number of possible mine types by a-priori knowledge.

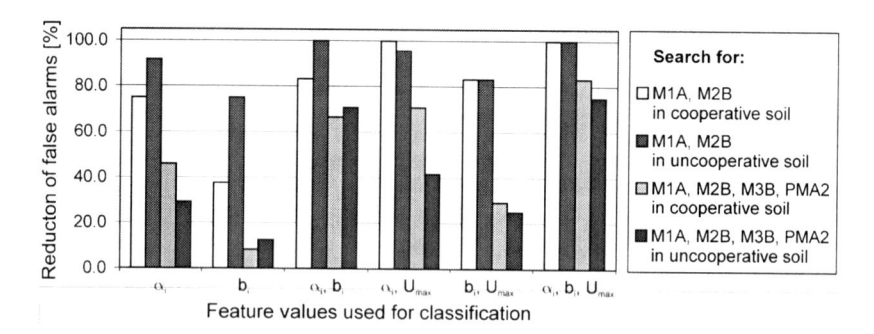

Fig. 6. Reduction of the false alarm rate by fuzzy classification, referred to a metal detector without classification: 24 clutter objects in 2 soils, feature combinations of α_i, b_i, U_{Max}.

4.2 Test Field from the JRC-Ispra (Italy)

The test lane of the Joint Research Centre in Ispra provides lanes with 7 different soils, from which we selected four [9].

- Lane 2: cleaned local soil (sand), $\kappa = 265 \cdot 10^{-6}$
- Lane 4: cleaned industrial soil (sand), $\kappa = 2475 \cdot 10^{-6}$
- Lane 5: soil from Piemond (silt), $\kappa = 145 \cdot 10^{-6}$
- Lane 7: volcanic soil from Naples, $\kappa = 5950 \cdot 10^{-6}$

The soils in lane 4 and 7 are classified as "uncooperative" because of their high susceptibility κ. But in fact only Lane 7 is "uncooperative" because soil 4 is very homogeneous and the soil effect is already well compensated by the differential coil arrangement of the metal detector.

In every lane we measured the signatures of 12 different clutter objects and 5 different mines in 4 depths from 0 to 15 cm. Previously, one exemplar of each mine type was measured under lab terms to generate the needed mine database with in-air-signatures.

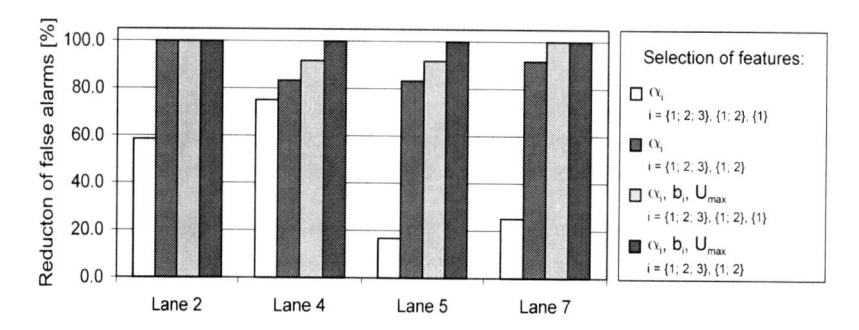

Fig. 7. Reduction of the false alarm rate by fuzzy classification, referred to a metal detector without classification: 12 clutter objects in 4 soils, feature combinations of α_i, b_i, U_{Max}

The left bares of each lane in Fig. 7 mark the result for the case in which only phase features are used. The false alarm rate is still high, because the number of usable signal components reduces for weak signatures because of the signal-to-noise ratio. That leads to a classification based on only one feature (1 phase angle) in the worst case.

Fixing the minimal number of components after soil compensation to three, the classification is based on a minimum of two phase features (2nd bars). But this method decreases also the detection probability because of the weaker soil reduction. Using instead all kinds of features (α_i, b_i and U_{max}) more than 90% of false alarms can be excluded without (3rd bar) and 100% with reduction of the detection depth (last bar).

5 Conclusions

A major problem of mine clearance using metal detectors is the high false alarm rate, which is caused primarily by soil in-homogeneity and harmless metal parts (so-called clutter). The key to the significant lowering of the false alarm rate is the usage of a-priori information about the types of mines in the field and the usage of space resolved multivariate MD data. The classification – mine of type x or clutter – of a signature, also in the presence of uncooperative soil, is possible with the use of feature-maintaining soil compensation.

From an N-parameter measurement (i.e. real and imaginary part of the voltage inducted in the sensing coil for two frequencies) up to N-1 parameter can be used to reduce N-1 disturbance influences. Therefore, the soil compensation is always a compromise between the reduction of soil influence and the number of usable parameters left for the classification. The position and spread of features in the feature space (Fig. 2) depends on the soil compensation parameters and also on the mine object itself. With help of the mine signature database, the membership functions used by the fuzzy classifier are adapted regarding to the soil compensation settings and the type of mines in the field. With π- and S-type membership functions for the different features (phase, amplitude, loop width), the fuzzy classifier reduces the false alarm rate up to 100%, if all extracted features are used. All mine signatures with a sufficient mine-to-soil signal ratio were detected without any false negative indication.

Further improvements are possible by using more than two excitation frequencies (with a reasonable spectral distance) to extract additional features and for a better soil compensation. Also the coil system should be optimized for the requirements of object classification problems by usage of a coil arrangement with a sharp rotation independent aperture function.

References

1. Ewald, H., Stieper, M.: Classification of eddy current signals using fuzzy logic and neural networks. In: Proc. SPIE, vol. 2947, p. 236 (1996)
2. Ewald, H., Leuschner, K.: Welding Seam Monitoring Using Fuzzy-Logic. In: Proceedings on the ANNIE 1999 Conference, St. Louis. Intelligent Engineering Systems Through Artificial Neural Networks, vol. 9, pp. 617–622. ASME Press, New York (1999)

3. Bruschini, C.: A Multidisciplinary Analysis of Frequency Domain Metal Detectors for Humanitarian Demining, PhD-Thesis, University Brussels (2002)
4. Krüger, H., Ewald, H.: New approach of signal processing for classification problems using a-priori information. In: Proceedings of IEEE-SENSORS 2009 - 8th IEEE Conference on Sensors, Christchurch (2009)
5. Libby, H.L.: Introduction to Electromagnetic Nondestructive Testing Methods. Wiley Interscience, New York (1971)
6. Krüger, H.: Imaging and classification of inductive metal detector data for humanitarian demining (in German: "Bildgebung und Klassifikation von Signalen induktiver Metalldetektoren zur Anwendung in der humanitären Minensuche"), PhD-Thesis, University of Rostock, published by MBV, Berlin (2010)
7. Krüger, H.: Solutions for 3D position referencing for handheld metal detectors used in humanitarian demining. Proceedings of IEEE-I2MTC 2010 - International Instrumentation and Measurement Technology Conference, Austin (2010)
8. Ewald, H.: 3-Dimensional Magnetic Leakage Field Sensor in Nondestructive Testing. In: Proceedings of IMTC-2003-Instrumentation and Measurement Technology Conference, Vail, CO, USA, pp. 1309–1311 (20-22, 2003)
9. Lewis, A.M., Bloodworth, T.J., Guelle, D.M., Littmann, F.R., et al.: Systematic Test & Evaluation of Metal Detectors (STEMD) - Interim Report - Laboratory Tests Italy (2006)

Velocity Measurement for Moving Surfaces by Using Spatial Filtering Technique Based on Array Detectors

Martin Schaeper and Nils Damaschke

University of Rostock, Institute of General Electrical Engineering, Albert-Einstein-Str. 2, 18059 Rostock, Germany
`Martin.Schaeper@uni-rostock.de`

Abstract. Spatial filtering velocimetry is an established measurement technique for process control e.g. measuring speed of moving surfaces. The technique uses optical gratings for filtering the velocity information. Usually optical transmission gratings as films or micro-lenticular arrays are used for generating the spatial filter. The article describes the spatial filtering technique by using structured detectors (CCD or CMOS-array), which reduces the optical setup significantly. The grating of the spatial filter is realized by weighting lines of pixels, possible for array detectors in different directions. After describing the technique and the realized system some measurement results are shown.

Keywords: spatial filter, structured detector, velocity measurement.

1 Introduction

Over the last decades a number of optical measurement techniques have been developed for the in-situ determination of velocities. Whereas Laser Doppler technique [1] and Particle Image Velocimetry (PIV) [2] generally used to characterize fluid velocities, the spatial filtering technique is mainly applied to surface velocity measurements. The original measurement effect, based on imaging of an inhomogeneous illuminated scene through an periodical optical grating, was already used in [3] and applied to an aerial camera system in [4]. Finally Ator 1963 [5] proposed a parallel slit reticle placed in the image plane, which is the basic configuration in spatial fitering devices. The moving spatially filtered image generates a temporally dependent oscillating signal. The signal frequency is proportional to the velocity of the imaged structure e.g. of an observed surface and can be measured by a photo detector, e.g. a standard photo diode.

Together with the development in optics and electronics a number of modifications enlarges the application field and improves the setup. By using two inverted spatial filters a offset free output signal can be generated [6]. Prism grating or lenticular arrays can be used for generating an optical grating [7]. A further development utilizes an fiber optical arrangement [8]. Beside using an optical grating and a photodiode, one dimensional spatial filter can also be realized by photodetector-arrays with specially adapted readout timing [9]. By weighting the pixels of a CCD or CMOS-array detector, a two-dimensional grating for a two dimensional spatial filter [10] can be realized. The different configurations open a wide application field, e.g. for fluid flow detection, for in plane vibration measurements, for velocity distributions of particles and for particle sizing.

Advantages of the spatial filtering technique are a robust optical setup, illumination by white light sources and ease adaptation to the process dynamics.

This article describes optical spatial filtering by using a structured array sensor.

2 Principle of Spatial Filtering Velocimetry

Optical spatial filters use the filtering effect of gratings in the optical path of an imaging system. The image of an observed scene is weighted by the spatial grating in the focal plane. The weighted information of the scene (image/grating) can be described by equation 1.

$$f(x,y,t) = b(x,y,t) \times a(x,y) \qquad (1)$$

The expression $b(x,y,t)$ represents the light-distribution of the image and $a(x,y)$ the periodic or non-periodic function of the optical grating. The spatial filtered intensity distribution $f(x,y,t)$ is added up in x- and y- direction respectively for example by using collective lenses for each direction. The result is the spatial filter signal $s(t)$ represented in equation 2.

$$s(t) = \int_{-\infty}^{+\infty}\int_{-\infty}^{+\infty} f(x,y,t) \, dx \, dy = \int_{-\infty}^{+\infty}\int_{-\infty}^{+\infty} \left[b(x,y,t) \times a(x,y) \right] dx \, dy \qquad (2)$$

By using a structured array detector instead of a single photodiode with a collecting lens the system design becomes simpler, because spatial filter and integration can be combined in the detector. For this configuration the hardware grating in the detectors path disappear. The pixels spacing of a standard CCD or CMOS sensor is already accurate enough to generate the optical spatial filter. The weighting of the pixels rows and columns (respectively to the x- and y- direction) is performed after the photo conversion in the array detector or after pixel acquisition. Therefore it becomes easy to implement different periods, grating structures or multiple gratings. Figure 1 shows the basic concept where a simple differential grating with the minimum period g_p of two pixels is used. The captured gray values of the pixels are summed up in every row and column. The generated values are weighted by the spatial filter function and the samples of the spatial filter signal $s(t)$ are generated by summation over all values for rows and columns respectively

$$s(t) = \sum_n B(n,t) A(n) \quad . \qquad (3)$$

Where n is the number of the line, $B(n,t)$ the summed gray values of each column or row and $A(n)$ the grating function.

The calculation has to be performed for each frame for each direction (x and y) to get two time dependent spatial filtering signals for x and y-direction. In case of structured detectors, orthogonal spatial filter structures are found to be the best by Bergeler [11]. The grating structures are not restricted to rectangular forms. Sinusoidal forms or even free definable non-periodic forms are also possible. So the grating function can be adapted to the inhomogeneous structure of the observed surface without changing the hardware setup of the spatial filter.

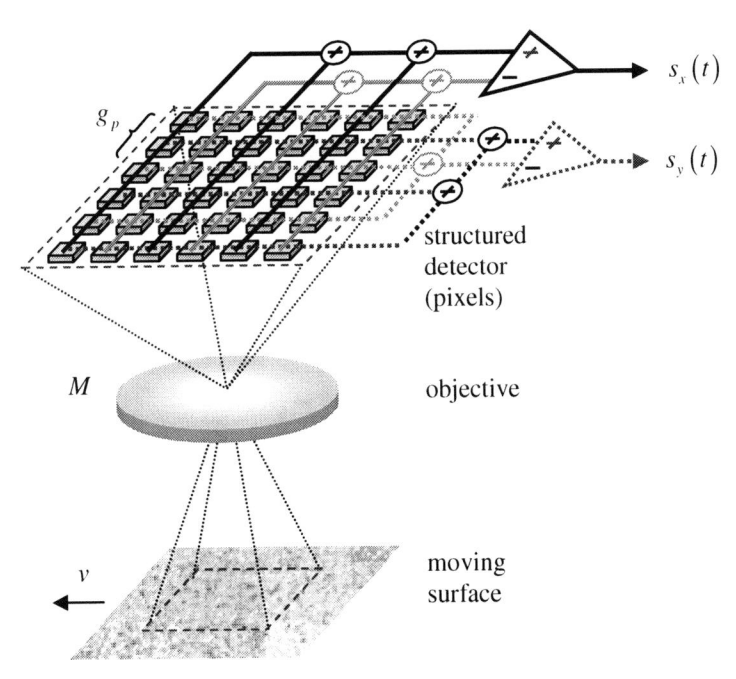

Fig. 1. Basic configuration of a spatial filtering system using a structured array detector. The shown weighting function is a differential type (+1,-1) for each direction.

The velocity v of the moving structure, e.g. an individual object or a surface, can be derived from the main frequency of the signal with respect to the magnification M of the imaging optics and the grating period g.

$$v = \frac{f_0 \, g}{M} \quad . \tag{4}$$

The grating period g is calculated by the pixel period g_p multiplied by the pixel pitch of the detector array. When using CCD or CMOS-array without shutter option, the sequential readout of the array must be considered. The so called rolling shutter induces, depending on the velocity direction, a relative compressing/dilation of the effective grating. The velocity correction is given in equation (5) and can be found in [10]:

$$v = \frac{v_{err} \, v_p}{v_p \pm v_{err}} \quad . \tag{5}$$

The velocity v_p is the effective read out speed of the pixel array and the sign depends on the moving direction relative to the read out direction.

3 Spatial Filtering System with Dedicated Structured Detector

A major problem of CCD- and CMOS-sensors is the low frame rate, because normally each gray value has to be read out sequentially. Possible ways to increase the frame rate are a reduced image size by defining a region of interest or to reduce the resolution by using pixel-binning. Both methods result in a reduced pixel number and limited sensor parameters. However, using a high speed profile sensor is a further solution to solve this problem. This used sensor has *256×256* pixels, but all gray values are summed up for every row and column internally. So the number of values to be transmitted for each frame is reduced to *2×256* pixel-line-values. Figure 2 shows the original application of the sensor for spot localization.

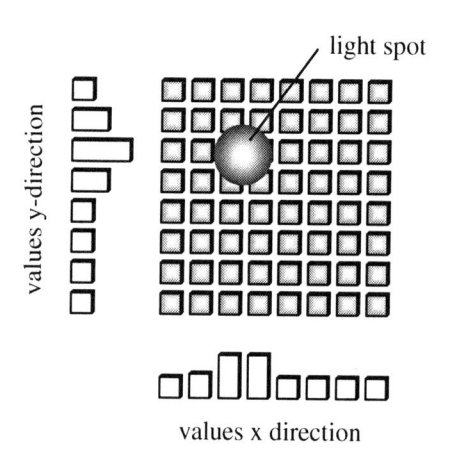

Fig. 2. Internal pre-processing used for detection of the position of a light spot. The values for x- and y-direction are summation of each pixel row and column.

The feature of internal summation increases the possible frame rate to *3200* frames (integrated line values) per second. Beside the readout time, the summation of each pixel-row and -column is also time consuming by generating the spatial filter signal from a sensor array. Also this signal generation step is integrated in the sensor design.

The applied sensor is controlled via a DSP-system (Digital-Signal-Processor). The gray values from the sensor are stored in the DSP by using direct memory access. The grating functions used to calculate the spatial filter signals are freely programmable via PC interface. After calculating a defined number of signal points, the power spectral density (PSD) distribution of a signal-section is calculated. The position of the main peak in the PSD is a measure of the dominant frequency in the signal and is converted to a velocity. The velocity samples are passed to the PC for visualization, calculation of statistics and storage.

4 Measurement Results for Speed of Solid Surface Motion

To demonstrate the feasibility of the sensor concept, two measurements will be presented. First, angle measurement by using a moving light spot will demonstrate the directional sensitivity. Second, the accuracy of the velocity determination will be shown.

4.1 Angle Measurement

Theoretically, the minimum resolvable angle and therefore the maximum resolution of the two orthogonal spatial filters are given by the grating period g_p. If the imaged surface inhomogeneities are moving parallel to the grating line there will be no measurement effect in the orthogonal direction. When a minimum of one grating period is crossed over the length of the spatial filter, the resulting spatial filtering signal will have one period which can be detected as a movement. So the maximum angle resolution $\alpha_{err\,max}$ can be found by:

$$\alpha_{err\,max} = \arctan\left(\frac{g_p}{length_p}\right). \quad (6)$$

For the demonstration measurements the sensor was rotated between 0° and 180° in 5° steps and the light spot was moved in front of the sensor in the same direction forwards and backwards.

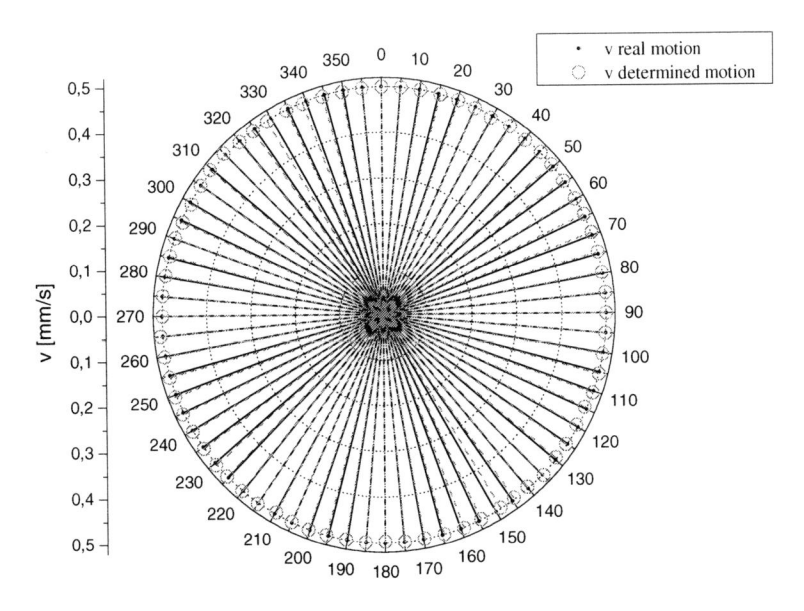

Fig. 3. Measurement results of rotation angle determination by the spatial filter system. A light spot was moved in front of the sensor.

Figure 3 shows a good agreement between the image motion and the determined angle and velocity. The standard derivation over all angles is about 0,426°. The theoretical maximum error from equation (6) is 0,45° with a grating period of $g_p = 2$ and the full length of the spatial filter of $length_p = 256$.

4.2 Speed Measurements of a Moving Surface

The setup for the velocity measurement on a solid surface is shown in figure 4. The constant surface movement is generated by a rotating disc, where the spatial filtering system observes the face side of the disc.

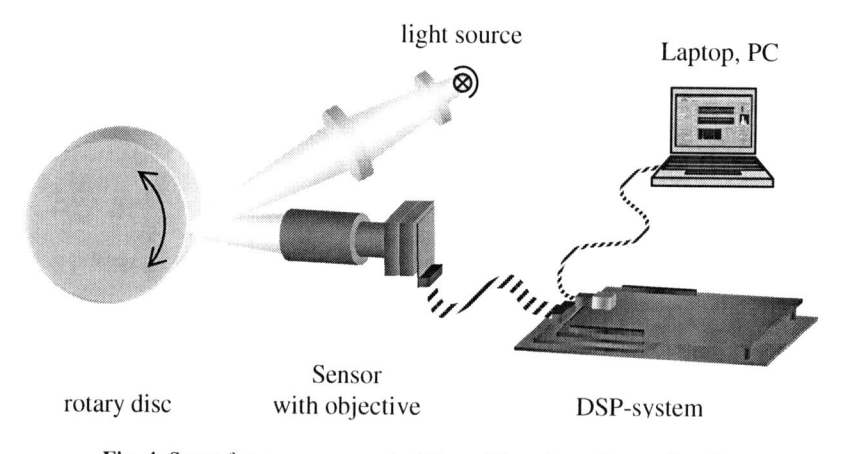

Fig. 4. Setup for measure speed of the solid surface of a rotating disc

The rotation speed of the disc was referenced by an encoder. A halogen lamp (*25W*) and collimation lenses provide incoherent illumination. The sensor was rotated by about *40°* compared to the velocity vector of the surface, consequently two component velocity measurement can be demonstrated.

Different velocities were tested with a measurement time of *15* seconds. The velocity varies from *100 mm/s* up to *500 mm/s* in steps of *100 mm/s*. Figure 5 shows the results of the measurements. The solid dots indicate the individual velocity values determined by the measurement system. Additionally the mean value (hollows square) and the standard deviation of the measurement are shown in Figure 5. The scatter of individual measurements increases with increasing velocity, but the relative deviation is constant with a maximum uncertainty of about 2.8%. Nevertheless, the accuracy and uncertainty of the velocity values also depend on the algorithm used to calculate the main frequency of the spatial filter signal. For the actual case the peak position in the power spectral density distribution was interpolated by a square fit for improving the spectral resolution. However this method has disadvantages when the raw signal includes phase jumps. In spatial filtering technique phase jumps occure when some in-homogeneities of the observed surface move in the field of view and some other leave the field of view. The phase jumps result in a phase noise and can cause cancellation of individual frequencies in the PSD nearby the peak frequency. Therefore, the accuracy of the interpolation of the spectrum is limited.

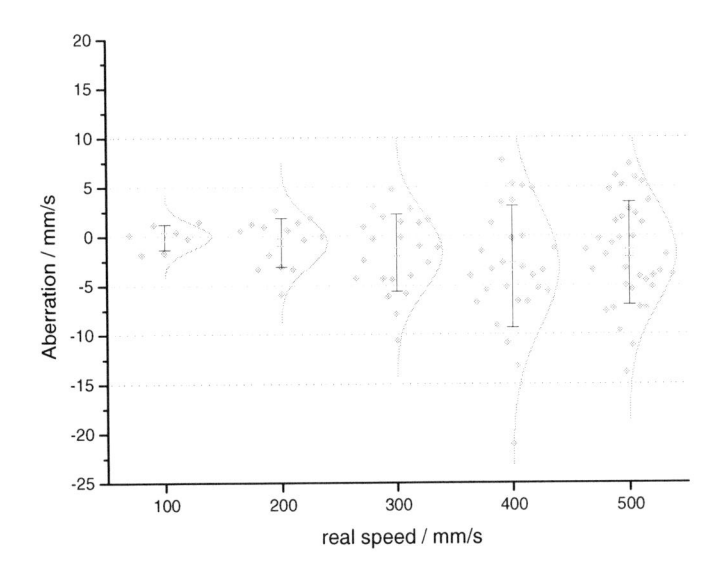

Fig. 5. Aberration of the determined velocity from the referenced velocity of the impulse generator

5 Conclusion

The paper presents a possibility for implementing an optical spatial filter by using a structured array detector, e.g. a CCD- or a CMOS-sensor. The advantage of using array detectors is the possibility to implement different grating structures like rectangular or sinusoidal grating without changing the hardware setup. By using a sensor array with on-chip integration a high frame rate is possible. The read out of the gray values and the data accumulation and calculation of the velocity can be done online by a digital signal processor.

Further application in the field of spatial filtering technique might be the interpolation of non-uniform grating forms as proposes in [12]. Here the flow profile of the velocity in a capillary is considered in the grating. Therefore the spatial filter correlates with the form of a velocity field.

References

1. Albrecht, H.-E., et al.: Laser Doppler and Phase Doppler Measurement Techniques. In: Adrian, R.J., et al. (eds.) Experimental Fluid Mechanics, Springer, Heidelberg (2003)
2. Raffel, M., Willert, C., Kompenhans, J.: Particle Image Velocimetry. In: Adrian, R.J., et al. (eds.) Experimental Fluid Mechanics. Springer, Heidelberg (1998)
3. Gerald, A.S.F.: Photo-electric system, United States (1935)
4. Hancock, J.D., Meinema, H.E.: Camera for aerial photography, United States (1946)
5. Ator, J.T.: Image-Velocity Sensing with Parallel-Slit Reticles. J. Opt. Soc. Am. 53, 1416–1419 (1963)

6. Aizu, Y., Asakura, T.: Principles and development of spatial filtering velocimetry. Applied Physics B: Lasers and Optics 43(4), 209–224 (1987)
7. Jakobsen, M.L., Yura, H.T., Hanson, S.G.: Speckles and their dynamics for structured target illumination: optical spatial filtering velocimetry. Journal of Optics A: Pure and Applied Optics 5(054001), 1464–4258 (2009)
8. Petrak, D., Rauh, H.: Micro-flow metering and viscosity measurement of low viscosity Newtonian fluids using a fibreoptical spatial filter technique. Flow Measurement and Instrumentation 20(2), 49–56 (2009)
9. Michel, K.C., et al.: A novel spatial filtering velocimeter based on a photodetector array. IEEE Transactions on Instrumentation and Measurement 47(1), 299–303 (1998)
10. Bergeler, S., Krambeer, H.: Novel optical spatial filtering methods based on two-dimensional photodetector arrays. Measurement Science and Technology 15(7), 1309–1315 (2004)
11. Bergeler, S.: Einsatz optoelektronischer Flächensensoren in der ein- und zweidimensionalen Ortsfiltertechnik. Universität Rostock, Rostock (2002)
12. Pau, S., Dallas, W.J.: Generalized spatial filtering velocimetry and accelerometry for uniform and nonuniform objects. Applied Optics 48(24), 4713–4722 (2009)

Towards Unified Performance Metrics for Multi-robot Human Interaction Systems

Jamil Abou Saleh and Fakhreddine Karray

University of Waterloo, PAMI Lab
Waterloo, ON, Canada
{jaabousa,karray}@uwaterloo.ca

Abstract. This paper intends to be a further step towards identifying common performance metrics for task-oriented human-robot interaction. We believe that within the context of human-robot interaction systems, both human and robot independent actions and joint interactions can significantly affect the quality of the accomplished task, thus proposing a generic performance metric to assess the performance of the human-robot team. In our previous work, and toward efficive modelling of such metric, we proposed a fuzzy temporal model to evaluate the human trust in automation which directly influences the interaction time that should be directly and indirectly dedicated toward interacting with the robot. Another fuzzy temporal-based model was also presented to evaluate the human reliability during interaction time. In this paper, the model is then extended to accomodate for multi-robot scenarios. Sequential and parallel robot cooperation schemes with varying levels of task dependency are considered.

Keywords: performance metrics, fuzzy logic, finite automata.

1 Introduction

Many systems have been implemented toward achieving effective human-machine collaboration or interaction, but running with the risk of being ignored if appropriate performance metrics are not in place. Therefore, presenting metrics that assess the performance of the human-robot system becomes very crucial. Mission effectiveness is one of the most popular and widely used metrics to evaluate the performance of human-robot teams [1]. However, such measures face several crucial problems; a robot might appear to be effective at a certain time, but might be making negative progress on the overall goal [2]. Situation awareness also finds itself as another emerging metric used in the literature to assess system performance. Endsley [3] defines situation awareness as *"the perception of the elements in the environment within a volume of time and space, the comprehension of their meaning and the projection of their status in the near future"*. Situation awareness, thus, requires task-specific metrics to be designed in order to assess its mentioned objectives.

The idea of developing a common toolkit of performance metrics and identifying measures that can be used to compare different research results has also been discussed by many researchers. Olsen and Goodrich discuss several interrelated performance metrics that can lead the design of human-robot interaction systems [2], [4]. They include:

- Neglect Tolerance (NT): NT is a measure of the time elapsed between the human instruction to the robot, and either a drop of the robot's performance below the effectiveness threshold, or the next human instruction in case the user detects and oversees a problem.
- Robot Attention Demand (RAD): RAD is a measure of how much time or fraction of the total task time must be spent by the user in interacting with the robot. RAD is defined as shown in equation 1, where IE represents the interaction effort.

$$RAD = \frac{IE}{IE + NT} \quad (1)$$

- Fan-out (FO): FO is a measure of how many robots a user can interact with simultaneously and effectively. It is defined as the total task time divided by the time spent interacting with one robot (RAD), as shown in equation 2.

$$FO = \frac{1.0}{RAD} = \frac{IE + NT}{IE} \quad (2)$$

- Interaction Effort (IE): IE is a measure of the time required to interact with a robot, however, defining this metric is not easy. Olsen suggests to experimentally measure NT and FO, and then approximate the IE using equation 2.

Although the set of metrics presented by Olsen and Goodrich is somewhat generic, it lacks focus on two essential factors: human trust in automation, and human reliability. Therefore, a metric framework that can be generalized should also involve the human trust and the human cognitive limitations in the human-robot interaction performance assessment loop [5], which were both addressed in our previous work [9], [10]. The purpose of this work is to present a further step toward a generalized common metric attempting to extend the model to acuommodate for multi-robot systems.

The remainder of the paper is organized as follow: Section 2 presents an overview of our previously proposed modeling of FO and RAD. A summary of the proposed human trust in automation and human reliability fuzzy temporal models is presented in section 3. Generalized mathematical models of our proposed metric to accommodate for multi-robot systems are presented in section 4. Simulations and experimental results are discussed in section 5. Finally, section 6 concludes this paper.

2 Effective Modeling of Fan Out Mteric

2.1 Fan Out and Human Reliability

Fan-out, a metric originally defined by Olsen and Goodrich, is a measure of how many robots with similar capabilities a human can simultaneously and effectively control and/or interact with. However, there are several limitations that make it difficult to measure this fan-out limit. The first limitation is caused by task saturation. The second critical limitation is based on the human cognitive and physical limitations. Several studies show human factors are responsible for 20 to 90 percent of the failures in many systems [6], [7], [8]. As a result, modelling this HR factor becomes a necessity, hence our new FO metric is defined as shown in equation 3, where HR (a value between 0 and 1) represents the human operator's ability to manage the increased task complexity.

$$FO = \frac{1.0}{RAD} \times HR \qquad (3)$$

2.2 Robot Attention Demand and Human Trust in Automation

Olsen defined robot attention demand (RAD) as a relationship between neglect tolerance (NT) and interaction effort (IE). Olsen, however, states that in this scenario, this metric is no more independent from the user, and hence the operator's trust in the robot's autonomous abilities becomes a critical issue that can highly influence such interaction process. In previous work, an alternative definition of the RAD as a function of both Direct Interaction Time (DIT) and Indirect Interaction Time (IIT) was presented, where the IIT is a direct consequence of trust. This is shown in equation 4, where Tr is the operator's trust in automation.

$$RAD = DIT + IIT = DIT + NT \times (1 - Tr) \qquad (4)$$

3 Proposed Human Trust in Automation and Human Reliability Fuzzy Temporal Model

3.1 Human Trust in Automation Model

Trust phenomenon is fuzzy and temporal by nature, thus, modelling this factor using only fuzzy logic might not give the anticipated results, and therefore, a finite fuzzy state machine is proposed. However, the larger the number of inputs to

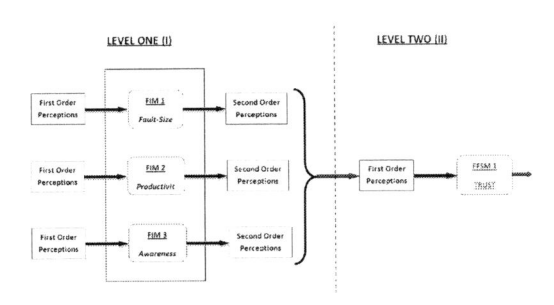

Fig. 1. Overall Architecture of the Proposed Trust Model

this model, the more complicated the knowledge base will be; consequently, identifying the crucial factors which contribute largely to building up trust becomes crucial. One very important observation is that such factors can be grouped under a few categories. This observation gave us the idea of building a two-level framework to identify and estimate the level of trust, as shown in figure 1. The second level relates trust to some **second order perceptions** that contribute directly and largely to the current level of trust. Level II is implemented using a finite fuzzy state machine (FFSM) which takes three inputs: fault size, productivity, and awareness, where each is modelled using three membership functions: Low, Medium, and High. The FFSM outputs the corresponding level of trust. Five states are used to model this phenomenon to ensure more accuracy and specificity (Very Low, Low, Medium, High, and Very High). The number of required rules, implemented in the form of *if-then* fuzzy rules, is: 3*3*3*5 = 135 rules. An Example Rule is: **IF** q(very low)[t-1], AND fault size is low, **AND** productivity is low, **AND** awareness is low, **THEN** q(low)[t].

Max-Min operation is used for determining the state activation level. The "AND" minimum operation is used to determine the firing strength of each fired rule, and the maximum firing strength that corresponds to the most dominant rule that reaches a specific state is considered as the state activation level. Then, we follow the structure defined by Sugeno-type fuzzy inference model, and map the states (Very Low, Low, Medium, High, Very High) to the zero-order consequences $(0.1, 0.3, 0.5, 0.7, 0.9)$. The crisp overall consequence is then generated using the weighted average method, as described in the equation 5.

$$\hat{c} = \frac{\sum_{i=1}^{n} w_i c_i}{\sum_{i=1}^{n} w_i} \qquad (5)$$

where w_i is the ith state firing strength, and c_i is the corresponding state consequent. Second-order perceptions inputted to the FFSM are themselves explained with lower-order perceptions, and hence should be properly modeled. Fault size, for example, is highly determined by three lower-order factors: fault frequency, fault cruciality, and fault recovery. These factors are not temporal, thus no FFSM is needed. As a result, a fuzzy Mamdani inferencing model is used. Three membership functions are chosen to represent each factor: Low, Medium, and High. The same applies to awareness. Three inputs mainly determine the value of this factor: machine awareness of its capabilities (MA), context awareness of the task (CA), and machine awareness of the human operator's cognitive and physical abilities and limitations (HA); each will be represented by three membership functions: Low, Medium, and High. Finally, as for productivity, two inputs mainly determine the value of this factor: task/goal successfulness and completion (TC), and task complexity and sophistication (TS). Each will be represented by three membership functions: Low, Medium, and High.

The idea behind building a two-level architecture to estimate the trust factor is very important as it dramatically decreases the complexity of the system and the size of the knowledge base. More details can be found in [9], [10].

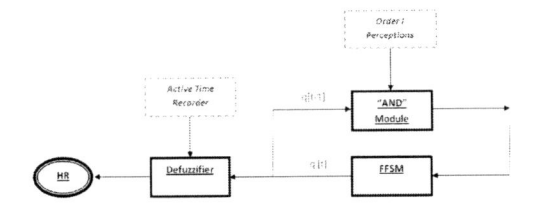

Fig. 2. Proposed Human Reliability Model Interconnection

3.2 Human Reliability Model

Previous discussions on human reliability (HR) demonstrate the crucial importance of modelling this factor when building an effective and generalized robotic metric framework. With respect to this work, we believe modelling human reliability should be also fuzzy. Since this factor is highly related to time and to its previous state, a FFSM is then required. The same description presented for level II in trust modelling is used for modelling human reliability. Many factors affect humans and their performance; some are psychological, physiological, or sociological, among others. However, in our work, we focused on those factors that arise from interacting directly with the robot, and future work will put more focus toward better modelling this somewhat complex phenomenon. In this context, three input variables are considered: (1) number of sub-tasks being simultaneously perceived (NS), (2) mental workload required during task completion (MW), and (3) external/internal burden (EB). Three membership functions are used to model each variable (Low, Medium, and High), and human reliability (HR) will be modelled using 5 states (Very Low, Low, Medium, High, Very High). However, an important factor that we should keep in mind is time. Although the above three variables play remarkable role in determining the human reliability, interaction time (or time while being active) is also crucial since human performance degrades with time even when the task is simple. Figure 2 shows the overall structure of the proposed human reliability module. Therefore, in the output function, which is also denoted as the defuzzifier module, and after obtaining or generating the output alphabet, we map the states (Very Low, Low, Medium, High, Very High) to the following consequences $(0.1-a\times t, 0.3-a\times t, 0.5-a\times t, 0.7-a\times t, 0.9-a\times t)$, where "a" is a subjective constant that varies from one person to another and represents the natural human degradation factor, and "t" represents the time factor. Weighted average method is then used to generate the crisp final consequence.

4 Metrics Generalization Models for Multi-robot Cognitive Systems

So far, FO metric has been introduced as a performance criteria to assess the performance of the human-robot system, where **one** robotic agent is involved

(More details can be found in our previous work [9], [10]. However, an important issue also ignored by Olsen and Goodrich, among other researchers, is a scenario where the robotic cognitive system is composed of more than one robot. Hence, a special consideration of multi-robot systems should be also addressed.

4.1 Sequential Execution of Tasks

In this scenario, N robots are cooperating with the guidance of a human user to finish a specific task. Doing so, only one robot is active at one point, while the other robots are idle. Therefore, the system FO is that of the active robot as shown in equation 6.

$$SystemFO = FO_1 \vee FO_2 \vee ... \vee FO_N \qquad (6)$$

where $FO(idlerobot) = 0$.

4.2 Parallel Execution of Tasks

Independent Execution of Tasks. In this scenario, the robots are active at the same point executing **independent** tasks in parallel, thus dependency-related issues can be ignored. Usually, in most cases the average of the N robots FO is indicative of the system FO, but in our scenario (and since robots do not contribute equally to the overall task completion) a weighted average method is applied as shown in equation 7, where W_i is the percent contribution of the i_{th} robot toward the final goal completion.

$$SystemFO = \sum_{i=1}^{N} W_i FO_i, where \sum_{i=1}^{N} W_i = 1 \qquad (7)$$

Dependent Execution of Tasks. This scenario is most complicated. Let us start by considering two robots that are active at the same point executing **dependent** tasks in parallel. When total (100 percent) task dependency is encountered, the system FO will be forced to follow the smallest robot FO, and the less dependency exists, the closer the system FO will be to the weighted average (the task independent scenario). Therefore, we can model the system FO as shown in equation 8, where d represents the percent task dependency between robots one and two.

$$SystemFO = d * min(FO_1, FO_2) + (1 - d) * \sum_{i=1}^{N} W_i FO_i \qquad (8)$$

To generalize the conclusion for N-robots, inter-task dependencies should be considered. For example, if we have three robots, there exist task dependencies between robot 1 and 2, robot 1 and 3, and robot 2 and 3. The system FO is shown in equation 9, where $d_{i,j}$ is the task percent dependency between robots i, and j, and FO_i is the Fan-out of i_{th} robot. Equation 9 consists of multiplication of two parts, where the second part (between brackets) is the **relative** weighted average FO of the i_{th} and j_{th} robots, and the first part of the equation considers how much contribution both robots make toward the final goal. However, when dealing with pair-wise dependencies, we have $\binom{N}{2}$ possibilities, hence the sum of all contribution weights corresponding to all possibilities will be $(N-1)$, hence explaining the division by $(N-1)$. System FO=

$$\sum_{i<j}^{N} \frac{W_i + W_j}{N-1} * \left\{ d_{i,j} * min(FO_i, FO_j) + (1 - d_{i,j}) * (\frac{W_i}{W_i + W_j} FO_i + \frac{W_j}{W_i + W_j} FO_j) \right\} \tag{9}$$

But one important issue is that inter-robot task dependencies might highly affect other dependencies in the system. Consequently, the abovementioned system FO is an upper bound for the real system FO which is lower bounded by the smallest robot FO. This is shown in equation 10.

$$min(FO_1, FO_1, ..., FO_N) < RealSystemFO < SysFO \tag{10}$$

5 Simulations and Experimental Results

In the following section, we discuss experimental and simulation results which intuitively support the importance of our proposed generic metric framework. Figure 3 shows the trust evolution in accordance with the temporal change of the three inferred second-order perceptions discussed in preceding section. At time equal to zero, the human trust in automation is assumed to be neutral. An important observation is the smoothness of change in the trust value, rather than an abrupt change. This is related to the fact that human trust is a step-by-step process, and strongly depends on its previous state. Results also show that when the fault size is low, and both awareness and productivity are high, the human trust in automation increases, and vice versa. Figure 4 shows the implications of the human trust in automation factor on both interaction time (IT) and free time (FT). Direct interaction time (DIT) is assumed to be 25 percent of the overall task time. Figure 4 shows that when the human trust in automation increases, the indirect interaction time (IIT) spent monitoring the

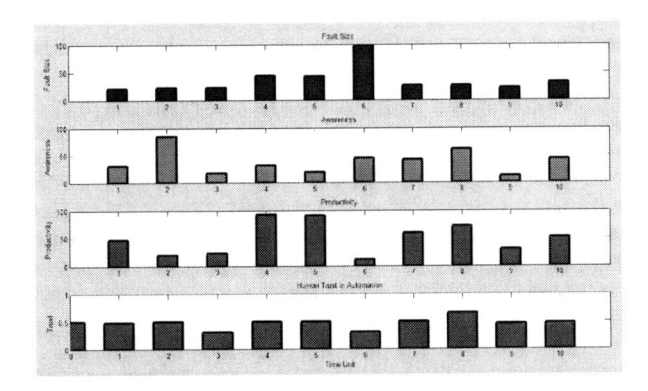

Fig. 3. Level II Trust Simulation Results -1-

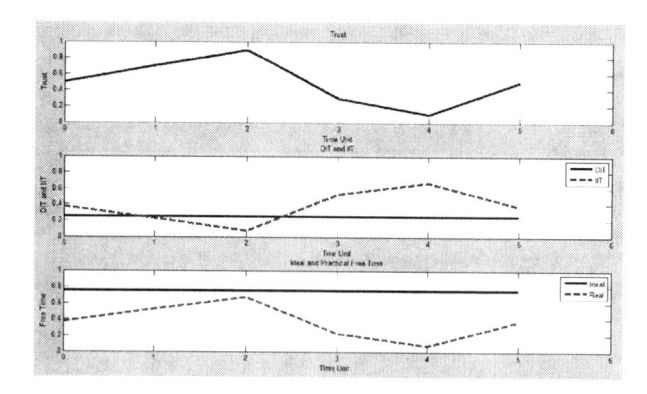

Fig. 4. Implication of Trust on Interaction Time and Practical Free Time

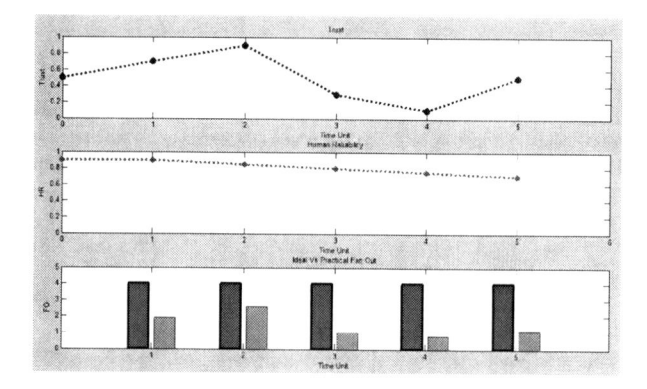

Fig. 5. FO Calculation

robot decreases, and hence the practical free time becomes closer to the ideal free time (obtained using only the DIT). Now, in order to see the implications of trust in automation and human reliability factors on the FO metric, figure 5 shows that when the trust is high (as in time t = 2) the practical FO is close to its ideal value (which is 4 assuming a DIT of 25 percent), while at time t = 4, trust is really low which significantly lowers the FO value.

Concerning our proposed generalized mathematical models for the multi-robot scenarios, Figure 6 shows the system FO for a two-robot system working sequentially toward achieving some tasks. The system FO is defined as the logical OR of the two robot's FO, where the FO of the idle robot is assumed to take a value of zero. On the other hand, figure 7 demonstrate a two-robot system, where it is assumed that both robots are active in parallel, and equally contribute to the final goal completion, conducting tasks that are 70 percent dependent. Results show that the system FO falls between the weighted average FO (representing the independent scenario) and the smallest FO (i.e full dependency scenario), but closer to the smallest FO because percent dependency is higher than 50 percent.

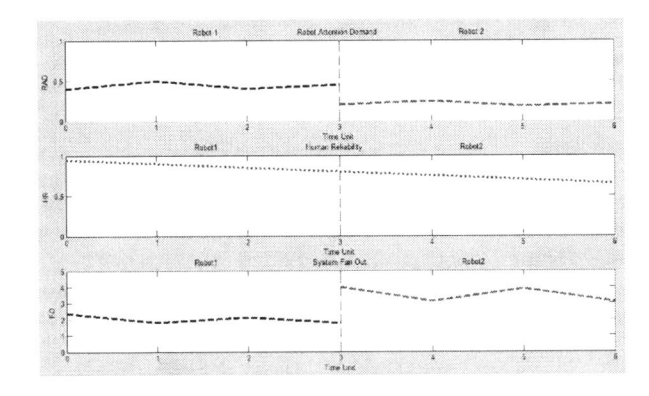

Fig. 6. Multi-Robot System: Sequential Execution of Tasks

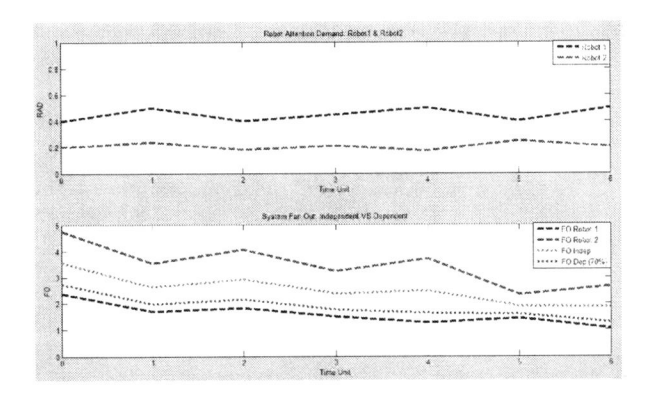

Fig. 7. Two-Robot System: Parallel Dependent Execution of Tasks

6 Conclusion

The main goal of any human-robot interaction system is to increase the effectiveness of the team in accomplishing some task. Therefore, designing a performance metric that can assess this effectiveness is crucial. We believe that such evaluation criteria should focus equally on human and machine performance. A two-level trust evaluation model which estimates the human trust in automation and a time-based human reliability assessment model were addressed. In this paper, the model is extended to accomodate for multi-robot systems. Sequential and Parallel task execution with varying levels of dependency are considered.

References

1. Donmez, B., Pina, P., Cummings, M.: Evaluation Criteria for Human-Automation Performance Metrics. In: PerMIS Conference Proceedings (2008)
2. Olsen, D.R., Goodrich, M.A.: Metrics for Evaluating Human-Robot Interactions. In: Performance Metrics for Intelligent Systems Workshop (2003)

3. Endsley, M.R.: Design and Evaluation of Situation Awareness Enhancement. In: Proceedings of the Human Factors Society 32nd Annual Meeting, vol. 1, pp. 97–101 (1988)
4. Goodrich, M.A., Olsen, D.R.: Seven Principles of Efficient Human Robot Interaction. In: Proceedings IEEE International Conference on Systems, Man and Cybernetics, pp. 3943–3948 (2003)
5. Pina, P., Cummings, M.L., Crandall, J.W., Penna, M.D.: Identifying Generalizable Metric Classes to Evaluate Human-Robot Teams. In: Workshop at the 3rd Annual Conference on Human-Robot Interaction (2008)
6. Santoni, C., Neto, J.A., Vieira, M.F., Scherer, D.: Proposing Strategies to Prevent the Human Error in Automated Industrial Environments. In: HCI International Conference (2009)
7. Williams, K.: Summary of Unmanned Aircraft Accident/Incident Data: Human Factors Implications. Technical Report (2004)
8. Dhillon, B.S.: Human Reliability and Error in Medical System. In: Series on Industrial And Systems Engineering - Medical (2003)
9. Abou Saleh, J., Karray, F.: Temporal Fuzzy Based Modeling as Applied to the Class of Man-Machine Interaction. In: IEEE International Conference on Fuzzy Systems (2010)
10. Abou Saleh, J., Karray, F.: Towards Generalized Performance Metrics for Human-Robot Interaction. In: International Conference on Autonomous and Intelligent Systems (2010)

Augmented HE-Book: A Multimedia Based Extension to Support Immersive Reading Experience

Abu Saleh Md Mahfujur Rahman, Kazi Masudul Alam, and Abdulmotaleb El Saddik

Multimedia Communications Research Lab,
University of Ottawa, Ottawa, ON, Canada
{kafi@mcrlab,malam@discover,abed@mcrlab}@uottawa.ca
http://mcrlab.uottawa.ca

Abstract. Traditionally we know that the use of various modalities such as audio-visual materials can influence the effective learning behaviour [6]. Can haptic rendering if combined with the audio-visual learning content have the similar effect? In order to examine that we present an intuitive approach of annotation based hapto-audio-visual interaction with the traditional digital learning materials such as eBooks. It leverages the haptic jacket interface to receive haptic emotive signals wirelessly in the form of patterned vibrations of the actuators and expresses the learning material by incorporating image based augmented display in order to pave ways for intimate reading experience in the popular eBook platform.

Keywords: Haptic, Multimedia, Annotation, E-Book, Mobile device, Augmented Rendering.

1 Introduction

Haptics is considered to be both independent and supplementary medium of communication channel [14] that mimics physical contacts. They are fundamental to mental and psychological development and hence their applications in various cognitive based applications have attracted attention of many researchers around the world [9]. Reading is a complex cognitive behaviour that intends to decode meanings from text by relating words with one existing knowledge base and understanding. According to Dale [6] effective learning styles are influenced by the use of modalities. He enumerated our learning phenomenon by 10% of what we read, 20% what we hear, 30% of what we see, 50% of what we hear and see, 70% of what we say and 90% of what we both say and do. Though these numerical labels are not widely accepted but the theory, "Cone of Learning" is more acknowledged in the research communities relating to analysis of the reading behaviours. Our ways of readings are continuously molded with the advent of new technological innovations, devices and platforms. For example, e-book is continuously challenging the existence of print book and at times e-books are replacing the former in the schools [13]. In our daily life, we are getting more and more involved to electronic materials than printed books as now-a-days most of the advanced handheld mobile devices provide e-book reading facilities [17]. Handheld mobile devices are also widely adopting vibrotactile actuators in order to compensate the lack of haptic feedbacks in their touch screen based interfaces [3]. With the growing popularity of the haptic sensory feedbacks in the mobile handheld devices [11], one of the challenge is now to find new avenues to intuitively incorporate the haptic-audio-visual feature (see, hear and do [6]) into mobile eBook applications in order to foster effective learning.

Our proposed Haptic E-Book (HE-Book) is a special type of e-book reader, which is capable of delivering vibrotactile feedbacks in the user's reading experience. For example. when a user is reading a terrible sea storm experience, HE-book attempts to provide content dependent vibrotactile haptic feedbacks by using both the haptic sofa[1] and the haptic jacket [16]. In order to better express the scene, it also presents an augmented depiction of the sea storm by using the mobile screen. We present a preliminary prototype exploring the suitability of haptic feedbacks in a mobile e-book device that generates tactile feedbacks when a user reads an annotated part of the e-book content. An overview of the system components is shown in Figure 1. We incorporated manual haptic annotation of the e-book content [15] with three different type of haptic stimulations from our previous work [16]. We evaluated user's acceptance of the system using standard usability test methods and found that there is a positive tendency of accepting such interactive experiences in e-book handheld devices.

Our contribution to this research article is three folds. First of all, we propose the annotation based haptic-audio-visual feedback integration in traditional digital learning contents. The haptic interaction is manually defined by analyzing the scenarios of the e-book content. Secondly, we present a generalized framework for such an HCI application that describes the haptic data annotation scheme and illustrate the various interaction mechanisms with annotated contents. The annotation method used in the prototype allows the creation of separate annotation files that could be stored in remote annotation repository. Third, we present an analysis of the prototype application with usability test results to address the influence of haptic rendering in the learning process by combining it with audio and visual learning materials. The remainder of this paper is organized as follows. In the beginning, in Section 2 we provide literature survey of existing or closely related applications. In Section 3, we illustrate various components of our system that facilitates the annotation based haptic experience from digital reading materials. Further in Section 4, we describe our analysis of the usability test results using the prototype system. We conclude the paper in Section 5 and state some possible future work directions.

---

[1] Haptic sofa, http://d-box.com/en/home-theatre/d-box-ready-seating/

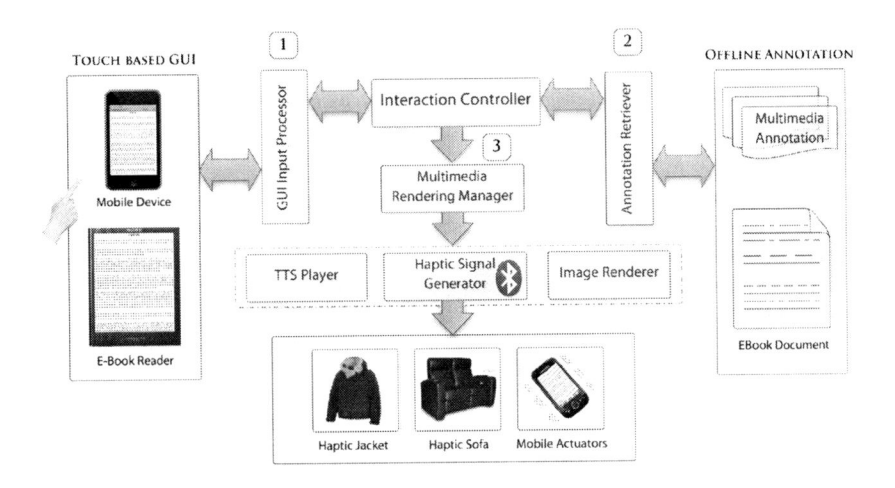

Fig. 1. High level architecture of the proposed system

2 Related Study

According to a research article by Harrison [10], french visionaries Robida and Uzanne stated the future of books in 1895. Their prognosis were that written book contents would soon be transferred into sound wave and be played in gramophone and subsequently gramophone would evolve to become pocket-sized player. In contemporary research, MagicBook [2] project was an early attempt of to bring augmented reality display in book reading experience, where a user could see 3D virtual models out of the pages. This scheme also supported multiple user immersion in the same scene. However, in order to replicate physical metaphors used in dealings to real book Chen et al. [5] introduced a new design of dual display for e-book reader and various supported interactions like folding, flipping, and fanning in their system.

In a long-term development of mixed reality book, Grasset et al. have focused on design and development of mixed reality applications in broader sense. They have discussed semantics of mixed reality book, design space and user experience with such type of interfaces [7]. Card et al. [4] have proposed physical behavioural simulation of a real book in the presentation of a 3D virtual book. They extended their prototype [12] to support virtual annotation or bookmarking. More recently, Welch et al. [18] extended the concept further to an immersive Virtual Reality book aimed at medical training for surgery. In another research, Gupta et al. explored integration of projected imagery with physical book in order to introduce tangible interface to multimedia data. They have used camera and projector pair to introduce a tracking framework, which monitors 3D position of planar pages when user turns book pages back and forth. Their book can be loaded with various multimedia contents like video, images, data etc. [8].

Several works on augmented books has explored relationship between real and digital books, various interaction techniques and technologies, audio-visual augmentation etc. But, none of them studied the opportunity of tactile feedback or haptics as an augmentation or annotation to the book content. In this paper we have introduced a new type of eBook, Haptic E-Book (HE-Book), which annotates eBook contents with various unique vibrotactile signals and later render the signals while user is reading that part of the book. To the best of our knowledge, haptic annotation to eBook contents is a new idea where a certain amount of research contribution is possible. For example, haptic annotation to augmented reality books can be an interesting genre to explore. In this paper we have presented our introductory prototype of HE-Book.

3 Haptic Ebook System

The HE-Book system consists of four major modules and an interaction controller as depicted in Figure 3. In Section 3.1 we present the *Annotation Manager* module that handles eBook annotation resolution and annotation retrieval operations. Next in Section 3.2 we describe *PDA GUI Client* module, which assists the annotation mapping service by locating target paragraphs based on the user finger touch detection and processing. *Multimedia Manager* module illustrated in Section 3.3. Operations of all the modules are centrally monitored and organized by the *Interaction Controller*, the description of which is presented in Section 3.4.

3.1 Annotation Manager

This component manages annotation generation and retrieval procedures. *Annotation Manager* is used to generate the *Annotation File* while *Annotation Mapping Service* is used to retrieve annotation blocks from the XML file.

Annotation Generator. Using the developed annotation scheme an author selected a paragraph of a selected page of an ebook and tag those by using various haptic and image properties. In haptic annotation scheme there were two supported devices and each device-supported list of supported vibration patterns. Those patterns were configured by using key control parameters for each haptic device. For image annotation steps, the editor provided a list of online image providers and an author chose one of them to search related annotation candidates for particular keywords. Further, the editor retrieved a list of possible images from the selected image providers and displayed them in an image container. Later, an author selected a relevant image from the list and thus image block was attached in the annotation file. The XML file was further stored to be processed by the HE-Book reader.

Advantage of seperate XML based annotation file was that a HE-Book user could decide to either use the multimedia extension or avoid it. For the annotated paragraph of an ebook we created an XML element *Page* which has attribute *ID* that denoted the unique number of the page. Every *Page* element was divided in *Para* which also had *ID* that denoted the paragraph no. Granularity level of our annotation ended at *Para* i.e. vibrotactile feedback would start and end in a paragraph level, which we termed as scene or paragraph based annotation. Under the *Para* we had elements *Haptic* and *Image*. *Haptic* was described using *Device Type*, *Pattern*, *Repeat* and *Delay* and *Image* had attributes *Src*, *URL*. Influenced by our past work [16] we manually annotated the ebook content with the previously defined *Touch, Fun, Hug, Tickle* type haptic feedbacks into our prototype HE-Book system.

Annotation Retriever. *Annotation Retriever* (Fig. 1) is responsible of searching the annotation file when it was necessary and retrieved the requested XML block. In a nutshell this was an XML parser. For a request of {*Page, Para*} pair, this XML parser would return corresponding *Haptic* , *Image* block from the XML file.

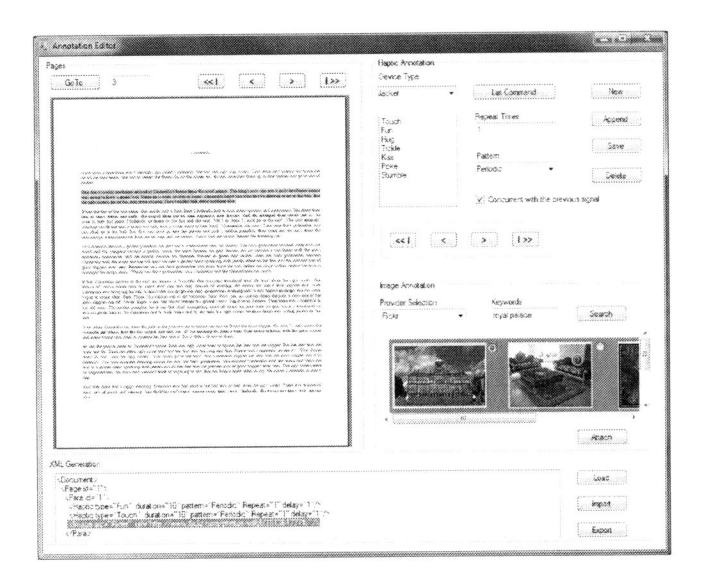

Fig. 2. Annotation editor for the e-book document (i.e. PDF)

3.2 Mobile GUI Client

The haptic eBook system works currently in a touch based device. The user can flip pages and scroll the paragraph of the page through the touch based interaction. In this case, the *eBook XML Reader* is deployed in an emulated touch mobile device, the standard touch SDK is used to obtain the screen coordinates by using the *GUI Interaction Listener*. The Listener uses the *Paragraph Locator* component in order to match the touched screen coordinate with the displayed eBook page content to deduce the paragraph that the user is currently pointing at. The eBook XML reader software continuously monitors the touched screen coordinate interaction. As soon as the user touches a paragraph of the page that has been previously annotated, the user is sent a motor vibration to confirm that a valid haptic annotation for the selected paragraph has been found. The GUI further initiates the haptic feedback retrieval process by sending the page ID and the corresponding paragraph ID to the *Interaction Controller* module. The Interaction Controller coordinates the retrieval operation and receives the appropriate haptic signal for the selected learning content.

3.3 Multimedia Manager

The Multimedia Manager module resides in the PDA Client and handles the hapto-audio-visual rendering. The rendering schemes are described as the following.

Haptic Renderer. Based on the page and paragraph information of *Touch Based GUI*, *Interaction Controller* gets haptic description from the *Annotation Retriever* submodule and generates the described haptic signal for the target haptic device. *Haptic Signal Generator* plays an important role for device specific haptic signal generation. As the target haptic device can be heterogeneous we considered device specific configuration. In our system architecture we have considered Bluetooth as a method of communication between our system and the corresponding haptic device. Such communication can also be extended to other possible personal network communication methods. In our prototype system we have used a Bluetooth enabled haptic jacket which has location based array of vibrotactile actuators. Location means arm, abdomen, shoulder, neck, chest etc. of human body.

Our haptic jacket consists of an array of vibrotactile actuators that are placed in different portions of the jacket and their patterned vibration can stimulate touch in the user's skin. Vibrotactile actuators communicate sound waves and create funneling illusion when it comes into physical contacts with skin. A series of small actuator motors are placed in a 2D plane in the jacket in a certain manner. In order to produce touch feeling the actuators are activated in a defined manner [1]. The jacket used in our prototype adhering to the aforementioned properties is discussed in detail in this work [16].

Text to Speech Interaction. With the vision of delivering audio feedback of the reading material to foster effective learning we incorporated Text to Speech (TTS) interaction scheme in the system. Using this interaction option a reader can choose male or female voice and issue commands to listen selected or all the remaining paragraphs of the book automatically. The real time mobile TTS processing engine produces audio stream from the eBook content so that the learner can choose to hear the reading material. While hearing the audio content the reader receives the haptic rendering and augmented images are delivered on the mobile screen to 'see' the learning material by using the annotation blocks in real time.

2D Image Rendering Approach. *The Imager Viewer* incorporates augmented visual feedback on the mobile screen so that a learner can 'see' images that are related to the learning content. The images are statically defined during the annotation steps for each paragraphs of a book. A semi-automated process allows the usage of the vast image database and makes those available during the reading process. Seeing an image of a snow storm while reading about it can help the user improve his/her reading and understanding abilities in order to help practice effective learning behaviour [6]. Additionally, the augmented visual display of the learning material can make the intuitive learning process more interactive and entertaining [2].

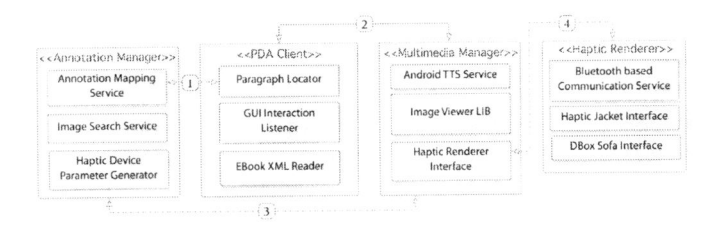

Fig. 3. HE-Book component architecture

3.4 Interaction Controller

Interaction Controller plays the central role to organize and synchronize necessary work flow in the haptic eBook reading system. The Interaction Controller polls to acquire user touch inputs after certain interval. As soon as a touch interaction is performed the Interaction Controller coordinates the screen coordinate based paragraph identification of the eBook document. The Annotation Manager module takes the page ID and the paragraph ID to retrieve the haptic signal associated with the paragraph and returns the signal data to the Interaction Controller. Further the Interaction Controller transmits the obtained haptic-audio-visual signal to the Multimedia Manager module where the signal is rendered and the user obtains the assigned haptic, auditory and visual feedback associated with the learning content. In addition to coordinating the aforementioned task *Interaction Controller* ensures that the system modules are not blocking or throttling the operation of the other modules. The Interaction Controller employs carrier-sensing algorithm to determine the active/idle states of the other modules.

4 Implementation and Results

We present the implementation details of the haptic eBook system in Section 4.1. Further in Section 4.2 we present user study of the implemented prototype system to determine the suitability of the proposed system.

4.1 Implementation Details

In order to use the Android 2.1 based mobile emulator to implement the mobile version of the haptic eBook reader prototype we incorporated the platform release that was published on January 2010. The SDK provided the basic packages to retrieve and display images on the mobile screen. The touch APIs were available both to determine the screen coordinate and mapping of the paragraph IDs of the touched texts from an eBook XML file. We used the Bluetooth Serial Port (SPP) driver with full support for VT-100/ANSI terminal emulation. Even though there was no direct support in the emulator for Bluetooth we developed a bare-bone reimplementation of the android Bluetooth API where we run a Bluetooth signal processor server (telnet to localhost, port 8199) on our PC to which the emulator connects. The developed packages were configured that extended the base classes in the *android.bluetooth* package to support easy discovery of the devices, creation of bluetooth services pool and connections. In order to use the simulator we called the $BluetoothAdapter.SetContext(this)$ during the initialization process of our activity/service. In general, the bluetooth emulator server accepted commands on the form: $\langle com_{ID} \rangle \diamond \langle param_{name} \rangle \ll param_{value} \rangle \cdots$. The command format was suitable for sending/receiving bluetooth signals to the haptic jacket's Parani ESD200 Bluetooth kit.

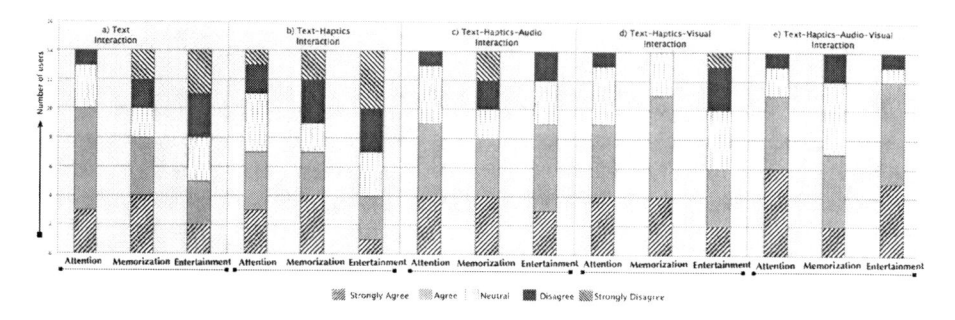

Fig. 4. Usability study of the HE-Book system. a) Likert scale presentation of the attention, memorization and entertainment metrics in a traditional text reading experience, b) The same study with haptic rendering added in the reading process, c) Incorporating haptic and audio playback simultaneously to determine the changes in the attention, memorization or entertainment factors, d) Haptic and visual feedbacks added in the reading process, e) Combining the hapto-audio-visual feedbacks in the paragraph reading experience to report the user scores.

4.2 User Study

In order to assess the influence of haptic rendering in the user's learning behaviour we set up five testbeds. The tests were took place at the university laboratory in a controlled environment with fourteen volunteers. In each test, the participating users were given a paragraph to read and answer three multiple choice type questions (MCQ) that analyzed their memory relating to the study material. The findings from each testbeds are depicted in Figure 4. The interaction modalities in each testbeds were targeted to determine the *Attention, Memorization and Entertainment* metrics in the user's reading activity. In the first test a) users were asked to read a paragraph consisting of textual information. In the second test, b) in addition to textual information we also annotated haptic data so that while reading certain section of the paragraph the user received touch/tickle type haptic feedback. Next in c) the paragraph was read out loud by using the TTS API and haptic feedbacks were rendered after certain parts of the paragraph were played-back. In experiment setup d) the paragraph was annotated with haptic feedbacks and augmented image was rendered on the mobile screen that explained the learning material. Last in setup e) the paragraph was annotated with hatpic, audio and augmented image material. The MCQs were designed to determine the user's memory relating to the learning contents.

In our experiments we deduced that the users seemed to score on average 5.7% more when they received haptic feedback in addition to other interaction modalities. Specially their score was significantly higher (14.2%) when haptic and augmented visual cues were combined during the reading activities. However, the results were not conclusive. Part of the reason for scoring higher while receiving haptic feedback could be due to the excitement that the users felt in receiving haptic feedbacks that relate with the learning materials. Also, visual feedbacks by itself improves memorization, hence the combination of haptic rendering modality with it conclusively do not prove that the former has a direct influence on the remembering process. However, empirically the experiment results convey the significance of haptic rendering in the user's reading experience.

5 Conclusion

Human computer interaction research has gained a pace with the advent of haptic devices. Our research goal was to explore the possibilities and opportunities this modality could bring to our traditional reading experience. In this regard, we introduced an intuitive eBook reader, which was capable of providing hapto-audio-visual modalities to a user's reading experience. We made five controlled usability tests to determine reader's experience to such multimodal reading activity in the domain of attention, memorization and entertainment. Although initial user feedback has been positive, it requires more laboratory and in school experiments to determine the significance of haptic modalities in the user's learning experience. We plan to conduct such experiments in our future endeavour in addition to improving the HE-Book system.

References

1. Barghout, A., Cha, J., El Saddik, A., Kammerl, J., Steinbach, E.: Spatial resolution of vibrotactile perception on the human forearm when exploiting funneling illusion. In: IEEE International Workshop on HAVE, pp. 19–23 (2009)

2. Billinghurst, M., Kato, H., Poupyrev, I.: The magicbook - moving seamlessly between reality and virtuality. IEEE Computer Graphics and Applications 21(3), 6–8 (2001)
3. Brewster, S.A., Chohan, F., Brown, L.M.: Tactile feedback for mobile interactions. In: Proceedings of ACM CHI, pp. 159–162. ACM Press/Addison-Wesley Publishing Co., San Jose, CA, USA (2007)
4. Card, S.K., Hong, L., Mackinlay, J.D., Chi, E.H.: 3book: A scalable 3d virtual book. In: Extended Abstracts of the 2004 Conference on Human Factors and Computing Systems (CHI), pp. 1095–1098. ACM Press, New York (2004)
5. Chen, N., Guimbretiere, F., Dixon, M., Lewis, C., Agrawala, M.: Navigation techniques for dual-display e-book readers. In: Proceeding of the SIGCHI Conference on Human Factors in Computing Systems, CHI 2008, pp. 1779–1788. ACM, USA (2008)
6. Dale, E.: Audio-Visual Methods in Teaching. Rinehart and Winstor (1969)
7. Grasset, R., Dunser, A., Billinghurst, M.: The design of a mixed-reality book: Is it still a real book? In: 7th IEEE/ACM International Symposium on Mixed and Augmented Reality, ISMAR 2008, p. 99–102 (2008)
8. Gupta, S., Jaynes, C.: The universal media book: tracking and augmenting moving surfaces with projected information. In: Proceedings of the 5th IEEE and ACM International Symposium on Mixed and Augmented Reality, pp. 177–180. IEEE Computer Society, Washington, DC, USA (2006)
9. Haans, A., IJsselsteijn, W.: Mediated social touch: a review of current research and future directions. Virtual Real. 9(2), 149–159 (2006)
10. Harrison, B.L.: E-books and the future of reading. IEEE Comput. Graph. Appl. 20, 32–39 (2000)
11. Hoggan, E., Brewster, S.A., Johnston, J.: Investigating the effectiveness of tactile feedback for mobile touchscreens. In: CHI 2008: Proceeding of the Twenty-Sixth Annual SIGCHI Conference on Human Factors in Computing Systems, pp. 1573–1582. ACM, New York (2008)
12. Hong, L., Chi, E.H., Card, S.K.: Annotating 3d electronic books. In: CHI 2005 Extended Abstracts on Human Factors In Computing Systems, pp. 1463–1466. ACM, USA (2005)
13. Mangent, A.: Hypertext fiction reading: haptics and immersion. Research in Reading 31(4), 404–419 (2008)
14. Park, T., Hwang, J., Hwang, W.: Facilitating the design of vibration for handheld devices. In: Jacko, J.A. (ed.) HCI International 2009. LNCS, vol. 5611, pp. 496–502. Springer, Heidelberg (2009)
15. Rahman, A.S.M.M., Cha, J., El Saddik, A.: Authoring edutainment content through video annotations and 3d model augmentation. In: IEEE International Conference on VECIMS, Hong Kong, China, pp. 370–374 (May 2009)
16. Rahman, A.S.M.M., Hossain, S.A., El Saddik, A.: Bridging the gap between virtual and real world by bringing an interpersonal haptic communication system in second life. In: IEEE International Symposium on Multimedia, Taiwan (November 2010)
17. Schilit, B.N., Price, M.N., Golovchinsky, G., Tanaka, K., Marshall, C.C.: The reading appliance revolution. Computer 32(1), 65–73 (1999)
18. Welch, G., State, A., Ilie, A., Low, K.L., Lastra, A., Cairns, B., Towles, H., Fuchs, H., Yang, R., Becker, S., Russo, D., Funaro, J., van Dam, A.: Immersive electronic books for surgical training. IEEE Multimedia 12(3), 22–35 (2005)

Human-Machine Learning for Intelligent Aircraft Systems

Stuart H. Rubin[1] and Gordon Lee[2]

[1] SPAWAR – SSC Pacific, 53560 Hull Street, San Diego, CA 92152-5001
[2] Dept. of Electrical and Computer Engineering, San Diego State University,
San Diego, CA 92182

Abstract. The solution for insuring the safety of tele-operated or fully unmanned autonomous systems (UASs) in the air space requires a) that the human remain in and on the loop to the maximal extent practical and b) that the UASs, which share the air space, have an intelligent backend for the processing of their sensory data. Moreover, it is necessary that this sensory processor be capable of generalizing and learning more than it was told in order that it properly handle situations not explicitly programmed for. Given the advent of advances in nanotechnology and microsystems, several research teams continue to investigate the integration of such technologies for single UASs and small swarms of UASs for military, commercial, and civilian applications. Our proposed technology can be readily adapted for transparent learning to serve as an assistant for human piloting as well as an emergency *intelligent* autopilot for all manner of piloted vehicles.

Keywords: CBR, Context-Free Grammar, Heuristics, KASER, Machine Learning.

1 Introduction

While any autonomous vehicle equipped with suitable sensors can be autonomously guided by an appropriate set of heuristics, the acquisition of such a base has proven to be elusive to date. This paper offers an explanation for the difficulty and a prescription for success. The problems encountered in developing a set of heuristics to guide an autonomous vehicle include: (a) rules are too crisp and not sufficiently flexible to cover real-world situations; (b) no one knows all of the rules needed in advance; (c) fuzzy logic only addresses quantitative aspects of the rules; whereas, the qualitative aspects are at least as important; (d) conventional rule acquisition is slow, error prone, and otherwise impractical: (e) neural networks or pattern classifiers cannot plan effectively; (f) neural networks and their derivatives cannot learn nearly fast enough to be used in conjunction with a human trainer; (g) cases cannot be autonomously adapted, which limits the applicability of conventional CBR here; and (h) not all cases can be predicted or incorporated *a priori*.

No one would be a pilot if he/she lacked common sense. One can learn how to fly, but without common sense, one would be preordained to fail. Yet, some researchers have sought to build UASs that are lacking in this most critical capability – presumably because they are of the mistaken belief that computers can never exhibit commonsense reasoning beyond that for which they are explicitly programmed. However, knowledge amplifiers, known as KASERs, have been designed [1], which allow for the symmetric expansion of a knowledge base – enabling heuristic programs to reason by analogy – just like human pilots.

While much research has been devoted to learning and machine intelligence, the field is still in its infancy. For example, a technology that will allow for the heuristic exploitation of information domain regularities to reduce the time required for knowledge acquisition while concomitantly resulting in an increase in the reliability of the acquired knowledge is still lacking. Furthermore, contemporary learning mechanisms such as neural architectures are inherently incapable of such performance. The objective of this paper is to present a new way of looking at learning and machine intelligence, which has applicability in many fields such as in robotics (e.g., UAVs), intelligent agents, data fusion, and cooperative sensing. Neural networks cannot compute *Modus Ponens*, have *NP-hard* learning algorithms where one or more hidden layers are required, and are subject to the theoretical limitations imposed by *Gödel's Incompleteness Theorem*. We propose to construct a new type of intelligence that allows fusion and transference of knowledge.

Several researchers have suggested that the learning process is primarily hindered by computational power; however the number of computational devices using embedded software is rapidly increasing and the functional capabilities are becoming increasingly complex. Embedded software often substitutes for functions previously realized in hardware such as custom ICs or the more economical, but slower gate arrays; for example, digital fly-by-wire flight control systems have superseded mechanical control systems in aircraft. Software also increasingly enables new functions, such as intelligent cruise control, pilot assistance, and collision avoidance systems in high-end military aircraft.

However, the great number of source lines of code itself is not a fundamental problem. The main difficulty stems from the ever-more complex interactions *across* software components and subsystems. All too often, coding errors only emerge after use. The software testing process must be integrated within the software creation process – including the creation of systems of systems in a spiral development. This follows because in theory, whenever software becomes complex enough to be capable of self-reference it can no longer be formally proven to be valid [2]. As software gets more complex, one might logically expect the number of components to grow with it. Actually, the exact opposite is true. Engineers are required to obtain tighter integration among components in an effort to address cost, reliability, and packaging considerations, so they are constantly working to decrease the number of software components, but deliver an ever-expanding range of capabilities. The door is thus opened for heuristic validation and this in turn implies software testing as a bona fide method for software synthesis. Unlike the conventional software development cycle, which may differ from the waterfall model because an allowance must be made for parallel development, heuristic validation can be modeled as in [3].

This paper is organized as follows: an introduction to the associate case-based reasoning as a learning mechanism is presented in Section 2. In Section 3, this software tool is applied to the problem of learning to fly a scaled helicopter. Finally, some concluding comments are made in Section 4.

2 An Associative Case-Based Reasoning Approach to Learning

Case-Based Reasoning (CBR) differs from KASER-based approaches in that knowledge is not represented in rules, but in examples. It reasons by comparing a new problem with a set of stored previous problems along with their solution. The solution to the new problem is obtained by retrieving similar problems from memory and adapting their associated solutions to apply to the new problem [4-5]. The main advantage that CBR has over reasoning with rules is that it is far easier to acquire a knowledge base (i.e., rules do not have to be rewritten – new cases are simply acquired). Thus, CBR is more amenable for use in problem domains that are not well understood.

A major problem in CBR, addressed here, pertains to the retrieval of cases that are sufficiently similar to a novel problem. The solutions of the retrieved cases need to be *adapted* to create a solution for the novel problem. The outstanding problem here is that no *similarity measure* can provide high values to cases that are similar to the novel problem in general [6]. Many approaches have been tried and some have succeeded in applying fuzzy logic to CBR; but, these approaches apply to quantitative possibilities and not to the constituent qualitative knowledge [7-8].

The real-time case retrieval and adaptation problem needs to be solved. Existing CBR systems are just too brittle [5]. KASERs can solve the retrieval and adaptation problem through generalization and analogy. However, this approach is hardware and search intensive, in part, because it is not fully user-interactive. User-interaction is so important that it must be considered in determining if the classic outstanding computer science problem; namely, if P = NP is solvable [9]. This paper offers initial results based upon a project that develops software associative memories (AMs) using context-free grammars (CFGs). Intelligent algorithms will be devised for performing iterative grammatical inference [10]. Moreover, UASs will be able to associatively retrieve *situational* and thus *actional* information through the fusion of heterogeneous properties. Our initial design will ascertain these properties through (iterative) human interaction, but they may also be automatically obtained through (iterative) interaction with various organic as well as possibly inorganic (backup) sensory systems. The underpinning science here is mathematically principled.

Algorithms are being developed that will greatly extend the effective size and ease of use of the case base. If the system is wrong, then the correct action is obtained from the user via a CFG and otherwise specified just as with a conventional CBR system. Creativity stems in part from disjunctive substitution. As a result, the system has not only acquired a new case, but iteratively redefined grammatical generalizations thereof. Real-time specification is made possible by chunking or *granular computing* [11]. Thus, machine learning has occurred at two levels. Formulas are being developed for determining the dynamic relative validity (possibility) of an applied case. Such analytics will be based on the cases previous success history and degree of adaptation and related statistics thereof.

An intelligent software system interacts with the user in two principal ways. First, it requests random knowledge be supplied where necessary. Second, it asks the user or knowledge engineer to confirm symmetric knowledge where presented. Note that supplying a selection from a pull-down menu is partially random and partially symmetric in its component tasks.

Clearly, if a knowledge engineer can supply the requested random or symmetric knowledge, then it is through the application of acquired knowledge. It follows that if that knowledge can be captured in a knowledge-based system(s), then the requested tasks can be automated. Furthermore, let our grammar-based system be used as the shell, which acquires the knowledge that the user or knowledge engineer would otherwise be requested. An interesting and inescapable conclusion follows. That is, the only thing not automated would be the acquisition of random knowledge by the auxiliary system(s). In other words, randomness can be defined along a continuum in degrees. What is being claimed is that a network of cooperating grammar-based systems requests knowledge that is random in proportion to the size of its collective randomized knowledge bases.

A consequence of Gödel's Incompleteness Theorem is that countably infinite truths are recursively enumerable, but not recursive [12]. Consider two total computable functions, f, and g. We say that they are total because they are defined over all members of the domain. They are said to be computable because they can be realized by some algorithm. Mutually symmetric domains can be characterized by rulebases consisting of fewer, shorter, and more reusable rules than are the more random pairs. The larger are the domains in the pairing, the more likely are they to embed symmetric constructs. We already know that the world is neither totally random, nor totally symmetric. Indeed, this follows from Gödel's Incompleteness Theorem. The degree of symmetry increases with scale. Were this not the case, then the universe would be random in the limit. Further, if the degree of symmetry did not change with scale, then the notion of chaos would be violated.

In other words, the degree of randomization possible is in proportion to the magnitude of the information, where there is no upper bound. Also, the processor time required to randomize such information is unbounded. Such absolute minimal entropy can of course never be achieved for it would violate the Incompleteness Theorem. However, there is nothing to preclude the construction of randomizing systems of any desired level of utility.

Transformational components need to be represented in a context-free grammar in such a manner as to be context sensitive in their use. For example, one software design may require a quicksort component because it involves a large data set, while another may require an insertion sort because it does not. Clearly, the suggested component can never converge in the absence of context. Furthermore, it is only appropriate to substitute one component for its siblings when that component is provably better. Such is the case when, for example, one component has been run through an optimizing compiler and the other has not.

Without breaking boundaries, the existential optimization is the same as the initial amended grammar. Interaction with the user will serve to further randomize this model. Randomization involves the substitution of existing component definitions into grammar strings. Clearly, this technique involves a heuristic search because the order of substitutions is critical.

The purpose of the inference engine is to select the next rule to be fired, if any, and update the blackboard, if applicable. Here is a high-level view of the applicable algorithm:

1) A most-specific match is sought for the context. Initially, this is done for the lowest subclass (i.e., the primitives). The most-specific question (i.e., if any), along with its previous response, is always first to be presented upon a match. The user or knowledge engineer may subsequently generalize this as necessary.

2) If no specific match can be found, then superclass (i.e., generalize) the instance. Thus, for example, 1.2.3 would be first generalized into 1.2.* Generalize the context from the right to the left; that is, from the most recent predicate to the least-recent predicate. Scan the entire grammar for a match before generalizing the next predicate. Again, seek a most specific match (i.e., longest predicate sequence).

3) Subclasses serve to delimit the search for a predicate when manually browsed. When automatically matching the context, they serve to delimit and otherwise order the space of functions that need be tested. Such processes could not be practically threaded in the absence of such constraints.

4) A special symbol(s) exists at the top of each generalization hierarchy, which provides the user or knowledge engineer with a means to insert knowledge (i.e., a new class or superclass) if nothing relevant is known. It is analogous in function to the idle process of an operating system.

If a component is capable of transforming other components, it is said to be *active*. The same component can be passive in one situation and active in another. Active components can also transform each other and even themselves.

Grammars that consist entirely of passive components allow for the design of relatively inefficient components. This follows because the user may select a locally correct component definition in a pull-down menu, but do so unaware that the resulting global design will be inefficient if that component is used in the local context. The inclusion of active components provides for the capture and reuse of the users expressed knowledge of optimization. Such optimizations can be applied to active components -- including, at least in theory, self-referential optimization. Thus, there need be no attendant inefficiency of scale if the grammar includes active components.

Clearly, randomization is ubiquitous. It must then be the case that the proper space-time tradeoff is defined by the domain to which the system is to be applied. Again, such systems are necessarily domain-specific in keeping with the dictates of the halting problem in computability theory. For example, a web address repeater should be composed entirely of passive components. On the other hand, a search for the deepest most random knowledge must be conducted by pure chance, strange though as it must sound. The great majority of other systems fall between these extremes and it is for those that these randomization techniques need to be detailed.

The associative case-based reasoning approach can thus be summarized as follows:

1. We can design a heuristic grammar to iteratively extract maximum granules.
2. The properties of a granule are determined by its constituent predicates.
3. The machine-generated mnemonic must be hierarchical, categorical, and lexicographic.

4. Grammatical disjuncts/conjuncts allow for creative substitution. This provides for exponential growth in their applicability.
5. Adaptive CBR implies a learning *semantic associative memory* (SAM). It is based on CFGs or some modification thereof for sets and sequences.
6. The antecedent grammar looks like SA → u v w A; A → a b, where A provides the domain context.
7. Then, the consequent is S → consequent1 or SA → consequent2. SA is more specific, where the A is acquired if a disjunct/conjunct proves wrong.
8. The idea is to a) use the CFGs to record all knowledge; b) use suffix symbols to specify domain; c) allow for heuristic randomization of the CFGs (within the grammars themselves) and across domains; and, d) be creative by allowing traversal up the CFG tree for a symbolic match.

3 Preliminary Results

The focus of this paper is to suggest an associative Case-Based Reasoning (CBR) technology which can be applied to a multitude of problems. In order to do this, consider the scenarios required to fly a helicopter. A pilot must consider several parameters such as flight paths, mission requirements, uncertain environments and of course survivability. Thus innovative concepts, technologies, and techniques are required to solve operational problems in the area of identifying unique, common/complementary, or exclusive maneuvers or rules for decisions governing sharing of aircraft control between mission objectives and Sense And Avoid (SAA) concerns.

Next, we present a Patent-Pending natural language processing and learning system, Cbfer, which applies the above principles to learn to understand natural language commands. It does not always correctly interpret the semantics, but it does learn to do so faster than it is taught (i.e., super-linear learning). The more coherent and populated the knowledge base, the greater the rate of super-linear learning. Here, we cold start the system with no knowledge and show how it learns, makes mistakes, and learns to correct those mistakes in a super-linear way.

Figure 1 shows that while flying may sound easy, it is a complex task, which requires the user to learn how to react and use the proportionate radio controls. This can often be counter intuitive.

Fig. 1. Learning to Fly with the Help of an Instructor [13]

In Figure 2, we start the system with no knowledge. Thus, we initially teach it. Next, when presented with a syntactically distinct, but semantically equivalent request the system states that it's semantics is unknown, considerably more likely to be in error than correct (i.e., with a possibility of 0.286 over [0, 1]), and yet properly mapped by the Cbfer algorithm. Note that just as it can learn to map distinct syntax to a common semantics, Cbfer can learn to correct a voice transcription.

```
Enter (multiline) context or "e" to exit:
How do I takeoff?

Enter (multiline) consequent, "u", return, return for UNKNOWN,
or return to search for a match of the context:
Position the helicopter with the nose in the wind.
Lift the helicopter to a height of 1m.

Words file successfully updated with 17 entries.

The CbferCases file was updated with 1 case(s).

Continue (c) or exit (e)?
c

Enter (multiline) context or "e" to exit:
What do I need to do to fly?

Enter (multiline) consequent, "u", return, return for UNKNOWN,
or return to search for a match of the context:

No exact crispy match of the context was found in the case base.
The best fuzzy match of the context was found at case number 0.
The possibility = 0.286 over [0, 1], or [no match, perfect match].

The best fuzzily-matched case antecedent was found to be:

How do I takeoff.
-eoa-

The associated consequent was found to be:

Position the helicopter with the nose in the wind
Lift the helicopter to a height of 1m
-eoc-

Is this the correct consequent for the current context (y)?
```

Fig. 2. The Initial Training

```
Enter (multiline) context or "e" to exit:
How do I fly?

Enter (multiline) consequent, "u", return, return for UNKNOWN,
or return to search for a match of the context:
Move the helicopter sideways 5m, stop, and back again in both directions.
You now get used to the controls of the helicopter without the training gear.

Words file successfully updated with 37 entries.

The CbferCases file was updated with 3 case(s).

Enter (multiline) context or "e" to exit:
It's flying so now what?

Enter (multiline) consequent, "u", return, return for UNKNOWN,
or return to search for a match of the context:

No exact crispy match of the context was found in the case base.
No fuzzy match of the context was found in the case base.

Enjoy yourself - it's later than you think!

Process completed.
```

Fig. 3. Non Sequitur

In Figure 3, we see that Cbfer often knows when it does not know something – a requirement that the military (e.g., DARPA) finds important. What is not shown here is that we will now provide the proper answer to the stated question, "It's flying so now what?" That answer will be, "Prepare to land the helicopter."

In Figure 4, the system finds that, "How do I land?" is likely (i.e., a possibility of 0.75 over [0, 1]) a good match for, "How do I fly?" After all, "fly" is a new word for it in a foreign language. Thus, it is taught the correct meaning of the query by supplying it with a correct consequent for this question.

```
Words file successfully read.
Cases file successfully read.

Enter (multiline) context or "e" to exit:
How do I land?

Enter (multiline) consequent, "u", return, return for UNKNOWN,
or return to search for a match of the context:

No exact crispy match of the context was found in the case base.
The best fuzzy match of the context was found at case number 2.
The possibility = 0.75 over [0, 1], or [no match, perfect match].
```

> The best fuzzily-matched case antecedent was found to be:
>
> How do I fly.
> -eoa-
>
> The associated consequent was found to be:
>
> Move the helicopter sideways 5m stop and back again in both directions
> You now get used to the controls of the helicopter without the training gear
> -eoc-
>
> Is this the correct consequent for the current context (y)?
> n
>
> Enter (multiline) consequent or "u" for UNKNOWN
> - do not enter a carriage return until then:
> Don't land the helicopter if it moves sideways, abort the landing in that case.
> Lift the helicopter to a height of 1m and land it.
>
> Words file successfully updated with 51 entries.
>
> The CbferCases file was updated with 5 case(s).
>
> Continue (c) or exit (e)?

Fig. 4. Process for Error Correction

Next, we provide an admittedly quick demo to show that it has retained a capability to disambiguate the queries as follows.

> Enter (multiline) context or "e" to exit:
> I want to takeoff.
>
> Enter (multiline) consequent, "u", return, return for UNKNOWN,
> or return to search for a match of the context:
>
> No exact crispy match of the context was found in the case base.
> The best fuzzy match of the context was found at case number 0.
> The possibility = 0.5 over [0, 1], or [no match, perfect match].
>
> The best fuzzily-matched case antecedent was found to be:
>
> How do I takeoff.
> -eoa-
>
> The associated consequent was found to be:

Position the helicopter with the nose in the wind
Lift the helicopter to a height of 1m
-eoc-

Enter (multiline) context or "e" to exit:
Show me how to fly.

Enter (multiline) consequent, "u", return, return for UNKNOWN,
or return to search for a match of the context:

No exact crispy match of the context was found in the case base.
The best fuzzy match of the context was found at case number 1.
The possibility = 0.25 over [0, 1], or [no match, perfect match].

The best fuzzily-matched case antecedent was found to be:

What do I need to do to fly.
-eoa-

The associated consequent was found to be:

Position the helicopter with the nose in the wind
Lift the helicopter to a height of 1m
-eoc-

Enter (multiline) context or "e" to exit:
Help me land.

Enter (multiline) consequent, "u", return, return for UNKNOWN,
or return to search for a match of the context:

No exact crispy match of the context was found in the case base.
The best fuzzy match of the context was found at case number 4.
The possibility = 0.25 over [0, 1], or [no match, perfect match].

The best fuzzily-matched case antecedent was found to be:

How do I land.
-eoa-

The associated consequent was found to be:

Don't land the helicopter if it moves sideways abort the landing in that case
Lift the helicopter to a height of 1m and land it
-eoc-

Is this the correct consequent for the current context (y)?
y

```
Words file successfully updated with 56 entries.

The CbferCases file was updated with 8 case(s).

Continue (c) or exit (e)?
e

I thoroughly enjoyed working with you!

Process completed.
```

Fig. 5. The Replay Demonstrates Semantic Understanding

Figure 5 completes our brief introduction to Cbfer. This software can also present questions and feedback the answers to these questions to augment the context. Only the basic features have been demonstrated in this paper.

4 Concluding Remarks

The Associate Case-Based Reasoning approach suggested in this paper provides the required science and technology for decision making by knowledge bases, decision aids, and supporting systems into a highly adaptive human-centric system. A novel associative case-based reasoning (CBR) technology is proposed, which defines an intelligent scalable system. It is not subject to overgeneralization in the way that neural networks are, which can lead to catastrophic failures. Indeed, its definition of intelligence differs from the traditional Turing definition in that it is objectively defined by the rate of acceleration in learning over time. Moreover, the proposed architecture will necessarily include some static sensor systems to further delimit the possibility for catastrophic events (e.g., a radar chip proximity-avoidance sensor, which we will not address further herein). As a consequence, the automatic programming of UAVs becomes distinctly practical.

References

[1] Rubin, S.H., Murthy, S.N.J., Smith, M.H., Trajkovic, L.: KASER: Knowledge Amplification by Structured Expert Randomization. IEEE Transactions on Systems, Man, and Cybernetics – Part B: Cybernetics 34(6), 2317–2329 (2004)
[2] Uspenskii, V.A.: Gödel's Incompleteness Theorem. Ves Mir Publishers, Moscow (1987) (Translated from Russian)
[3] Rubin, S.H., Lee, G.: Intelligent Guidance of an Unmanned Helicopter. In: Learning, Planning and Sharing Robot Knowledge Seminar, Dagstuhl (2010),
http://drops.dagstuhl.de/portals/index.php?semnr=10401
[4] Rubin, S.H. (ed.): Special Issue of ISA Trans. on Artificial Intelligence for Engineering, Design and Manufacturing (1992)
[5] McSherry, D.: Increasing Dialogue Efficiency in Case-Based Reasoning without Loss of Solution Quality. In: IJCAI 2003, Acapulco, Mexico, pp. 121–126 (2003)

6. Pal, S.K., Dillon, T.S., Yeung, D.S. (eds.): Soft Computing in Case Based Reasoning. Springer-Verlag Inc., London (2001)
7. Zadeh, L.A.: From Computing with Numbers to Computing with Words – From Manipulation of Measurements to Manipulation of Perceptions. IEEE Trans. Ckt. Syst. 45, 105–119 (1999)
8. Rubin, S.H.: Computing with Words. IEEE Trans. Syst. Man Cybern. 29, 518–524 (1999)
9. Fortnow, L.: The Status of the P versus NP Problem. Comm. ACM 52, 78–86 (2009)
10. Solomonoff, R.: A Formal Theory of Inductive Inference. Inform. Contr. 7, classic paper, 1–22, 224–254 (1964)
11. Pedrycz, W., Rubin, S.H.: Data Compactification and Computing with Words. Int. J. Engineering Applications of Artificial Intelligence (2009)
12. Chaitin, G.J.: Randomness and Mathematical Proof. Scientific American 232, 47–52 (1975)
13. http://www.rchelicopterweb.com/LearningToFly/LearningToFly.htm

Haptic Data Compression Based on Curve Reconstruction

Fenghua Guo[1,2], Yan He[1], Nizar Sakr[2], Jiying Zhao[2], Abdulmotaleb El Saddik[2]

[1] School of Computer Science and Technology, Shandong University, P.R. China
feguo@site.uottawa.ca, liu_hui@mail.sdu.edu.cn
[2] School of Information Technology and Engineering, University of Ottawa, Canada
nsakr@discover.uottawa.ca, jyzhao@site.uottawa.ca,
elsaddik@mcrlab.uottawa.ca

Abstract. In this paper, the problem of haptic data compression is addressed. The algorithm partitions haptic data samples into subsets while relying on knowledge from human haptic perception. A geometric distance-based approach is used to reduce the number of haptic data subsets. In particular, to improve approximation precision, each haptic data subset is fitted by a quadratic curve. Accordingly, rather than directly using the original haptic data, only the coefficients of the quadratic curves are stored or transmitted. Experiments are performed to compare the suggested curve reconstruction method with the more common linear method. The results prove the effectiveness of the proposed approach in improving data reduction rate and approximation precision.

Keywords: Haptic data compression, human haptic perception, curve reconstruction.

1 Introduction

These days, haptics has been recognized as being compelling to further augment human-to-human and human-to-machine interaction. Especially in telepresence and teleaction (TPTA) systems, the haptic modality plays a central role. Telepresence systems enable a user to be immersed into environments that are distant, inaccessible (including virtual environments), macro- or nano-dimensions, or hazardous [1]. Some example applications include tele-surgery through simulation for medical training, tele-rehabilitation which deals with the delivery of rehabilitation services over telecommunication networks, and more generally, tele-education and tele-entertainment [2]. Haptic data, similar to other types of media such as audio and video, can be processed in online or offline conditions depending on the desired application. In online (networked) conditions, hard delay constraints have to generally be considered to maintain the stability of a networked haptic system. Conversely, in offline haptic data processing conditions, delay constraints and other distortions potentially introduced in network systems (packet loss, jitter, etc.) are not of fundamental importance.

Many haptic applications involve the transmission, storage, and retrieval of haptic data, which depict trajectory, cutaneous and kinesthetic information (i.e., force, position, orientation, torque, velocity, etc). The multidimensional, high-frequency, and data intensive nature of haptic media motivates the research of haptic data compression. Examples of applications in which haptic data compression is of great importance include: The compression of voluminous haptic data files typically produced during a haptic session (i.e., offline compression). In particular, a haptic training session where the user learns to manipulate certain virtual objects can be stored to be later analyzed, or even replayed (e.g., for medical training purposes) [3, 1]. Additionally, in networked haptic (telehaptic) environments (i.e., online compression), it is highly desirable to use compression techniques to reduce haptic data traffic and improve system performance while maintaining a high-quality telehaptic user experience. The focus of this paper is on offline haptic data compression.

In this paper, a novel haptic data compression algorithm is presented. Based on knowledge from human haptic perception, the algorithm partitions haptic data samples into subsets. A geometric distance approach is explorer rather than the more common algebraic distance, to reduce the number of haptic data subsets. Moreover, in order to improve approximation precision, each haptic data subset is fitted by a quadratic curve. Accordingly, only the coefficients of the quadratic curves are stored or transmitted rather than the entire original haptic signal. The rest of the paper is organized as follows. In Section 2 the related work is discussed. In Section 3 the proposed haptic data compression and reconstruction technique is presented. Section 4 presents the experimental results, whereas the conclusion and some directions for future work are highlighted in Section 5.

2 Related Work

Data compression for efficient storage and transmission has been investigated extensively by the information theory and source coding communities. However, the compression and communication of haptic data has not been a field of intense research so far. The related research papers discussed here are those primarily intended for offline haptic data compression. In [4], Kron *et al.* discussed a low-delay coding technique that relies on Differential Pulse Code Modulation (DPCM) together with fixed quantization and Huffman coding. A similar technique was also introduced in [5], however in this case, adaptive DPCM was used, where the quantization step-size is increased or decreased during periods of fast or slow haptic motion respectively. In [6], Shahabi *et al.* presented a method in which a combination of adaptive sampling and adaptive DPCM is performed for the reduction of data samples during storage or transmission of haptic sensory readings. In [7], Borst made use of both lossy uniform and nonuniform quantization of first-order differences in order to investigate the possible data reduction that can be achieved before compression artifacts are haptically sensed. In more recent work [8], Sakr *et al.* presented an autoregressive model for the prediction of both, haptic movement and force information; the model's application for haptic data compression is also discussed. In [9], Sakr *et al.* presented an architecture for lossy and lossless compression of haptic media files. The architecture's primary components consist of the haptic prediction model discussed in [9], as well as a run-length coder and a Huffman-based entropy coder. In [3], Kammerl *et al.* proposed an offline coding technique for haptic media that is primarily intended for haptic playback applications.

Their method relies on human haptic perception (the deadband principle that will be discussed in Section 3). Kammerl *et al.*'s technique is similar to [9] as it consists of a hybrid compression algorithm that combines the concept of the Just Noticeable Differences with predictive and entropy coding in order to encode the required perceptually-significant haptic data samples, using only a few number of bits. Specifically, prediction is performed using a basic first-order linear predictor. The prediction errors are initially encoded using a deadband-based differential coder before being further compressed using adaptive arithmetic coding.

3 Haptic Data Compression and Reconstruction

In this section, the proposed haptic data compression method is presented. The objective is to enable a high data reduction performance and approximation precision, while preserving a high-quality haptic experience during playback of the compressed haptic data streams. The suggested data prediction and reconstruction strategy relies on linear prediction combined with quadratic curve fitting. Knowledge from human haptic perception is incorporated into the architecture to assess the perceptual quality of the compressed haptic signals.

3.1 Haptic Data and Curve Reconstruction

In many science and engineering problems there is often a need to fit a curve or a number of curves to a regularly or irregularly spaced set of points [10]. The points typically correspond to a series of measurements of a physical phenomenon. The objective in this process may be to find patterns using the points and describe the phenomenon analytically. Curve reconstruction has been studied considerably in approximation theory and geometric modeling and many books on the subject have been published [11, 12, 13, 14, 15]. There are many techniques to fit a curve to a set of points by minimizing an error criterion [15, 16, 17, 18, 19]. Haptic data, which are acquired from a series of

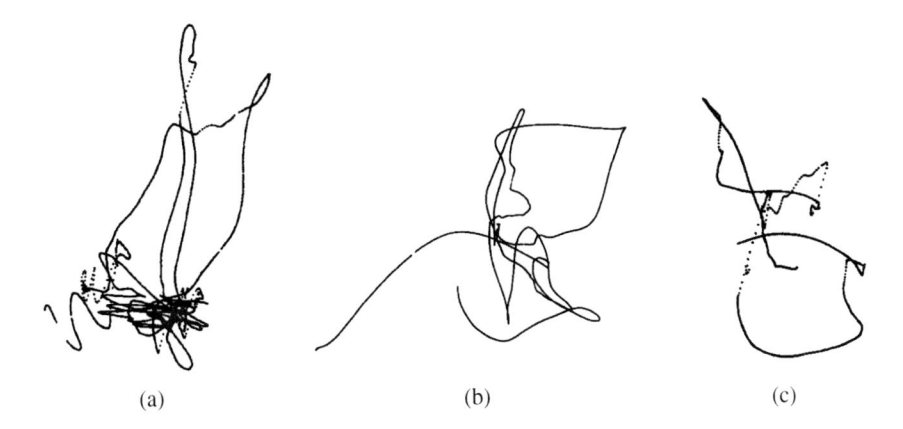

Fig. 1. Haptic data

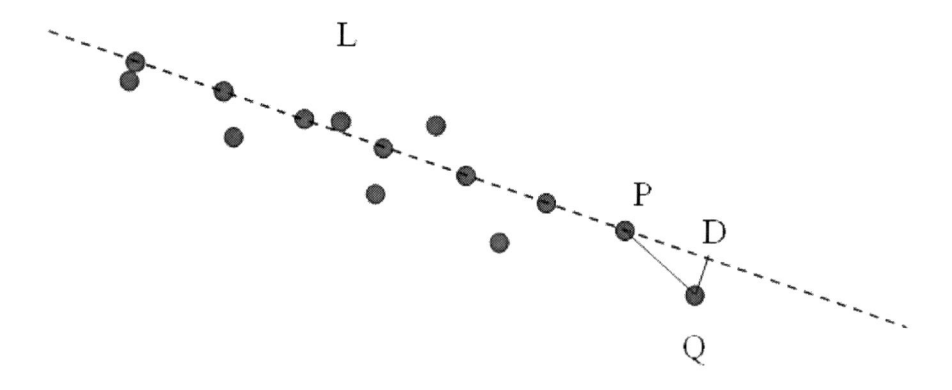

Fig. 2. A visual illustration of the geometric and algebraic distances

sensory measurements acquired at different instances in time, are discrete and ordered in nature. The haptic sensory readings correspond to movement and force data acquired upon user interactions with virtual or remote environments. Haptic signals are generally nonlinear and continuous in nature as illustrated in Fig. 1; in this case, however, only position signals are shown. Therefore, the use of curve reconstruction is a natural choice for haptic data compression. In this paper, quadratic curve data fitting is used to improve approximation precision.

3.2 Haptic Perceptibility and Deadband Approach

The suggested data reduction method relies on the limitations of human haptic perception. In particular, human haptic perception is analyzed using Weber's law of Just Noticeable Differences (JND). The JND consists of the minimum amount of change in stimulus intensity which results in a noticeable variation in sensory experience. This relation can be expressed as

$$\frac{\Delta I}{I} = K \qquad (1)$$

where I is the stimulus intensity, ΔI is the so-called difference threshold or the JND and K is a constant called the Weber fraction. Generally, the JND for human haptic perception ranges from 5% to 15% [20, 21]. This suggests that, if a change in haptic force (or movement) magnitude is less than the JND, the user would not perceive a force-feedback (or movement variation). Weber's law provides a very simple mathematical model to characterise human haptic perception. It essentially provides an approximate model that allows the detection of perceptible changes in haptic signals. In the haptic data compression literature, a haptic perception threshold is generally referred to as a *deadband* [22]. Generally, the deadband principle states that haptic signal changes (e.g., due to prediction) do not need to be stored or transmitted, unless they exceed a certain perceptual threshold. For example, in [23, 24], a method that relies on a linear predictor and knowledge from human haptic perception (and the deadband principle) is presented. Specifically, haptic data samples are only transmitted (or stored) if their prediction falls

Fig. 3. (a) Algebraic distance-based linear prediction; (b) a quadratic curve is fitted to each subset partitioned by the geometric distance-based linear prediction

outside a tolerable error boundary (deadband violation). However, a possible drawback of the methods so far explored in the haptic compression literature [23, 24, 22] is their use of the algebraic distance when evaluating prediction errors. Fig. 2 provides a visual distinction between algebraic and geometric distances. Specifically, point Q denotes the original sample, line L is the output of the linear predictor, point P is its predicted point (that falls on line L). The distance between Q and P which is in fact measured as $\| P - Q \|$ corresponds to the algebraic distance between P and Q. Generally, as haptic data represent highly nonlinear signals, the linear predictor can not predict the haptic signal accurately. Moreover, the calculated algebraic distance (from P to Q) does not represent the true minimal distance between Q and the potential predictor output, which in fact corresponds to the (geometric) distance from Q to the prediction line L (point D).

Consequently, given the same perception threshold and using the geometric distance approach (instead of the common algebraic distance) a haptic data reduction performance improvement can be expected. For example, as illustrated in Fig. 3, considering the same haptic dataset, perception threshold, and using the same linear prediction method, the output generated using the geometric distance clearly provides a better and smoother result. Specifically, when the geometric distance is considered, the haptic signal is partitioned into 4 subsets. Conversely, when relying on the algebraic distance, the linear prediction technique partitions the data into 8 subsets. Moreover, in the proposed method, once the haptic data are partitioned using the geometric distance, the computed subset is fitted by a quadratic curve to further improve approximation precision.

3.3 The Proposed Algorithm

A detailed description of the proposed algorithm is as follows. First, the algorithm constructs a linear predictor which is then used to divide a haptic signal into subsets, i.e., haptic data samples in the same subset are those that can be predicted (within a tolerable perceptual error ε) using the linear predictor. Then, samples in each subset are fitted by a quadratic curve. The procedure of the proposed method is divided into the following steps. Given a predefined perception threshold ε and a haptic data set, repeat the following three steps until there are no more haptic data samples left (these steps are repeated for each subset).

1- Given the ordered haptic data samples set, M (M is the filter length, $M \geq 3$) data points $P_0, P_1, \ldots, P_{M-1}$ are considered in sequence. Based on the M samples, a parametric line $P(u) = a_0 + ua_1$ is constructed, where a_0 and a_1 are the coefficients of the line, u is a parameter that denotes the index of each sample in the sequence. Equation $P(u)$ acts as the linear predictor.
2- For the successive sample Q in the haptic data set do the following:
   If (the perpendicular distance from Q to line $P(u)$ is less than the tolerable distortion ε)[1];
   Then Q can be approximated by line P(u), goto step 2;
   Else Q can not be approximated by line $P(u)$; Q will be the first point of the next point subset which will be approximated by a new line.
3- All the data points which can be approximated by a line $P(u)$ are fitted by a parametric quadratic curve $P(u) = a_0 + ua_1 + u^2 a_2$, where a_0, a_1, a_2 are the coefficients of the curve, u is a parameter that denotes the index of each sample in a subset sequence. Then store (or transmit) only a_0, a_1, a_2. The haptic subset can be reconstructed by simply using the curve coefficients a_0, a_1, a_2.

In order to obtain precise linear predictions and parametric quadratic curve estimations, the least-squares method is used. The parametric line $P(u) = a_0 + ua_1$ that would best fit the original haptic data samples $P = [P_0, P_1, \ldots, P_{M-1}]^T$ in the least squares sense is derived. The equation is then formulated as follows:

$$\phi A = P, \qquad (2)$$

where,

$$\phi = \begin{bmatrix} 1 & u_0 \\ 1 & u_1 \\ \vdots & \vdots \\ 1 & u_{M-1} \end{bmatrix}, \quad A = \begin{bmatrix} a_0 \\ a_1 \end{bmatrix}, \quad P = \begin{bmatrix} P_0 \\ P_1 \\ \vdots \\ P_{M-1} \end{bmatrix},$$

A is the unknown to be solved. $u_0, u_1, \ldots, u_{M-1}$ are the parameters corresponding to the haptic data $P = [P_0, P_1, \ldots, P_{M-1}]^T$. In addition, the filter length M has been selected such that a compromise between reliable velocity estimation and fast computation is achieved. Median filtering is also used to eliminate any fine irregularities and outliers remaining in the computed velocity (first derivative) measures, while preserving the true signal discontinuities [24]. The solution to (2) in the least-squares sense can be found to be $A = (\phi^T \phi)^{-1} \phi^T P$.

---

[1] It should be mentioned that this condition is only used when position data are considered. Weber's law is used when force data are processed; i.e., the condition is replaced with *If* (the perpendicular distance from Q to line $F(u)$ is less than $K \cdot F(u)$).

Similarly, the quadratic curve $P(u) = a_0 + ua_1 + u^2 a_2$ that would best fit the original haptic data samples $P = [P_0, P_1, \ldots, P_{k-1}]^T$ in the least squares sense is also derived (k is the length of the subset). The equation is then formulated as follows:

$$\phi' A' = P, \qquad (3)$$

where,

$$\phi' = \begin{bmatrix} 1 & u_0 & u_0^2 \\ 1 & u_1 & u_1^2 \\ \vdots & \vdots & \vdots \\ 1 & u_{k-1} & u_{k-1}^2 \end{bmatrix}, \quad A' = \begin{bmatrix} a_0 \\ a_1 \\ a_2 \end{bmatrix}, \quad P = \begin{bmatrix} P_0 \\ P_1 \\ \vdots \\ P_{k-1} \end{bmatrix},$$

A' is the unknown to be solved. $u_0, u_1, \ldots, u_{k-1}$ are the parameters corresponding to the haptic data $P = [P_0, P_1, \ldots, P_{k-1}]^T$. Similarly to the linear case, the solution to (3) in the least-squares sense is $A' = (\phi'^T \phi')^{-1} \phi'^T P$.

4 Results

The experimental data consists of haptic movement (position) and force data produced from the haptic interaction with a virtual environment. The haptic device is a PHANToM Omni (SensAble Technologies, Inc., Woburn, MA, USA), whereas the graphical display is a haptic-enabled virtual environment developed for training novice surgeons to perform cataract eye surgeries [25]. In addition, position and force data acquisition is performed at 1 kHz, as this can be found to be necessary to create compelling and stable force feedback. In Fig. 4 and Table 1, a comparison of the proposed method with the general linear method is provided when 3D position data are considered, using a filter length $M = 10$, and different perception thresholds ε. Similarly, in Fig. 5 and table 2, a comparison of the proposed method with the general linear method is provided when 3D force data are considered using also a filter length $M = 10$, and different Weber fractions K (deadband).

In Tables 1 and 2, the data reduction rate is expressed in terms of the number of curve segments computed using the proposed method over the number of linear segments computed using the general linear approach. In particular, the number of stored

Table 1. Comparison of the proposed method with the general linear method when position data are considered

Position Deadband (ε)	Curve Segments/Linear Segments	Curve MSE/Linear MSE
0.8	39/61	$2.85 \cdot 10^{-2}/1.22 \cdot 10^{-1}$
0.4	67/88	$2.19 \cdot 10^{-3}/2.88 \cdot 10^{-2}$
0.2	96/127	$3.43 \cdot 10^{-4}/7.08 \cdot 10^{-3}$
0.1	135/177	$1.31 \cdot 10^{-4}/1.85 \cdot 10^{-3}$
0.05	185/238	$5.00 \cdot 10^{-5}/4.76 \cdot 10^{-4}$
0.025	260/324	$2.16 \cdot 10^{-5}/1.55 \cdot 10^{-4}$

(a) $\varepsilon = 0.8$

(b) $\varepsilon = 0.4$

(c) $\varepsilon = 0.2$

(d) $\varepsilon = 0.1$

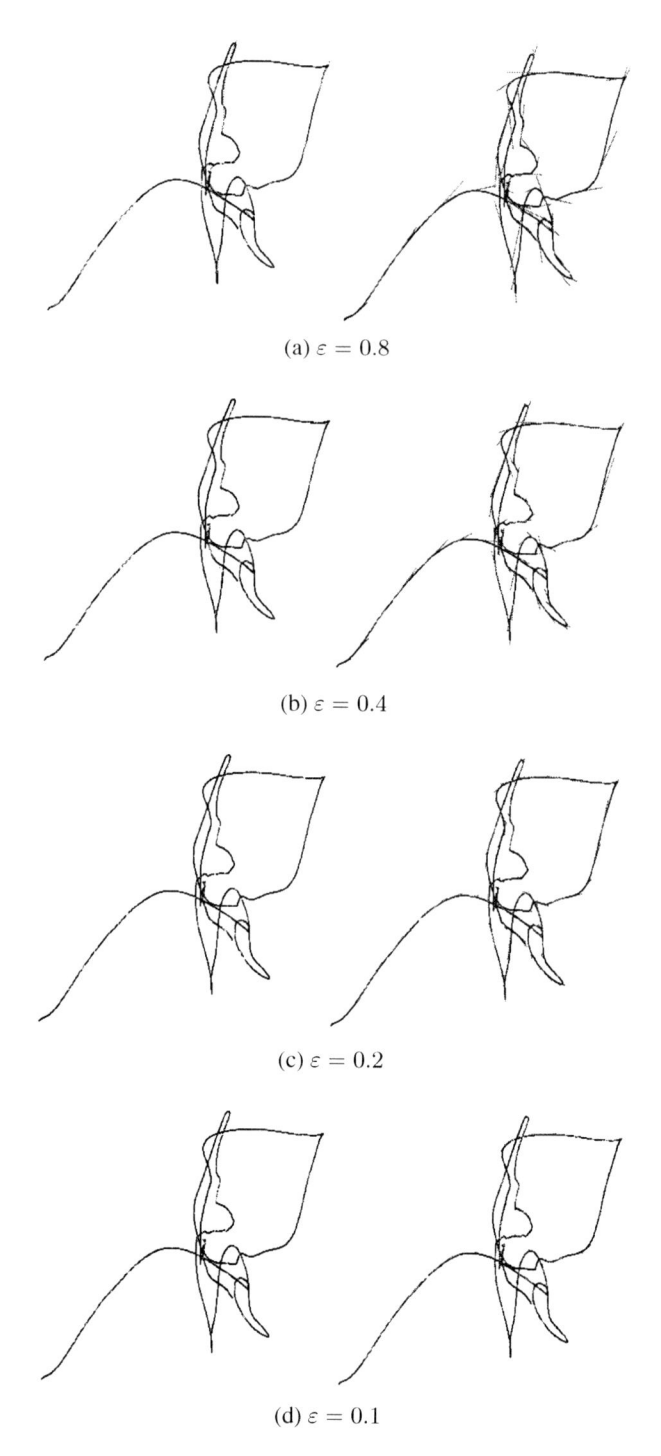

Fig. 4. Plots comparing the output of the proposed data reduction approach (left side of each subplot) with the general linear method (right side of each subplot) when position data are considered

(a) $K = 15\%$

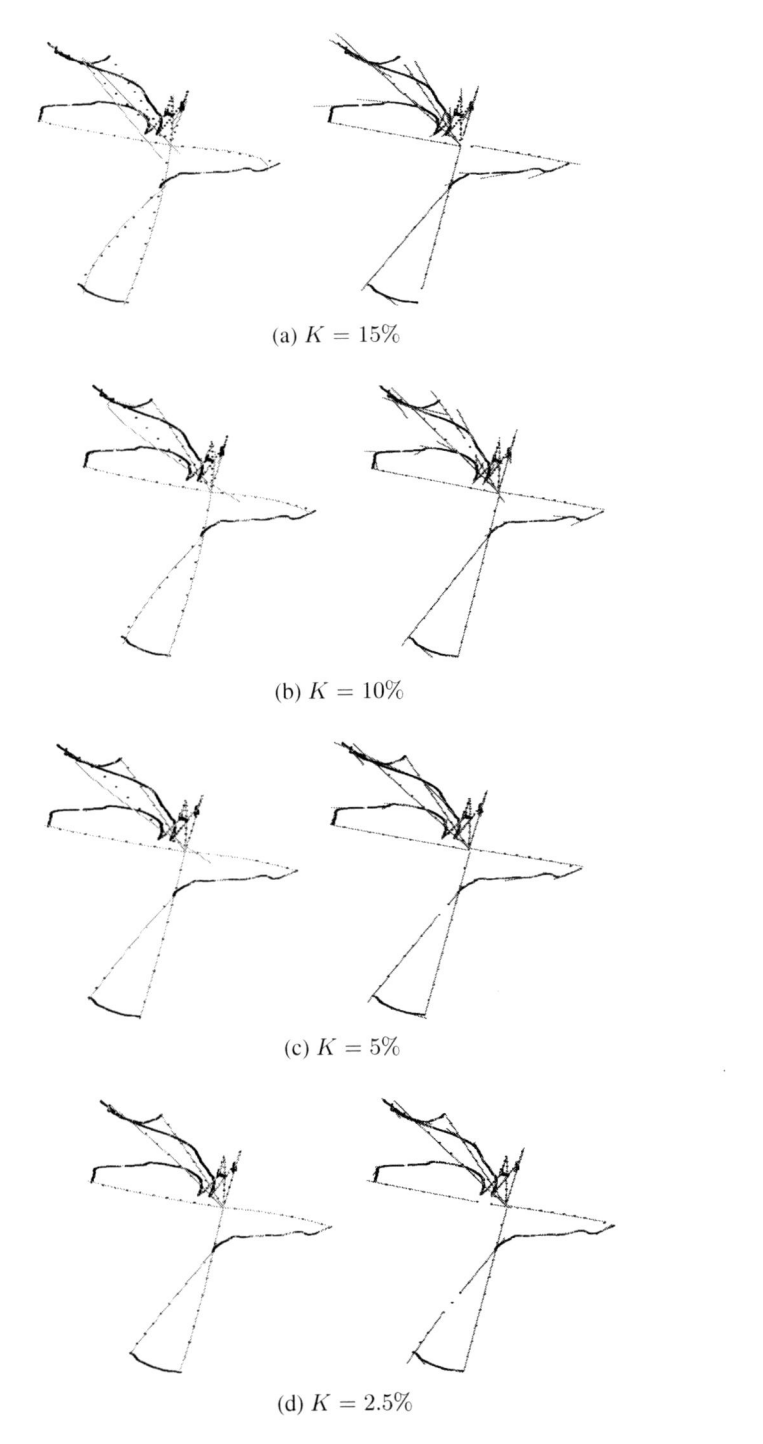

(b) $K = 10\%$

(c) $K = 5\%$

(d) $K = 2.5\%$

Fig. 5. Plots comparing the output of the proposed data reduction approach (left side of each subplot) with the general linear method (right side of each subplot) when force data are considered

Table 2. Comparison of the proposed method with the general linear method when force data are considered

Force Deadband (K)	Curve Segments/Linear Segments	Curve MSE/Linear MSE
15%	23/36	$1.04 \cdot 10^{-2}/4.90 \cdot 10^{-2}$
10%	27/44	$3.33 \cdot 10^{-3}/2.46 \cdot 10^{-2}$
5%	37/54	$1.06 \cdot 10^{-3}/7.40 \cdot 10^{-3}$
2.5%	53/71	$1.57 \cdot 10^{-4}/1.49 \cdot 10^{-3}$
1%	77/93	$8.90 \cdot 10^{-5}/2.37 \cdot 10^{-4}$
0.5%	101/123	$3.00 \cdot 10^{-5}/6.05 \cdot 10^{-5}$

curve coefficients = stored curves×3; this is due to the fact that a parametric quadratic curve is described using three coefficients. The number of stored coefficients of linear segments = stored lines×2; this is because a parametric line is described using only two coefficients. From Table 1 and Fig. 4, it can be clearly seen that for different position perception thresholds (ε) the proposed geometric distance/quadratic curve fitting strategy outperforms the general linear approach. First, the number of curves to be transmitted/stored are fewer, and second, the reconstructed compressed curves have consistently a lower mean squared error ($MSE = \frac{1}{N}\sum_{i=0}^{N-1}(\| p(i) - \hat{p}(i) \|)^2$) and are noticeably smoother when the proposed method is considered. Similarly, from Table 2 and Fig. 5, it is can be observed that for different force deadband values (K) the proposed geometric distance/quadratic curve fitting approach outperforms the general linear method. First, the number of curves to be transmitted/stored are fewer, and second, the reconstructed compressed curves have lower MSE values and are clearly smoother when the proposed data reduction algorithm is exploited.

5 Conclusion

In this paper, an offline haptic data reduction algorithm was proposed. The algorithm uses the geometric distance and quadratic curve fitting to improve data reduction rate and signal approximation precision. Furthermore, the limitations of human haptic perception are considered in the method to insure that distortions introduced by the compression scheme remain imperceptible to the user. The experimental results proved the effectiveness of the proposed approach. Future work consists of further testing this algorithm using more datasets generated from various haptic applications. Finally, the data reduction method will be extended to operate in online conditions, i.e., to reduce haptic data packets in telehaptic applications.

References

1. Steinbach, E., Hirche, S., Kammerl, J., Vittorias, I., Chaudhari, R.: Haptic data compression and communication for telepresence and teleaction. IEEE Signal Processing Magazine 28(1), 87–96 (2011)

2. Sakr, N., Zhou, J., Georganas, N.D., Zhao, J., Petriu, E.M.: Robust perception-based data reduction and transmission in telehaptic systems. In: Proc. of World Haptics, pp. 214–219 (March 2009)
3. Kammerl, J., Steinbach, E.: Deadband-based offline-coding of haptic media. In: Proc. of ACM International Conference on Multimedia, pp. 549–558 (2008)
4. Kron, A., Schmidt, G., Petzold, B., Zah, M.F., Hinterseer, P., Steinbach, E.: Disposal of explosive ordnances by use of a bimanual haptic telepresence system. In: Proc. of IEEE International Conference Robotics and Automation, pp. 1968–1973 (April 2004)
5. McLaughlin, M.L., Hespanha, J.P., Skhatme, G.S.: Lossy compression of haptic data. In: McLaughlin, M. (ed.) Touch in Virtual Environment: Haptics and the Design of Interactive Systems. IMSC Press, Prentice Hall (2002)
6. Shahabi, C., Ortega, A., Kolahdouzan, M.R.: A comparison of different haptic compression techniques. In: Proc. of IEEE International Conference on Multimedia and Expo, pp. 657–660 (August 2002)
7. Borst, C.W.: Predictive coding for efficient host-device communication in a pneumatic force-feedback display. In: Proc. of Joint Eurohaptics Conference and Symposium on Haptic Interfaces for Virtual Environment and Teleoperator Systems, pp. 596–599 (March 2005)
8. Sakr, N., Georganas, N.D., Zhao, J., Shen, X.: Motion and force prediction in haptic media. In: Proc. of IEEE International Conference on Multimedia and Expo, pp. 2242–2245 (July 2007)
9. Sakr, N., Georganas, N.D., Zhao, J., Shen, X.: Towards an architecture for the compression of haptic media. In: Proc. of IEEE Int. Conf. on Virtual Environments, Human-Computer Interfaces and Measurement Systems, pp. 13–18 (June 2007)
10. Ardeshir Goshtasby, A.: Grouping and parameterizing irregularly spaced points for curve fitting. ACM Transactions on Graphics 19(3), 185–203 (2000)
11. Beach, R.C.: An Introduction to the Curves and Surfaces of Computer-Aided Design. Van Nostrand Reinhold Co., New York (1991)
12. Farin, G.: Curves and Surfaces for Computer Aided Geometric Design. Academic Press Prof., Inc., San Diego (1990)
13. Faux, I.D., Pratt, M.J.: Computational Geometry for Design and Manufacture. Ellis Horwood, Upper Saddle River (1979)
14. Hagen, H.: Curve and Surface Design. SIAM, Philadelphia (1992)
15. Lancaster, P., Salkauskas, L.: Curve and Surface Fitting. Academic Press, Inc., New York (1986)
16. Cohen, E., O'dell, C.L.: A data dependent parametrization for spline approximation. In: Lyche, T., Schumaker, L.L. (eds.) Mathematical Methods in Computer Aided Geometric Design, pp. 155–166. Academic Press Prof., Inc., San Diego (1989)
17. Fritsch, F.N., Carleson, R.E.: Monotone piecewise cubic interpolation. SIAM Journal on Numerical Analysis 17(2), 238–246 (1980)
18. Kosters, M.: Curvature-dependent parametrization of curves and surfaces. Computer Aided Desgn 23(8), 569–578 (1991)
19. Sarkar, B., Menq, C.-H.: Parameter optimization in approximating curves and surfaces to measurement data. Computer Aided Desgn 8(4), 267–290 (1991)
20. Jones, L.A., Hunter, I.W.: Human operator perception of mechanical variables and their effects on tracking performance. ASME Advances in Robotics 42, 49–53 (1992)
21. Burdea, G.C.: Force and touch feedback for virtual reality. Wiley, New York (1996)
22. Hinterseer, P., Hirche, S., Chaudhuri, S., Steinbach, E., Buss, M.: Perception-based data reduction and transmission of haptic data in telepresence and teleaction systems. IEEE Transactions on Signal Processing 56(2), 588–597 (2008)

23. Sakr, N., Zhou, J., Georganas, N.D., Zhao, J., Shen, X.: Prediction-based haptic data reduction and compression in tele-mentoring systems. In: Proc. of IEEE Instrumentation and Measurement Technology Conference, pp. 1828–1832 (May 2008)
24. Sakr, N., Zhou, J., Georganas, N.D., Zhao, J.: Prediction-based haptic data reduction and transmission in telementoring systems. IEEE Transactions on Instrumentation and Measurement 58(5), 1727–1736 (2009)
25. Shen, X., Zhou, J., Hamam, A., Nourian, S., El-Far, N.R., Malric, F., Georganas, N.D.: Haptic-enabled tele-mentoring surgery simulation: design, implementation and evaluation. IEEE Multimedia 15(1), 64–76 (2008)

Diastolic Timed Vibrator: Applying Direct Vibration in Diastole to Patients with Acute Coronary Ischemia during the Pre-hospitalization Phase

Farzad Khosrow-khavar[1], Marcin Marzencki[1,2], Kouhyar Tavakolian[1,2], Behrad Kajbafzadeh[1,2], Bozena Kaminska[2], and Carlo Menon[1]

[1] MENERVA Group, Simon Fraser University, Burnaby, BC
[2] CIBER Lab, Simon Fraser University, Burnaby, BC
cmenon@sfu.ca

Abstract. A biomedical device that applies vibration to the chest of a patient during the pre-hospitalization phase of acute coronary ischemia is introduced. This device improves the blood flow in order to remove the thrombosis. In order to help the heart pumping mechanism, vibration is applied during diastole only and hence this device is called a Diastolic Timed Vibrator (DTV). We introduce two algorithms (slope-lookup and slope-energy) to detect diastole by using an electrocardiogram signal available in most ambulances and pre-emergency units. The robustness of the algorithms were examined by considering many types of arrhythmias that might occur during acute ischemia. We show that the slope-energy is a robust method and can be used to detect diastole interval in most types of arrhythmias.

Keywords: Diastolic Timed Vibrator(DTV), Thrombosis, ECG, Diastole Detection, Hamilton and Tompkins filter.

1 Introduction

According to the world health organization, ischemic heart disease accounts for 12.5 percent of annual death rate in the world [1]. The main reason for ischemia is formation of blood clot or thrombus in the blood vessel that reduces the blood flow and can have short and long term effects. Depending on the location and size of the blood clot, a heart attack or stroke can occur. Heart attack or Myocardial Infarction (MI) occurs when the blood clot prevents the flow of blood to the heart leading to destruction of the heart muscles. On the other hand, if the blockage occurs in the artery that supplies blood to the brain, stroke occurs.

Blood clotting is a complicated process which takes place during an injury to body tissues. This is very important protective mechanism. However, pathophysiological processes could potentially generate different sizes of blood clots [3]. The process of break-down or dissolution of blood clot has always been a challenge for researchers. The suggested solutions range from using non-invasive methods such as injection of thrombolic drugs [4] and applying low frequency ultrasound [5–8] to using invasive methods such as angioplasty.

Two important studies by Pozen, et al. illustrate the importance of pre-hospitalization phase of treatment of ischemic heart disease [2]. According to these studies, majority of acute ischemic deaths occur prior to the patient arriving to the hospital. Consequently, we propose the design of a non-invasive device that applies vibration during the diastolic cardiac cycle directly to the chest of the patient. We refer to this device as the Diastolic Timed Vibrator or DTV in short. Depending on the clot size and density, the applied vibration could potentially either directly disintegrate the clot or improve the blood flow that leads to removal of the clot [9]. As it was previously shown, this method can be used either in conjunction with thrombolytic agent or as a stand-alone method [9].

While our method applies direct vibrations to the chest, other types of vibration methods have been utilized. Non-invasive ultrasonic methods use waves with different frequencies and intensities to enhance thrombolysis for therapeutic applications. Such methods apply frequencies from 48 KHz upto 1 MHz with various intensities [5–8]. Furthermore, many research have shown the positive effect of mechanical vibrations in increasing the blood flow in various parts of the body [11–14].

The paper is organized as follows: In section 2, the principle operation of the DTV is explained. Section 3 illustrates a detailed implementation of our device. In section 4, we show the effectiveness of the diastole detection algorithm when applied to different types of cardiac rhythms.

2 Principle of DTV

The human heart could be represented as two synchronized pumps that circulates blood to the rest of the body. Each side of the heart consists of two chambers: atrium and ventricle. The right side receives the de-oxygenated blood from the rest of the body and pumps it to the lung whereas the left side receives the oxygenated blood from the lungs and pump it to the rest of the body. The pumping mechanism is periodic and is referred to as the cardiac cycle. During this period, the ventricle goes through two phases: systole and diastole. During diastole, the ventricles relax and blood starts to fill in slowly while the valves between atria and ventricle slowly open due to pressure difference. Just at the end of this cycle, the atria start to contract and push more blood in ventricle. During the systole, the ventricle contracts and pumps blood to the rest of the body.

As it was mentioned before, It has been shown that applying external vibrations during the diastole can essentially increase the blood flow both in humans and animals [9, 11]. Furthermore, studies have shown that applying low frequency vibrations can lead to recanalization of thrombosed flow system held at atrial like pressure [12–14].

The cardiac cycle is synchronized by electrical activity of the heart muscles. By placing electrodes on human body, this electrical activity can be recorded.

This type of recording is referred to as Electrocardiogram (ECG). Figure. 1 shows a sample ECG signal. The different segments of the ECG are annotated as P, QRS and T wave. As it is also shown in Fig. 1, the P segment corresponds to the depolarization of the atrial muscle, the QRS complex to the depolarization of the ventricle and the T wave to the relaxation and the re-polarization of the ventricle. Diastole starts from the middle of the T-wave to the middle of the R-wave [10, 11].

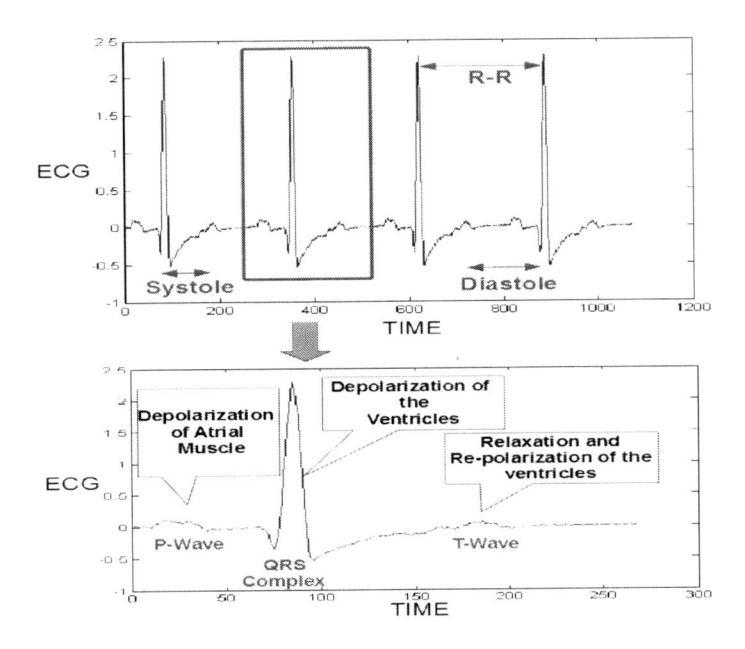

Fig. 1. The Cardiac Cycle and ECG Segments

Our device consists of a commercial massager that is controlled and synchronized with the diastole period. During this interval, the heart is relaxing and the induced vibrations does not have any adverse effect on the heart.

In order to examine the DTV, both normal and abnormal heart rhythms (arrhythmias) were applied to the device. In order to be successful, our device should be able to detect the diastole/systole interval in order to limit the vibration during diastole only. The heart rhythms used are listed in Table.1.

3 Implementation of DTV

3.1 System Implementation

Figure. 2 shows various sub-systems used for implementation of DTV. For the experimentation phase, a Fluke PS410 ECG simulator was used to synthetically

Table 1. Heart Rhythm and Arrhythmia Types

Normal ECG
Premature Atrial Contraction (PAC)
Premature Nodal Contraction (PNC)
Premature Ventricular Contraction (PVC)
Bigeminy Rhythm
Trigeminy Rhythm
Ventricular Tachycardia
Supraventricular Tachycardia
Atrial Tachycardia
First/Second/Third Degree Block Rhythm
Right Bundle Branch Block (RBBB)
Left Bundle Branch Block (LBBB)
ST Elevation
Pacemaker Rhythm

generate ECG signals. This device is capable of generating 35 different types of arrhythmias and heart rhythms with different beat rates. After the experimentation phase, we will use an ECG system that connects directly to the human chest.

The acquired ECG signal is in millivolt range and thus should be amplified in order to have enough resolution required for the analog to digital conversion. Accusync72 is a commercial system (certified under FDA) capable of amplification as well as detection of the diastole. The amplified signal/diastole-detected output wave of the Accusync72 is fed into the DAQ card.

The diastole detection algorithm that we implemented as a Labview module is compared with the Accusync72 output. The Swan vibrator is modified by adding a relay that can turn the motor on during the diastole and turn it off during the systole. The PC software controls vibration during the diastole by switching the relay. Furthermore, a feedback PID controller is used to set the frequency of vibration by controlling the input voltage. By doing so, we can find the best frequency or range of frequencies that achieves the best result. This is accomplished by controlling the DC power-supply which uses the RS-232 serial protocol as the means of communication. The feedback is accomplished by acquiring data from the accelerometer mounted on the vibrator.

3.2 Software Implementation

There has been extensive amount of research in extracting the QRS segment of the ECG signal. The survey by Kohler et al. summarizes all these algorithms in detail [15]. These implementations range from using simple, filter based, to very complex machine learning algorithms. We used a modified version of Hamilton and Tompkins algorithm [16, 17]. Figure.3 shows the block diagram of the Hamilton and Tompkins filter.

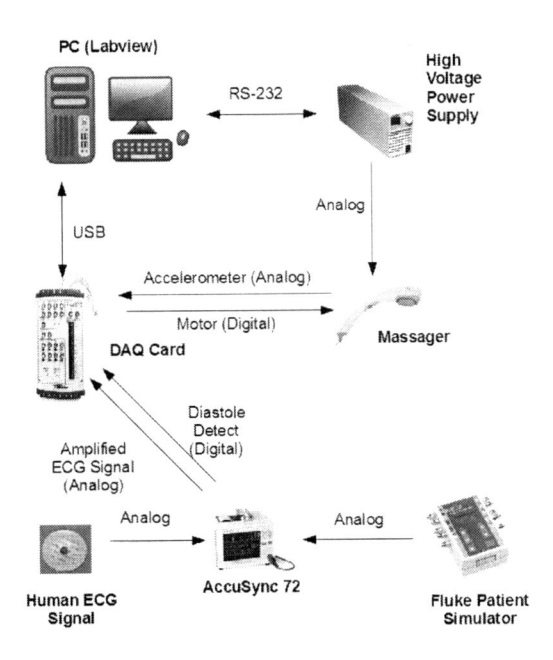

Fig. 2. The Overall System Set-up for Diastolic Timed Vibrator

A band-pass (combination of a low-pass (LP) and a high-pass (HP)) filter extracts the segment of the signal where most of the energy resides. In the next stage, the signal is differentiated to find the slope and then is squared to make it positive. The integrator sums the area of the positive wave form. The width of the integrator (32 in Fig.3) is chosen carefully to be long enough to consider abnormally long QRS sectors and short enough so that it does not overlap with the T-wave [17].

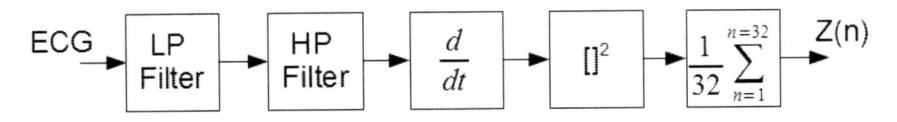

$$ECG \rightarrow \boxed{\text{LP Filter}} \rightarrow \boxed{\text{HP Filter}} \rightarrow \boxed{\frac{d}{dt}} \rightarrow \boxed{[\,]^2} \rightarrow \boxed{\frac{1}{32}\sum_{n=1}^{n=32}} \rightarrow Z(n)$$

Fig. 3. The Block Diagram of the Hamilton and Thompkins [17]

The research by Andrew Hoffman et al. shows that the duration of the systole starts on the onset of the R-wave and ends in the middle of the T-wave where the diastole starts [11]. We realized that the output of Hamilton and Tompkins filter matches very well with the duration of the systole. During the onset of the R-wave, the slope of the filter is very steep which can be used to detect the R-wave (we refer to it as slope-method). During the systole, the minimum value of samples are higher than certain threshold. At the end of the systole (that corresponds to the period between the middle and the end of the T-wave), the value of the filter falls under this threshold. In order to exclude outliers, the average of 8 consecutive samples were used and consequently was used to compare with the fixed threshold (we refer to this as energy-method).

We propose two algorithms to calculate the duration of the diastole/systole period. In the first method, the slope method is used to detect the R-wave and a lookup table is used to detect the systole/diastole period. This method is referred to as slope-lookup method. In the second approach, the slope method is used to detect the R-wave and the average energy method is used to calculate the systole period. This method is referred to as slope-energy method.

4 Experimental Result

The lookup table for the slope-lookup method is summarized in Table.2. The data is obtained from the normal ECG's of the Fluke patient simulator with various beat rates. This type of patient simulator is used for training purposes in hospitals and has a very realistic ECG that emulates a real patient. The systole periods were calculated from the onset of the R-wave to the middle of the T-wave [11]. If the exact value of the beat rate was not found in the table, a linear interpolation between two consecutive entries were used to extract the diastole and systole period. The fixed threshold for the slope-energy method was found to be 0.25.

Figure.4 - Figure.6 show the result of both the slope-lookup and slope-energy method for normal ECG with 80 BPM, normal ECG with 220 BPM, Premature Ventricular Contraction (PVC), Premature Atrial Contraction (PAC), bigeminy, trigeminy, pacemaker rhythm, Right Bundle Branch Block (RBBB), Left Bundle Branch Block (LBBB), ST elevation with +0.6 mV, ST elevation with -0.6mV, tachycardia and supraventricular tachycardia. Each figure also shows

Table 2. Lookup Table for R-R , Diastole and Systole Time Intervals

Beat Per Minute	R-R Period [sec]	Diastole [sec]	Systole [sec]
30	2.0	1.59	0.41
40	1.5	1.16	0.34
60	1.0	0.69	0.31
80	0.75	0.44	0.31
100	0.60	0.34	0.26
120	0.50	0.23	0.27
140	0.43	0.20	0.23
160	0.38	0.18	0.20
180	0.33	0.19	0.14
200	0.30	0.18	0.12
220	0.27	0.13	0.14
240	0.25	0.10	0.15
260	0.23	0.09	0.14
280	0.21	0.08	0.13
300	0.20	0.07	0.13

Diastolic Timed Vibrator 361

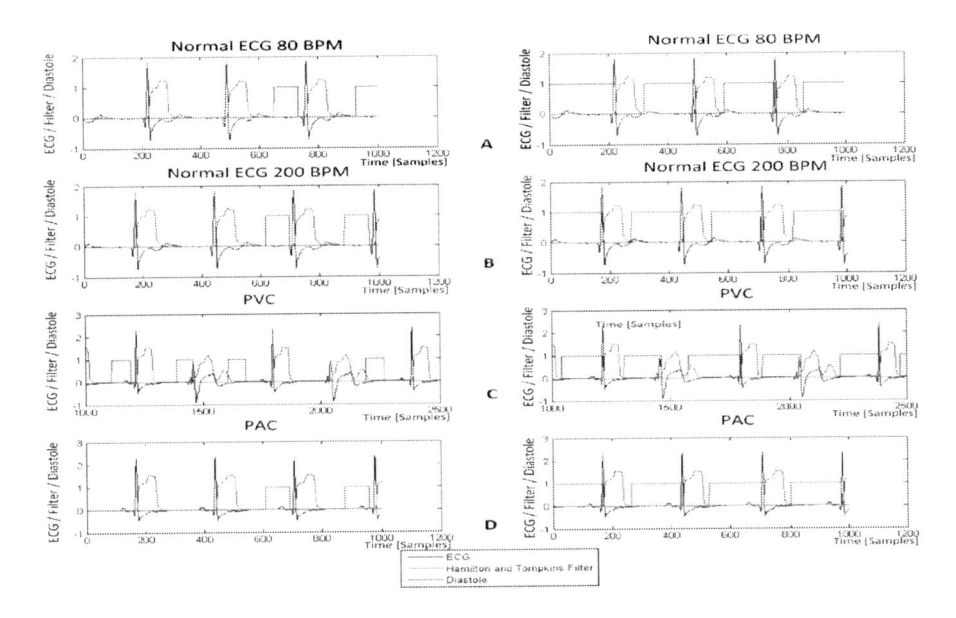

Fig. 4. DTV Operation using Slope-Lookup (Left) and Slope-Energy (Right) for A. Normal ECG with 80 BPM, B. Normal ECG with 220 BPM, C. Premature Ventricular Contraction (PVC) and D. Premature Atrial Contraction (PAC).

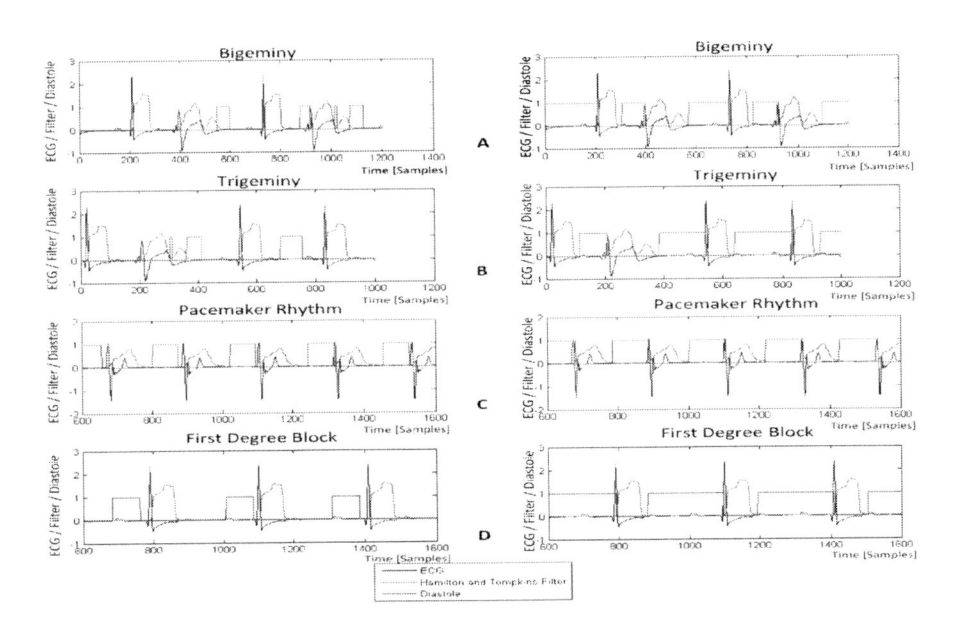

Fig. 5. DTV Operation using Slope-Lookup (Left) and Slope-Energy (Right) for A. Bigeminy, B. Trigeminy, C. Pacemaker Rhythm and D. First Degree Block.

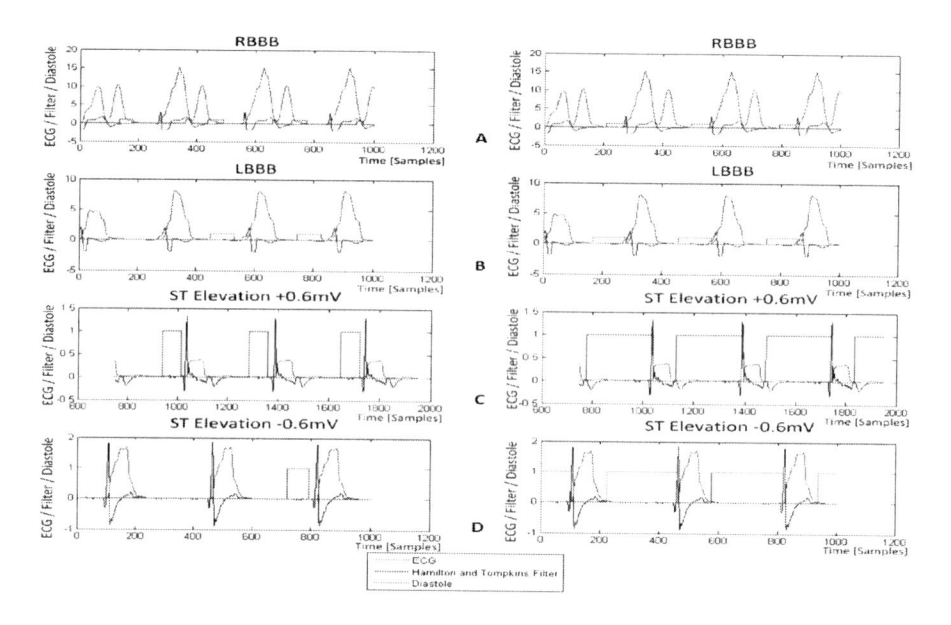

Fig. 6. DTV Operation using Slope-Lookup (Left) and Slope-Energy (Right) for A. Right Bundle Branch Block (RBBB), B. Left Bundle Branch Block (LBBB), C. ST Elevation +0.6mV and D.ST Elevation -0.6 mV.

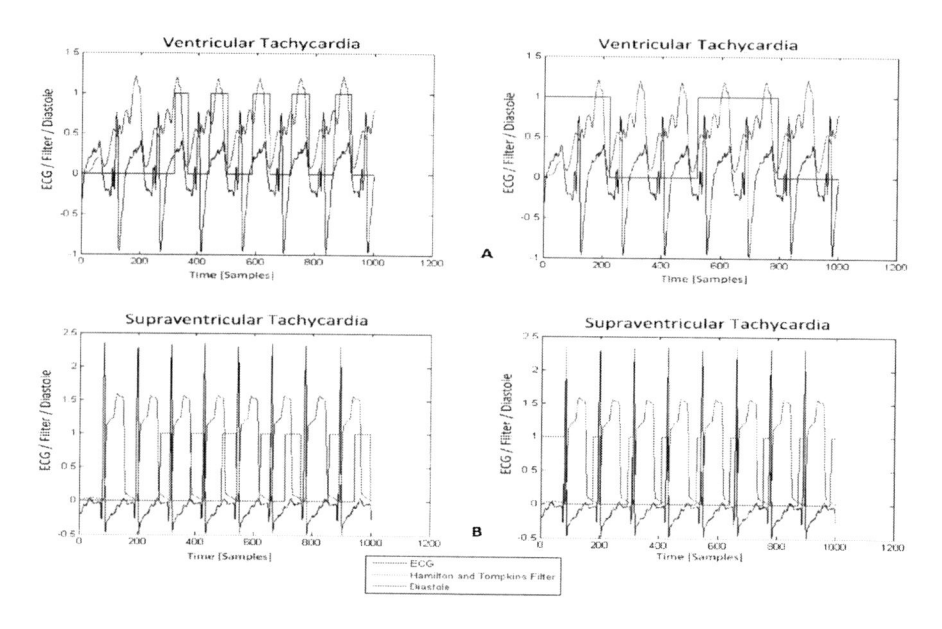

Fig. 7. DTV Operation using Slope-Lookup (Left) and Slope-Energy (Right) for A. Ventricular Tachycardia and B. Supraventricular Tachycardia.

the output of the Hamilton and Tompkins algorithm as well as the diastole detection intervals. Since the lookup table is obtained from the normal ECG case, the slope-lookup also underestimates for the cases where the diastole period is greater than the normal case. The example of this case can be observed in the pacemaker rhythm (Fig.5-B) and in the ST elevation (Fig.6-C and Fig.6-D). Furthermore, the slope-lookup method also underestimates in the cases of varying beat rate. For instance, we can observe how the variation of beat rate at around sample 1700 affects the underestimation of the slope-lookup method in the the PVC case (Fig.4-C). In contrast, the slope-energy method performs efficiently for most of the arrhythmias except for the situations where the QRS complex of two consecutive cycles are too close to each other. This method can't identify diastole in the case of ventricular tachycardia, supraventricular and data rates higher than 200 BPM.

Table.3 compares the two aforementioned algorithms. In the table, "Efficient" indicates up to 80 percent , "Underestimate" indicates between 50 to 80 percent, and "Not Efficient" lower than 50 percent of correct accurate diastole detection. Since the beat rates of above 200 BPM are rare, the slope-energy method is a better alternative and could be used in conjunction with slope-lookup for beat rates higher than 200 BPM.

Table 3. Comparison of Slope-Lookup and Slope-Energy Method

Heart Rhythm Types	Slope-Lookup Method	Slope-Energy Method
Normal Beat Rate	Efficient	Efficient up to 200 BPM
PVC	Underestimate	Estimate
PAC	Underestimate	Efficient
PNC	Underestimate	Efficient
Bigeminy	Not Efficient	Efficient
Trigeminy	Not Efficient	Efficient
Ventricular Tachycardia	Not Efficient	Not Efficient
Supraventricular Tachycardia	Efficient	Underestimate
Atrial Techycardia	Not Efficient	Efficient
First Degree Block	Not Efficient	Efficient
Second Degree Block	Not Efficient	Efficient
Third Degree Block	Not Efficient	Efficient
RBBB	Not Efficient	Not Efficient
LBBB	Underestimate	Underestimate (less)
ST Elevation +0.6mV	Not Efficient	Efficient
ST Elevation -0.6mV	Not Efficient	Efficient
Pacemaker Rhythm	Underestimate	Underestimate (less)

5 Conclusion

In this paper, we showed how a Diastolic Timed Vibrator can be used during acute coronary ischemia. DTV applies vibration during diastole to the patient chest. Since the heart operates irregularly during ischemia, many different types of heart rhythms and arrhythmias with different beat rates were considered. We illustrated two algorithms for detection of diastole by using ECG. We observed that the slope-energy is a robust method that can be used for most types of arrhythmias encountered during ischemia.

6 Future Work

Even though the slope-energy algorithm detects the diastole/systole intervals successfully in most of the arrhythmias and heart rhythms, we need to further investigate the algorithm efficiency with real ischemic patients. Furthermore, we have to research about the effectiveness of our device to such patients. Moreover, the position and the efficient frequency or range of frequencies of vibration is still an open research area and should be addressed in the future research.

In order to make this device clinically useful, our research team is going to pursue both animal and clinical studies. It is our aim to use this device in ambulances and pre-emergency units to treat patients with acute coronary ischemia.

References

1. World Health Organization (2004). Annex Table 2: Deaths by Cause, Sex and Mortality Stratum in WHO Regions, Estimates for 2002, The world health report (2004)
2. Pozen, M., Fried, W., Voigt, D.D., Studies, D.D.: of Ambulance Patients with Ischemic Heart Disease. American Journal of Public Health 67(6), 532–553 (1977)
3. Dressler, D.K.: Acute Coronary Syndromes, Ischemic Stroke, Pulmonary Embolism, and Disseminated Intravascular Coagulation. AACN Advanced Critical Care 20(2), 166–176 (2009)
4. Wardlaw, J.M., Zoppo, G., Yamaguchi, T., Berge, E.: Thrombolysis for Acute Ischaemic Stroke. Cochrane Database of Systematic Reviews (2003)
5. Nedelmann, M., Brandt, C., Schneider, F., Eicke, B., Kempski, O., Krummenauer, F., Dieterich, M.: Ultrasound-induced Blood Clot Dissolution without a Thrombolytic Drug is More Effective with Lower Frequencies. Journal of Cerebrovascular Diseases (2005)
6. Charles, W.F., Valentina, N.S.: Ultrasound and Thrombolysis. Journal of Vascular Medicine, 181–187 (2001)
7. Sara, W., Hans, W., Ha/Kan, E., Kjell, H., Hans, B.: Effect of Externally Applied Focused Acoustic Energy on Clot Disruption in Vitro. Journal of Clinical Science, 67–71 (1999)
8. Azita, S., Kim, R.V., Doulas, R.H.: Effect of Modulated Ultrasound Parameters on Ultrasound-induced Uhrombolysis. Journal of Physics and Biology, 6837–6847 (2008)
9. Koiwa, Y., Ohyama, T., Takagi, J., Kikuchi, T., Honda, H., Hoshi, N., Takishima, T.: Clinical Demonstration of Vibration-induced Depression of Left Ventricular Function. Tohoku Journal of Experimental Medicine 159(3), 247–248 (1989)
10. Purves, W.k., Sadava, D., Orians, G.H., Heller, H.C. (eds.): Life, the Science of Biology, 7th edn. Sinauer Associates and W. H. Freeman (2003)

11. Yohannes, F., Hoffmann, A.: Non-invasive Low Frequency Vibration as a Potential Emergency Adjunctive Treatment for Heart Attack and Stroke. An In-vitro Flow Model. Journal of Thrombolysis, 251–258 (2008)
12. Janssen, P.M.L., Honda, H., Koiwa, Y., Shirato, K.: The Effect of Diastolic Vibration on the Relaxation of Rat Papillary Muscle, Journal of Cardiovascular Research, 344–350 (1996)
13. Lohman, E.B. , Petrofsky, J.S., Maloney-Hinds, C., Betts-Schwab, H., Thorpe, D.: The Effect of Whole Body Vibration on Lower Extremity Skin Blood Flow in Normal Subjects. Journal of Medical Science Monitor (2007)
14. Boyle, L.J.: The Effects Of Whole Body Vibration and Exercise on Fibroanalysis In Men. Thesis, Ball State University (2009)
15. Kohler, B., Henning, C., Orglmeister, R.: The Principle of QRS Detection, Review and Comparing Algorithms for Detecting the Important ECG Waveform. IEEE Engineering in Medicine and Biology (2002)
16. Hamilton, P.S., Tompkins, W.J.: Quantitative Investigation of QRS Detection Rules Using the MIT/BIH Arrhythmia Database. IEEE Transaction Biomedical Engineering, 1157–1165 (1986)
17. Suri, J., Spaan, J.A.E., Krishnan, S.M.: Advances in Cardiac Signal Processing, 1st edn. Springer, Heidelberg (2010)

Low Noise CMOS Chopper Amplifier for MEMS Gas Sensor

Jamel Nebhen[1,2], Stéphane Meillère[1], Mohamed Masmoudi[2],
Jean-Luc Seguin[1], and Khalifa Aguir[1]

[1] Aix-Marseille Université, IM2NP-CNRS-UMR 6242, Avenue Escadrille Normandie Niemen
- Case 152, 13397 Marseille Cedex 20, France
[2] EMC Research Group-National Engineering, school of Sfax, Electrical Engineering
Department, Route de Soukra Km 2.5, BP. 1173 – 3038, Sfax, Tunisia
{jamel.nebhen,stephane.meillere,
jean-luc.seguin,khalifa.aguir}@im2np.fr,
Mohamed.masmoudi@enis.rnu.tn

Abstract. We describe in this paper a low-noise, low-power and low-voltage analog front-end amplifier dedicated to high resistive gas sensor detection. A mobile sensor system for very low level signals such as gas spikes detection is required to implement with a scaled CMOS technology. For a key circuit of these systems, a Chopper Stabilization Amplifier (CHS) which suppresses DC offset and 1/f noise figure of MOS devices is commonly used.

Keywords: Gas sensor, low-noise, low-power, chopper modulation, analog integrated circuits.

1 Introduction

Typical sensor output signals are in the microvolt range and have bandwidths ranging from DC up to a few kilohertz. One of the main challenges in weak sensor signals data acquisition systems are low-frequency 1/f noise and DC-offset [1]. To achieve the sub-microvolt level both for offset and noise, the CHS has been found a prime candidate to meet these requirements [2][3]. CHS is employed in the amplification stage to eliminate the non-ideal low frequency effects, such as the flicker noise and DC-offset voltage [4]. The purpose of this CHS is aimed to eliminate the flicker noise and DC-offset voltage from Gas-signal analog-front-end interface [5].

2 Basic Principle

Chopper Stabilized Amplifiers are low-noise continuous-time amplifiers useful for amplifying dc and very low frequency signals, mainly used for instrumentation application such as Gas sensor detection. Often the design objective is to reach the microvolt level for both offset and noise, with a bandwidth limited to a few hundred Hz while maintaining the low power consumption. Chopper Stabilized Amplifiers are best suited for low-power, portable, very low-noise, very small offset and offset drift, high performance applications such as Gas sensors.

The CHS technique uses an AC carrier to amplitude modulate the input signal. The principle of chopper amplification is illustrated in Fig.1 [6] with input V_{in}, output V_{out}, and A is the gain of a linear amplifier. The signal $m_1(t)$ and $m_2(t)$ are modulating and demodulating carriers with period $T=1/f_{chop}$ where f_{chop} is the chopper frequency. Also, V_{OS} and V_N denote deterministic DC-offset and 1/f noise. It is assumed that the input signal is band limited to half of the chopper frequency f_{chop} so no signal aliasing occurs. Basically, amplitude modulation using a square-wave carrier transposes the signal to higher frequencies where there is no 1/f noise, and then the modulated signal is demodulated back to the base band after amplification.

For the periodic carrier with a period of T and 50% duty cycle, its Fourier representation is [1]:

$$m(t) = 2 \sum_{k=1}^{\infty} \frac{\sin(\frac{k\pi}{2})}{(\frac{k\pi}{2})} \cos(2\pi f_{chop} kt) \qquad (1)$$

Its k-th Fourier-coefficients, Mk, have the property:

$$M_0 = M_{\pm 2} = M_{\pm 4} \cdots = 0 \qquad (2)$$

The modulated signal is the product of the initial signal and equation (1). The spectrum of the product $V_{in} \cdot m_1(t)$ in Fig. 1 shows that the signal is transposed to the odd harmonic frequencies of the modulating signal. After amplification, the modulated signal is then demodulated by multiplying $m_2(t)$ to obtain [1]:

$$V_d(t) = 4AV_{in}(t) \left[\sum_{k=1}^{\infty} \frac{\sin\left(\frac{k\pi}{2}\right)}{\left(\frac{k\pi}{2}\right)} \cos(2\pi f_{chop} kt) \right] * \left[\sum_{l=1}^{\infty} \frac{\sin\left(\frac{l\pi}{2}\right)}{\left(\frac{l\pi}{2}\right)} \cos(2\pi f_{chop} lt) \right] \qquad (3)$$

To recover the original signal in amplified form, the demodulated signal is applied to a low-pass filter with a cut-off frequency slightly above the input signal bandwidth. In this case, a half of the chopper frequency. Noise and offset are modulated only once.

If $S_N(f)$ denotes the power spectral density (PSD) of the noise and offset, then the PSD of $(V_{OS} + V_N) \cdot m_2(t)$ is [1]:

$$S_{CS}(f) = \sum_{k=-\infty}^{\infty} |M_{2k+1}|^2 S_N(f - \frac{2k+1}{T})$$

$$= \left(\frac{2}{\pi}\right)^2 \sum_{k=-\infty}^{\infty} \frac{1}{(2k+1)^2} S_N(f - \frac{2k+1}{T}) \qquad (4)$$

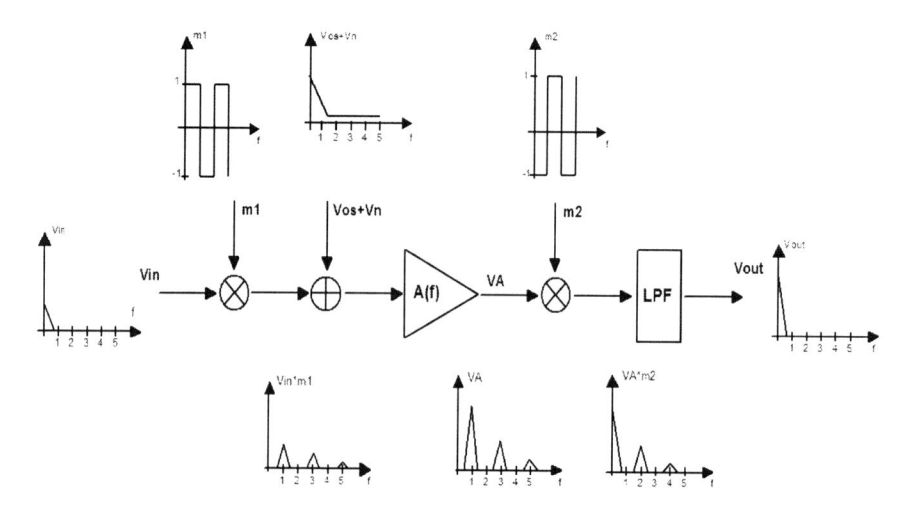

Fig. 1. The Chopper amplification principle [6]

So noise and offset are translated to the odd harmonic frequencies of the modulating signal, leaving the chopper amplifier ideally without any offset or low-frequency noise.

Assume the input signal Vin is a DC signal, if the amplifier has an infinite bandwidth and no delay, the signal at its output, V_A, is simply the same square wave with an amplitude A·Vin and the signal after demodulation is again a DC signal of value A·Vin. The amplifier output signal $V_A(t)$ is now a sinewave corresponding to the fundamental component of the chopped DC signal with an amplitude $(4/\pi)(A \cdot Vin)$. The output Vout of the second modulator is then a rectified sinewave containing even-order harmonic frequencies components.

The output will have to be low-pass filtered to recover the desired amplified signal. After low-pass filtering, the DC value is $(8/\pi^2)(A \cdot Vin)$, thus an approximately 20% degradation on DC gain. So a larger bandwidth of the main amplifier results in a bigger DC gain.

3 Design Consideration and Simulation Results

The schematic of a classical OTA-Miller integrated circuit implemented with CMOS transistors is illustrated in Fig.2 [7].

It consists of a basic differential pair implemented with PMOS transistors (M_1 and M_2), which has a single-ended current source as active load implemented with NMOS transistors (M_3 and M_4). This stage is biased with the current mirror formed with PMOS transistors M_5 and M_6, whose reference current source is I_{REF}=200 nA. The second stage is a basic common source amplifier with an NMOS transistor (M_7) acting as amplifier and a PMOS transistor (M_8) acting as a current source load. We assign the values indicated in Table I to the transistor sizes such as length L and width W. This OTA-Miller is designed to drive a load capacitor, C_L, of 5 pF. It is biased with voltage sources $V_{DD} = -V_{SS} = 1.25$ V. The process parameters for the transistors used in this work correspond to the AMS 0.35 µm Technology. The design of our CMOS OTA-Miller integrated circuit starts by defining the design specifications in terms of the performance parameters of interest, such that a high open lop voltage gain, |Av|, a high phase margin, PM, a low slew rate, SR, and a low power dissipation, PD.

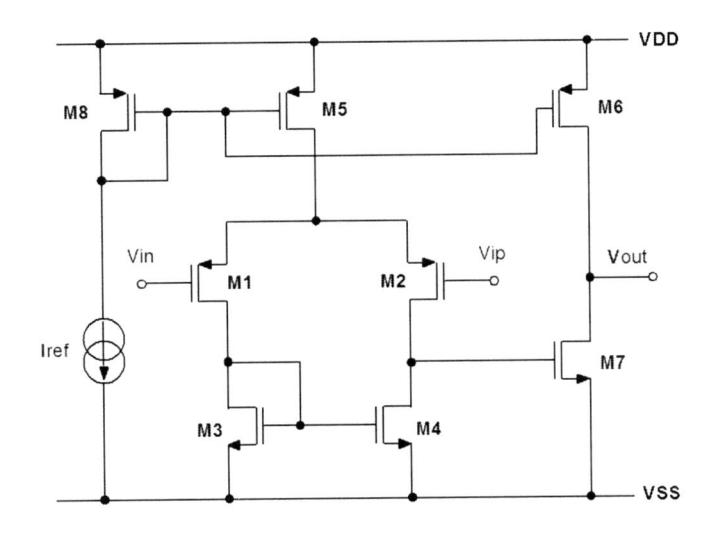

Fig. 2. Schematic of a classical CMOS OTA-Miller

Table 1. Transistor sizes of the OTA-Miller

Transistors	Type	Size W/L (µm/µm)
M1-M2	pmos	30.5/0.5
M3-M4	nmos	0.5/1
M7	nmos	4.5/1
M5-M6-M8	pmos	3/1

The results of circuit simulation for the OTA-Miller are shown in Fig.3.

We simulate an initial |Av| of 25.75 dB at low frequencies, a 1.208 MHz frequency at unity DC-gain and 159 kHz for a -3dB cut off frequency. On the other hand, the Phase Margin in Fig.4, PM, is approximately PM = 46°.

The simulated positive Slew Rate, SR+, is 0.822 mA/µs, while the SR- is -0.383 mA/µs. A high common-mode rejection ratio, CMRR, of about 94 dB and a high power-supply rejection ratio, PSRR, of about 88 dB are measured. Finally, the Total Power Dissipation, TPD, at the starting point is 1.28 µW. The architecture that has been used to implement the 2nd order band-pass filter is Sallen-Key Topology. This was chosen because of its simplicity. The Active-RC Sallen-Key filters have a range

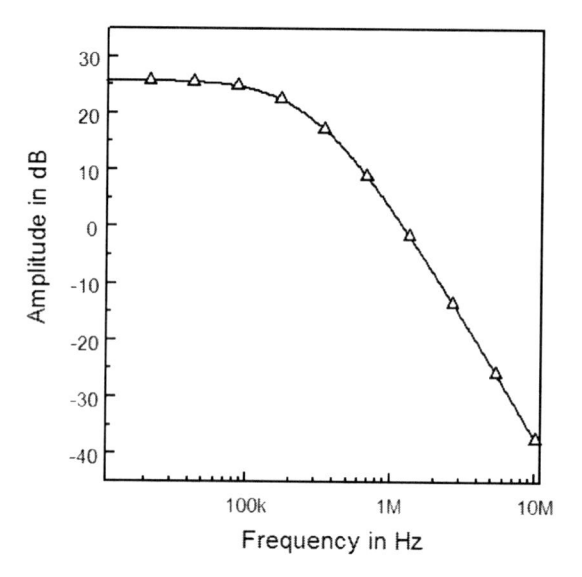

Fig. 3. DC gain

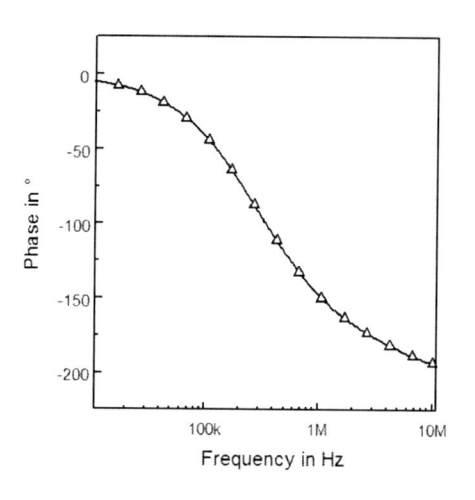

Fig. 4. Phase margin

of advantages when used for lower order of the filter: have excellent linearity, have low power dissipation and are easy to design and analyze. A circuit diagram for second order Sallen-Key band-pass filter is shown in Fig.5. The results of circuit maker simulation for the 2nd order Active-RC Sallen-Key band-pass filter are shown in Fig.6. The filter design has passband frequency of 40 kHz, a center frequency of about 203 kHz, a passband gain of about 12.2, a quality factor Q of 5, a low cut-off frequency of 185 kHz and a high cut-off frequency of 225 kHz. The output noise PSD

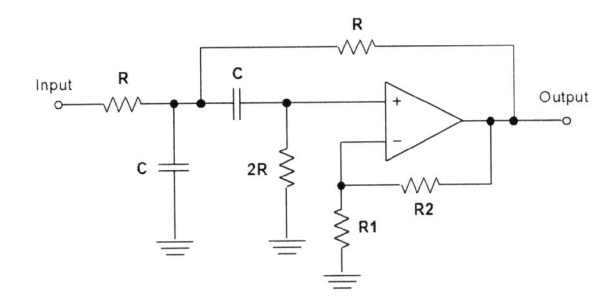

Fig. 5. Second order Sallen-Key band-pass filter

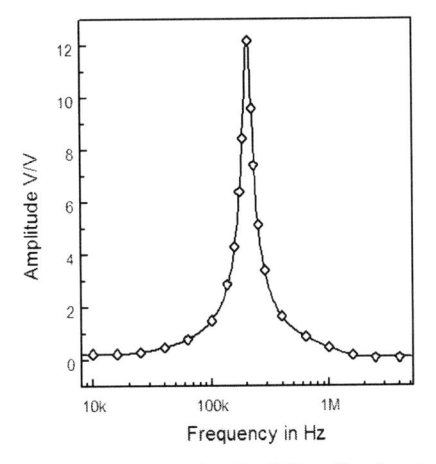

Fig. 6. Response of the second order Sallen-Key band-pass filter

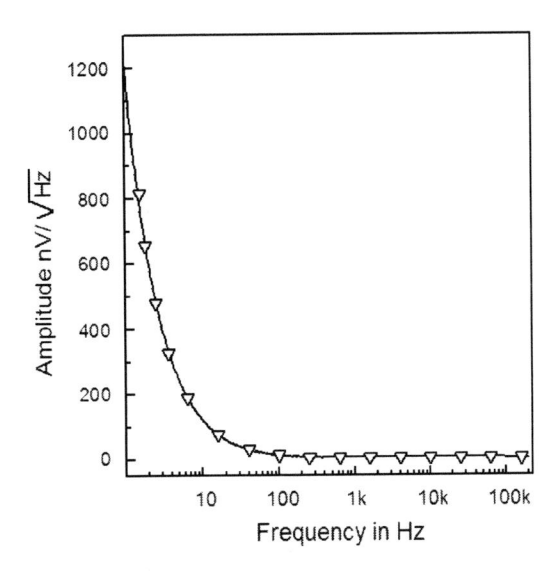

Fig. 7. Input-referred voltage noise spectrum

of the filter is depicted in Fig. 7. The effective noise floor referred to the input is $12.7\,nV/\sqrt{Hz}$. The process parameters for the transistors used in this work correspond to the AMS 0.35 µm technology. All the design specifications for the OTA-Miller were optimized while the sizes of CMOS transistors remain in the valid range for the selected technology.

4 Design Comparison

In this section, we compared a low-power low-noise CMOS amplifier for neural recording applications [8] and a 2 µW $100\,nV/\sqrt{Hz}$ chopper stabilized instrumentation amplifier for chronic measurement of neural field potentials [9] to our low noise micro-power chopper amplifier for high resistive MEMS Gas sensor [10].

The performance comparison between the proposed amplifier and published amplifiers is summarized in Table II. In the proposed amplifier, the low power consumption and the low equivalent input referred noise are considered. From the table II we can see the proposed amplifier achieved relatively low equivalent input referred noise and low consumption compared to published CHS in [8] and in [9]. In both published amplifiers, if the power consumption decreases the equivalent input referred noise increase. The contrary is true. The power consumption is inversely proportional to the equivalent input referred noise. However, This CHS achieves a good trade-off between power consumption and input noise.

Table 2. Comparison of the reference amplifier and this work

	This work	Ref. [8]	Ref. [9]
Supply voltage (V)	2.5	3.3	5
Equivalent input referred noise ($nV/\sqrt{Hz}$)	12.7	21	100
Power consumption (µW)	1.25	80	2
Application	MEMS Gas sensor	EEG amplifier	EEG amplifier

5 Conclusion

A fully integrated low voltage analog front-end preamplifier dedicated to implantable gas sensors is presented. The chopper modulation technique was chosen to reduce the low-frequency noise and DC offset. It is an efficient way to implement sensors that transduce very weak input signals. The amplifier is designed in AMS 0.35 µm technology at a 2.5 V power supply. It consumes a 1.25 µW. It achieves a noise floor of $12.7\,nV/\sqrt{Hz}$, a high CMRR of about 94 dB and a high PSRR of about 88 dB. It can be used in the amplification bloc and in the Low Pass Filter bloc simultaneously to obtaining a good trade-off between power consumption and input noise. The circuit features very low offset, low noise, and low power consumption.

Acknowledgment

This work was done thanks to financial support of the Franco-Tunisian Integrated Action of the French Ministry of Foreign and European Affairs and the Ministry of Higher Education, Scientific Research and Technology of Tunisia (project grant 09G1126).

References

1. Enz, C.C., Temes, G.C.: Circuit techniques for reducing the effects of op-amp imperfections: autozeroing, correlated double sampling, and chopper stabilization. Proceedings of the IEEE 84(11), 1584–1614 (1996)
2. Menolfi, C., Huang, Q.: A fully integrated, untrimmed CMOS instrumentation amplifier with submicrovolt offset. IEEE J. Solid-State Circuits 34, 415–420 (1999)
3. Hu, Y., Sawan, M.: CMOS front-end amplifier dedicated to monitor very low amplitude signal from implantable sensors. Analog Int. Circ. Signal Proc. 33, 29–41 (2002)
4. Nielsen, J.H., Bruun, E.: A CMOS low-noise instrumentation amplifier using chopper modulation. IEEE Analog Integrated Circuits and Signal Processing 42(1), 65–76 (2005)
5. Wu, R., Makinwa, K.A.A., Huijsing, J.H.: A Chopper Current-Feedback Instrumentation Amplifier with a 1mHz 1/f Noise Corner and an AC-Coupled Ripple Reduction Loop. J. Solid-State Circuits 44(12), 3232–3243 (2009)
6. Menolfi, C., Huang, Q.: A Low-Noise CMOS Instrumentation Amplifier for Thermoelectric Infrared Detectors. IEEE J. Solid-State Circuits 32(7), 968–976 (1997)
7. Pérez-Acosta, L.N., Rayas-Sánchez, J.E., Martínez-Guerrero, E.: Optimal design of a classical CMOS OTA-Miller using numerical methods and SPICE simulations. In: XIII International Workshop Iberchip (IWS 2007), Lima, Peru, pp. 387–390 (March 2007)
8. Harrison, R.R., Charles, C.: A low-power low-noise CMOS amplifier for neural recording applications. IEEE Journal of Solid-State Circuits 38, 958–965 (2003)
9. Denison, T., Consoer, K., Santa, W., Avestruz, A., Cooley, J., Kelly, A.: A 2 μW 100 nV/rtHz Chopper-Stabilized Instrumentation Amplifier for Chronic Measurement of Neural Field Potentials. IEEE Journal of Solid-State Circuits 42(12), 2934–2945 (2007)
10. Guérin, J., Bendahan, M., Aguir, K.: A dynamic response model for the WO3-based ozone sensors. Sensors and Actuators B 128-2, 462–467 (2008)

Modeling Reliability in Wireless Sensor Networks

Ahmed Ibrahim Hassan[1], Maha Elsabrouty[2], and Salwa Elramly[3]

[1] Ericsson Egypt Ltd.
[2] Egypt-Japan University for Science and Technology, Egypt
[3] Ain-Shams University, Egypt
ahmed.ibraheem@ericsson.com, maha2000_eg@yahoo.com,
salwa_elramly@eng.asu.edu.eg

Abstract. Energy efficient and reliable data forwarding becomes important if resources are limited such as in Wireless Sensor Networks. In this paper, we discuss how the error rate associated with a link affects the overall probability of reliable delivery, and consequently the energy associated with the reliable transmission of a single packet. The analysis includes both fixed-power and variable-power scenarios along with the End-to-End Retransmission (EER) and Hop-by-Hop Retransmission (HHR) techniques. In the EER case, a threshold value for the packet error rate at which both scenarios will result in the same energy costs is defined. The relation between this threshold value and the difference in number of hops between both scenarios is also derived. The simulation results finally confirm the theoretical model.

Keywords: Wireless Sensor Networks, Reliability, Retransmission Techniques.

1 Introduction

Wireless Sensor Networks are potentially one of the most important technologies of this century. Recent technological advances have led to the emergence of small, low-power devices that integrate sensors and actuators with limited on-board processing and wireless communication capabilities. Pervasive networks of such sensors and actuators open new vistas for constructing complex monitoring and control systems, ranging from habitat monitoring, target tracking, home automation, ubiquitous sensing for smart environments, construction of safety monitoring, and inventory tracking [1].

The key challenge in sensor networks is to maximize the lifetime of sensor nodes that rely on limited battery power. Therefore, computational operations of nodes and communication protocols must be made as energy efficient as possible. Data-centric routing approaches have been proposed in the literature to provide more energy-efficient routing as opposed to the traditional end-to-end routing schemes known as address-centric approaches [1-4]. A comparison between both implementations is given in Table 1.

Whereas to compare between two data-centric implementations, points of comparison may include: structure of aggregation tree (planar, cluster, etc), direction of diffusion (push, pull, etc) and type of applications (event-driven, sink-initiated, etc) [1, 5].

Table 1. Data-Centric versus Address-Centric Implementations

Heading level	Data-Centric Implementation	Address-Centric Implementation
Communication Mode	Neighbor-to-Neighbor (Hop-by-Hop) Communication	End-to-End Communication
Localized Algorithm	Yes (no need to maintain overall network topology)[1]	No (rely on global topological knowledge)
Routing Decisions	Based on the contents of the payloads of packets[2]	Based on packets destination addresses

Energy-aware routing protocols typically compute the shortest-cost path, where the cost associated with each link is some function of the transmission (and/or reception) energy associated with the corresponding nodes. To adapt such minimum cost route determination algorithms (such as Dijkstra's or the Bellman–Ford algorithm) for energy-efficient reliable routing, the link cost must now be a function of not just the associated transmission energy, but the link error rates as well [6]. In this research, we analyze the consequences of this behavior with different packet retransmission techniques.

Reliability, as one of the performance measures of routing algorithms for Wireless Sensor Networks, should not be mixed with information accuracy. The later is questioned when data aggregation is used. Data aggregation is the combination of data from different sources according to a certain aggregation function and is an enabler for data-centric routing. It can be categorized into two classes: lossless and lossy [3]. With *lossless aggregation*, all detailed information is preserved. Whereas *lossy aggregation* may discard some detailed information and/or degrade data quality for more energy savings.

The rest of the paper is organized as follows. In section 2, the related work will be described. To understand the effect of introducing link errors to our model, the error-free transmission case is studied first in section 3. In section 4, two generally retransmission techniques are examined. Both sections 3 and 4 represent the analytical model of the work. In section 5, the simulation results for the energy costs of different scenarios will be discussed and compared to the theoretical results. Finally, section 6 will conclude the paper with the possible directions for future work.

2 Related Work

To route data efficiently in Wireless Sensor Networks, various routing protocols have been proposed. Data-Centric routing is a common routing approach in Wireless Sensor Networks [2, 7, 8, 9]. Factors affecting the performance of data-centric routing,

---

[1] Local communication implies that, as far as a node knows, the data that is received from a neighbor came from that neighbor. This can be energy efficient in highly dynamic networks when changes in topology need not be propagated across the network.

[2] Nodes store and interpret tasks/data rather than simply forwarding them along.

such as the number and placement of sources and the communication network topology, are investigated in [2].

Directed diffusion [7] is a network layer protocol based on data-centric routing. It can be classified as pull-diffusion mechanism. In directed diffusion, the sink broadcasts interests to all sensor nodes in the network. Each sensor node keeps the interest in its local cache and uses the gradient fields within the interest descriptors to recognize the most suitable routes to the sink. These established routes are then used by source nodes to forward the sensed data (events) to the sink.

Robustness is addressed in directed diffusion by enabling *local repair* of failed or degraded paths. Causes for failure or degradation include node energy depletion, and environmental factors affecting communication (*e.g.*, obstacles, rain fade). When the quality of the link between the source and an intermediate node degrades and events are frequently corrupted, the intermediate node negatively reinforces the direct link to the source. This will eventually lead to the discovery of one empirically good path.

Reliable routing has been also tackled in [8] where a new routing scheme, known as EARS; Energy-efficient And Reliable routing Scheme, is proposed. EARS relies on interaction between routing and MAC layers, with the overall goal of achieving energy efficiency and reliability through cross layer optimization.

In EARS, data are forwarded to that neighbor node which has lesser value of the Radio-aware Metric C_{lq}. Each time a Request to Send (RTS) packet is sent to a selected neighbor node. If RTS request is successful, data are routed to that neighbor node, otherwise another neighbor node is selected among the candidate neighbor nodes, which has comparatively less value of the C_{lq}. In this way a route to the corresponding neighbor is selected until the source or the sink is reached.

In this work, modeling of reliable routing in Wireless Sensor Networks is presented. Two scenarios are studied; fixed-power scenario and variable-power scenario. Routing in variable-power scenario, in which each sensor node (source) adjusts its transmission power based on the distance between itself and the receiver (data recipient/sink), has been proposed in a previous work [10]. Simulation results show that more energy savings are achieved in the variable-power scenario than in fixed-power scenario where transmission power is constant and is independent of the characteristics of the link between the transmitter and the receiver.

3 Error-Free Transmission

In this section, the energy costs of the error-free transmission are presented. The communication model used throughout the analysis can be explained as follows [6]. Let us consider a sender (S) and a receiver (R) separated by a distance D. let N represent the total number of hops between S and R, so that $N - 1$ represents the number of forwarding nodes i: $i = \{2, ..., N\}$, with node i referring to the $(i - 1)$th intermediate hop in the forwarding path. Node 1 refers to S and node $N + 1$ refers to R. This is illustrated in Figure 1.

In this case, the total energy spent in simply transmitting a packet once (without considering whether or not the packet was reliably received) from the sender to the receiver over the $N - 1$ forwarding nodes is:

N hops

(S) ────●────●────●────●── ─ ──●──── (R)
1   i = 2 i = 3 i = 4   i = N   N+1
       └──── N-1 forwarding nodes ────┘
←──────────────── D ────────────────→

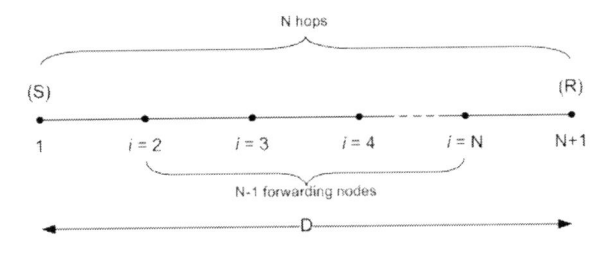

Fig. 1. Communication Model

$$E_t = \sum_{i=1}^{N} E_{i,i+1}$$

In the fixed-power scenario, $E_{i,i+1}$ is independent of the link characteristics; in the variable-power scenario, $E_{i,i+1}$ is a function of the distance between nodes i and $i+1$. Thus:

Fixed-Power Scenario   **Variable-Power Scenario**

$$E_f = \sum_{i=1}^{N_f} \alpha R^K = \alpha R^K N_f \quad (1) \qquad E_v = \sum_{i=1}^{N_v} \alpha D_{i,i+1}^K \quad (2)$$

Where,
  R is the communication radius[3]
  $D_{i,i+1}$ is the distance between nodes i and $i+1$ ($D_{i,i+1} \leq R$)
  K is the coefficient of channel attenuation ($K \geq 2$)
  α is a proportionality constant.

From (1) it is obvious that if links are considered error-free, then minimum hop paths are the most energy-efficient for the fixed-power case. While to understand the trade-offs associated with the choice of the number of hops in the variable-power case, N_v, it is assumed that each of the hops is of equal length D/N_v. In that case, E_v in (2) is given by:

$$E_v = \sum_{i=1}^{N_v} \frac{\alpha D^K}{N_v^K} = \frac{\alpha D^K}{N_v^{K-1}} \quad (3)$$

From (3) it is easy to see that, in the absence of transmission errors, paths with a large number of small hops are typically more energy efficient in the variable-power case. These results are illustrated in Figure 2.

It has been confirmed experimentally in [10] that, for a wide range of communication radius and number of sources, E_v is always smaller than E_f while N_v is always larger than N_f.

---

[3] All nodes are assumed to be able to communicate with any other nodes that are within some distance called the communication radius.

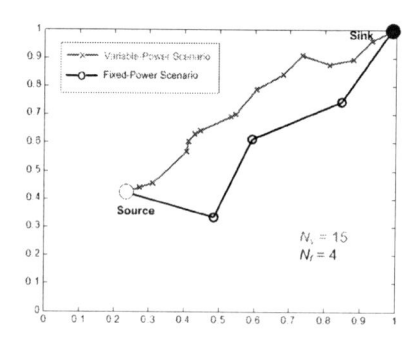

Fig. 2. Fixed-Power Path versus Variable-Power Path

Definition 1: Let us define the variable-power gain, G, obtained by using the variable-power scenario as opposed to the fixed-power scenario as follows:

$$G = \frac{E_f}{E_v} = \frac{R^K N_f}{\sum_{i=1}^{N_v} D_{i,i+1}^K} \qquad (4)$$

and $G \geq 1$

The value of G depends on the number of sources and the communication radius.

4 Retransmission Techniques

In the presence of link errors, none of the choices explained in section 3 may give optimal energy efficient paths. In this section, two generally packet retransmission techniques that are used to overcome link errors are examined; namely, the End-to-End Retransmission (EER) and the Hop-by-Hop Retransmission (HHR) techniques [6]. This should not be mixed with the communication modes that are used in combination with data-centric/address-centric implementations given in Table 1.

4.1 End-to-End Retransmission (EER)

$p = 1-$ *the probability of an error-free transmission* $= 1-(1-p_{link})^N$

The number of transmissions (including retransmissions) necessary to ensure the successful transfer of a packet between S and R is then a geometrically distributed random variable X, such that [6]:

$$\Pr\{X = k\} = p^{k-1} \times (1-p), \forall k$$

The *mean* number of individual packet transmissions for the successful transfer of a single packet is thus:

$$\text{Mean} = \frac{1}{1-p}$$

Since each such transmission uses total energy given above by (1) and (2), the total expected energy required in the reliable transmission of a single packet is given by:

Fixed-Power Scenario

$$E_f(EER) = E_f \cdot \frac{1}{1-p} = \alpha R^K N_f \cdot \frac{1}{1-p}$$

$$\therefore E_f(EER) = \frac{\alpha R^K N_f}{(1-p_{link})^{N_f}} \quad (5)$$

Variable-Power Scenario

$$E_v(EER) = E_v \cdot \frac{1}{1-p} = \sum_{i=1}^{N_v} \alpha D_{i,i+1}^K \cdot \frac{1}{1-p}$$

$$\therefore E_v(EER) = \frac{\sum_{i=1}^{N_v} \alpha D_{i,i+1}^K}{(1-p_{link})^{N_v}} \quad (6)$$

Definition 2: Let us define a threshold value for p_{link} such that:

$$E_f(EER) = E_v(EER) \text{ at } p_{link} = p_{link}(threshold)$$

$$\therefore \frac{\alpha R^K N_f}{(1-p_{link}(threshold))^{N_f}} = \frac{\sum_{i=1}^{N_v} \alpha D_{i,i+1}^K}{(1-p_{link}(threshold))^{N_v}}$$

Using the definition of G in (4) we get:

$$p_{link}(threshold) = 1 - (G)^{-\frac{1}{\Delta N}} \quad (7)$$

Where,

$$\Delta N = N_v - N_f, \quad \text{and } \Delta N \geq 0$$

It can be noted that:

For $p_{link} < p_{link}(threshold)$, $E_f(EER) > E_v(EER)$

For $p_{link} > p_{link}(threshold)$, $E_f(EER) < E_v(EER)$

4.2 Hop-by-Hop Retransmission (HHR)

For the HHR model, the number of transmissions on each link is *independent of the other links* and is geometrically distributed. The total energy cost for the HHR case with N intermediate nodes and having a link packet error rate of p_{link} is:

$$E_t(HHR) = \sum_{i=1}^{N} E_{i,i+1} \cdot \frac{1}{1-p_{i,i+1}}$$

Fixed-Power Scenario **Variable-Power Scenario**

$$E_f(HHR) = \sum_{i=1}^{N_f} \frac{\alpha R^K}{1-p_{link}} = \frac{\alpha R^K N_f}{1-p_{link}} \qquad (8) \qquad E_v(HHR) = \sum_{i=1}^{N_v} \frac{\alpha D_{i,i+1}^K}{1-p_{link}} \qquad (9)$$

Applying the same assumption in (3), that each hop is of distance D/N_v, to the variable-power equation in (9) we get:

$$E_v(HHR) = \frac{1}{1-p_{link}} \cdot \sum_{i=1}^{N_v} \frac{\alpha D^K}{N_v^K} = \frac{\alpha D^K}{N_v^{K-1} \cdot (1-p_{link})} \qquad (10)$$

From (8), (9) and (10), it can be noticed that the analysis of the hop-by-hop retransmission technique is similar to the analysis of the error-free transmission. Furthermore, similar results are obtained in regards to the relation between the energy costs and the number of hops, i.e. $E_f(HHR)$ decreases as N_f decreases whereas $E_v(HHR)$ decreases as N_v increases. Obviously, energy costs are higher in HHR case than in error-free transmission providing that $p_{link} > 0$.

5 Simulation Results

In this section, the simulation results are presented and compared to the theoretical results obtained in sections 3 and 4. For simplifying the analysis, a single-source model is assumed. Routing is thus treated as the *shortest-path problem* in graphs. When two nodes wish to communicate, a *minimum-weight path (shortest path)* connecting the corresponding pair of nodes is selected. The simulation parameters used are shown in Table 2.

Table 2. Simulation Parameters

Number of Nodes	100
Dimensions	square of unit size
Communication radius (R)	0.3
Coefficient of channel attenuation (K)	2
Proportionality constant (α)	1

The first two parameters represent the same setup used in [2]. The value of the communication radius, R, is chosen to be 0.3 to ensure that no experiment may result in unconnected graphs. In our study, we assume free space model with no obstacles, i.e. $K = 2$. The proportionality constant is assumed to be 1.

5.1 Variable-Power Gain

The value of the variable-power gain, G, defined in (4) is calculated in the error-free transmission case, i.e. $p_{link} = 0$, by running 100 experiments. Each consists of random placement of 100 nodes including the sink node in a square of unit size. The shortest-path between the source and the sink in fixed-power and variable-power scenarios is obtained using Dijkstra's algorithm. The average is then calculated to represent the energy costs of each scenario. The value of G is computed to be: 3.5.

5.2 Energy Costs versus Link Error Rate in EER

For the end-to-end retransmission case, energy costs of both scenarios are calculated at different values of link error rate, plink, which is varied from 0% to 30% with step of 1%. In this example, ΔN is fixed at a value of 6 ($N_v = 9$ and $N_f = 3$). The threshold value of the link error rate, plink (threshold), is obtained experimentally from the intersection of the fixed-power curve with the variable-power curve. Whereas the theoretical value of plink (threshold) is obtained by solving equation (7) at $\Delta N = 6$. Both values are shown in Figure 3.

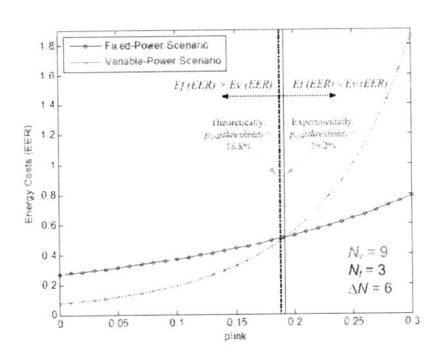

Fig. 2. Energy Costs versus Link Error Rate in EER

5.3 Threshold Link Error Rate versus ΔN in EER

The relation between the threshold value of the link error rate and the difference in number of hops between the variable-power scenario and the fixed-power scenario, ΔN, is given by equation (7). To verify this relationship, the same experimental setup discussed above is used. In this case, 300 experiments were run at different values of p_{link}. At each individual value of p_{link}, the value(s) of ΔN at which the energy costs of the EER variable-power start to exceed that of the EER fixed-power (or vice versa) was recorded. The results are plotted in Figure 4.

5.4 Error-Free Transmission versus EER and HHR

In this section, the energy costs of the error-free transmission are compared to that of different retransmission techniques. Similar to section 5.2, results are calculated at

$N_v = 9$ and $N_f = 3$. Figure 5 shows fixed-power scenario represented by equations (1), (5) and (8) for error-free transmission, EER and HHR respectively. Figure 6 shows variable-power scenario represented by equations (2), (6) and (9) for error-free transmission, EER and HHR respectively.

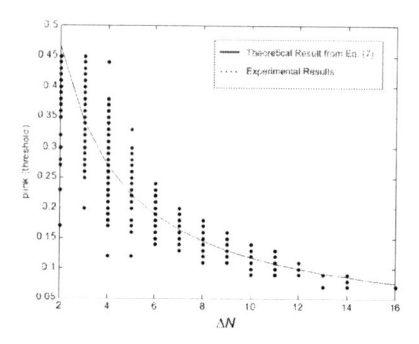

Fig. 4. Threshold Link Error Rate versus ΔN in EER

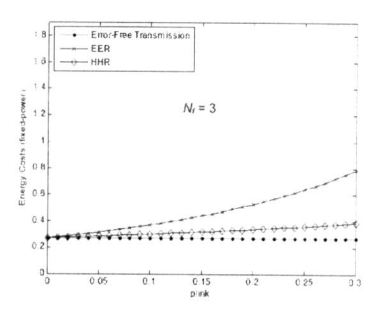

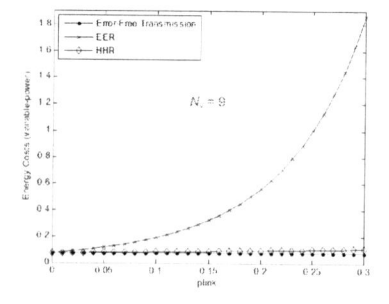

Fig. 5. Fixed-Power Scenario

Fig. 6. Variable-Power Scenario

In both scenarios the energy costs of HHR case are less than the EER case.

6 Conclusion and Future Work

In this paper, we have modeled and analyzed the performance of reliable routing in Wireless Sensor Networks in both fixed and variable power scenarios. A formula for the threshold value of the link error rate, at which the energy costs of the EER variable-power scenario is equal to that of the EER fixed-power scenario, was derived. A comparison between the simulation results of the error-free transmission, EER and HHR was also presented.

Throughout the simulation, the communication radius was fixed at a constant value. Modeling the effect of the communication radius is a topic for future study. Moreover, the analysis has focused on the case where there is a single source. It is reasonable to ask what would happen if there were additional sources. Extending the analysis to include multiple-sources scenario is part of future research.

References

1. Stojmenovic, I.: Handbook of Sensor Networks: Algorithms and Architectures. John Wiley & Sons, Inc., New Jersey (2005)
2. Krishnamachari, B., Estrin, D., Wicker, S.: Modelling Data-Centric Routing in Wireless Sensor Networks. In: Proc. IEEE INFOCOM (June 2002)
3. Intanagonwiwat, C., Estrin, D., Govindan, R., Heidemann, J.: Impact of Network Density on Data Aggregation in Wireless Sensor Networks. In: Proc. 22nd International Conference on Distributed Computing Systems (November 2001)
4. Mikami, S., Aonishi, T., Yoshino, H., Ohta, C., Kawaguchi, H., Yoshimoto, M.: Aggregation Efficiency-Aware Greedy Incremental Tree Routing for Wireless Sensor Network. IEICE Trans. Commun. E89-B(10), 2741–2751 (2006)
5. Chen, Y., Shu, J., Zhang, S., Liu, L., Sun, L.: Data Fusion In Wireless Sensor Networks. In: Proc. Second International Symposium on Electronic Commerce and Security (2009)
6. Glisic, S.G.: Advanced Wireless Networks 4G Technologies. John Wiley & Sons Ltd., England (2006)
7. Intanagonwiwat, C., Govindan, R., Estrin, D.: Directed Diffusion: A Scalable and Robust Communication Paradigm for Sensor Networks. In: Proc. of the Sixth Annual ACM/IEEE International Conference on Mobile Computing and Networking (Mobicom 2000), Boston, Massachusetts (August 2000)
8. Tariq, M., Kim, Y.P., Kim, J.H., Park, Y.J., Jung, E.-H.: Energy Efficient and Reliable Routing Scheme for Wireless Sensor Networks. In: Proc. IEEE Int. Conference on Communication Software and Networks, China, pp. 181–185 (February 2009)
9. AL-khdour, T.A., Baroudi, U.: A Generalized Energy-Aware Data Centric Routing for Wireless Sensor Network. In: Proc. IEEE Int. Conference on Signal Processing and Communications, Dubai, United Arab Emirates, pp. 117–120 (November 2007)
10. Hassan, A.I., Elsabrouty, M., El-Ramly, S.: Energy-Efficient Routing in Variable-Power Wireless Sensor Networks. In: Proc. 27th National Radio Science Conference, Egypt (March 2010)

Fading Compensation for Improved Indoor RF Tracking Based on Redundant RSS Readings

Andreas Fink and Helmut Beikirch

Department of Computer Science and Electrical Engineering,
Rostock University, A.-Einstein-Str. 2, 18059 Rostock, Germany
{andreas.fink,helmut.beikirch}@uni-rostock.de
http://www.ief.uni-rostock.de

Abstract. In partially hazardous environments a high reliable tracking of humans is of interest. For example, the monitoring of maintenance staff in the underground longwall mining – in particular close to heavy weight self-advancing hydraulic shields – is crucial to safe human lives. A centroid location estimation technique based on received signal strength (RSS) readings offers a well known and low-cost tracking solution in such a rough environment where many other systems with optical, magnetical or ultrasound sensors fail. A redundant RF communication concept using spatial and frequency diversity compensates the influence of multipath fading and leads to a more precise and high available tracking. The experimental results on a motion test track show that an improved tracking performance can be reached with low infrastructural costs.

Keywords: Diversity Combining, Indoor Tracking, ISM Radio, Multipath Fading, Received Signal Strength.

1 Introduction

The longwall mining technique is used for highly productive underground coal mining where massive shearers cut coal from a wall face, which falls onto a conveyor belt for removal. Heavy weight hydraulic shields are placed in a long line in order to support the roof of the coalface and maintaining a safe working space along the face for the miners. The automatic motion of the hydraulic shields need to be stopped when a miner is localized in front of them. The precision of the miner's tracking system determines the size of the security zone and the amount of shields which are involved. Besides the precision also the availability of the tracking system plays a significant role for the availability and efficiency of the longwall mining system, since the safety critical system architecture requires an immediate idle state of the system.

The basis for a reliable tracking system is a high available and precise localization. The different localization methods can be classified according to their sensor signal type (cf. Fig. 1). The environment of an underground longwall mining system does not permit a global radio frequency (RF) localization via GPS. The use of directional sensors like infra-red (IR), ultrasound, optical and magnetic systems is limited to line-of sight (LOS) scenarios. Due to the mining infrastructure components (e.g. hydraulic shields or coal shearer, cf. Fig. 2) non-line-of sight (NLOS) conditions have to be assumed.

Classification of Localization Techniques

Sensor Type	Estimation End	Signal Metric	Metric Processing
• RF (Radio Frequency)	• Network-Based (Location-Aware)	• AOA (Angle of Arrival)	• Triangulation
• IR (Infra-Red)	• Handset-Based (Location-Support)	• RSS (Received Signal Strength)	• Pattern Matching
• Ultrasound		• TOA / TDOA (Time of Arrival / Time Difference of Arrival)	
• Optical			
• Magnetic			

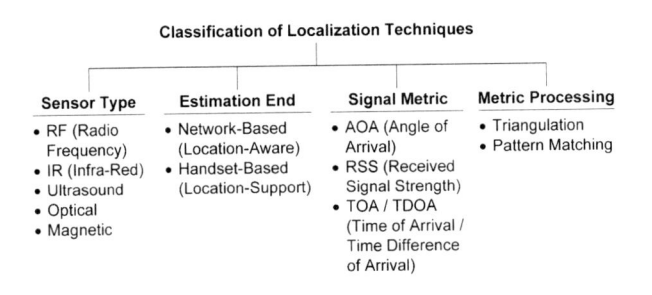

Fig. 1. Classification of localization techniques. Source: [1].

A local RF system with a multilateration of several distance approximations between an unknown node (blind node, BN) and fixed anchor nodes (reference nodes, RNs) is applicable to find out the position of a miner equipped with a BN. In principle the location estimation can be realized via analyzing signal propagation delays (Time of Arrival - TOA, Time Difference of Arrival - TDOA) or received signal strength (RSS) readings. The different methods are regarded in [2] as well as the general characteristics of localization systems. Further, a taxonomy of the systems takes place based on the criteria accuracy, scaling, costs and limitations.

RSS-based localization techniques are an easy method to predict the position of the mobile BN since the data acquisition puts low demands on the hardware infrastructure. Suitable estimation algorithms calculate a position information with a few computations (e.g. centroid localization), so common and proven embedded systems with low-power microcontrollers can be used. For TOA or TDOA techniques with the use of RF signals high-precision timer modules are required [3].

For an RSS-based indoor localization system it is recommended to concentrate on the proper acquisition of the RSS values since their interpretation has the main influence on the tracking system's reliability. RSS is notorious for being a noisy signal. For a ranging-based localization it is difficult to rely on the raw RSS values only. For a correct interpretation of the values according to a path loss model additional measures are required, since the influence of large-scale and small-scale fading effects leads to bad distance estimations. [4], [5]

The small-scale fading due to reflection, diffraction and scattering on obstacles is the main issue. When the transmitter and / or receiver are moved in the order of the wavelength (e.g. $\approx 12.5\,\text{cm}$ for $2.4\,\text{GHz}$) the RSS may vary by three or four orders of magnitude ($30..40\,\text{dBm}$). Therefore it is challenging to approximate the distance between transmitter and receiver with the RSS only. In some cases
– when the RSS of a single channel falls under the receiver sensitivity level

– the signals are completely wiped out by a destructive interference and the transmitted data gets lost. A detailed description of fading principles is given in [6], [7] and [8].

A common solution to compensate the small-scale fading behavior of radio channels is a diversity scheme with the combining of the received data of more than one communication channel. The diversity schemes are subdivided according to their physical domain into time, frequency, spatial and angle diversity. Time diversity with a repeated transmission often is not an acceptable solution – especially for real-time communication systems. For a reliable real-time communication we investigate the simultaneous use of spatial and frequency diversity with an RSS-based selection combining.

In section 2, the architecture of the RF tracking system with the used indoor propagation model and the diversity transceiver is explained. In section 3, we validate the tracking system performance by experimental results of a dynamic tracking measurement on a motion test track. In the last section 4, the results are discussed and investigated in terms of an outlook for further system developments.

2 Tracking System Architecture

In previous research we have carried out localization measurements for the availability and precision of a two-dimensional localization system in an obstructed factory building [9]. For the tracking of humans in a longwall mining environment we assume the one-dimensional scenario shown in Fig. 2.

Fig. 2. Infrastructure components in a longwall mining application. Source: Own elaboration.

The miner with the BN can move along the path between the shields and the conveyor belt. The shields are equipped with a certain number of RNs which are interconnected with a wired bus interface (e.g. CAN). The tracking procedure and structure is given in Fig. 3. The tracking algorithm can be subdivided into the components data acquisition, data preprocessing and location estimation.

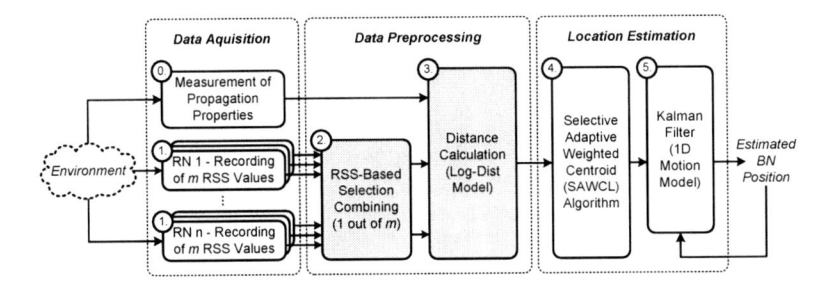

Fig. 3. Structure of the RSS-based tracking system. Source: Own elaboration.

The acquisition of the RSS values is realized by a periodic redundant data transmission between the BN and the RNs. Between the BN and one RN eight communication channels are available with our used diversity concept. We use four spatial divided 868 MHz channels and four spatial divided 2.4 GHz channels. The detailed diversity scheme is presented in section 2.2. The redundant RSS values from all RNs are collected by a data concentrator (e.g. industrial PC).

The distance approximation according to a specific indoor propagation model as well as the selection combining of the redundant RSS values are implemented on the data concentrator. The distance approximation is discussed in detail in section 2.1 – together with the used indoor propagation model.

The location estimation of the BN uses the Selective Adaptive Weighted Centroid Localization (SAWCL) algorithm. The computed distances between the RNs and the BN are converted into weights. The unknown BN position is given by the weighted centroid of the known RN positions. Details on the SAWCL algorithm can be found in [9].

For a tracking of the BN the connection of subsequent location estimations is used. The estimations are processed with a Kalman filter according to a one-dimensional motion model. The corresponding equations for the prediction and update phase can be derived from [10]. Although, the location estimation process with the Kalman filter and the SAWCL algorithm offer a lot of fading compensation opportunities, in this paper we concentrate on the impact of the data preprocessing stage. The path loss model and the diversity concept are the issues and are highlighted in the process structure in Fig. 3.

2.1 Indoor RF Propagation Model

Without any disturbances (free space propagation) the distance-depending path loss shows a logarithmic dropping of power with linear increasing distance according to the Log-distance path loss model. With (1) the average path loss $\overline{PL}(d)$ (in dBm) over a distance d is given by the reference path loss $PL(d_0)$ over a reference distance d_0 and the environment-specific propagation coefficient n.

$$\overline{PL}(d) = PL(d_0) + 10n \log\left(\frac{d}{d_0}\right). \qquad (1)$$

The value of $PL(d_0)$ is influenced by the characteristics of the analog RF front-end. The effective radiated power (ERP) of an RF transceiver depends on the output power of the IC, the RF matching network and the antenna. An 868 MHz SRD transceiver with an output power of $+1\,\text{dBm}$ gives a $PL(d_0)$ of $-79\,\text{dBm}$ while the 2.4 GHz ISM transceiver with an output power of $+10\,\text{dBm}$ gives a $PL(d_0)$ of $-67\,\text{dBm}$.

The value of n is influenced by the specific environmental propagation conditions and the used frequency. In [6] values for n between 1.8 and 3.2 are given for obstructed indoor environments and frequencies between 900 MHz and 4.0 GHz.

In Fig. 4 the average path loss $\overline{PL}(d)$ for these two values of n are shown on the left as a function of the distance between transmitter (TX) and receiver (RX). The corresponding probability density functions (PDFs) for the distributions of $PL(d)$ at the receiver ($PL(d_{RX})$ – indicated by the measured RSS value) are shown on the right.

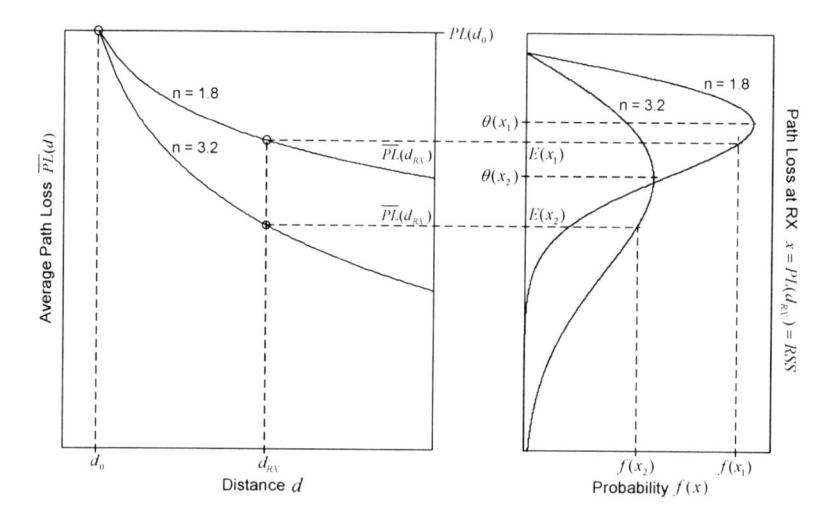

Fig. 4. Log-normal path loss model (left) and PDFs (right) for two different fading channel estimations ($n = 1.8$ and $n = 3.2$). Source: Own elaboration.

The fading follows a Rayleigh distribution. The PDF of the distribution is defined as follows:

$$f(x,\theta) = \begin{cases} \frac{xe^{-\frac{x^2}{2\theta^2}}}{\theta^2} &, x \geq 0 \\ 0 &, x < 0 \end{cases}, \qquad (2)$$

where the maximum likelihood estimator of θ is defined as

$$\theta = \sqrt{\frac{1}{2N}\sum_{i=0}^{N} x_i^2}. \qquad (3)$$

The expectation value $E(X_r)$ of the Rayleigh distribution is defined as

$$E(X_r) = \theta \sqrt{\frac{\pi}{2}}. \tag{4}$$

The distance dependent path loss from (1) is an average value and therefore not suitable to describe a real channel. For obstructed indoor environments a zero-mean Gaussian random variable X_σ with standard deviation σ is added to the average path loss:

$$PL(d) = PL(d_0) + 10n \log\left(\frac{d}{d_0}\right) + X_\sigma. \tag{5}$$

X_σ can be described with the standard deviation σ of the Rayleigh distribution which is defined as follows:

$$\sigma = \sqrt{\frac{4-\pi}{2}\theta^2}. \tag{6}$$

With (1) and a reference path loss $PL(d_0)$ at a distance $d_0 = 1$ m the distance between transmitter and receiver can be calculated with

$$d = 10^{\left(\frac{PL(d_{RX}) - PL(d_0)}{10n}\right)}, \tag{7}$$

where $PL(d_{RX})$ is the path loss measured at the receiver. Rearranged to n we get

$$n = \frac{PL(d_{RX}) - PL(d_0)}{10 \log d}. \tag{8}$$

2.2 Diversity RF Transceiver Module

To investigate the optimum antenna spacings we proposed a prototype diversity platform with four uncorrelated channels and flexible antenna spacings in a previous work [11]. In Fig. 5 our miniaturized diversity concept with eight redundant channels using proprietary low-power RF transceivers with chip antennas and fixed antenna spacings is shown.

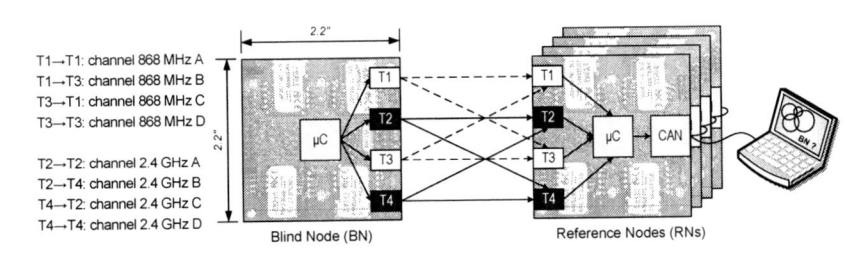

Fig. 5. Overview of the radio tracking system and diversity concept with eight RF channels. Source: Own elaboration.

The concept uses spatial and frequency diversity simultaneously to decrease the probability of signal dropouts, improve the distance estimation of the localization system, and also increase the systems availability. An RSS-based selection combining is realized on the receiver. The influence of the number of uncorrelated channels on the path loss distribution is indicated with the PDFs in Fig. 6.

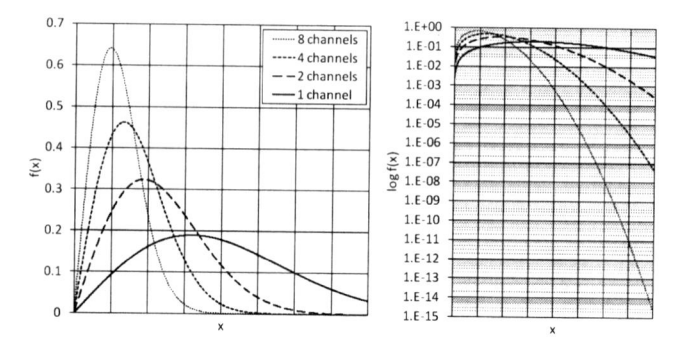

Fig. 6. PDFs of RSS-based selection combining for various numbers of channels (left: linear scale, right: logarithmic scale). Source: Own elaboration.

Let assume a probability of 10^{-1} for a specific signal dropout of a single channel (e.g. by 20 dBm). With the selection combining of two channels this probability is reduced to 10^{-2}. With the diversity configuration from Fig. 5 using eight redundant channels the dropout probability is reduced to 10^{-8}.

3 Experimental Results

To figure out the impact of diversity for the tracking performance we performed two different experiments. In a path loss measurement we investigate the environment-specific propagation coefficient n for the two different frequency bands and the impact of spatial diversity on the fading compensation. The validation of the tracking improvement due to the fading compensation is realized with a dynamic measurement on a motion test track.

The test bed for the measurements is shown in Fig. 7 on the left. The BN performs periodic movements on a motion test track according to the motion profile shown on the right. The duration of one movement from position A to B and back to A is 65 s. For an explicit multipath propagation we installed metallic reflecting walls next to the track.

3.1 Path Loss Measurement

For the test environment we investigated the value of n for different channel configurations. The receiver is placed at $x = 0.0$ m. The transmitter starts at B ($x = 1.4$ m). The transmitter sends out a training packet every 0.25 s while it

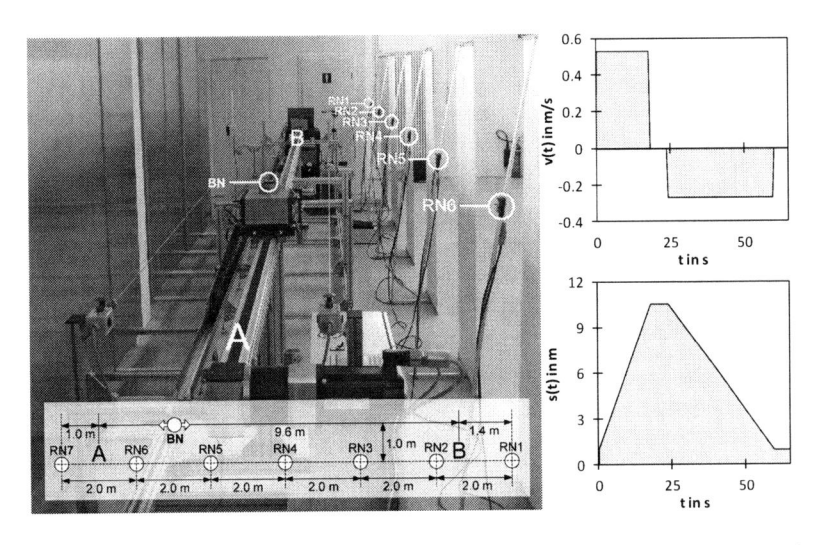

Fig. 7. Measurement setup on a motion test track in an obstructed test hall (speed and position profiles on the right show one A-B-A motion cycle, $T = 65\,\text{s}$). Source: Own elaboration.

moves straightforward to A ($x = 11.0\,\text{m}$) with a speed of approximately $0.27\,\text{m/s}$ (cf. motion profile in Fig. 7). We collect the RSS from the receiver for eight redundant channels. For channel A of the 868 MHz band and the 2.4 GHz band the values are shown in the path loss diagrams in Fig. 8 on the left. For each RSS value we compute n according to (8). With all 150 values of a channel we compute the maximum likelihood estimation θ for the Rayleigh distribution according to (3). With θ and (4) we get the expectation value for the path loss coefficient n. For channel A of the 868 MHz band we get a coefficient of 2.31, the value for channel A of the 2.4 GHz band is 2.62. On the right in Fig. 8 the path loss curves for an RSS-based selection combining of four spatial divided channels are shown for each frequency band. Comparing the path loss curves for one single channel with the combined values, the fading compensation gets obvious. When there are a lot of signal dropouts and even signal losses for the single channels, the combined signals show only a few signal dropouts and even no signal losses for 2.4 GHz.

The PDFs of n are shown in Fig. 9 for all four channels and the combining of two and four channels. The values of n for 868 MHz range between 2.18 and 2.93. For 2.4 GHz the values are between 1.91 and 3.02. The probability of certain signal dropouts is indicated by the probability of a specific value for n. The free space propagation without interferences corresponds to $n = 1$. An $n < 1$ indicates a positive interference of multipath signals. The PDFs can be used to compare the diversity gain of the used concept with the theoretical values from Fig. 6. For a comparison we assume the worst case scenario for a single channel. Here again we assume a signal dropout probability of 10^{-1} which corresponds to an n of 5.70 for 868 MHz (channel B) and 5.80 for 2.4 GHz (channel C).

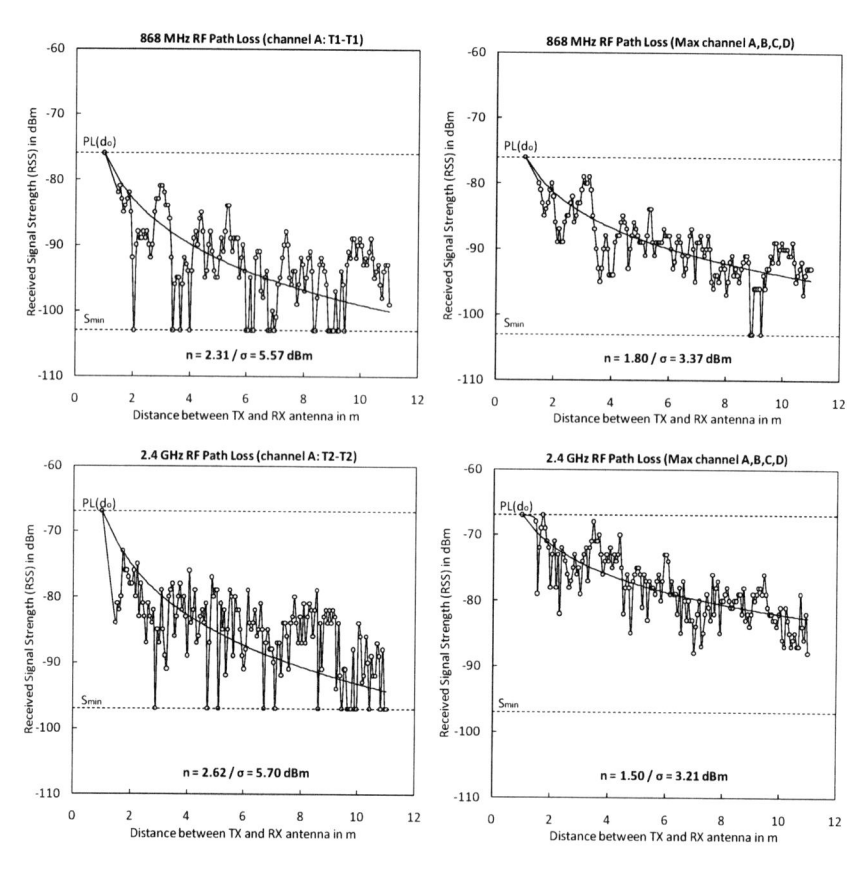

Fig. 8. Path loss measurement for 868 MHz and 2.4 GHz over a distance of 11 m (150 RSS samples for each frequency, TX speed = 0.27 m/s, 5 Hz update rate). Source: Own calculation.

With a selection combining of four channels this probability can be reduced to 1.2×10^{-2} for 868 MHz and 2.0×10^{-3} for 2.4 GHz. The diversity gain of the real path loss is slightly lower than the theoretical values. Especially for 868 MHz the spacing between the spatial divided antennas is not optimal and thus the channels show a weak correlation. For 2.4 GHz the spacing is nearer to the optimum and thus the channels are less correlated and the diversity gain is higher. A derivation of the optimum antenna spacing is given in [12].

3.2 Tracking Performance

The precision of the tracking on the motion test track (cf. Fig. 7) was elaborated for different diversity configurations. In Fig. 10 the trajectories of the BN on the motion test track and the absolute location estimation error for a complete motion cycle (A-B-A) are shown. Different configurations are compared to point out the influence of the diversity on the accuracy of the position estimation.

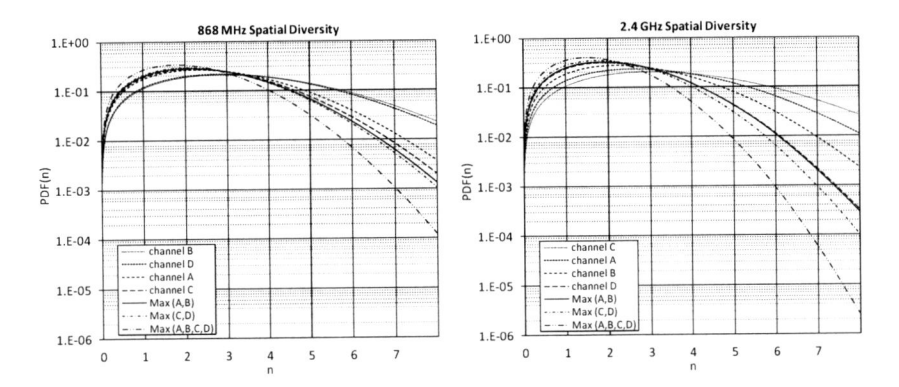

Fig. 9. Probability density function (PDF) of propagation coefficient n for 868 MHz and 2.4 GHz path loss measurements. Source: Own calculation.

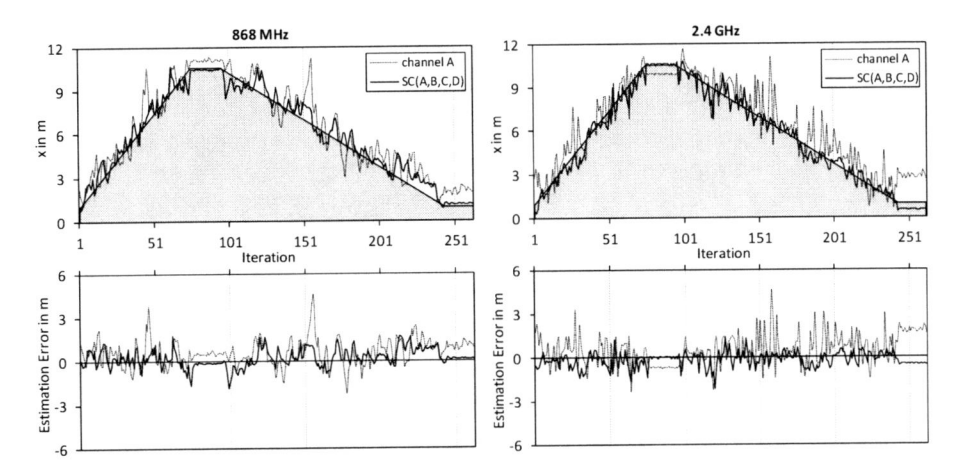

Fig. 10. Estimated tracks and location estimation error for 868 MHz and 2.4 GHz measurements on a motion test track. Source: Own calculation.

The selection combining (SC) of the four channels reduces the error for both, the 868 MHz and the 2.4 GHz band. The improvement is slightly higher for the 2.4 GHz band.

To have a closer look at the estimation error we compare the cumulative distribution functions (CDFs) of the selection combining of all channels with the single channels in Fig. 11. The corresponding values for the maximum, median and 95th percentile of the error are given in Table 1. The median error of all configurations is below 0.74 m. For the reference application with the tracking of human beings it is necessary to have a look at the maximum estimation error. Here the improvement by the diversity concept with the selection combining of four channels is pointed out. The maximum error is reduced by more than 43 % / 31 % for the 868 MHz / 2.4 GHz band.

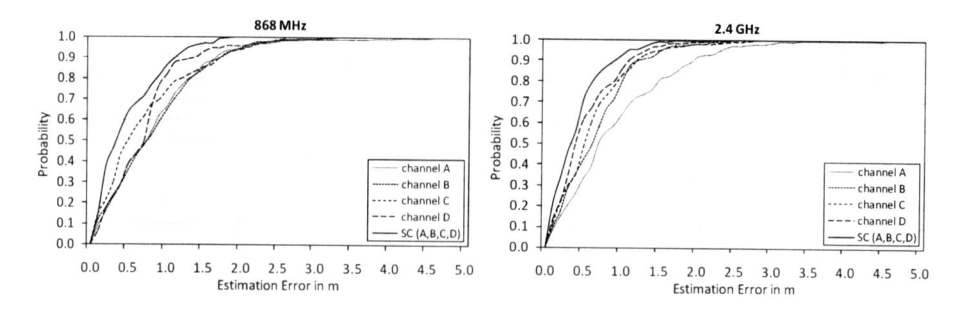

Fig. 11. CDFs for the location estimation error of various configurations of the 868 MHz and 2.4 GHz tracking measurements. Source: Own calculation.

Table 1. Performance comparison of different configurations (LEE - location estimation error in meters, PER - packet error ratio for transmission between BN and all RNs)

RF band	868 MHz					2.4 GHz				
channel	A	B	C	D	SC(A,B,C,D)	A	B	C	D	SC(A,B,C,D)
LEE_{med}	0.74	0.69	0.62	0.73	0.50	0.53	0.39	0.71	0.33	0.33
σ_{LEE}	0.67	0.73	0.47	0.63	0.64	0.48	0.42	0.53	0.43	0.35
$LEE_{95\%}$	1.96	2.29	1.51	2.01	2.08	1.49	1.30	1.74	1.29	1.09
LEE_{max}	4.54	4.56	2.43	3.25	2.58	2.78	2.42	3.18	1.85	2.17
PER	0.11	0.15	0.08	0.12	0.01	0.08	0.04	0.07	0.01	0.00

4 Conclusion and Future Work

The experimental results for the tracking on the motion test track show that the accuracy of a single channel RSS-based localization is not sufficient for a tracking of humans in a longwall mining application. With a maximum location estimation error of 4.56 m the size of the security zone and the amount of involved shields lead to an inefficient mining strategy. The experimental results show that diversity is a good idea to reduce the LEE in a significant way. The best results for the accuracy of the localization is reached with a selection combining of the four 2.4 GHz channels. The maximum LEE is reduced to 2.17 m. In this case only a few shields have to be stopped and the mining system reaches a higher efficiency compared to the single channel system.

The results from a path loss measurement match the theoretical Rayleigh model assumption. Without diversity the 2.4 GHz path loss coefficient is slightly higher than it is for the 868 MHz propagation. This can be ascribed to the higher losses and stronger multipath effects with an increasing frequency. The improved availability due to diversity is indicated by a reduced PER for the selection combining of four channels. The diversity gain indicates a more optimum spatial channel separation for 2.4 GHz, although there is also a significant improvement for 868 MHz.

For a more sophisticated evaluation of the tracking system we are going to introduce a further measurement with the controlled inclusion of external interferences (e.g. broadband WLAN channel) to point out the influence of the frequency diversity. The tracking system should be tested in a real scenario of an underground longwall mining application. Therefore, we are currently working on the intrinsic safety of the infrastructure components.

References

1. Guvenc, I.: Enhancements to RSS Based Indoor Tracking Systems Using Kalman Filters. In: GSPx & International Signal Processing Conference (2003)
2. Hightower, J., Boriello, G.: A Survey and Taxonomy of Location Systems for Ubiquitous Computing. Technical report, University of Washington, Department of Computer Science and Engineering (2001)
3. Jang, W.S., Skibniewski, M.J.: A Wireless Network System for Automated Tracking of Construction Materials on Project Sites. Technical report, University of Maryland, Department of Civil and Environmental Engineering (2008)
4. Elnahrawy, E., Li, X., Martin, R.: The Limits of Localization Using Signal Strength: A Comparative Study. In: Sensor and Ad Hoc Communications and Networks, pp. 406–414 (2004)
5. Whitehouse, K., Karlof, C., Culler, D.: A Practical Evaluation of Radio Signal Strength for Ranging-based Localization. SIGMOBILE Mob. Comput. Commun. Rev. 11, 41–52 (2007)
6. Rappaport, T.S.: Wireless Communications Principles and Practice. Prentice Hall PTR, Englewood Cliffs (2002)
7. Haykin, S., Moher, M.: Modern Wireless Communications. Pearson, Prentice Hall (2005)
8. Molisch, A.: Wireless Communications. John Wiley & Sons, Chichester (2005)
9. Fink, A., Beikirch, H., Voss, M.: Improved Indoor Localization with Diversity and Filtering based on Received Signal Strength Measurements. International Journal of Computing 9, 9–15 (2010)
10. Welch, G., Bishop, G.: An Introduction to the Kalman Filter. Technical report (2006)
11. Beikirch, H., Voss, M., Fink, A.: Redundancy Approach to Increase the Availability and Reliability of Radio Communication in Industrial Automation. In: 14th IEEE Conference on Emerging Technologies & Factory Automation (ETFA), pp. 1682–1685 (2009)
12. Vedral, A., Kruse, T., Wollert, J.: Development and Performance Evaluation of an Antenna Diversity Module for Industrial Communication based on IEEE 802.15.4. In: 12th IEEE Conference on Emerging Technologies & Factory Automation (ETFA), pp. 177–186 (2007)

Redundant Reader Elimination Approaches for RFID Networks

Nazish Irfan[1], Mustapha C.E. Yagoub[1], and Khelifa Hettak[2]

[1] School of Information Technology and Engineering (SITE),
University of Ottawa, Ottawa, Canada
{nirfan,myagoub}@site.uottawa.ca
[2] Communication Research Center,
3701 Carling Avenue,
Ottawa, Canada
khelifa.hettak@crc.gc.ca

Abstract. Radio Frequency Identification (RFID) systems, due to recent technological advances, have been used for various advantages in industries like production facilities, supply chain management etc. However, sometimes this implies a dense deployment of readers to cover the working area. Without optimizing reader's location and number, many of them will be redundant, reducing the efficiency of the whole RFID system. There are many algorithms proposed by researchers to solve redundant reader problem, but all the algorithms are based on omnidirectional reader antenna pattern, which is not practical. Therefore, the approach proposed in this paper was first applied to RFID readers using omni-directional antenna and thus, extended to directional reader antennas. Simulation results demonstrate that our approach for omnidirectional reader antenna outperforms other well-known techniques like Redundant Reader Elimination (RRE) algorithm, Layered Elimination Optimization (LEO) and LEO+RRE while preserving the coverage ratio quite close to those obtained by RRE, LEO and LEO+RRE. Then, we extended the concept to real directional reader antenna pattern using genetic algorithm to optimize the antenna beam to eliminate redundant readers.

Keywords: RFID, antenna, redundant reader, tilt, genetic algorithm.

1 Introduction

Radio Frequency Identification (RFID) is based on radio communication for tagging and identifying an object. Over the last few years, RFID has drawn a great deal of attention and it is now widely believed that RFID can bring revolutionary changes. Some of the major retailers have already invested significantly in RFID and mandated their manufacturers to place tags on cases and pallets, which resulted in mass production of inexpensive RFID tags [1].

In recent years, more efforts have been made to implement RFID-applications in inventory control and logistics management. Applications like inventory detection and automated product receiving in supply chain management require

RFID readers to read tags anywhere within a large geographic area. Since the range of reader to tag communication is very limited, readers must be deployed in high densities over the entire area [2]. Therefore, the deployment of RFID system has generated the RFID network planning problem that needs to be solved for large scale deployment. However, RFID network planning is one of the most challenging problems since it has to meet many requirements for efficient operation of RFID system [3]. This dense deployment of RFID systems in large scale results in unwanted effects. In fact, when multiple readers share the same working environment and communicate over shared wireless channels, a signal from one reader may reach other readers and can cause interference among readers. Hence, the effect of reader interference on the RFID interrogation range should be analyzed before any large scale deployment of readers in a RFID system [4]. Also, unnecessary readers in the network may consume power, which can be wasteful. Therefore, finding redundant readers is of great importance for an optimal deployment of a RFID network. This ensures user that the minimum number of readers is used to cover all the tags in a specified zone.

The problem of redundant reader elimination has been studied extensively in [5,6,7]. In this paper, we proposed two approaches to find redundant readers, first based on omni-directional reader antenna pattern and second based on real directional reader antenna pattern. In the second approach, we have demonstrated that practical measured/simulated data from commercial or inhouse antennas can be efficiently utilized to eliminate redundant readers from RFID networks. The redundant reader elimination techniques can be then used for efficient deployment of readers in large warehouse, manufacturing units in production lines and large retail stores.

2 Related Work

During the last decade, the RFID collision problem has been extensively covered in literature. It can be categorized as reader-to-reader interference or reader-to-tag interference. The reader collision problem not only results in incorrect operation but also results in reduction of overall read rate of the RFID system [4,8,9]. There are many algorithms in literature, which cover reader collision problem [2,8,9,10].

Another approach to avoid collision is to reduce the number of redundant readers in the RFID network. The RRE algorithm [5] is based on greedy method. The main idea of this method is to record "tag count" i.e. the number of tags a reader covers into RFID tags memory. The reader which has the maximum tag count will be the holder of the corresponding tag. This procedure iterates the above steps until all the tags in a network are assigned to readers. Finally, readers with no tags assigned are eliminated as redundant readers. In [6], the authors demonstrated that RRE procedures fail to eliminate redundant readers for some specific topologies of RFID networks. Thus, they introduced the LEO algorithm, the fundamental approach of this procedure being "first read first own". In a RFID network, all the readers send command signal to RFID tags in their coverage zones to get the record of the tags. The reader that first sent its signal is the owner of the tag. If this tag already has another owner (reader) ID, then its ID cannot be changed. Finally, the readers in the network with no tags in their coverage are eliminated as redundant readers. The authors have also demonstrated that LEO and RRE can be combined for better performance. In [7], the authors proposed an algorithm which takes advantage of the concept of neighboring reader density to assess the priority of reading. A reader having higher priority among its neighbor will own the tags. Finally, any reader owing no tag is eliminated as redundant. In this algorithm, the priority value of a reader depends on the number of its neighboring readers.

Redundant reader elimination techniques mentioned above are all based on omni-directional reader antenna, which is not the case for real antennas designed for RFID systems. Moreover, in [11], the authors mentioned that antennas having an omni-directional radiation pattern should be avoided and wherever possible directional antennas should be used because directional antennas have advantages of less disturbance to the radiation pattern and to the return loss. Also, omni-directional antennas have short range and no particular advantage in terms of signal to noise ratio (SNR). Therefore, for more realistic applications, redundant reader elimination should be based on directional reader antenna pattern.

In antenna, tilt is an important design parameter for network planning when considering coverage vs. capacity for cell planning as well as tuning network. Antenna tilting of base station is a very common technique for improving cell isolation and / or increasing covers in cellular networks. Tilt angle optimization can be achieved electrically, mechanically or by a combination of both [12]. It is reported that both coverage and quality performances are very sensitive to antenna tilt variations [13]. There are many papers [12,14,15] in literature, which have studied the effect of antenna tilt control on radio network planning and cellular network. Authors have shown that significant reductions in path-loss delay spread, transmitted power and system interference can be achieved when suitable height and antenna tilt are selected. It is reported that in a dense network, the importance of antenna down tilt becomes more significant. Therefore, in this paper we have presented an redundant reader elimination approach, based on antenna tilt optimization using genetic algorithm, which can be effectively utilized with any commercial or in-house directional antenna radiation pattern data.

3 Genetic Algorithm

Genetic algorithm is an optimization and search technique based on the principles of genetics and natural selection. The method was developed by John Holland [16]. Initially the population is generated randomly. The fitness values of all chromosomes are evaluated by calculating the objective function in a decoded form (phenotype). From the population, a particular group of chromosomes (parents) are selected to generate the offspring by the defined genetic operations. The fitness of the offspring is evaluated in a similar fashion to their parents. Based on certain replacement strategy, the chromosomes in the current population are replaced by their offspring. The above GA cycle is repeated until a desired termination criterion is reached. Finally, the best chromosomes in the final population can become a highly evolved solution of the population. There are various techniques that are employed in the GA process for encoding, fitness evaluation, parent selection, genetic operation and replacement.

4 First Approach (Omni-directional Reader Antenna)

In any arbitrary RFID network, any reader which covers more tags and has fewer number of neighbor readers must be given priority. A reader with more neighbors has higher probability of getting its operation interfered by the neighbor readers. The proposed algorithm assigns weights to each reader based on the number of neighbors it has and the number of tags it covers to ensure that best possible readers are selected to help efficient working of any RFID network.

Some of the assumptions for the proposed techniques are:

1. Reader coordinates are easily available.
2. Coverage information i.e. number of tags each reader has covered in initial round can easily be obtained by data processing subsystem.

It can be noted that the second assumption of collecting coverage information i.e. total number of tags covered by each reader at central host system does not require any new setup to RFID systems. Since such processing system is already included in existing RFID setup, this assumption adds no extra cost to it. Since proposed work is based on the number of neighbors to a reader, the neighbor criterion is defined as: Reader A is neighbor to reader B if ($d > 0\, meter$ & $d < 6\, meters$), where d is the distance between readers A and B.

Total weight assigned to a reader is a function of cost functions and multiplication factor. Cost function of a reader is defined in terms of its coverage and number of its neighbors. Cost function of a reader due to its coverage and number of neighbors is given by (1) and (2) respectively.

$$f_c = \frac{coverage(r)}{\alpha \times max(coverage(\mathbf{R}))} \quad (1)$$

$$f_n = 1 - \frac{(neighbor(r))}{max(neighbor(\mathbf{R}))} \quad (2)$$

where, r defines each individual reader in a network and $\mathbf{R}$ defines the list of all readers in the network with their individual tag counts and neighbor counts, respectively. The user-defined multiplication factor α, usually between $(1-3)$, is used so that the cost functions due to coverage and neighbors are in proportion and can influence each other.

$$TW_{reader} = l_c \times f_c + l_n \times f_n \quad (3)$$

where l_c and l_n are the load factors assigned to a cost function of a reader for coverage f_c and number of neighbors f_n, respectively. Load factors l_c and l_n are user-defined and satisfy the criterion that $l_c + l_n = 1$.

Basic operation of the proposed work can be summarized as follows:

1. All readers in the RFID network send commands to all the tags in their interrogation zone.
2. Each reader coverage information is sent to the central host station i.e. how many tags (with IDs) each reader has read.
3. For each tag of the RFID network, the proposed algorithm checks how many readers have read it. Further, algorithm compares the weights of readers who have read the tag. Reader having the maximum weight will own the tag.
4. All the readers of the network with no tag assigned to them are eliminated as redundant readers.

After eliminating the redundant readers, algorithm switches to its second part which is optimization of the network. In the optimization mode, the algorithm picks a reader from the remaining readers based on minimum coverage and maximum neighbors and eliminates it. The algorithm again assigns weights based on the number of readers left and total tags covered by the remaining readers using (1), (2) and (3). Further, it follows step 2 of its operation to reorder the readers based on total weights assigned to each reader. The procedure iterates till all the readers have a number of neighbors equals or less than 3.

5 Second Approach (Directional Reader Antenna)

In this section we present a proof of concept for redundant reader elimination technique using real directional reader antennas. Figure 1 shows the typical

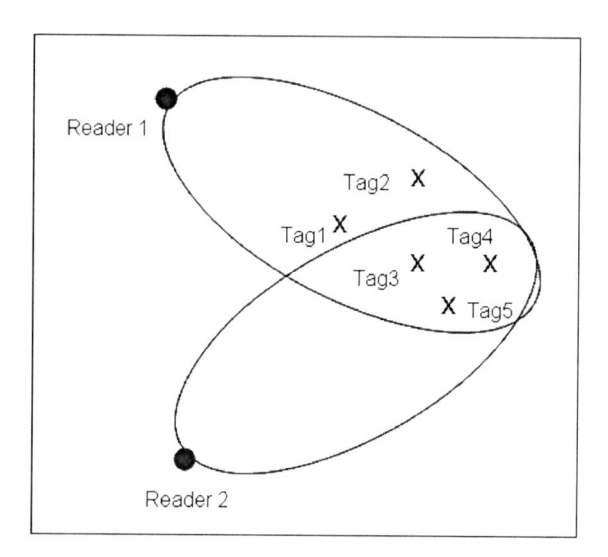

Fig. 1. Typical redundant reader topology for directional antenna in RFID network

scenario where a reader can optimize the coverage by adjusting its tilt angle. This scenario is common to redundant reader elimination setup in RFID networks with antenna readers having directional radiation patterns. From this figure, we see that reader 1 after optimizing its tilt angle can cover tags 1-5 and similarly reader 2 covers tags 3-5. Since reader 1 covers all the tags which reader 2 also covers then reader 2 can be eliminated safely without compromising the network coverage.

6 Simulation Setups and Results

6.1 Simulation Setups for First Approach

In this section, we will evaluate the performance of our first approach i.e. redundant reader elimination for omni-directional reader antenna. We selected 5 different experimental areas and randomly located readers and tags for each setup. When random locations for readers and tags in a RFID network have been generated, it was taken care that no reader or tag was located at same position.

Next, we will discuss the performance of our algorithm and compare our results with those obtained by the state-of-the-art approaches listed above. Initially, we compared the maximum coverage i.e. the maximum number of tags covered by all remaining readers in the RFID network. This step was undertaken to ensure that the coverage attained by the proposed work is in close relation with RRE, LEO and LEO+RRE algorithms.

After coverage evaluation, the number of redundant readers eliminated by each algorithm has been compared in this work. Figure 2 shows that the number of redundant readers eliminated by the proposed procedure outperforms the other compared algorithms.

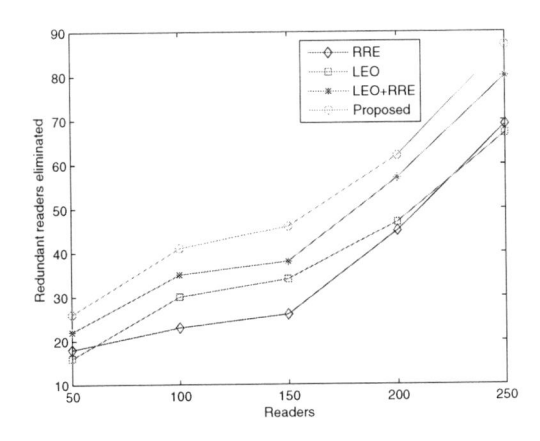

Fig. 2. Performance comparison of the proposed technique vs. RRE, LEO and LEO+RRE

Note that all the existing redundant reader elimination techniques [5,6,7] have many read-write operations. The procedure LEO presented in [6] has minimum write operations (reader writing or updating information on the tag) and the density based procedure [7] has the maximum write operations. Compared to other algorithms, our work has no write operations and has only one read operation. More write operation procedures require more resources and eventually add to the cost of operation of RFID systems.

Since the existing algorithms [5,6,7] require write operations, these procedures are only suitable for tags which have both read and write options. Since tags that support read-write operations have extra-functionality, it adds to the cost of RFID systems. On the other hand, the proposed work is suitable for any type of tags because it requires no write operation.

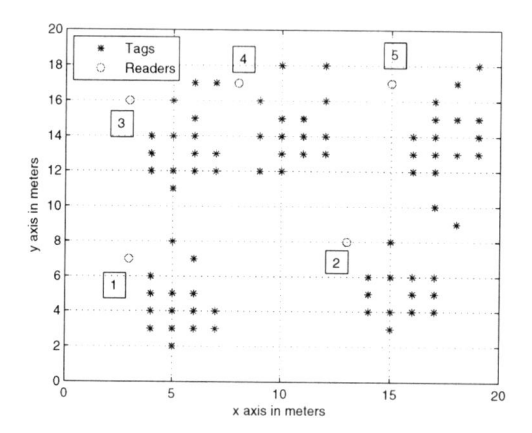

Fig. 3. RFID network topology of tilt angle optimization

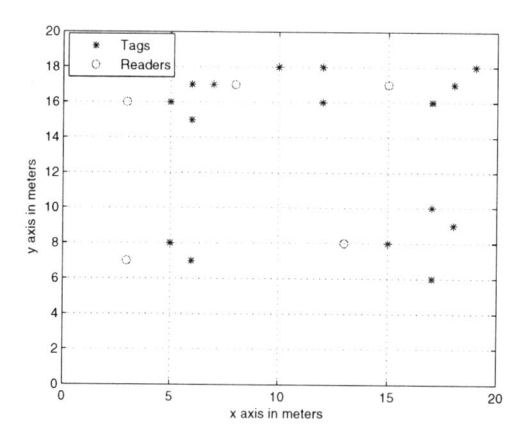

Fig. 4. Tags covered by reader with not tilt angle optimization

Procedure RRE focuses on maximum number of tags each reader covers for its operation. Similarly, algorithm LEO works with first read first own and density based procedure focuses on the number of neighbors each reader has for its operation. On the other hand, our work takes both total tags covered by each reader, i.e individual coverage and the number of neighbors each reader has for its operation. The advantage of this approach is that it not only eliminates the maximum possible redundant readers by attaining appreciable coverage but also optimizes the RFID network for efficient operation. In the proposed technique, maximum neighbors in any reader in a RFID network is set equal or less than 3. We have taken 3 as a lower bound number to ensure that the coverage of tags in a RFID network is not reduced below appreciable limit.

6.2 Simulation Setup for Second Approach

To further demonstrate the efficiency of our proposed approach, we implemented an experimental setup with area $20 \times 20m$, 5 readers and 75 tags. Figure 3 shows the proposed experimental setup. As the test device, the Intermec RFID reader IA33A circularly polarized panel was adopted and the Intermec UHF tag was used [17].

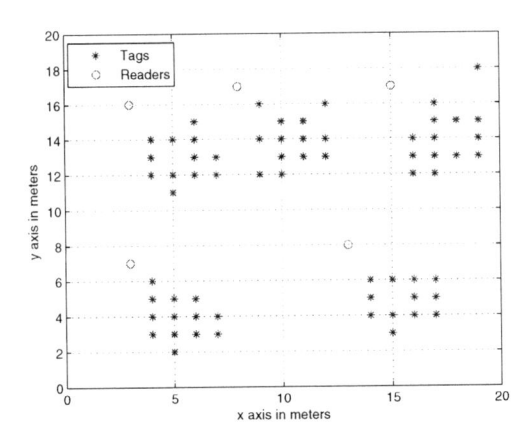

Fig. 5. Tags covered by readers with tilt angle optimization

Table 1. Number of tags covered before and after optimizing the tilt angle

Reader	Tags covered before	Tags covered after	Optimized tilt angle
1	2	13	-21.74 °
2	4	12	-23.50 °
3	4	12	-23.94 °
4	3	13	-23.73 °
5	3	14	-23.33 °

In this work, a continuous Genetic algorithm encoding scheme was adopted and the Roulette wheel fitness technique was used to evaluate the objective function. Genetic algorithm parameters used for optimization are mutation rate=0.2 and selection=0.5. In this experimental setup, the reader tilt angle was optimized. Figure 4 shows the number of tags covered by each reader when reader tilt angle is not optimized. It can observed that out of 75 tags only 16 tags were covered by 5 readers. On the other hand, as shown in Figure 5, after optimizing the reader tilt angle, 64 tags were covered by 5 readers. Table 1 summarizes the results for both experiments.

These results prove that the optimal placement of the beam angle of a directional antenna can greatly increase the coverage. The optimization helps in deciding on the optimal number of reader placement thereby eliminating the redundant reader and reducing the overall cost of the RFID system. This experimental setup demonstrates that this technique can be effectively used in eliminating redundant readers from the RFID network.

7 Conclusion

In this paper, we presented a redundant reader elimination techniques for both omni-directional and directional reader antennas. Algorithm for omni-directional reader antenna optimizes the RFID network by giving importance to a reader who has fewer number of neighbors and more coverage. This work can be used with any arbitrary RFID network. With our simulation results, it is demonstrated that the proposed algorithm outperforms other state of the art techniques presented in literature such as RRE, LEO and LEO+RRE. It eliminates more redundant readers than RRE, LEO and LEO+RRE by keeping the coverage close to that of RRE, LEO and LEO+RRE. Similarly, simulation results of our approach for directional reader antenna demonstrates that the proposed approach can effectively works as a redundant reader elimination tool for reader with practical directional radiation pattern in RFID network.

References

1. Sarma, S.E.: Towards the five-cent tag. Technical Report MIT-AUTOID-WH-006, MIT Auto ID Center (2001)
2. Waldrop, J., Engels, D.W., Sarma, S.E.: A MAC for RIFD reader networks. IEEE Wireless Communications and Networking (WCNC) 3, 1701–1704 (2003)
3. Chen, H., Zhu, Y.: RFID networks planning using evolutionary algorithms and swarm intelligence. In: 4th International Conference on Wireless Communications, Networking and Mobile Computing (WiCOM)), pp. 1–4 (2008)
4. Engels, D.W.: The reader collision problem. White Paper MIT-AUTOID-WH-007, MIT Auto ID Center (2001)
5. Carbunar, B., Ramanathan, M.K., Koyuturk, M., Hoffmann, C., Grama, A.: Redundant-reader elimination in RFID systems. Second Annual IEEE Communications Society Conference on Sensor and Ad Hoc Communications and Networks (SECON), pp. 176–184 (2005)

6. Hsu, C.H., Chen, Y.M., Yang, C.T.: A layered optimization approach for redundant reader elimination in wireless rfid networks. In: IEEE Asia-Pacific Services Computing Conference, pp. 138–145 (2007)
7. Yu, K.M., Yu, C.W., Lin, Z.Y.: A density-based algorithm for redundant reader elimination in a RFID network. In: 2nd International Conference on Future Generation Communication and Networking, vol. 1, pp. 89–92 (2008)
8. Birari, S.M., Iyer, S.: Mitigating the reader collision problem in RFID networks with mobile readers. In: 13th IEEE International Conference on Networks, vol. 1, pp. 463–468 (2005)
9. Shis, D.H., Sun, P.L., Yen, D.C., Huang, S.M.: Taxonomy and survey of RFID anti-collision protocols. J. Computer Communications 29, 2150–2166 (2006)
10. Hwang, K., Kim, I., Ecom, K.T., DiCa, D.S.: Distributed tag access with collision-avoidance among mobile RFID readers. In: EUC Workshops, pp. 413–422 (2006)
11. Foster, P., Burberry, R., Antenna, R.A.: Antenna problems in RFID systems. In: IEEE Colloquium on RFID Technology, pp. 3/1–3/5 (1999)
12. Athley, F., Johansson, M.N.: Impact of electrical and mechanical antenna tilt on LTE downlink system performance. In: IEEE 71st Vehicular Technology Conference (VTC 2010), pp. 1–5 (2010)
13. Dinan, E., Kurochkin, A.: The effects of antenna orientation errors on UMTS network performance. In: IEEE 17th International Symposium on Personal, Indoor and Mobile Radio Communications, pp. 1–5 (2006)
14. Niemela, J., Isotalo, T., Lempiainen, J.: Optimum antenna downtilt angles for macro cellular WCDMA network. EURASIP Journal on Wireless Communications and Networking 2005(5), 816–827 (2005)
15. Wadhwa, M., Song, M., Rali, V., Shetty, S.: The impact of antenna orientation on wireless sensor network performance. In: 2nd IEEE International Conference on Computer Science and Information Technology, pp. 143–147 (2009)
16. Haupt, R.L., Haupt, S.E.: Practical genetic algorithms, second edition, 2nd edn. A John Wiley and Sons Inc., Hoboken (2004)
17. Intermec, http://www.intermec.com/products/rfid/antennas

NoC-MPSoC Performance Estimation with Synchronous Data Flow (SDF) Graphs

Kamel Smiri[1,2] and Abderrazak Jemai[1,3]

[1] LIP2 Laboratory, University of Tunis El Manar, Faculty of Sciences of Tunis,
Campus Universitaire, 2092 Manar 2, Tunisia
[2] University of Kairouan, ISMAI- Kairouan, Assad Iben Fourat - 3100 Kairouan, Tunisia
smiri_kamel@yahoo.fr
[3] University of Carthage, INSAT, B.P. 676, 1080 Tunis Cedex, Tunisia
abderrazak.jemai@insat.rnu.tn

Abstract. Multi-Processor Systems-on-Chip (MPSoC) are going to be the leading hardware platform featured in embedded systems, if they aren't already. This article deals with the performance estimation problem on these systems. We present in this paper, a new approach of performance estimation of migration software task to hardware component for MPSoC systems with Synchronous Data Flow (SDF) Graphs. This approach is structured on four steps: 1) annotation Kahn Process Network (KPN) model, 2) transformation the annotated KPN model to SDF model, 3) synthesis under constraints and 4) comparison of results. We have using the SDF3 tool to determine performance estimation of migration software task to hardware component. Experiments on MJPEG decoder are made to illustrate the efficiency of our approach of performance estimation.

Keywords: KPN, NoC-MPSoC, software-to-hardware migration, multimedia application, SDF graphs, SDF3 tool.

1 Introduction

Multimedia applications generally have important computation needs, and induce big transfers of data. Furthermore, they are composed of several independent tasks that can run in an autonomous way as soon as they have the appropriate input data at their input points. To fulfill that need for computation, while ensuring correct performances, designers often have recourse to software-to-hardware migration of tasks. But when implementing applications on MPSoC systems, migrating software tasks to hardware tasks is a costly process. It is time, men and money consuming.

One of the ways to avoid such expenses is performance estimation at design level. The general problematic needed to be answered is how to enable a designer to estimate the performances of an application destined to run on an MPSoC platform, at design level.

We present in this paper a new approach of performance estimation of migration software task to hardware component for MPSoC systems with Synchronous Data Flow (SDF) Graphs. The chosen application to illustrate our approach is the MJPEG decoder that is characterized by its high treatment density and important data exchange.

This paper is organized as follows. Section 2 presents the related work on performance estimation of MPSoC systems. Section 3 describes our approach of performance estimation of migration software task to hardware component with SDF Graphs. Section 4 presents the SDF3 tool, which was used to analyze performance. Section 5 presents the experimentation of MJPEG Decoder. The article ends with section 6 that comes as a conclusion and a glimpse of the work perspectives.

2 Related Work

Several research projects already treat the subject of performance estimation of migration on MPSoC systems. Indeed, recently, research is more and more concerned with the use of the Synchronous Data Flow (SDF) graph in the MPSoC design. Jerraya [1] points out that NoC-MPSoC design is the subject of more than 60 research projects over the world.

For their part, Kumar and al. [2] have proposed a solution to analyze applications that are modeled by SDF graphs and executed on MPSoC platform through the use of Resource Manager (RM). RM is a task which controls and directs the use of resources. RM is responsible for resources access (critical or not), and optimization of their usage. The designer reserves for the RM a whole execution node (CPU, memory, bus ...) which increases the cost of the total MPSoC system.

Lahiri and all [3] presented methodologies and algorithms to design communication architectures in order to achieve the desired performance on MPSoC systems. They developed a tool named Communication Analysis Graph (CAG), which relies on SDF graphs for MPSoC performance analysis and constraint satisfaction.

Wiggers and all [4] proposed a solution that consists in mapping purely software tasks and their communication channels on the target processors. They exploited the SDF graphs to compare the throughput obtained with the target throughput of the application.

Finally, S. Stuijk [5] proposed a design flow for mapping multimedia applications on NoC-MPSoC platforms.

3 Approach of Performance Estimation of Migration Software Task to Hardware Component with SDF Graphs

We propose in Fig. 1 a approach for estimating the performance, which is structured on the following four steps: (1) annotation Kahn Process Network (KPN) model (for the execution time of each task, size of data exchanged, number of times executed, percentage of occupation by the processor...), (2) transformation the annotated KPN model to SDF model, (3) synthesis under constraints and (4) comparison of results.

3.1 Annotation KPN Model

The first step is to annotate the KPN [12] model of parallel application by the execution times and the sizes of data exchanged between tasks. These metrics are

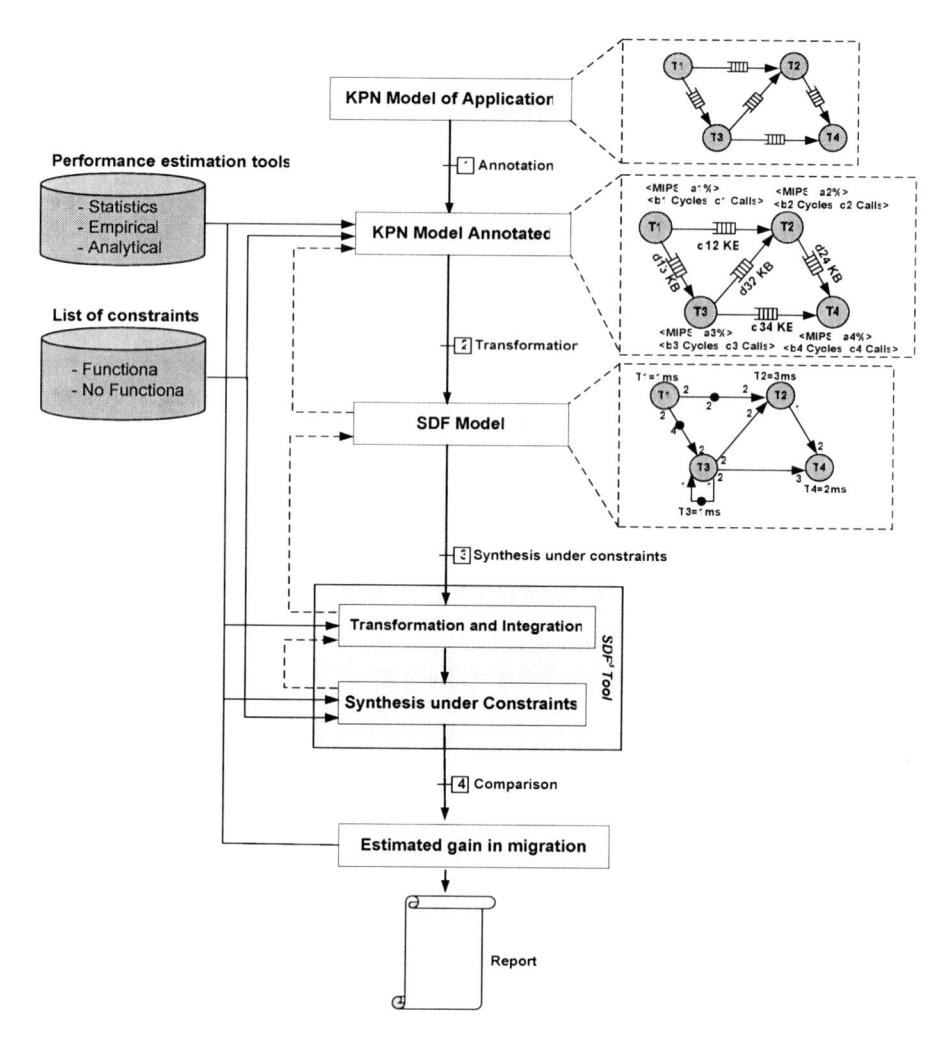

Fig. 1. Approach of performance Estimation of migration software task to hardware component with SDF Graphs in MPSoC systems

obtained by the performance estimation tools such as simulation (CASS, SystemC, ...), profiling (kprof, gprof, Callgrind, memprof, ...) and analytical methods (RapidRMA, TimeWize, SDF3, ...).

3.2 Transformation the Annotated KPN Model to SDF Model

The second step is to transform the annotated KPN parallel model to SDF model. This transformation must necessarily depend of application and hardware architecture of the target platform. The parameters that must be modeled by the SDF graph are: 1) the type of communication system (Bus or NoC), 2) the internal architecture of each tile (CPU type, Super Scalar, VLIW ...) and 3) the structure of memory (local, shared, single-port, multi-port...).

3.3 Synthesis under Constraints

This third step is the subject of the synthesis under constraints. First, we formalize the functional constraints (ex. number of frames per seconds) imposed by the application and no-functional constraints imposed by the target platform: number of CPU frequency, type of processor (VLIW, superscalar...), type of communication system, memory architecture (local to an execution node, shared, universal access port, multi-port access), type of memory (ROM, RAM, ...).

Then, we summarize the performance under constraints. Subsequently, a study of migration tasks will be carried out to estimate the application performance in various migration scenarios. We reconfigure the tool SDF3 (SDF For Free), which is offered by Stuijk, Geilen and Basten in [5][8], to realize this step of synthesis under constraints. SDF3 is free software available for free download at http://www.es.ele.tue.nl/sdf3 [9].

3.4 Comparison of Results

The last step is to compare the estimated performance gains with SDF graphs to those measured by simulation in [6] and draw appropriate conclusions.

4 The SDF3 Tool and Communication Protocols

4.1 Description of SDF3 Tool

The SDF3 tool has been developed by Stuijk, Geilen and Basten, from the Eindhoven University of Technology [8][11]. It is an open-source tool and can thus be downloaded for free at http://www.es.ele.tue.nl/sdf3/download.php [9]. The main purpose of SDF3 is performance estimation of applications on NoC-MPSoC systems, through static analysis of SDF graphs. The tool is written in C++. It can be used through both an API and a set of commands. Two of these commands have been used during this work. The first is the sdf3analysis command, which allows, among others, throughput, connexity and deadlocking analysis of SDF graphs. The second is the sdf3flow command, which executes a design flow called the predictable design flow [5], given (1) the application's SDF graph, (2) the destination NoC-MPSoC platform, and (3) a throughput constraint that formalizes the targeted performances. The input and output formats of the sdf3flow command is the XML format, described by an XML DTD (Document Type Definition).

The predictable design flow contains four phases. Phase 1 aims at determining the memory that will be allocated for each actor and channel of the application graph on the tiles of the target platform. Phase 2 computes the minimal bandwidth needed by the application and the maximal latency on the NoC. Phase 3 binds the actors and channels of the application graph to the tiles of the NoC-MPSoC target platform. The arbitration is based on TDMA wheels for each processor. Phase 4 schedules the communications on the NoC, given the actors and channels mappings.

When it is successful, the output of the predictable design flow is a NoC-MPSoC configuration that satisfies the throughput constraint. The configuration contains a mapping of the actors and channels of the SDF graph to the tiles and connections of the NoC-MPSoC destination platform. The mapping is modeled by applying transformations to the application graph, in order to model the binding of actors and channels to tiles and connections. The SDF3 tool has a conservative approach, which means that it will assume the worst-case scenario at any stage of the analysis.

4.2 Communication Protocols Modeling with SDF Graphs

Following the migration of a software task in hardware, the communication protocol must change and adapt. We adopted the solutions proposed in [8] to model these changes in the SDF graph. Communication protocols used: Communication Soft/Hard HS and communication Hard / Soft HS.

4.2.1 The SH1 Communication Protocol
Let A and B be two actors from a given SDF application graph (2.a). A and B are connected by a channel d. If A is migrated to a hardware task, the communication between A and B becomes a hardware-to-software communication. In the application graph, the SH1 communication protocol is modeled by the sub-graph represented in Fig. 2. (a) **Initial sub-graph, (b) Sub-graph after SH1 modeling, (c) Sub-graph after HS1 modeling.**

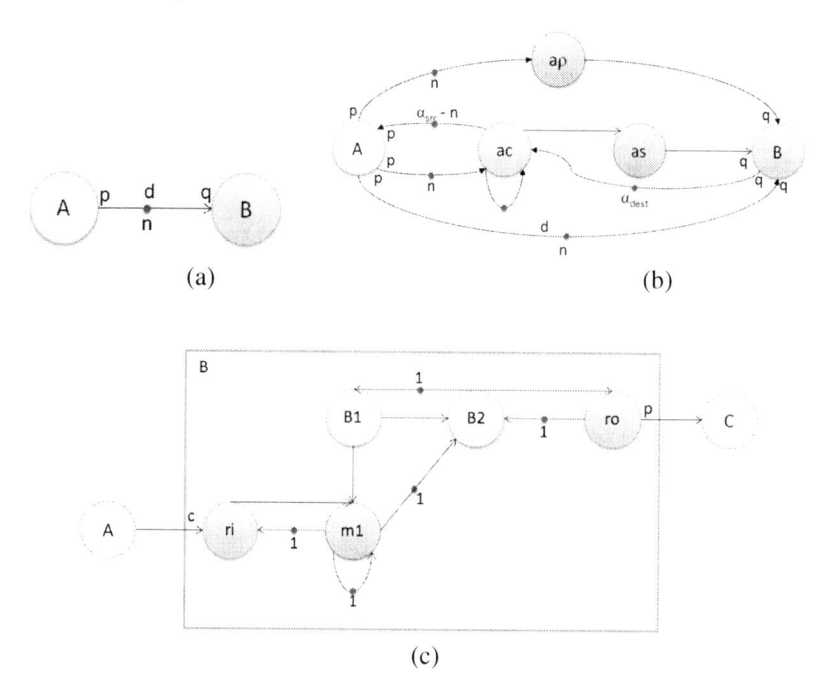

Fig. 2. (a) Initial sub-graph, (b) Sub-graph after SH1 modeling, (c) Sub-graph after HS1 modeling

The ac actor models the time necessary to send all the tokens from the actor A to the actor B. The ap actor models the maximal latency of the NoC. The as actor models the unavailability of the B actor. Indeed, in the worst-case scenario, the B actor has consumed all the TDMA (TDMA Time Division Multiple Access) slices allocated for it, so it has to wait for its processor's TDMA wheel to complete its current turn.

4.2.2 The HS1 Communication Protocol

Let's assume the same starting situation as in the previous case. If B is migrated to a hardware task, the communication between A and B becomes a software-to-hardware communication. It is modeled by the following SDF sub-graph ().

The hardware task, A, writes its results in its local memory. The B actor has to access that memory as a remote memory. It is split into two actors B1 and B2. B1 models the time necessary to send a request to the remote memory. The B2 actor models the initial B actor. The m1 actor models the time needed to access the remote memory and send the required data over the NoC.

5 Case Study: MJPEG DECODER

5.1 Description of MJPEG Decoder

The MJPEG codec processes video sequences that are composed of separate JPEG images. These images are 128x128 pixels big, and split in 8x8 pixels macro-blocks. The first task is the Huffman decoding. It is followed by a reverse zigzag ordering, and a reverse quantization. Then comes the reverse discrete cosine transform. The two final tasks are color conversion from the YUV color space to the RGB color space, and the reordering of the pixels in order to adapt them to the output device. These two tasks are not JPEG tasks but are needed when for output purposes. Table 1 presents the task done by each actor of the application graph. To make comparison, we are based on the results of simultaneously presented by our team and presented in [6]. These simulation results are presented in Table 2.

Table 1. Actor/Task mapping

Actor	Task
VLD	Huffman decoding
IZZ	Inverse zigzag reordering
IQ	Inverse quantization
IDCT	Inverse discrete cosine transform
LIBU	Pixel reorganization and adapting to the output device

Table 2. Simulation results of MJPEG decoder

Migration scenario	Performance (fps)
None	12,07
IDCT	11,49
IDCT and VLD	24,07

5.2 Validation of Approach

5.2.1 Annotation KPN Model

Fig. 3 show the result of the first step in our approach to performance estimation. During this stage, we annotate the parallel model KPN (Kahn Process Network) of MJPEG decoder by the execution time of tasks that are executed on ISS (Instruction Set Simulator) of target platform. And also we have annotated the KPN model by the sizes of data exchanged for image 128x128 pixels. The functional constraint imposed by the application is to have a decoding speed of 25 frames / second. The no-functional constraints imposed by the target platform are: 2 CPU MIPS3000, 50 MHz, CPU type superscalar, NoC communication system, local memory and uni-port architecture.

5.2.2 Transformation the Annotated KPN Model to SDF Model

The second phase of our approach is the transformation the annotated KPN model to SDF model. Fig. 4 shows the SDF model of the MJPEG decoder. Actors IZZ, IQ, IDCT consume and produce at each of their activations, a macro-block. The actor LIBU reconstructs an entire image and adapts to the output device, so it needs the 256 macro-blocks of that image.

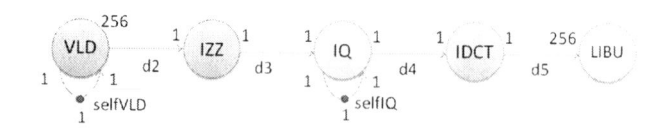

Fig. 3. SDF model of MJPEG decoder

5.2.3 Synthesis under Constraints with SDF3 Tool

This section is the third step of the approach, which is the step synthesis under constraints. Before running the SDF3 tool, some information needs to be collected: Actors execution times, Token sizes and Throughput constraint.

a) Actors execution times

The first information needed to be collected is the individual execution times for actors, expressed in abstract time units. The sdf3flow command uses an abstract time-unit.

Table 3. Simulation results of MJPEG decoder

Actor	Cycles per image	Cycles per macroblock
VLD	1.140.800	-
IZZ	446.400	1743
IQ	644.800	2518
IDCT	2.281.600	8912
LIBU	248.000	-

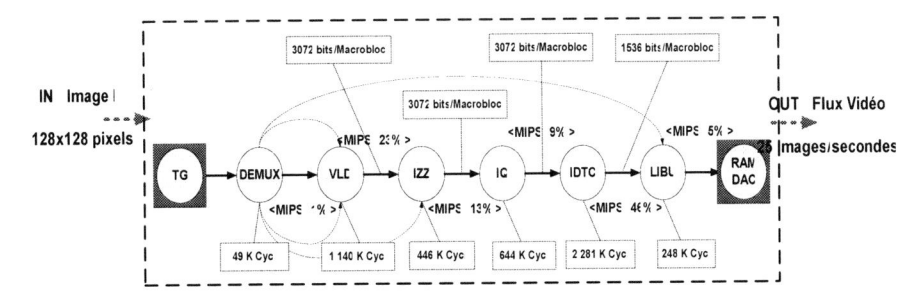

Fig. 4. Annotation of the KPN model of the MJPEG decoder

(a)

(b)

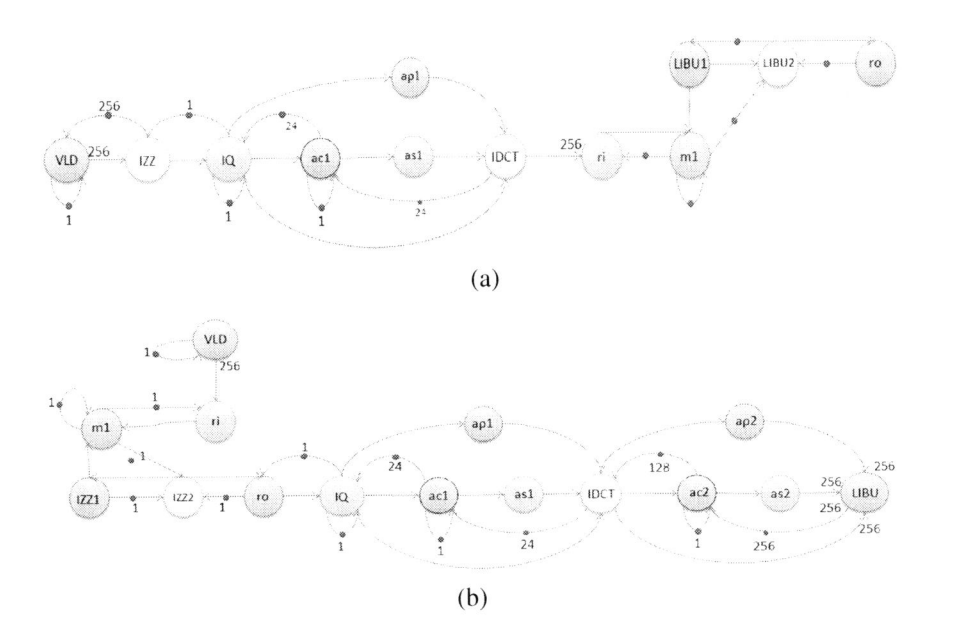

Fig. 5. a) : Application graph after modeling the IDCT migration, b) Application graph after modeling of the VLD migration

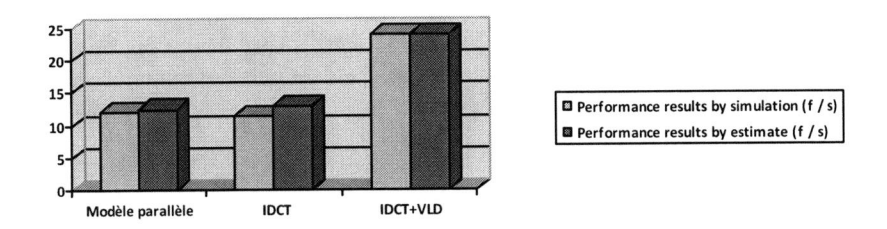

Fig. 6. Comparison of estimated results (SDF) to the measured results (simulation)

The time unit used in this work is the processor cycle. There are two methods for obtaining actor execution times in cycles. The first is the use of WCET tools, which process worst-case execution times for pieces of source code, or binaries. The second is simulation, which was used in this work. Table 3 presents the individual execution times for the actors of the MJPEG decoder.

b) Token sizes

In the worst case possible, the decoder has to process JPEG images with a 4:4:4 resolutions, which means that no sub-sampling is used. Therefore, the luminance Y is coded with the same resolution as the blue and red chrominance. For each 8x8 pixels macro-block, the decoder will then have to process 192 samples (64 x 3).

In the frequency domain, samples are coded on 2 octets because their interval is [-1024, 1024]. In the spatial domain, samples are coded on one single octet because their interval is [0,255]. The token sizes for each channel of the application graph are presented in Table 4.

Table 4. Token sizes for the application graph

Channel	d2	d3	d4	d5
Size in bits (per MB)	3072	3072	3072	1536

c) Throughput constraint

The throughput constraint has to be expressed in iterations per time-unit. If the target performance is n fps, and the processors of the target platform have a frequency of 50 MHz, which implies a cycle time of 20 ns per cycle, then the constraint is as follows:

Constraint = $n \times 20 \times 10^{-9}$ iterations / time-unit (E1)

5.2.4 Comparison of Results

Fig. 5.a represents the application graph which was applied the required transformations to model the SH1communication protocol between the IQ and IDCT actors, and the HS1 communication between the IDCT and LIBU actors. Fig. 5.b represents the application graph which was applied the required transformation to model the HS1 communication between the VLD and IZZ actors. Fig. 6 shows the comparison of estimated results (SDF) to the measured results (simulation). We find that our approach to performance analysis based SDF graphs is very interesting because it can quickly determine the source of problems (i.e. the data rate, consuming tasks of processing...). Moreover, this approach is more precise (running time) compared with simulation results.

6 Conclusion

We present in this paper, a new approach of performance estimation of migration software task to hardware component with Synchronous SDF Graphs. This approach is structured on four steps: which is structured on the following four steps: 1) annotation KPN model (for the execution time of each task, size of data exchanged, number of times executed, percentage of occupation by the processor...), 2) transformation the annotated KPN model to SDF model, 3) synthesis under constraints and 4) comparison of results. Our methodology can help the designer to choose one or more tasks to migrate. However, it can provide a solution to problem of hardware/software partitioning. We find that our approach is very interesting because it can quickly determine the source of problems.

References

1. Jerraya, A.A.: System compilation for MPSoC based on NoC. In: 8th International Forum on Application-Specific Multi-Processor SoC, MPSoC (2008)
2. Kumar, B., Mesman, B., Theelen, H., Corporaal, Y.: Analyzing composability of applications on MPSoC platforms. Journal of Systems Architecture (2008) ISSN 1383-7621
3. Lahiri, K., Raghunathan, A., Dey, S.: Efficient Exploration of the SoC Communication Architecture Design Space. IEEE Computer Aided Design (2000)
4. Wiggers, M.H., Kavaldjiev, N., Smit, G.J.M., Jansen, P.G.: Architecture Design Space Exploration for Streaming Applications Through Timing Analysis. Communicating Process Architectures (2005)
5. Stuijk, S.: Predictable Mapping of Streaming Applications on Multiprocessors, PhD thesis, Eindhoven University of Technology (2007)
6. Smiri, K., Jemai, A., Ammari, A.-C.: Evaluation of the Disydent Co-design Flow for an MJPEG Case Study. International Review on Computers and Software, I.RE.CO.S. (2008)
7. Smiri K., Jemai A., Migration Methodology in MPSoC Based on Performance Analysis via SDF Graph. In: Advances in Computational Sciences and Technology (ACST), pp. 149-166, (2009)
8. Stuijk, S., Geilen, M., Basten, T.: SDF3: SDF for free. In: ACSD 2006, pp. 276–278 (2006)
9. Eindhoven University of Technology, SDF3 official website (2009), http://www.es.ele.tue.nl/sdf3/
10. Leupers, R., Castrillon, J.: MPSoC Programming using the MAPS Compiler. In: ASP-DAC, Taipei (2010)
11. Stuijk, S., Geilen, M., Basten, T.: Exploring trade-offs in buffer requirements and throughput constraints for synchronous dataflow graphs. In: Proceedings of the 43rd annual Design Automation Conference DAC 2006 (2006) ISBN: 1-59593-381-6
12. Kahn, G.: The semantics of a simple language for parallel programming. In: Information Processing 74, Stockholm, Sweden, pp. 471–475 (1974)

Author Index

Abida, Kacem 231
Abida, Wafa 231
Aguir, Khalifa 366
Ahn, Hyeong-Joon 83
Alam, Kazi Masudul 321
Alemzadeh, Milad 222

Bayezit, Ismail 124
Beikirch, Helmut 384
Biglarbegian, Mohammad 93
Boubaker, Olfa 199

Cruz, Joseph Sun de la 167

Damaschke, Nils 303

Elramly, Salwa 374
Elsabrouty, Maha 374
El Saddik, Abdulmotaleb 134, 156, 177, 321, 343
Ewald, Hartmut 294

Farrokhsiar, Morteza 112
Fiaidhi, Jinan 241
Fidan, Baris 124
Fink, Andreas 384
Firouzi, Hadi 273

Gueaieb, Wail 21, 146
Guo, Fenghua 343

Hassan, Ahmed Ibrahim 374
He, Xiaochuan 146
He, Yan 343
Hettak, Khelifa 396
Huang, Guang-Bin 253, 263
Hudson, Jonathan A. 51
Huissoon, Jan P. 102

Irfan, Nazish 396
Islam, S. 134, 156, 177

Javed, Mohammad Azam 284
Jemai, Abderrazak 406

Jeon, Soo 83
Jnifene, Amor 11

Kajbafzadeh, Behrad 355
Kamel, Mohamed S. 1
Kaminska, Bozena 355
Karray, Fakhri 1, 222, 231, 311
Keshmiri, Soheil 31
Khajepour, Amir 102
Khamis, Alaa 1
Khosrow-khavar, Farzad 355
Khoury, Richard 212, 222
Krüger, Hendrik 294
Kulić, Dana 167

Lan, Yuan 253
Lanteigne, Eric 11
Lee, Gordon 331
Li, Howard 41
Lin, Zhiping 263
Liu, P.X. 134, 156, 177

Marzencki, Marcin 355
Masmoudi, Mohamed 366
Meech, John 63
Meillère, Stéphane 366
Melek, William W. 93, 284
Menon, Carlo 355
Miah, M. Suruz 21
Miao, Yun-Qian 1
Mkaouar, Hanéne 199
Mohammed, Sabah 241

Najjaran, Homayoun 112, 273
Nebhen, Jamel 366

Owen, William 167, 284

Pan, Yajun 51
Parreira, Juliana 63
Paull, Liam 41
Payandeh, Shahram 31

Rahman, Abu Saleh Md Mahfujur 321
Reis, Luís Paulo 73
Rubin, Stuart H. 331

Sadeghian, Alireza 93
Saeedi G., Sajad 41
Sakr, Nizar 343
Saleh, Jamil Abou 311
Schaeper, Martin 303
Seguin, Jean-Luc 366
Seo, Jaho 102
Servos, Daniel 241
Seto, Mae L. 41, 51
Smiri, Kamel 406
Soh, Yeng Chai 253
Spike, Jonathan 284

Tavakolian, Kouhyar 355
Teófilo, Luís Filipe 73

Waslander, Steven 284
Wu, Dongrui 189

Yagoub, Mustapha C.E. 396

Zhang, Rui 253
Zhao, Jiying 343
Zhou, Hongming 263
Zhu, Jiayi 146
Zong, Weiwei 263